Quantum Mechanics in a Nutshell

Quantum Mechanics in a Nutshell

Gerald D. Mahan

PRINCETON UNIVERSITY PRESS • PRINCETON AND OXFORD

Published by Princeton University Press, 41 William Street,
Princeton, New Jersey 08540

In the United Kingdom: Princeton University Press,
6 Oxford Street, Woodstock, Oxfordshire OX20 1TW

Library of Congress Cataloging-in-Publication Data

Mahan, Gerald D.

Quantum mechanics in a nutshell / Gerald D. Mahan.

p. cm.—(In a nutshell)

Includes bibliographical references and index.

ISBN 978-0-691-13713-1 (cloth : alk. paper) 1. Photons—Scattering.
2. Photon emission. 3. Scattering (Physics) 4. Quantum theory. I. Title.

QC793.5.P428M34 2009

530.12—dc22 2008017144

British Library Cataloging-in-Publication Data is available

This book has been composed in Scala

Printed on acid-free paper. ∞

press.princeton.edu

Printed in the United States of America

10 9 8 7 6 5 4 3 2 1

Contents

Preface

This manuscript is a textbook for a graduate course in quantum mechanics. I have taught this course 15–20 times and gradually developed these notes. Orginally, I used as a text *Quantum Mechanics* by A.S. Davydov. When that fine book went out of print, I wrote these notes following a similar syllabus. It contains much new material not found in older texts.

The beginning chapters follow a traditional syllabus. Topics include solving Schrödingers equation in one, two, and three dimensions. Approximate techniques are introduced such as (1) variational, (2) WKBJ, and (3) perturbation theory. Many examples are taken from the quantum mechanics of atoms and small molecules. Solid-state examples include exchange energy, Landau levels, and the quantum Hall effect. Later chapters discuss scattering theory and relativistic quantum mechanics. The chapter on optical properties includes both linear and nonlinear optical phenomena. Each chapter concludes with numerous homework problems.

Preliminary versions of these lectures have been handed to several generations of graduate students. Their feedback has been invaluable in honing the material.

Quantum Mechanics in a Nutshell

1 Introduction

1.1 Introduction

Quantum mechanics is a mathematical description of how elementary particles move and interact in nature. It is based on the wave–particle dual description formulated by Bohr, Einstein, Heisenberg, Schrödinger, and others. The basic units of nature are indeed particles, but the description of their motion involves wave mechanics.

The important parameter in quantum mechanics is Planck's constant $h = 6.626 \times 10^{-34}$ J s. It is common to divide it by 2π, and to put a slash through the symbol: $\hbar = 1.054 \times 10^{-34}$ J s. Classical physics treated electromagnetic radiation as waves. It is particles, called *photons*, whose quantum of energy is $\hbar\omega$ where ω is the classical angular frequency. For particles with a mass, such as an electron, the classical momentum $m\vec{v} = \vec{p} = \hbar\mathbf{k}$, where the wave vector k gives the wavelength $k = 2\pi/\lambda$ of the particle. Every particle is actually a wave, and some waves are actually particles.

The *wave function* $\psi(\mathbf{r}, t)$ is the fundamental function for a single particle. The position of the particle at any time t is described by the function $|\psi(\mathbf{r}, t)|^2$, which is the probability that the particle is at position $\mathbf{r}$ at time t. The probability is normalized to one by integrating over all positions:

$$1 = \int d^3r |\psi(\mathbf{r}, t)|^2 \tag{1.1}$$

In classical mechanics, it is assumed that one can know exactly where a particle is located. Classical mechanics takes this probability to be

$$|\psi(\mathbf{r}, t)|^2 = \delta^3(\mathbf{r} - \mathbf{v}t) \tag{1.2}$$

The three-dimensional delta-function has an argument that includes the particle velocity $\mathbf{v}$. In quantum mechanics, we never know precisely where to locate a particle. There is always an uncertainty in the position, the momentum, or both. This uncertainty can be summarized by the *Heisenberg uncertainty principle*:

$$\Delta x \Delta p_x \geq \hbar \tag{1.3}$$

Table 1.1 Fundamental Constants and Derived Quantities

Name	Symbol	Value
Electron mass	m	$9.10938215 \times 10^{-31}$ kg
Electron charge	e	$1.602176487 \times 10^{-19}$ C
Planck's constant	h	$6.62606896 \times 10^{-34}$ J s
	$\hbar = h/2\pi$	$1.054571628 \times 10^{-34}$ J s
Boltzmann's constant	k_B	$1.3806504 \times 10^{-23}$ J/K
Light speed	c	299,792,458 m/s
Atomic mass unit	AMU	$1.660538782 \times 10^{-27}$ kg
Bohr magneton	μ_B	$927.400915 \times 10^{-26}$ J/T
Neutron magnetic moment	μ_n	$-0.99623641 \times 10^{-26}$ J/T
Bohr radius	a_0	$0.52917720859 \times 10^{-10}$ m
Rydberg energy	E_{Ry}	13.605691 eV
Fine structure constant	α	$7.2973525376 \times 10^{-3}$
Compton wavelength	λ_c	$2.463102175 \times 10^{-12}$ m
Flux quantum	$\phi_0 = h/e$	$4.13566733 \times 10^{-15}$ T/m^2
Resistance quantum	h/e^2	25,812.808 Ω

Source: Taken from NIST web site http://physics.nist.gov/

where Δx is the uncertainty in position along one axis, Δp_x is the uncertainty in momentum along the same axis, and ħ is Planck's constant h divided by $2\pi(\hbar = h/2\pi)$, and has the value $\hbar = 1.05 \times 10^{-34}$ joules-second. Table 1.1 lists some fundamental constants.

1.2 Schrödinger's Equation

The exact form of the wave function $\psi(\mathbf{r}, t)$ depends on the kind of particle, and its environment. Schrödinger's equation is the fundamental nonrelativistic equation used in quantum mechanics for describing microscopic particle motions. For a system of particles, Schrödinger's equation is written as

$$i\hbar \frac{\partial \psi}{\partial t} = H\psi \tag{1.4}$$

$$H = \sum_j \left[\frac{p_j^2}{2m_j} + U(\mathbf{r}_j, s_j) \right] + \sum_{i>j} V(\mathbf{r}_i - \mathbf{r}_j) \tag{1.5}$$

The particles have positions $\mathbf{r}_i$, momentum $\mathbf{p}_j$, and spin s_j. They interact with a potential $U(\mathbf{r}_j, s_j)$ and with each other through the pair interaction $V(\mathbf{r}_i - \mathbf{r}_j)$. The quantity H is the Hamiltonian, and the wave function for a system of many particles is $\psi(\mathbf{r}_1, \mathbf{r}_2, \mathbf{r}_3, \cdot, \mathbf{r}_N; s_1, s_2, \ldots, s_N)$.

The specific forms for H depends on the particular problem. The relativistic form of the Hamiltonian is different than the nonrelativistic one. The relativistic Hamiltonian is discussed in chapter 11. The Hamiltonian can be used to treat a single particle, a collection of identical particles, or different kinds of elementary particles. Many-particle systems are solved in chapter 9.

No effort is made here to justify the correctness of Schrödinger's equation. It is assumed that the student has had an earlier course in the introduction to modern physics and quantum mechanics. A fundamental equation such as eqn. (1.4) cannot be derived from any postulate-free starting point. The only justification for its correctness is that its predictions agree with experiment. The object of this textbook is to teach the student how to solve Schrödinger's equation and to make these predictions. The students will be able to provide their own comparisons to experiment.

Schrödinger's equation for a single nonrelativistic particle of mass m, in the absence of a magnetic field, is

$$i\hbar \frac{\partial \psi}{\partial t} = H\psi \tag{1.6}$$

$$H = \frac{p^2}{2m} + V(\mathbf{r}) \tag{1.7}$$

The potential energy of the particle is $V(\mathbf{r})$. This potential is usually independent of the spin variable for nonrelativistic motions in the absence of a magnetic field. Problems involving spin are treated in later chapters. When spin is unimportant in solving Schrödinger's equation, its presense is usually ignored in the notation: the wave function is written as $\psi(\mathbf{r})$.

In quantum mechanics, the particle momentum is replaced by the derivative operator:

$$\mathbf{p} \rightarrow -i\hbar\vec{\nabla} \tag{1.8}$$

$$H = -\frac{\hbar^2 \nabla^2}{2m} + V(\mathbf{r}) \tag{1.9}$$

Schrödinger's equation (1.4) is a partial differential equation in the variables $(\mathbf{r}, t)$. Solving Schrödinger's equation for a single particle is an exercise in ordinary differential equations. The solutions are not just mathematical exercises, since the initial and boundary conditions are always related to a physical problem.

Schrödinger's equation for a single particle is always an artificial problem. An equation with $V(\mathbf{r})$ does not ever describe an actual physical situation. The potential $V(\mathbf{r})$ must be provided by some other particles or by a collection of particles. According to Newton's third law, there is an equal and opposite force acting on these other particles, which are also reacting to this mutual force. The only situation in which one particle is by itself has $V = 0$, which is a dull example. Any potential must be provided by another particle, so

Schrödinger's equation is always a many-particle problem. Nevertheless, there are two reasons why it is useful to solve the one-particle problem using classical potentials. The first is that one has to learn using simple problems as a stepping stone to solving the more realistic examples. Secondly, there are cases where the one-particle Schrödinger's equation is an accurate solution to a many-particle problem: i.e., it describes the relative motion of a two-particle system.

1.3 Eigenfunctions

In solving the time-dependent behavior, for the one-particle Schrödinger's equation (1.8), an important subsidiary problem is to find the eigenvalues ε_n and eigenfunctions $\phi_n(\mathbf{r})$ of the time-independent Hamiltonian:

$$H\phi_n(\mathbf{r}) = \varepsilon_n \phi_n(\mathbf{r}) \tag{1.10}$$

There is a silly convention of treating "eigenfunction" and "eigenvalue" as single words, while "wave function" is two words. The name wave function is usually reserved for the time-dependent solution, while eigenfunction are the solutions of the time-independent equation. The wave function may be a single eigenfunction or a linear combination of eigenfunctions.

The eigenfunctions have important properties that are a direct result of their being solutions to an operator equation. Here we list some important results from linear algebra: The Hamiltonian operator is always Hermitian: $H^\dagger = H$.

- Eigenvalues of Hermitian operators are always real.
- Eigenfunctions with different eigenvalues are orthogonal:

$$[\varepsilon_n - \varepsilon_m] \int d^3 r \phi_n^*(\mathbf{r}) \phi_m(\mathbf{r}) = 0 \tag{1.11}$$

  which is usually written as

$$\int d^3 r \phi_n^*(\mathbf{r}) \phi_m(\mathbf{r}) = \delta_{nm} \tag{1.12}$$

  These two statements are not actually identical. The confusing case is where there are several different states with the same eigenvalue. They do not have to obey eqn. (1.12), but they can be constructed to obey this relation. We assume that is the case.

- The eigenfunctions form a complete set:

$$\sum_n \phi_n^*(\mathbf{r}) \phi_n(\mathbf{r}') = \delta^3(\mathbf{r} - \mathbf{r}') \tag{1.13}$$

These properties are used often. Orthogonality is important since it implies that each eigenfunction $\phi_n(\mathbf{r})$ is linearly independent of the others. Completeness is important, since any function $f(\mathbf{r})$ can be uniquely and exactly expanded in terms of these eigenfunctions:

$$f(\mathbf{r}) = \sum_n b_n \phi_n(\mathbf{r}) \tag{1.14}$$

$$b_n = \int d^3 r f(\mathbf{r}) \phi_n(\mathbf{r}) \tag{1.15}$$

The function of most interest is the wave function $\psi(\mathbf{r}, t)$. It can be expanded exactly as

$$\psi(\mathbf{r}, t) = \sum_n a_n \phi_n(\mathbf{r}) e^{-i\varepsilon_n t/\hbar} \tag{1.16}$$

$$i\hbar \frac{\partial \psi}{\partial t} = \sum_n a_n \varepsilon_n \phi_n(\mathbf{r}) e^{-i\varepsilon_n t/\hbar} = H\psi \tag{1.17}$$

The coefficients a_n depend on the initial conditions. They depend on neither $\mathbf{r}$ nor t. One example is when the system is in thermal equilibrium. If the particles obey Maxwell-Boltzmann statistics, the coefficients are

$$|a_n|^2 = \exp[-\beta(\varepsilon_n - \Omega)], \quad \beta = \frac{1}{k_B T} \tag{1.18}$$

and Ω is the grand canonical potential. Another example occurs during an experiment, where the system is prepared in a particular state, such as an atomic beam. Then the coefficients a_n are determined by the apparatus, and not by thermodynamics.

The wave function has a simple physical interpretation. The probability $P(\mathbf{r}, t)$ that the particle is at the position $\mathbf{r}$ at the time t is given by the square of the absolute magnitude of the wave function:

$$P(\mathbf{r}, t) = |\psi(\mathbf{r}, t)|^2 \tag{1.19}$$

In quantum mechanics, there is no certainty regarding the position of the particle. Instead, the particle has a nonzero probability of being many different places in space. The likelihood of any of these possibilities is given by $P(\mathbf{r}, t)$. The particle is at only one place at a time.

The normalization of the wave function is determined by the interpretation of $P(\mathbf{r}, t)$ as a probability function. The particle must be someplace, so the total probability should be unity when integrated over all space:

$$1 = \int d^3 r P(\mathbf{r}, t) = \int d^3 r |\psi(\mathbf{r}, t)|^2 \tag{1.20}$$

The normalization also applies to the wave function. The eigenfunctions are also orthogonal, so that

$$1 = \int d^3 r |\psi(\mathbf{r}, t)|^2 = \sum_{n,m} a_n^* a_m e^{it(\varepsilon_n - \varepsilon_m)/\hbar} \int d^3 r \phi_n^* \phi_m \tag{1.21}$$

$$= \sum_n |a_n|^2 \tag{1.22}$$

The summation of the expansion coefficients $|a_n|^2$ must be unity. These coefficients

$$P_n = |a_n|^2, \ \sum_n P_n = 1 \tag{1.23}$$

are interpreted as the probability that the particle is in the eigenstate $\phi_n(\mathbf{r})$.

The average value of any function $f(\mathbf{r})$ is obtained by taking the integral of this function over all of space, weighted by the probability $P(\mathbf{r}, t)$. The bracket notation denotes the average of a quantity:

$$\langle f\rangle(t) = \int d^3r f(\mathbf{r}) P(\mathbf{r}, t) \tag{1.24}$$

For example, the average potential energy $\langle V\rangle$ and the average position $\langle r\rangle$ are

$$\langle V\rangle = \int d^3r V(\mathbf{r}) P(\mathbf{r}, t) \tag{1.25}$$

$$\langle \mathbf{r}\rangle = \int d^3r \mathbf{r} P(\mathbf{r}, t) \tag{1.26}$$

A similar average can be taken for any other function of position.

There is no way to take an average of the particle velocity $\mathbf{v} = \dot{\mathbf{r}}$. Since $P(\mathbf{r}, t)$ does not depend on $\dot{\mathbf{r}}$, there is no way to average this quantity. So $\langle \dot{\mathbf{r}}\rangle$ does not exist. Instead, the average velocity is found by taking the time derivative of the average of $\mathbf{r}$, such as

$$\frac{\partial}{\partial t}\langle \mathbf{r}\rangle = \int d^3r \mathbf{r} \frac{\partial}{\partial t} P(\mathbf{r}, t) \tag{1.27}$$

$$= \int d^3r \mathbf{r}\left[\psi^* \frac{\partial \psi}{\partial t} + \psi \frac{\partial \psi^*}{\partial t}\right] \tag{1.28}$$

Now use Schrödinger's equation and its complex conjugate, to find

$$\frac{\partial \psi}{\partial t} = \frac{-i}{\hbar}\left[-\frac{\hbar^2\nabla^2}{2m} + V\right]\psi \tag{1.29}$$

$$\frac{\partial \psi^*}{\partial t} = \frac{i}{\hbar}\left[-\frac{\hbar^2\nabla^2}{2m} + V\right]\psi^* \tag{1.30}$$

which is used in $\partial\langle \mathbf{r}\rangle/\partial t$:

$$\frac{\partial}{\partial t}\langle \mathbf{r}\rangle = \frac{i\hbar}{2m}\int d^3r \mathbf{r}[\psi^*\nabla^2\psi - \psi\nabla^2\psi^*] \tag{1.31}$$

$$= \frac{i\hbar}{2m}\int d^3r \mathbf{r}\vec{\nabla}\cdot[\psi^*\vec{\nabla}\psi - \psi\vec{\nabla}\psi^*] \tag{1.32}$$

The terms containing the potential energy V canceled. The equivalence of the last two expressions is found by just taking the derivative in the last expression. Each term in brackets generates a factor of $(\vec{\nabla}\psi^*)\cdot(\vec{\nabla}\psi)$, which cancels.

An integration by parts gives

$$\frac{\partial}{\partial t}\langle \mathbf{r}\rangle = -\frac{i\hbar}{2m}\int d^3r[\psi^*\vec{\nabla}\psi - \psi\vec{\nabla}\psi^*]\cdot\vec{\nabla}\mathbf{r} \tag{1.33}$$

If $\mathbf{A}$ is the quantity in brackets, then $(\mathbf{A}\cdot\vec{\nabla})\mathbf{r} = \mathbf{A}$, so the final expression is

$$\frac{\partial}{\partial t}\langle \mathbf{r}\rangle = \frac{\hbar}{2mi}\int d^3r[\psi^*\vec{\nabla}\psi - \psi\vec{\nabla}\psi^*] \tag{1.34}$$

The integrand is just the definition of the particle current:

$$\mathbf{j}(\mathbf{r},t)=\frac{\hbar}{2mi}[\psi^{*}\vec{\nabla}\psi-\psi\vec{\nabla}\psi^{*}] \tag{1.35}$$

$$\frac{\partial}{\partial t}\langle\mathbf{r}\rangle=\int d^{3}r\mathbf{j}(\mathbf{r},t) \tag{1.36}$$

The function $\mathbf{j}(\mathbf{r}, t)$ is the particle current, which has the units of number of particles per second per unit area. If it is multiplied by the charge on the particle, it becomes the electrical current density $J = ej$, with units of amperes per area.

In the integral (1.34), integrate by parts on the second term in brackets. It then equals the first term:

$$\frac{\partial}{\partial t}\langle\mathbf{r}\rangle=\frac{1}{m}\int d^{3}r\psi^{*}\left(\frac{\hbar}{i}\vec{\nabla}\right)\psi=\frac{\langle p\rangle}{m} \tag{1.37}$$

The momentum operator is $\mathbf{p}=\hbar\vec{\nabla}/i$, and the integral is the expectation value of the momentum. In quantum mechanics, the average value of the velocity is the average value of the momentum divided by the particle mass.

The expectation value of any derivative operator should be evaluated as is done for the momentum: the operator is sandwiched between ψ^* and ψ under the integral:

$$\langle O\rangle=\int d^{3}r\psi^{*}(\mathbf{r},t)O(\mathbf{r})\psi(\mathbf{r},t) \tag{1.38}$$

Other examples are the Hamiltonian and the z-component of angular momentum:

$$\langle H\rangle=\int d^{3}r\psi^{*}(\mathbf{r},t)\left[-\frac{\hbar^{2}\nabla^{2}}{2m}+V(\mathbf{r})\right]\psi(\mathbf{r},t) \tag{1.39}$$

$$\langle L_{z}\rangle=\frac{\hbar}{i}\int d^{3}r\psi^{*}(\mathbf{r},t)\left[x\frac{\partial}{\partial y}-y\frac{\partial}{\partial x}\right]\psi(\mathbf{r},t) \tag{1.40}$$

Once the wave function is known, it can be used to calculate the average value of many quantities that can be measured.

The last relationship to be proved in this section is the equation of continuity:

$$0=\frac{\partial}{\partial t}\rho(\mathbf{r},t)+\vec{\nabla}\cdot\mathbf{j}(\mathbf{r},t) \tag{1.41}$$

where $\rho(\mathbf{r}, t) \equiv P(\mathbf{r}, t)$ is the particle density and $\mathbf{j}(\mathbf{r}, t)$ is the particle current. The continuity equation is proved by taking the same steps to evaluate the velocity:

$$\frac{\partial\rho}{\partial t}=\frac{\partial}{\partial t}\psi^{*}\psi=\psi^{*}\frac{\partial\psi}{\partial t}+\psi\frac{\partial\psi^{*}}{\partial t} \tag{1.42}$$

Use the above expressions (1.29, 1.30) for the time derivative of the wave functions. Again the potential energy terms cancel:

$$\frac{\partial\rho}{\partial t}=\frac{i\hbar}{2m}\vec{\nabla}\cdot\left[\psi^{*}\vec{\nabla}\psi-\psi\vec{\nabla}\psi^{*}\right]=-\vec{\nabla}\cdot\mathbf{j}(\mathbf{r},t) \tag{1.43}$$

which proves the equation of continuity. Schrödinger's equation has been shown to be consistent with the equation of continuity as long as the density of particles is interpreted to be $\rho(\mathbf{r}, t) = |\psi(\mathbf{r}, t)|^2$, and the current is eqn. (1.35).

1.4 Measurement

Making a measurement on a particle in a quantum system is a major disruption of the probability distribution. Suppose we have N identical particles, for example, atoms, in a large box. They will be in a variety of energy states. It is not possible to say which atom is in which state. If $\phi_n(\mathbf{r})$ is an exact eigenstate for an atom in this box, the wave function of an atom is

$$\psi(x, t) = \sum_n a_n \phi_n(\mathbf{r}) \exp[-itE_n/\hbar] \tag{1.44}$$

The amplitudes a_n, when squared ($P_n = |a_n|^2$), determine the probability of an atom being in the state with energy E_n.

Suppose we do a measurement on a single atom, to determine its energy. One might drill a small hole in the side of the box, which allows one atom to escape at a time. If one measures the energy of that particular atom, one will find a definite value E_j. The result of the measurement process is that one state, out of the many possible states, is selected.

Suppose that one measures the energy of the same atom at a later time. If its flight path has been undisturbed, one will again find the same energy E_j. After the first measurement of energy, the particle continues its motion starting from the eigenstate $\phi_j(\mathbf{r})$, and not from the distribution of eigenstates in $\psi(\mathbf{r}, t)$. Of course, the first measurement of energy may disrupt the flight path a bit, according to the uncertainly relation, so the second measurement may give a slightly different energy. The important point is that measurement disrupts the statistical distribution, and imposes a new initial condition for the particle motion.

1.5 Representations

In the early days of quantum mechanics, Schrödinger and Heisenberg each presented versions that appeared to be different. There was a lively debate on the virtues of each version, until it was shown by Dirac and Jordan that the two are mathematically identical. This history is well described by von Neumann. The two theories do not look identical, at least superficially. The two versions are described here briefly, and an equally brief explanation is given as to why they are identical.

Schrödinger's version is the one discussed so far in this chapter, and which is mainly treated in the early parts of this book. The Heisenberg version, which stresses operators rather than wave functions, is introduced in later chapters. Both versions are used extensively in modern physics, and both are important.

The measurable quantities in physics are always expectation values: they are the average quantities. If $F(\mathbf{r}, \mathbf{p})$ is some function of position $\mathbf{r}$ and momentum $\mathbf{p}$, then its expectation value is

$$\langle F\rangle(t) = \int d^3r\psi^*(\mathbf{r}, t)F(\mathbf{r}, -i\hbar\nabla)\psi(\mathbf{r}, t) \tag{1.45}$$

The bracket on the left is a shorthand notation for taking the average over $\mathbf{r}$, as indicated on the right-hand side of the equation. The Schrödinger and Heisenberg versions of quantum mechanics are equivalent because they always give the same expectation values. The same result is obtained even as a function of time, so the two versions give the same result for the time derivatives of $\langle F\rangle(t)$.

1.5.1 *Schrödinger Representation*

This representation has two important features.

1. Wave functions $\psi(\mathbf{r}, t)$ depend on time, and the time development is governed by Schrödinger's equation:

$$i\hbar\frac{\partial\psi}{\partial t} = H\psi(\mathbf{r}, t) \tag{1.46}$$

$$\psi(\mathbf{r}, t) = e^{-iHt/\hbar}\psi(\mathbf{r}, t=0) \tag{1.47}$$

   The second equation is a formal solution to eqn. (1.46).

2. Operators do not depend on time. Operators such as $\mathbf{r}$, $\mathbf{p}$, H, L_z are time independent.

In the Schrödinger representation, only the wave functions depend on time. The time derivative of an expectation value, such as (1.45), is

$$\begin{aligned}\frac{\partial}{\partial t}\langle F\rangle(t) = \int d^3r\Bigg[&\left(\frac{\partial\psi^*(\mathbf{r}, t)}{\partial t}\right)F(\mathbf{r}, -i\hbar\nabla)\psi(\mathbf{r}, t)\\ &+ \psi^*(\mathbf{r}, t)F(\mathbf{r}, -i\hbar\nabla)\left(\frac{\partial\psi(\mathbf{r}, t)}{\partial t}\right)\Bigg]\end{aligned} \tag{1.48}$$

The derivatives are evaluated using eqn. (1.46), giving

$$\frac{\partial}{\partial t}\langle F\rangle(t) = \frac{i}{\hbar}\int d^3r[(H\psi^*)F\psi - \psi^*F(H\psi)] \tag{1.49}$$

In the first term on the right, the positions of H and ψ^* can be interchanged. Recall that $H = p^2/2m + V$. The scalar V can be interchanged. The kinetic energy term is $-\hbar^2\nabla^2/2m$, which can be interchanged after a double integration by parts on the $\mathbf{r}$ variable. Then the time derivative is written as a commutator:

$$[H, F] = HF - FH \tag{1.50}$$

$$\frac{\partial}{\partial t}\langle F\rangle(t) = \frac{i}{\hbar}\int d^3r\psi^*[H, F]\psi \tag{1.51}$$

The above equation can be summarized as

$$\frac{\partial}{\partial t}\langle F\rangle(t) = \left\langle \frac{i}{\hbar}[H, F] \right\rangle \tag{1.52}$$

The identical equation is found in the Heisenberg representation, as described below.

1.5.2 Heisenberg Representation

This representation has several important features:

1. Wave functions are independent of time $\psi(\mathbf{r})$.
2. Operators are time-dependent according to the prescription

$$F(\mathbf{r}, \mathbf{p}, t) = e^{iHt/\hbar} F(\mathbf{r}, \mathbf{p}) e^{-iHt/\hbar} \tag{1.53}$$

3. The expectation values are given by

$$\langle F\rangle(t) = \int d^3 r \psi^*(\mathbf{r}) F(\mathbf{r}, \mathbf{p}, t)\psi(\mathbf{r}) \tag{1.54}$$

$$= \int d^3 r \psi^*(\mathbf{r}) e^{iHt/\hbar} F(\mathbf{r}, \mathbf{p}) e^{-iHt/\hbar}\psi(\mathbf{r}) \tag{1.55}$$

The latter definition is formally identical to (1.45). The time derivative is also identical:

$$\frac{\partial}{\partial t} F(\mathbf{r}, \mathbf{p}, t) = \frac{i}{\hbar} e^{iHt/\hbar}[HF - FH] e^{-iHt/\hbar} \tag{1.56}$$

$$= \frac{i}{\hbar}[H, F(\mathbf{r}, \mathbf{p}, t)] \tag{1.57}$$

This equation is the fundamental equation of motion in the Heisenberg version of quantum mechanics. The focus of attention is on the operators and their development in time. Once the time development is determined, by either exact or approximate techniques, the expectation value of the operator may be evaluated using the integral in eqn. (1.55). The solution to the time development in (1.57) involves the solution of operator expressions. The Heisenberg version of quantum mechanics, with its emphasis on the manipulation of operators, appears to be very different than the Schrödinger's equation (1.29), which is a differential equation in $\mathbf{r}$-space.

Yet the two representations do make identical predictions. For the average of the time derivative of F, the Heisenberg representaton gives, from (1.57),

$$\left\langle \frac{\partial F}{\partial t} \right\rangle = \frac{i}{\hbar} \int d^3 r \psi^*(\mathbf{r}) e^{itH/\hbar}[H, F(\mathbf{r}, \mathbf{p})] e^{-itH/\hbar}\psi(\mathbf{r}) \tag{1.58}$$

This result is identical to the Schrödinger result in (1.51), just as the average value of F is the same as (1.45) and (1.55). The simularities of the two approaches is more apparent when using eqn. (1.47). The Hermitian conjugate of this expression is

$$\psi^\dagger(\mathbf{r}, t) = \psi^*(\mathbf{r}, t) e^{itH/\hbar} \tag{1.59}$$

and the expectation value of F in the Schrödinger representation is

$$\langle F\rangle = \int d^3r\psi^\dagger(\mathbf{r},t)F(\mathbf{r},p)\psi(\mathbf{r},t) \tag{1.60}$$

This Schrödinger result is identical to the Heisenberg result in (1.55). The two representations make identical predictions for the average values of operators and for all values of time. The two versions of quantum mechanics are equivalent in predicting measureable quantitites.

1.6 Noncommuting Operators

The important variables of classical mechanics usually exist as operators in quantum mechanics. A partial list of such operators is momentum $\mathbf{p}$, energy H, and angular momentum L_z. An important consideration between pairs of operators is whether they commute or do not commute.

An example of a pair of noncommuting operators in one dimension is the position x and momentum $p = -i\hbar d/dx$. Take some arbitrary function $f(x)$, and operate on it by the two combinations

$$xpf = -i\hbar x\frac{df}{dx} \tag{1.61}$$

$$pxf = -i\hbar\frac{d}{dx}(xf) = -i\hbar\left(f + x\frac{df}{dx}\right) \tag{1.62}$$

Subtract these two results:

$$(xp - px)f = i\hbar f = [x, p]f \tag{1.63}$$

The bracket notation for the commutator is used in the second equality: $[A, B] \equiv (AB - BA)$. This notation was introduced in the prior section. The commutatior $[x, p]$, when operating on any function of x, gives $i\hbar$ times the same function of x. This result is usually written by omitting the function $f(x)$:

$$[x, p] = i\hbar \tag{1.64}$$

The presence of such a function is usually implied, even though it is customarily omitted. The position x and momentum p do not commute. They would be said to commute if $[x, p] = 0$.

A theorem in linear algebra states that if two operators commute, then it is possible to construct simultaneous eigenfunctions of both operators. For example, if two operators F and G commute, then it is possible to find eigenstates ϕ_n such that

$$F\phi_n = f_n\phi_n \quad G\phi_n = g_n\phi_n \tag{1.65}$$

where (f_n, g_n) are eigenvalues. The converse is also true, in that if two operators do not commute, then it is impossible to find simultaneous eigenvalues from the same eigenfunction.

It is impossible to find simultaneous eigenfunctions of x and p. There is no function ϕ_n that has the dual property that $x\phi_n = x_n\phi_n$, $p\phi_n = p_n\phi_n$. One can find an eigenfunction of x but it is not an eigenfunction of p, and vice versa. The statement that one cannot define simultaneous eigenvalues of x and p causes the uncertainly principle that $\Delta x \Delta p \geq \hbar$. A similar uncertainly principle exists between any pair of noncommuting operators.

The most important operator in quantum mechanics is the Hamiltonian H. The eigenfunctions of H are used extensively in wave functions and elsewhere. It is useful to ask whether these eigenfunctions are also exact eigenfunctions of other operators, such as momentum and angular momentum $\vec{L}$. The answer depends on whether these other operators commute with H. Angular momentum often commutes with H, and then one can construct simultaneous eigenstates of H and $\vec{L}$. The momentum p only occasionally commutes with H. Since p and $\vec{L}$ never commute, one can never find simultaneous eigenstates of all three operators.

An example is useful. Consider the Hamiltonian in three dimensions. The potential $V(r) = 0$.

$$H = \frac{p^2}{2m} = -\frac{\hbar^2 \nabla^2}{2m} \tag{1.66}$$

One choice of eigenfunction is the plane wave state $\phi(\mathbf{k}, \mathbf{r}) = Ae^{i\mathbf{k}\cdot\mathbf{r}}$, where A is a normalization constant. The eigenvalue of H is

$$H\phi(\mathbf{k}, \mathbf{r}) = -\frac{\hbar^2 \nabla^2}{2m}(Ae^{i\mathbf{k}\cdot\mathbf{r}}) = \varepsilon_k \phi(\mathbf{k}, \mathbf{r}), \quad \varepsilon_k = \frac{\hbar^2 k^2}{2m} \tag{1.67}$$

The plane wave solution is also an eigenfunction of the momentum operator, but not of the angular momentum:

$$\mathbf{p}\phi(\mathbf{k}, \mathbf{r}) = -i\hbar\nabla\phi(\mathbf{k}, \mathbf{r}) = \hbar\mathbf{k}\phi(\mathbf{k}, \mathbf{r}) \tag{1.68}$$

$$L_z\phi(\mathbf{k}, \mathbf{r}) = -i\hbar\left(x\frac{\partial}{\partial y} - y\frac{\partial}{\partial x}\right)\phi = \hbar(xk_y - yk_x)\phi(\mathbf{k}, \mathbf{r}) \tag{1.69}$$

The plane wave is an example of a simultaneous eigenfunction of H and $\mathbf{p}$.

Another choice of eigenfunction for H is the product of a spherical Bessel function $j_\ell(kr)$ and spherical harmonic angular function $Y_\ell^m(\theta, \phi)$ in spherical coordinates:

$$\psi_{\ell m}(kr) = j_\ell(kr) Y_\ell^m(\theta, \phi) \tag{1.70}$$

Readers unfamiliar with these functions should not worry: they are explained in chapter 5. This function is an exact eigenfunction of H and L_z, but not of momentum:

$$H\psi_{\ell m}(kr) = \varepsilon_k \psi_{\ell m}(kr) \tag{1.71}$$

$$L_z\psi_{\ell m}(kr) = \hbar m \psi_{\ell m}(kr) \tag{1.72}$$

Here is an example of simultaneous eigenfunctions of H and L_z.

The Hamiltonian of a "free particle," which is one with no potential ($V = 0$), has a number of eigenfunctions with the same eigenvalue ε_k. The eigenfunctions in (1.70), for different values of (ℓ, m), where $(0 \geq \ell < \infty, -\ell \leq m \leq \ell)$, all have the same eigenvalue ε_k. Any linear combination of eigenstates with the same eigenvalues are still eigenfunctions of H. The plane wave state $\exp(i\mathbf{k} \cdot \mathbf{r})$ is a particular linear combination of these states. Some linear combinations are eigenfunctions of $\mathbf{p}$, while others are eigenfunctions of L_z. Since $\mathbf{p}$ and L_z do not commute, there are no eigenfunctions for both operators simultaneously.

References

1. J. von Neumann, *Mathematical Foundations of Quantum Mechanics* (Princeton, University Press, Princeton, NJ, 1955) Translated from the German by R.T. Beyer
2. P.A.M. Dirac, *The Principles of Quantum Mechanics* (Oxford, University Press, Oxford, UK, 1958)

Homework

1. Prove that

$$e^L a e^{-L} = a + [L, a] + \frac{1}{2!}[L, [L, a]] + \frac{1}{3!}[L, [L, [L, a]]] + \cdots$$

where (a, L) are any operators.

2. If F is any operator that does not explicitly depend on time, show that $\partial\langle F\rangle/\partial t = 0$ in an eigenstate of H with discrete eigenvalues.

2 One Dimension

It is useful to study one-dimensional problems, since there are many more exact solutions in one dimension compared to two and three dimensions. The study of exact solutions is probably the best way to understand wave functions and eigenvalues. This chapter gives some examples of exact solutions. There are many more exactly solvable examples than the few cases given here.

A person with a knowledge of quantum mechanics should, if presented with some potential $V(x)$, be able to give a freehand sketch of the wave function $\psi(x)$. This sketch need not be precise, but should oscillate in the right regions, and decay exponentially in the other regions. A major purpose of the next two chapters is to develop in the student this level of intuition. The present chapter discusses exact solutions, while the next chapter gives approximate solutions. One dimension is a good place to sharpen one's intuitive understanding of wave functions.

There are other reasons to investigate one-dimensional behavior. Many physical systems exhibit quasi-one-dimensional properties. Another reason is that the mathematics of one and three dimensions are very similar. Two dimensions is different.

2.1 Square Well

The easiest solutions of Schrödinger's equation are those where the potential $V(x)$ has only constant values V_i. Several such cases are solved in this section, all under the general title of *square well*. They are characterized by having the potential $V(x)$ consist of constant values V_i in all regions of space.

The first case is a particle in a box (figure 2.1). The potential $V(x)$ has a constant value V_0 inside of the box of length L and infinite values everywhere else:

$$V(x) = V_0 \quad 0 < x < L \tag{2.1}$$

$$V(x) = \infty \quad \text{elsewhere} \tag{2.2}$$

Take $V_0 = 0$, which defines the energy zero to be the bottom of the box.

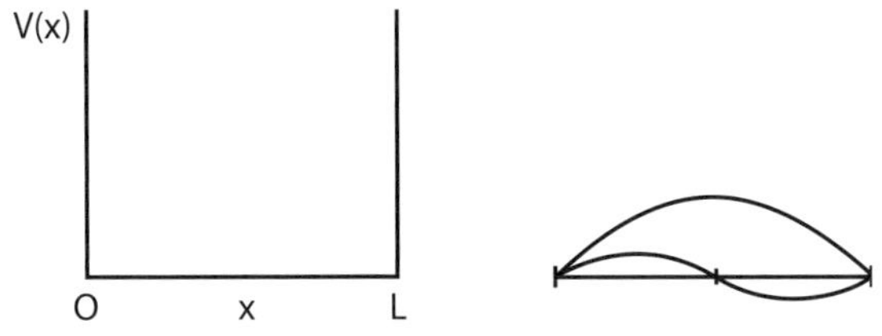

FIGURE 2.1. (a) $V(x)$ for a particle confined to a box $0 \le x \le L$. (b) The lowest two eigenfunctions.

Inside of the box, where $V(x) = 0$, Schrödinger's equation is

$$(H-E)\psi(x) = 0 = -\left(\frac{\hbar^2}{2m}\frac{d^2}{dx^2} + E\right)\psi(x) \tag{2.3}$$

The first constraint is that E has to be positive. This assertion can be proven using mathematics, where it can be shown that the differential equation (2.3) has no solution for $E < 0$ for the boundary conditions stated below. The physics statement is that Schrödinger's equation has no eigenvalues E_n that are lower in value than the smallest value of $V(x)$. All eigenvalues must be greater than the smallest value of $V(x)$. This constraint applies to all problems in all dimensions.

Since $E > 0$, it is used to define the wave vector:

$$k = \sqrt{\frac{2mE}{\hbar^2}} \tag{2.4}$$

Schrödinger's equation (2.3) is now a familiar differential equation

$$\left(\frac{d^2}{dx^2} + k^2\right)\psi(x) = 0 \tag{2.5}$$

with the solution

$$\psi(x) = C_1 \sin(kx) + C_2 \cos(kx) \tag{2.6}$$

The constants $C_{1,2}$ are chosen using the boundary and normalization conditions.

The boundaries of the wave function are at the ends of the box: $(x = 0, L)$. The divergent potential at these points $(V = \infty)$ forces $\psi(x)$ to vanish at $x = 0$ and at $x = L$. The argument requiring that $\psi(0) = 0 = \psi(L)$ has two parts.

1. The energy E should be finite and given by the expression

$$E = \int_{-\infty}^{\infty} dx\psi^*(x)H\psi(x) \tag{2.7}$$

where the limits of integration span $-\infty < x < \infty$. The Hamiltonian H contains the potential energy $V(x)$. Since $V(x) = \infty$ for $x < 0$, and for $x > L$, the integral will diverge over these intervals unless $\psi(x)^2 V(x) \neq \infty$, which forces $\psi(x) = 0$. So the wave function is zero outside of the box.

2. The wave function is continuous everywhere. This contraint is due to the nature of the second-order differential equation:

- If $\psi(x) = 0$ for $x > L$ then it is zero at the point $x = L$: $\psi(L) = 0$.
- If $\psi(x) = 0$ for $x < 0$ then it is zero at the point $x = 0$; $\psi(0) = 0$.

We conclude that the important boundary conditions in the case of a box with infinite walls is $\psi(0) = \psi(L) = 0$.

Since $\psi(x = 0) = 0$, then the coefficient C_2 in (2.6) is zero: The term $\cos(kx)$ does not vanish at $x = 0$, so its coefficient must be zero. There remains the solution

$$\psi(x) = C_1 \sin(kx) \tag{2.8}$$

The function $\sin(kx)$ does vanish at $x = 0$. It must be forced to also vanish at $x = L$. Only certain selected values of k obey the condition that $\sin(kL) = 0$. These are

$$kL = n\pi \tag{2.9}$$

$$k_n = \frac{n\pi}{L} \tag{2.10}$$

$$\psi(x) = C_1 \sin\left(\frac{n\pi x}{L}\right) \tag{2.11}$$

$$E_n = \frac{\hbar^2 k_n^2}{2m} = \frac{\hbar^2}{2m}\left(\frac{n\pi}{L}\right)^2 \tag{2.12}$$

$$= n^2 E_1, \quad E_1 = \frac{\hbar^2 \pi^2}{2mL^2} \tag{2.13}$$

The minimum eigenvalue is E_1. The boundary conditions permit only selected values of k_n and E_n. The allowed values are called *eigenvalues*. The corresponding functions are *eigenfunctions*:

$$\psi_n(x) = C_1 \sin(k_n x) \tag{2.14}$$

The coefficient C_1 is a normalization constant. It is chosen so $|\psi_n(x)|^2$ has unit value when integrated over the range of values where it is nonzero:

$$1 = \int_0^L dx |\psi_n(x)|^2 = C_1^2 \int_0^L dx \sin^2(k_n x) \tag{2.15}$$

Change the integration variable to $\theta = k_n x$ and the integral goes over the interval $0 < \theta < n\,\pi$

$$1 = C_1^2 \frac{L}{n\pi} \int_0^{n\pi} d\theta \sin^2(\theta) \tag{2.16}$$

The average value of $\sin^2(\theta)$ between zero and $n\pi$ is one-half, so the integral equals $n\pi/2$:

$$1 = C_1^2 \frac{L}{2} \tag{2.17}$$

$$C_1 = \sqrt{\frac{2}{L}} \tag{2.18}$$

$$\psi_n(x) = \sqrt{\frac{2}{L}} \sin\left(\frac{n\pi x}{L}\right) \tag{2.19}$$

The last equation gives the eigenfunction with proper normalization.

Another configuration for the same problem is to displace the box and have it span $-L/2 < y < L/2$. In the present problem, that can be accomplished by letting $y = x - L/2$:

$$\psi_n(y) = \sqrt{\frac{2}{L}} \sin\left[k_n\left(y + \frac{L}{2}\right)\right]$$

$$\sin\left[k_n y + \frac{n\pi}{2}\right] = \sin(k_n y)\cos\left(\frac{n\pi}{2}\right) + \cos(k_n y)\sin\left(\frac{n\pi}{2}\right) \tag{2.20}$$

But $\cos(n\pi/2) = 0$ if n is odd and $\sin(n\pi/2) = 0$ if n is even. One can write the eigenfunction when the box is $-L/2 < y < L/2$:

$$\psi_n(y) = \begin{cases} \sqrt{\frac{2}{L}} \sin\left(\frac{n\pi y}{L}\right) & n \text{ is even} \\ \sqrt{\frac{2}{L}} \cos\left(\frac{n\pi y}{L}\right) & n \text{ is odd} \end{cases} \tag{2.21}$$

The potential is symmetric in the present example: $V(-y) = V(y)$. When the potential is symmetric, the eigenfunctions are either of even or odd parity.

- Even parity eigenfunctions have $\psi(-y) = \psi(y)$. Expand this relation in a Taylor series about $y = 0$, and find that

$$0 = \left(\frac{d\psi(y)}{dy}\right)_{y=0} \tag{2.22}$$

  This equation is useful for finding the even-parity eigenfunctions of symmetric potentials. In the present example, they are the cosine functions.

- The odd-parity case has $\psi(-y) = -\psi(y)$. At $y = 0$ this identity gives $\psi(0) = -\psi(0)$, which requires that $\psi(0) = 0$. This latter identity is useful for finding odd-parity eigenfunctions. In the present problem they are the sine functions.

We prefer to write the particle in the box as spanning $0 < x < L$ since one only has to write down a single set of eigenfunctions.

A third example is a problem with three different regions of constant $V(x)$:

$$V(x) = \begin{cases} \infty & x < 0 \\ -V_0 & 0 < x < a \\ 0 & a < x \end{cases} \tag{2.23}$$

This potential is shown in figure 2.2. The quantity V_0 is positive, so the potential energy is negative for $0 < x < a$. The zero of energy is taken as the region where $x > a$.

- *Bound states* are defined as having an eigenvalue $E < 0$. For these states the wave function is localized near the region where the potential energy has negative values. In making the statement that $E < 0$, zero energy is chosen as the value of $V(x)$ far from the region of negative potential.

- *Continuum states* are defined as those with $E > 0$. They have an eigenfunction that extends over all space $x > 0$.

Bound states are discussed first. There will be a finite number of them for most potentials. This number could be zero—there may not be any bound states if the attractive

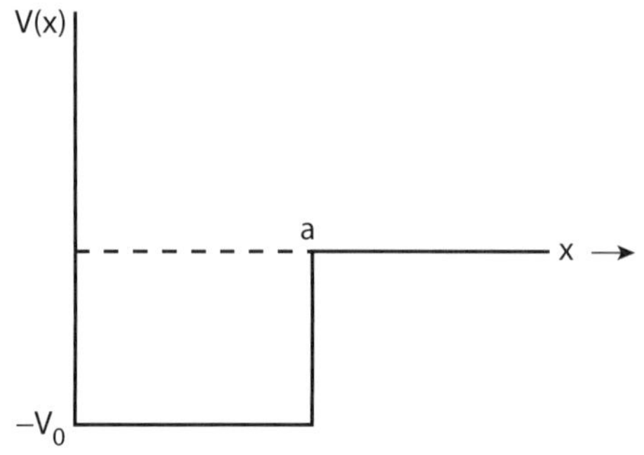

FIGURE 2.2. $V(x)$ for the second problem.

regions are weak. All bound-state energies are within $-V_0 < E < 0$, where again $-V_0$ is the minimum value of the potential energy.

The potential $V(x)$ is piecewise continuous. The mathematical procedure is to solve Schrödinger's equation first in each region of space. In each region there are two independent solutions of the second-order differential equation. Each is multiplied by a constant such as C_1 or C_2. The constants are determined by matching the solution at the boundaries between the regions.

- If the potential step is infinite, as in the preceding example, the eigenfunction must vanish at that step.
- If the potential step is finite at a point, the eigenfunction is continuous at that point.
- The first derivative of the eigenfunction is also continuous at each point, except those points x_0 with a delta-function potential $[V(x) = A\delta(x-x_0)]$. The derivative at a delta-function point is derived below.

For the bound state, there are two regions where the eigenfunction is nonzero.

- $0 < x < a$ has

$$0 = \left(\frac{d^2}{dx^2} + p^2\right)\psi(x), \quad p^2 = \frac{2m}{\hbar^2}(V_0 + E) > 0 \tag{2.24}$$

$$\psi(x) = C_1 \sin(px) + C_2 \cos(px) \tag{2.25}$$

The last equation is the most general solution to the differential equation.

- $a < x$ is a region where the eigenfunction decays exponentially:

$$0 = \left(\frac{d^2}{dx^2} - \alpha^2\right)\psi(x), \; \alpha^2 = -\frac{2m}{\hbar^2}E > 0 \tag{2.26}$$

$$\psi(x) = C_3 e^{-\alpha x} + C_4 e^{\alpha x} \tag{2.27}$$

These are the two most general solution of the differential equation. The latter solution can also be written as

$$\psi(x) = C_5 \sinh(\alpha x) + C_6 \cosh(\alpha x) \tag{2.28}$$

The two forms (2.27) and (2.28) are entirely equivalent. The first one (2.27) is preferable for the present problem.

The first boundary condition is that $\psi(x=0)=0$ since the potential becomes infinite at this point. The boundary condition forces $C_2=0$ since $\cos(0)\neq 0$. The second boundary condition is that the eigenfunction must vanish far from the region of negative potential. For $x>a$, the term $\exp(\alpha x)$ diverges at large values of x. This term must be omitted, so set $C_4=0$. Therefore, the eigenfuncion has the functional form

$$\psi(x)=\begin{cases} 0 & x<0 \\ C_1\sin(px) & 0<x<a \\ C_3 e^{-\alpha x} & a<x \end{cases} \tag{2.29}$$

The final step is to match these eigenfunctions at the point $x=a$. Matching the eigenfunction and its derivative gives two equations:

$$C_1\sin(pa)=C_3 e^{-\alpha a} \tag{2.30}$$

$$pC_1\cos(pa)=-\alpha C_3 e^{-\alpha a} \tag{2.31}$$

The equation for the eigenvalue is found by dividing these two equations, which cancels the constants $C_{1,3}$:

$$\frac{\tan(pa)}{p}=-\frac{1}{\alpha} \tag{2.32}$$

$$\tan\sqrt{\frac{V_0+E}{E_a}}=-\sqrt{\frac{V_0+E}{-E}},\quad E_a=\frac{\hbar^2}{2ma^2} \tag{2.33}$$

A bound state exists for every value of $-V_0<E<0$ that satisfies this equation. The equation has a minus sign. Since the square root is always positive, the minus sign comes from the tangent function. It has a negative value for arguments larger than $\pi/2$. This contraint gives the minimum eigenvalue:

$$\sqrt{\frac{V_0+E}{E_a}}>\frac{\pi}{2} \tag{2.34}$$

$$E>-V_0+\frac{\pi^2}{4}E_a \tag{2.35}$$

There are no bound states if the right-hand side of this equation is positive.

Another way to write the eigenfunction equation (2.32) is to define the *strength of the potential*:

$$g^2=\frac{V_0}{E_a}=\frac{2mV_0a^2}{\hbar^2} \tag{2.36}$$

It is a dimensionless quantity. Also define a dimensionless binding energy ε using $E=-\varepsilon V_0$. Since $-V_0<E<0$ then $1>\varepsilon>0$. Now the eigenvalue equation is

$$\tan\sqrt{g^2(1-\varepsilon)}=-\sqrt{\frac{1-\varepsilon}{\varepsilon}} \tag{2.37}$$

The allowed values of ε depend only on the single parameter g.

Above we found the minimum binding energy. Now let us find the minimum value of the potential strength g_c to have a bound state. For $g < g_c$ there are no bound states, while for $g > g_c$ there are one or more bound states. The critical value of g is found easily by observing that at critical binding $\varepsilon \to 0$, so that the right-hand side of (2.37) goes to infinity. As $\varepsilon \to 0$, then

$$\tan(g_c) = -\infty, \; g_c = \frac{\pi}{2}(2n+1) \tag{2.38}$$

So the minimum value of the potential strength is $g_c = \pi/2$. The minimum value of V_0 for the existence of a bound state is

$$V_0 > \frac{\pi^2}{4} E_a \tag{2.39}$$

This value applies only to the case that the potential at one end of the box goes to infinity. For a general value of $g > g_c$, one can find the bound states by solving the eigenvalue equation using graph paper or a computer.

Continuum states are those with positive energy ($E > 0$). They describe states that are not bound. Here we find the continuum states for the above potential energy. One still solves the two equations in (2.24)–(2.26), except now $\alpha^2 = -k^2 < 0$. The quantity p^2 is still useful. The general solution in the two regions is

$$\psi(x) = \begin{cases} a_1 \sin(px) + a_2 \cos(px) & 0 < x < a \\ b_1 \sin(kx) + b_2 \cos(kx) & a < x \end{cases} \tag{2.40}$$

$$k = \sqrt{\frac{2mE}{\hbar^2}} \tag{2.41}$$

Again there are constant coefficients (a_1, a_2, b_1, b_2) that depend on the boundary conditions and on the matching at $x = a$. Again set $a_2 = 0$ to satisfy the boundary condition that the eigenfunction vanishes at the point $x = 0$, where the potential becomes infinite. The second equation is expressed in terms of two coefficients (b_1, b_2). Although it is perfectly correct to express the eigenfunction this way, as a sum of sine and cosine functions, the conventional way in physics is to express the result in terms of two other constants. One of these is the amplitude D, while the other is called the *phase shift* $\delta(k)$:

$$\psi(x) = \begin{cases} a_1 \sin(px) & 0 < x < a \\ D \sin(kx + \delta) & a < x \end{cases} \tag{2.42}$$

It is emphasized that the two forms for the eigenfunction, for $x > a$, are identical:

$$\psi(x) = b_1 \sin(kx) + b_2 \cos(kx) = D \sin(kx + \delta) \tag{2.43}$$

The right-hand expression is expanded using the trigonometric identity for the sine of the sum of two angles:

$$\sin(kx+\delta) = \sin(kx)\cos(\delta) + \cos(kx)\sin(\delta) \tag{2.44}$$

$$b_1 = D\cos(\delta), \quad b_2 = D\sin(\delta) \tag{2.45}$$

$$D = \sqrt{b_1^2 + b_2^2}, \quad \tan(\delta) = \frac{b_2}{b_1} \tag{2.46}$$

These formula establish the equivalence of the two forms for the eigenfunction. There are always two independent constants in a second-order differential equation. In the preferred form, they are D and δ.

The constants a_1 and D are normalization coefficients. They are real, as shown below. So the entire continuum wave function is real. This feature is a consequence of forcing the eigenfunction to vanish at some point, here at $x=0$.

If the value of $V_0=0$, then $p=k$ and the solution has $D=a_1$, $\delta=0$. The presence of any potential will cause the phase shift δ to have a value different than zero. The phase shift gets its name since it causes the phase of the wave function $(kx+\delta)$ to shift its value. Of course, if the shift is a multiple of π then there is no observable effect on the particle density $|\psi(x)|^2$. In chapter 10 we will learn that the phase shift of a potential is all that can be determined in an elastic scattering measurement.

Now return to solving for the continuum eigenfunction. The two forms for the eigenfunction, for $0<x<a$ and $a<x$, are matched in value and slope at $x=a$:

$$a_1 \sin(pa) = D\sin(ka+\delta) \tag{2.47}$$

$$a_1 p\cos(pa) = kD\cos(ka+\delta) \tag{2.48}$$

Divide these two equations to eliminate (a_1, D) and derive the equation for the phase shift:

$$\frac{\tan(pa)}{p} = \frac{\tan(ka+\delta)}{k} \tag{2.49}$$

$$\delta(k) = -ka + \arctan\left[\frac{k}{p}\tan(pa)\right] \tag{2.50}$$

The only quantities remaining to be determined are the amplitudes (a_1, D). Together they provide the normalization of the eigenfunction. This normalization is treated in a later section.

A numerical example is presented for the potential shown in figure 2.2. Take m to be the mass of the electron, $a=2.0$ Å, and $V_0=4.0$ eV. The potential strength in (2.36) is $g=2.05$, which is above the value for critical binding ($g_c=\pi/2$). There is one bound state. Solving eqn. (2.37) gives $\varepsilon=0.119$. The bound-state energy is $E=-0.48$ eV, which is near to the top of the potential well. Figure 2.3 shows the unrenormalized eigenfunction ($C_1=1$) for the bound state $\psi_B(x)$ as the solid line. Finding the correct normalization coefficient is a homework assignment. The eigenfunction $\psi_B(x)$ rises linearly from the origin, has a single maximum, and then decays to zero outside of the potential well.

Also shown, as the dashed line, is the continuum-state eigenfunction $\psi_k(x)$ with kinetic energy $E=1.0$ eV. This curve is also normalized arbitrarily ($a_1=1$). For values of x outside of the potential well, the continuum eigenfunction $\psi_k(x)$ oscillates according to

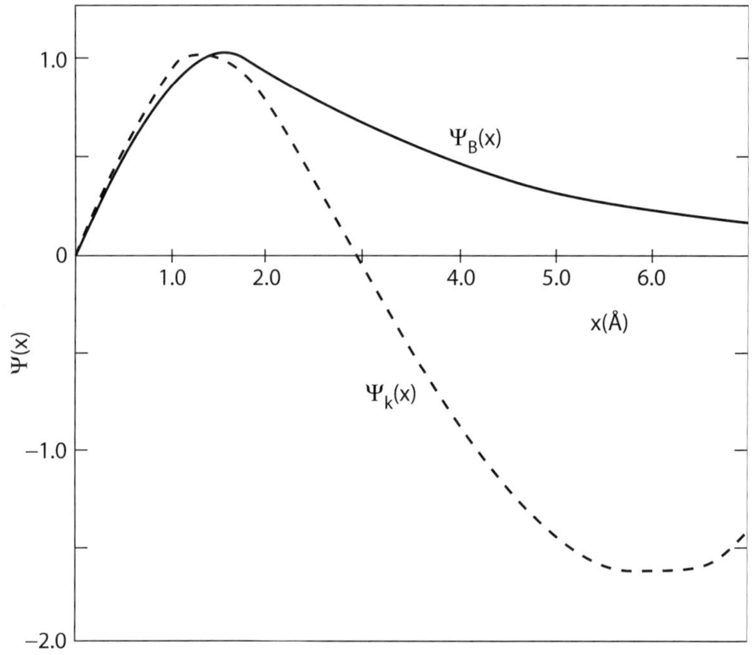

FIGURE 2.3. The solid line is the bound-state eigenfunction, while the dashed line is the continuum-state eigenfunction for $E = 1.0$ eV.

the simple formula of $\psi_k = D\sin(kx + \delta)$. Notice that the continuum eigenfunction changes sign quickly so that it satisfies the requirement that it be orthogonal to the bound state:

$$0 = \int_0^{\infty} dx \psi_B(x)\psi_k(x) \tag{2.51}$$

All solutions of the Schrödinger's equation must be orthogonal to all others with the same potential energy function.

The last example in this section on constant potentials is when a particle reflects at a barrier. Figure 2.4 shows a potential system $V(x) = V_0\Theta(x)$, where the Heaviside function $\Theta(x)$ is zero for $x < 0$ and is one for $x > 0$. $V_0 > 0$, so the potential is positive. A particle enters from the left with energy $E = \hbar^2k^2/2m > 0$ and amplitude I. Upon getting to the barrier at $x = 0$, some of the wave reflects with amplitude R. The behavior for $x > 0$ depends on whether $E > V_0$ or $E < V_0$. Here it is assumed that $E > V_0$, so the there is a wave on the right with wave vector p and amplitude T:

$$\psi(k, x) = \begin{cases} Ie^{ikx} + Re^{-ikx} & x < 0 \\ Te^{ipx} & x > 0 \end{cases} \tag{2.52}$$

$$p^2 = \frac{2m}{\hbar^2}(E - V_0), \quad k^2 = \frac{2mE}{\hbar^2} \tag{2.53}$$

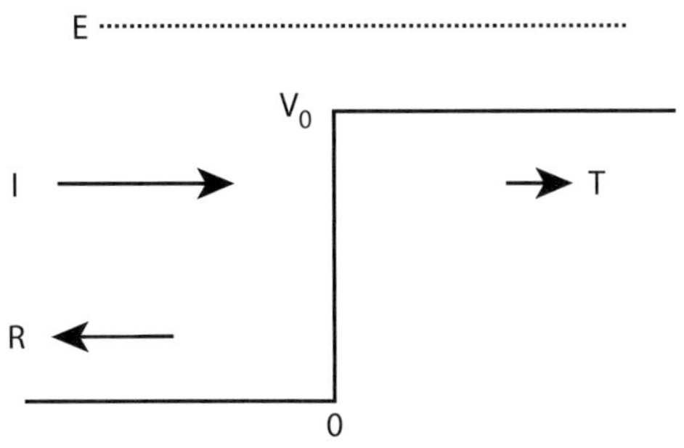

FIGURE 2.4. A particle entering from the left reflects from the barrier at $x = 0$.

The amplitude I is assumed known, and the values of R and T are found by matching the eigenfunction and its slope at $x=0$:

$$I+R=T \tag{2.54}$$

$$ik(I-R)=ipT \tag{2.55}$$

$$R=I\frac{k-p}{k+p} \tag{2.56}$$

$$T=I\frac{2k}{k+p} \tag{2.57}$$

The first two equations are from the matching conditions, and the last two are their solution. The eigenfunction is completely determined.

The current operator in one dimension is given in chapter 1:

$$j=\frac{\hbar}{2mi}\left[\psi^*\frac{d\psi}{dx}-\psi\frac{d\psi^*}{dx}\right] \tag{2.58}$$

The phrase "current operator" has two possible meanings. One is the flux of particles (j), so it has the units in one dimension of particles per second. The second possible meaning is the electrical current ($J=ej$), which has the units of amperes. These two definitions differ only by a prefactor, which is the charge e of the particle. Multiplying the particle current by the charge gives the electrical current. The above formula is for the particle current.

Let us apply this formula to the three terms in the eigenfunction:

- The incoming particles have a term Ie^{ikx} and a particle current

$$j_i=\frac{\hbar k}{m}|I|^2=v_k|I|^2, \quad v_k=\frac{\hbar k}{m} \tag{2.59}$$

- The reflected wave has a term Re^{-ikx} and a current

$$j_r=-\frac{\hbar k}{m}|R|^2=-v_k|R|^2 \tag{2.60}$$

- The transmitted wave Te^{ipx} has the particle current

$$j_t=\frac{\hbar p}{m}|T|^2=v_p|T|^2 \tag{2.61}$$

- The outgoing flux equals the incoming flux j_i

$$j_t - j_r = \frac{\hbar k}{m}|I|^2 \frac{(k-p)^2 + 4kp}{(k+p)^2} = \frac{\hbar k}{m}|I|^2 \tag{2.62}$$

The flux of the particles is conserved in the process of having the wave scatter from the potential step.

Another way to evaluate the current is to calculate the total particle current in the region $x < 0$:

$$j_< = \frac{v_k}{2}[(I^* e^{-ikx} + R^* e^{ikx})(Ie^{ikx} - Re^{-ikx}) \tag{2.63}$$

$$+ (Ie^{ikx} + Re^{-ikx})(I^* e^{-ikx} - R^* e^{ikx})]$$

$$= v_k[|I|^2 - |R|^2] = j_i + j_r \tag{2.64}$$

The current on the left is the sum of the incoming and reflected currents. The cross terms such as I^*R cancel.

Most potential functions, such as $V(x)$ in one dimension or $V(\mathbf{r})$ in three dimensions, neither create nor absorb particles. For these simple potentials, the current of particles is conserved. Just as many particles enter the system as leave. This sum rule is a useful check on the correctness of the eigenfunction.

Consider again the transmission and reflection of a particle from the barrier in figure 2.4. Now have the incident particle from the right, so the eigenfunction when $E > V_0$ has the form

$$\psi(k, x) = \begin{cases} I'e^{-ipx} + R'e^{ipx} & x > 0 \\ T'e^{-ikx} & x < 0 \end{cases} \tag{2.65}$$

and (k, p) have the same meaning as before. The amplitude I' is assumed known, and the values of R' and T' are found by matching the eigenfunction and its slope at $x = 0$:

$$I' + R' = T' \tag{2.66}$$

$$-ip(I' - R') = -ikT' \tag{2.67}$$

$$R' = I'\frac{p-k}{k+p} \tag{2.68}$$

$$T' = I'\frac{2p}{k+p} \tag{2.69}$$

It is interesting to compare these reflection and transmission amplitudes with those coming from the other direction. In both cases set the incident amplitudes to unity ($I = I' = 1$). Then we find

$$R' = -R, \qquad T' = \frac{p}{k}T \tag{2.70}$$

There is a simple relation between the amplitudes for the two directions.

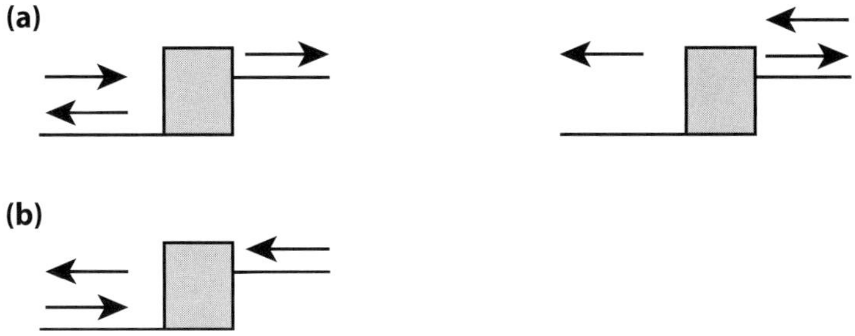

FIGURE 2.5. (a) The incident and reflected amplitudes for waves coming from either direction to an arbitrary barrier, represented as a box. (b) The time-reversed wave from the left.

The above derivation can be extended to the transmission and reflection of a barrier of any shape or thickness. Figure 2.5a shows the incident and transmitted waves for a barrier of any shape, which is represented by a gray box.

1. The first figure shows an incident wave from the left. It has unit amplitude, and reflection (R_L) and transmission (T_{LR}) amplitudes:

$$\psi_L = \begin{cases} e^{ikx} + R_L e^{-ikx} & x<0 \\ T_{LR} e^{ipx} & x>0 \end{cases} \tag{2.71}$$

2. The second figure shows an incident wave from the right. It has unit amplitude, and reflection (R_R) and transmission (T_{RL}) amplitudes:

$$\psi_R = \begin{cases} e^{-ipx} + R_R e^{ipx} & x>0 \\ T_{RL} e^{-ikx} & x<0 \end{cases} \tag{2.72}$$

In figure 2.5b, take the complex conjugate of the wave coming from the left. Since Schrödinger's equation is invariant under the complex conjugate operation, this should also be a valid physical eigenfunction. In fact, the complex conjugate is the time-reversal operation: the complex conjugate of $\exp(ikx)$ is $\exp(-ikx)$. A wave going to the right has been changed to a wave going to the left:

$$\psi_L^* = \begin{cases} e^{-ikx} + R_L^* e^{ikx} & x < 0 \\ T_{LR}^* e^{-ipx} & x > 0 \end{cases} \tag{2.73}$$

This state has an interesting feature: there is no wave leaving on the right. Both incident waves in part (a) have a wave going to the right. Although (b) has two incoming waves, none is leaving to the right. The two contributions must cancel. We can construct the time-reversed state by multiplying the two eigenfunctions in figure 2.5a by R_L^* and T_{LR}^* respectively:

$$\psi_L^* = R_L^* \psi_L + T_{LR}^* \psi_R \tag{2.74}$$

$$= \begin{cases} R_L^*(e^{ikx} + R_L e^{-ikx}) + T_{LR}^* T_{RL} e^{-ikx} & x<0 \\ T_{LR}^*(e^{-ipx} + R_R e^{ipx}) + R_L^* T_{LR} e^{ipx} & x>0 \end{cases} \tag{2.75}$$

$$= \begin{cases} R_L^* e^{ikx} + (|R_L|^2 + T_{LR}^* T_{RL}) e^{-ikx} & x < 0 \\ T_{LR}^* e^{-ipx} + (R_R T_{LR}^* + R_L^* T_{LR}) e^{ipx} & x > 0 \end{cases} \tag{2.76}$$

Compare these coefficients with those of the time-reversed state in (2.73) and find

$$1 = |R_L|^2 + T_{LR}^* T_{RL} \tag{2.77}$$

$$0 = R_L^* T_{LR} + R_R T_{LR}^* \tag{2.78}$$

Two other identities are found from the conservation of current:

$$k(1 - |R_L|^2) = p |T_{LR}|^2 \tag{2.79}$$

$$p(1 - |R_R|^2) = k |T_{RL}|^2 \tag{2.80}$$

The only solution to all of these equations is

$$T_{RL} = \frac{p}{k} T_{LR} \tag{2.81}$$

$$R_R = -R_L^* \frac{T_{LR}}{T_{LR}^*} \tag{2.82}$$

Time-reversal arguments have derived general relations between the reflected and transmitted wave amplitudes while approaching a barrier from either direction. They also agree with the simple example in (2.70). The last relationship gives $|R_R| = |R_L|$. The reflection amplitude is the same for the two sides of the barrier.

2.2 Linear Potentials

In one dimension, the linear potential has the form

$$V(x) = Fx \tag{2.83}$$

where F is a constant. There are numerous examples where this potential is applicable. One is a particle in a gravitational field near the surface of the earth, where $F = mg$. Another case is when a particle of charge q is in a constant electric field E, and $F = -qE$. Other cases are encountered later.

Schrödinger's equation is

$$0 = \left[-\frac{\hbar^2}{2m} \frac{d^2}{dx^2} + Fx - E \right] \psi(x) \tag{2.84}$$

It is convenient to use dimensionless variables. A unit of length is

$$x_0 = \left(\frac{\hbar^2}{2m|F|} \right)^{1/3} \tag{2.85}$$

$$z = \frac{x - E/F}{x_0} \tag{2.86}$$

The variable z is dimensionless. The form of Schrödinger's equation depends on the sign of the field:

$$0 = \left(\frac{d^2}{dz^2} - z\right)\psi(z), \quad F > 0 \tag{2.87}$$

$$0 = \left(\frac{d^2}{dz^2} + z\right)\psi(z), \quad F < 0 \tag{2.88}$$

Schrödinger's equation in (2.84) is converted into a simple equation (2.87). This latter equation is called Airy's equation. There are two forms, depending on whether F is positive or negative. However, by changing z to $-z$ in the second equation, it becomes identical to the first. So it is necessary to solve only one case. The other is obtained by changing the sign of z.

Since Airy's equation is a second-order differential equation, it has two solutions, called $Ai(z)$ and $Bi(z)$:

$$\psi(z) = C_1 Ai(z) + C_2 Bi(z) \tag{2.89}$$

The constants $C_{1,2}$ are again determined by the boundary conditions. A differential equation such as (2.84) is a mathematical problem that is incompletely specified: the boundary conditions are needed to completely determine the eigenfunction.

The two Airy functions can be expressed as Bessel's functions of index one-third. An alternative and equivalent procedure is to express them as integrals over a variable t:

$$Ai(z) = \frac{1}{\pi}\int_0^\infty dt \cos\left(\frac{t^3}{3} + zt\right) \tag{2.90}$$

$$Bi(z) = \frac{1}{\pi}\int_0^\infty dt \left[e^{-t^3/3 + zt} + \sin\left(\frac{t^3}{3} + zt\right)\right] \tag{2.91}$$

We prove that one of them, say $Ai(z)$, does satisfy the differential equation (2.87):

$$\frac{d^2}{dz^2} Ai(z) = -\frac{1}{\pi}\int_0^\infty dt t^2 \cos\left(\frac{t^3}{3} + zt\right) \tag{2.92}$$

$$\left[\frac{d^2}{dz^2} - z\right] Ai(z) = -\frac{1}{\pi}\int_0^\infty dt (t^2 + z) \cos\left(\frac{t^3}{3} + zt\right) \tag{2.93}$$

$$= -\frac{1}{\pi}\int_0^\infty dt \frac{d}{dt} \sin\left(\frac{t^3}{3} + zt\right) = 0 \tag{2.94}$$

The last line shows that the integrand is an exact differential, which vanishes at both end points. The value of the sine function as $t \to \infty$ is not obvious, since it oscillates. However, it is taken to be zero anyway.

Figure 2.6 shows a plot of the potential $V = Fx$ on top, and the two Airy functions below. Given a value of energy E, the turning point b is where $V(b) = E$, $b = E/F$. Turning points separate the regions where $V > E$ from those with $V < E$. In Airy's equation, the turning point separates $z < 0$ from $z > 0$. Figure 2.6 shows that $Ai(z)$ decays to zero for $z > 0$, while it oscillates for $z < 0$. $Bi(z)$ grows without limit for $z > 0$, and also oscillates for $z < 0$. Their asymptotic forms are needed later and are listed here. Define the variable $\rho = \left(\frac{2}{3}\right) z^{3/2}$

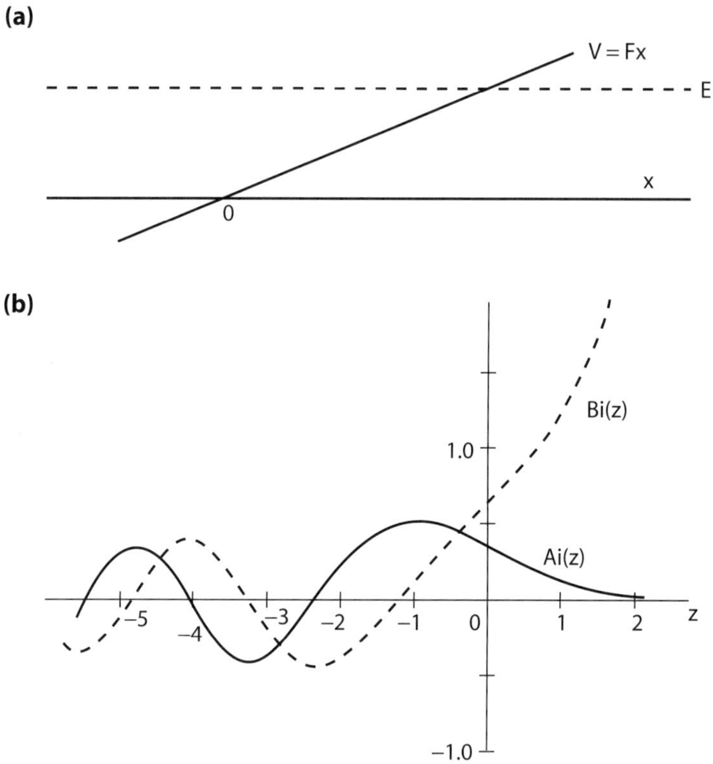

FIGURE 2.6. (a) The solid line is $V = Fx$ and the dashed line is E. They cross at the point $z = 0$. (b) The solid line is $Ai(z)$ and the dashed line is $Bi(z)$.

$$\lim_{z\to\infty}\begin{cases} Ai(z) = \dfrac{e^{-\rho}}{2\sqrt{\pi\sqrt{z}}}\left[1+O\left(\dfrac{1}{\rho}\right)\right] \\ Bi(z) = \dfrac{e^{\rho}}{\sqrt{\pi\sqrt{z}}}\left[1+O\left(\dfrac{1}{\rho}\right)\right] \end{cases} \tag{2.95}$$

$$\lim_{z\to\infty}\begin{cases} Ai(-z) = \dfrac{\sin(\rho+\pi/4)}{\sqrt{\pi\sqrt{z}}}\left[1+O\left(\dfrac{1}{\rho}\right)\right] \\ Bi(-z) = \dfrac{\cos(\rho+\pi/4)}{\sqrt{\pi\sqrt{z}}}\left[1+O\left(\dfrac{1}{\rho}\right)\right] \end{cases} \tag{2.96}$$

The interesting feature of Airy functions is that they depend on a single variable z. There is no other eigenvalue, such as k. Changing the energy E just moves the turning point, and the eigenfunction just slides over in x-space, without changing shape. Similarly, changing the field F just expands or compresses the scale over which the function is graphed. Changing the sign of F just changes z to $-z$.

Now consider Schrödinger's equation to be a physics problem. If the potential $V(x) = Fx$ exists throughout all space, the solution $Bi(z)$ must be discarded by setting $C_2 = 0$, since it diverges on the positive side of the turning point. Most physics problems contain just the term $Ai(z)$, since it is well behaved at large value of z.

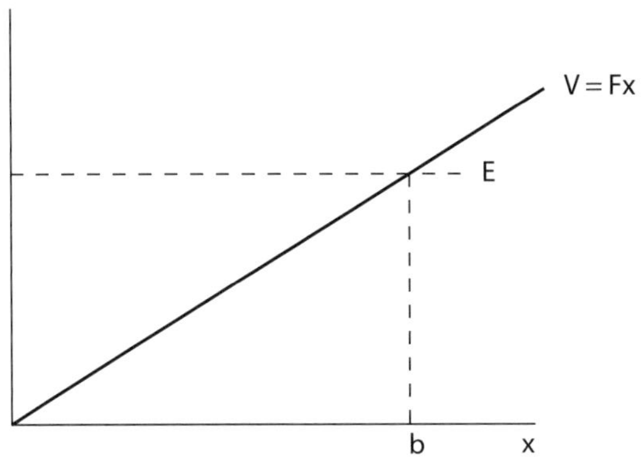

FIGURE 2.7. The potential $V(x)$ is linear for $x > 0$ and ∞ for $x < 0$.

As an example of the use of Airy functions, consider the potential

$$V(x) = \begin{cases} \infty & x < 0 \\ Fx & x > 0 \end{cases} \tag{2.97}$$

The potential is linear to the right of the origin $x = 0$ and diverges to the left, as shown in figure 2.7. The boundary condition at $x = 0$ forces the eigenfunction to vanish at that point:

$$\psi(x) = C_1 Ai\left(\frac{x - E/F}{x_0}\right) \tag{2.98}$$

$$\psi(0) = 0 = C_1 Ai\left(\frac{-E}{x_0 F}\right) \tag{2.99}$$

Figure 2.6 shows that $Ai(z)$ oscillates for negative values of z. Each point at which it crosses zero (i.e., each node) is a solution to the eigenvalue equation. Some mathematical tables give values for the zeros z_n of Airy functions. The allowed eigenvalues are

$$E_n = -x_0 F z_n \tag{2.100}$$

where $z_1 = -2.3381$, $z_2 = -4.0879$, $z_3 = -5.5206$, etc. Since $z_n < 0$, then $E_n > 0$. The constant C_1 is a normalization constant that is derived in a later section.

2.3 Harmonic Oscillator

Consider a classical system of a mass m connected to a spring of constant K. The spring obeys Hooke's law that the force is $F = -Kx$, for a displacement x, and the potential energy is $V(x) = Kx^2/2$. The Hamiltonian for this system is

$$H = \frac{p^2}{2m} + \frac{K}{2}x^2 \tag{2.101}$$

This Hamiltonian is called the harmonic oscillator. It is an important system in quantum mechanics, since it is applied to many problems in physics.

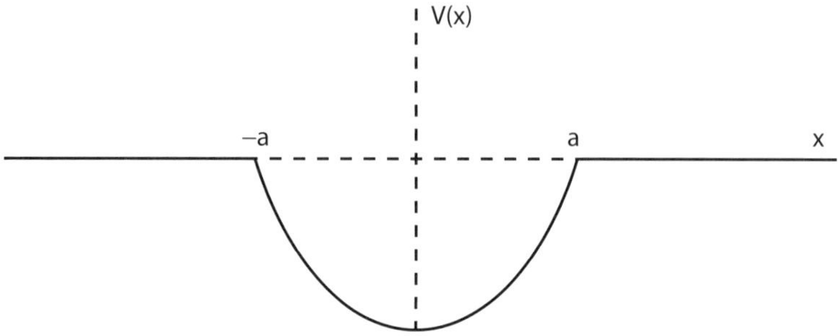

FIGURE 2.8. $V(x)$ with a segment having $V \propto x^2$.

At this point in my lecture, I always ask the student to name the polynomial that is the solution to this Hamiltonian. Many eager students volunteer the answer that it is a Hermite polynomial. This answer is usually incorrect.

The most general solutions to the differential equation

$$0 = \left[-\frac{\hbar^2}{2m}\frac{d^2}{dx^2} + \frac{K}{2}x^2 - E \right] \psi(x) \tag{2.102}$$

are parabolic cylinder functions. There are two of them, since the equation is second order. They are not given here since we will not use them, but they are given in books on mathematical physics. Solutions involving parabolic cylinder functions are required when the potential $V(x) \propto x^2$ exists only over a finite segment of space. For example,

$$V(x) = \frac{K}{2}(x^2 - a^2), \quad -a < x < a \tag{2.103}$$

$$= 0, \quad |x| > a \tag{2.104}$$

This potential is shown in figure 2.8. The two parabolic cylinder functions are needed for $-a < x < a$ and plane waves are needed elsewhere.

The harmonic oscillator Hamiltonian has a solution in terms of Hermite polynomials when the potential $V \propto x^2$ exists for all space—for all values of x. In this case, it is useful to define a frequency ω and length x_0:

$$\omega = \sqrt{\frac{K}{m}}, \quad x_0 = \sqrt{\frac{\hbar}{m\omega}} = \left(\frac{\hbar^2}{mK}\right)^{1/4} \tag{2.105}$$

It is convenient to rewrite eqn. (2.102) in dimensionless units by defining a dimensionless distance as $\xi = x/x_0$ and a dimensionless energy as $\varepsilon = 2E/\hbar\omega$. Multiply eqn. (2.102) by $-2/\hbar\omega$ and find

$$0 = \left[\frac{d^2}{d\xi^2} - \xi^2 + \varepsilon \right] \psi(\xi) \tag{2.106}$$

This equation has only bound states, with discrete eigenvalues. There are no continuum states. The eigenvalues and eigenvectors are

$$\varepsilon_n = 2n+1, \quad E_n = \hbar\omega\left(n+\frac{1}{2}\right) \tag{2.107}$$

$$\psi_n(\xi) = N_n H_n(\xi) e^{-\xi^2/2} \tag{2.108}$$

$$N_n = [\sqrt{\pi}(n!)2^n]^{-1/2} \tag{2.109}$$

where n is a nonnegative integer. Below are given the Hermite polynomials of lowest order, a recursion relation, and other relations:

$$H_0(\xi) = 1, \quad H_1 = 2\xi, \quad H_2 = 4\xi^2 - 2 \tag{2.110}$$

$$\frac{1}{2} H_{n+1} = \xi H_n - n H_{n-1} \tag{2.111}$$

$$\frac{d}{d\xi} H_n(\xi) = 2\xi H_n - H_{n+1} = 2n H_{n-1} \tag{2.112}$$

$$H_n(\xi) = (-1)^n e^{\xi^2} \frac{d^n}{d\xi^n} e^{-\xi^2} \tag{2.113}$$

$$e^{2z\xi - z^2} = \sum_{n=0}^{\infty} \frac{z^n}{n!} H_n(\xi) \tag{2.114}$$

The last equation is called a generating function.

The eigenfunction is normalized according to

$$\delta_{n\ell} = \int_{-\infty}^{\infty} d\xi \psi_n(\xi) \psi_\ell(\xi) \tag{2.115}$$

The functions $\psi_n(\xi)$ are dimensionless. Sometimes it is useful to have the dimensional functions of x, which are called $\phi_n(x)$. They are normalized acccording to

$$\phi_n(x) = \frac{1}{\sqrt{x_0}} \psi_n(x/x_0) \tag{2.116}$$

$$\delta_{n\ell} = \int_{-\infty}^{\infty} dx \phi_n(x) \phi_\ell(x) \tag{2.117}$$

This completes the discussion of eigenfunctions and eigenvalues.

Generating functions are useful for evaluating integrals of harmonic oscillator functions. They will be used to prove (2.115):

$$I_{n\ell} = \int_{-\infty}^{\infty} d\xi \psi_n(\xi) \psi_\ell(\xi) = N_n N_\ell \int_{-\infty}^{\infty} d\xi H_n(\xi) H_\ell(\xi) e^{-\xi^2} \tag{2.118}$$

The integral on the right is evaluated using generating functions. Multiply together the two generating functions below and then integrate over all $d\xi$:

$$e^{2z\xi - z^2} = \sum_n \frac{z^n}{n!} H_n(\xi) \tag{2.119}$$

$$e^{2y\xi - y^2} = \sum_\ell \frac{y^\ell}{\ell!} H_\ell(\xi) \tag{2.120}$$

$$\int_{-\infty}^{\infty} d\xi e^{-\xi^2} e^{2z\xi - z^2} e^{2y\xi - y^2} = \sum_{n\ell} \frac{z^n y^\ell}{n!\ell!} \int_{-\infty}^{\infty} d\xi H_n(\xi) H_\ell(\xi) e^{-\xi^2} \tag{2.121}$$

The integral on the left is a simple Gaussian, and is evaluated easily:

$$e^{-z^2-y^2}\int_{-\infty}^{\infty} d\xi e^{-\xi^2+2\xi(z+y)} = \sqrt{\pi}e^{-z^2-y^2+(z+y)^2} = \sqrt{\pi}e^{2zy}$$

$$= \sqrt{\pi}\sum_{k=0}^{\infty}\frac{(2zy)^k}{k!} \tag{2.122}$$

Compare the two series on the right side of eqns. (2.121) and (2.122). They must be equal. The series in (2.122) has only terms with identical exponents for z and y. The series in (2.121) must have the same feature, so the integral vanishes unless $n = \ell$. Equating each term in the two series gives the identity:

$$\frac{z^n y^\ell}{n!\ell!}\int_{-\infty}^{\infty} d\xi H_n(\xi)H_\ell(\xi)e^{-\xi^2} = \sqrt{\pi}\delta_{n\ell}\frac{(2zy)^n}{n!} \tag{2.123}$$

Canceling the factors of $(zy)^n$ we get

$$\int_{-\infty}^{\infty} d\xi H_n(\xi)H_\ell(\xi)e^{-\xi^2} = \sqrt{\pi}2^n n!\delta_{n\ell} = \frac{\delta_{n\ell}}{N_n^2} \tag{2.124}$$

This derivation shows that $I_{nm} = \delta_{nm}$ in eqn. (2.118), and also derives the value of the normalization constant N_n. Generating functions are useful for evaluating a variety of integrals involving Hermite polynomials.

The next quantity to evaluate is the integral of x with two different eigenfunctions:

$$\int_{-\infty}^{\infty} dx\phi_n(x)x\phi_\ell(x) = \langle n|x|\ell\rangle \tag{2.125}$$

This integral is called the matrix element of x. On the right is the Dirac notation for this matrix element. The two angular brackets indicate the two eigenfunctions. Dirac took the word "bracket" and divided it into "bra" $\langle n|$ and "ket" $|\ell\rangle$. For example, in this notation the orthogonality relation (2.117) is

$$\langle n|\ell\rangle = \delta_{n\ell} \tag{2.126}$$

The matrix element (2.125) occurs in many problems, and it is a good exercise to evaluate it now. The first step is to rewrite the recursion relation (2.111) as

$$\xi H_n(\xi) = \frac{1}{2}H_{n+1}(\xi) + nH_{n-1}(\xi) \tag{2.127}$$

The Hermite polynomials are converted to eigenfunctions by multiplying the above equation by the factor $N_n \exp(-\xi^2/2)$. This gives $\xi\psi_n(\xi)$ on the left of the equal sign. On the right, convert N_n to $N_{n\pm1}$ by adding the appropriate factors:

$$\xi\psi_n(\xi) = \frac{1}{\sqrt{2}}\left[\sqrt{n+1}\psi_{n+1}(\xi) + \sqrt{n}\psi_{n-1}(\xi)\right] \tag{2.128}$$

$$\xi|n\rangle = \frac{1}{\sqrt{2}}[\sqrt{n+1}|n+1\rangle + \sqrt{n}|n-1\rangle] \tag{2.129}$$

The two equations are identical, and the second line is in Dirac notation. Now it is simple to evaluate the matrix element in (2.125):

$$\langle n|x|\ell\rangle = x_0\langle n|\xi|\ell\rangle = \frac{x_0}{\sqrt{2}}\langle n|\left[\sqrt{\ell+1}|\ell+1\rangle + \sqrt{\ell}|\ell-1\rangle\right]$$
$$= \frac{x_0}{\sqrt{2}}\left[\sqrt{n}\delta_{n,\ \ell+1} + \sqrt{n+1}\delta_{n,\ \ell-1}\right] \tag{2.130}$$

The matrix element of x between two harmonic oscillator states n and ℓ has nonzero values only when $n=\ell\pm1$. The matrix element is proportional to the unit of length x_0.

The matrix element of x^k can be evaluated by iterating the above process. For example, if $k=2$, the matrix element of x^2 is found from

$$\xi^2|n\rangle = \frac{\xi}{\sqrt{2}}[\sqrt{n+1}|n+1\rangle + \sqrt{n}|n-1\rangle] \tag{2.131}$$
$$= \frac{1}{2}[\sqrt{(n+1)(n+2)}|n+2\rangle + \sqrt{n(n-1)}|n-2\rangle$$
$$+ (2n+1)|n\rangle] \tag{2.132}$$
$$\langle n|x^2|\ell\rangle = \frac{x_0^2}{2}[\sqrt{n(n-1)}\delta_{n,\ \ell+2} + \sqrt{(n+1)(n+2)}\delta_{n,\ \ell-2}$$
$$+ (2n+1)\delta_{n\ell}] \tag{2.133}$$

This completes the discussion of the matrix elements of x.

It is equally useful to know the matrix elements of the momentum operator p between two harmonic oscillator states:

$$\langle n|p|\ell\rangle = \frac{\hbar}{ix_0}\left\langle n\left|\frac{d}{d\xi}\right|\ell\right\rangle \tag{2.134}$$

The derivative of the harmonic oscillator function involves the derivative of the Hermite polynomial:

$$\frac{d\psi_n}{d\xi} = N_n\frac{d}{d\xi}\left(H_n(\xi)e^{-\xi^2/2}\right) \tag{2.135}$$
$$= N_n e^{-\xi^2/2}\left(-\xi H_n + \frac{d}{d\xi}H_n\right) \tag{2.136}$$

Evaluate the two terms in parentheses using eqns. (2.111) and (2.112) and find

$$\frac{d\psi_n}{d\xi} = N_n e^{-\xi^2/2}\left[nH_{n-1}(\xi) - \frac{1}{2}H_{n+1}(\xi)\right] \tag{2.137}$$
$$\frac{d}{d\xi}|n\rangle = \frac{1}{\sqrt{2}}[\sqrt{n}|n-1\rangle - \sqrt{n+1}|n+1\rangle] \tag{2.138}$$

The derivative in eqn. (2.134) is

$$\langle n|p|\ell\rangle = \frac{\hbar}{ix_0\sqrt{2}}[\sqrt{n+1}\delta_{n.\ell-1} - \sqrt{n}\delta_{n,\ell+1}] \tag{2.139}$$

The matrix element of p^2 is found by iterating this process:

$$\frac{d^2}{d\xi^2}|n\rangle = \frac{1}{2}[\sqrt{(n+1)(n+2)}|n+2\rangle + \sqrt{n(n-1)}|n-2\rangle - (2n+1)|n\rangle]$$

$$\langle m|p^2|n\rangle = \frac{1}{2}\left(\frac{\hbar}{ix_0}\right)^2 [\sqrt{(n+1)(n+2)}\delta_{m=n+2} + \sqrt{n(n-1)}\delta_{m=n-2} - (2n+1)\delta_{nm}] \tag{2.140}$$

The matrix elements of x^2 and p^2 are needed to evaluate the matrix element of the Hamiltonian (2.101):

$$\langle n|H|\ell\rangle = \frac{\langle n|p^2|\ell\rangle}{2m} + \frac{K}{2}\langle n|x^2|\ell\rangle \tag{2.141}$$

The dimensional prefactors in both terms are identical:

$$\frac{\hbar^2}{2mx_0^2} = \frac{\hbar^2}{2m}\left(\frac{m\omega}{\hbar}\right) = \frac{1}{2}\hbar\omega \tag{2.142}$$

$$\frac{1}{2}Kx_0^2 = \frac{K}{2}\sqrt{\frac{\hbar^2}{mK}} = \frac{\hbar}{2}\sqrt{\frac{K}{m}} = \frac{1}{2}\hbar\omega \tag{2.143}$$

Using eqns. (2.133) and (2.140) gives

$$\langle n|H|\ell\rangle = \hbar\omega\left(n + \frac{1}{2}\right)\delta_{n\ell} \tag{2.144}$$

The terms with $n = \ell \pm 2$ canceled to zero. There remains only the term with $n = \ell$. This answer is expected since $|\ell\rangle$ is an eigenstate of the Hamiltonian:

$$H|\ell\rangle = \hbar\omega\left(\ell + \frac{1}{2}\right)|\ell\rangle \tag{2.145}$$

These results for the harmonic oscillator are used many times in the remainder of the text.

2.4 Raising and Lowering Operators

The raising and lowering operators for the harmonic oscillator are introduced in this section. These two operators are defined, and many of the results of the prior section are expressed in terms of operators. There are two main reasons for introducing them. First, the Heisenberg approach to quantum mechanics, wherein problems are solved by manipulating operators, is a valid and important method. Advanced theoretical physics tends to use the operator formalism, and the student should get used to it. The second reason for introducing the raising and lowering operators for the harmonic oscillator is that they are very important in physics. For example, chapter 8 discusses the quantum approach to the electromagnetic field, where it is shown that photons obey harmonic oscillator statistics. The operator formalism is important and that of the harmonic oscillator is the most important of all.

Raising operators are also called *creation operators*, and lowering operators are called *destruction operators*. The title of this section could have been "Creation and Destruction Operators."

Start by recalling two equations from the prior section, (2.129) and (2.138):

$$\xi\psi_\ell(\xi) = \frac{1}{\sqrt{2}}[\sqrt{\ell}\psi_{\ell-1}(\xi) + \sqrt{\ell+1}\psi_{\ell+1}(\xi)] \tag{2.146}$$

$$\frac{d}{d\xi}\psi_\ell(\xi) = \frac{1}{\sqrt{2}}[\sqrt{\ell}\psi_{\ell-1}(\xi) - \sqrt{\ell+1}\psi_{\ell+1}(\xi)] \tag{2.147}$$

These two equations are first added and then subtracted to obtain two new equations:

$$\frac{1}{\sqrt{2}}\left(\xi + \frac{d}{d\xi}\right)\psi_\ell = \sqrt{\ell}\psi_{\ell-1}(\xi) \tag{2.148}$$

$$\frac{1}{\sqrt{2}}\left(\xi - \frac{d}{d\xi}\right)\psi_\ell = \sqrt{\ell+1}\psi_{\ell+1}(\xi) \tag{2.149}$$

Define the raising ($a^\dagger$) and lowering (a) operators as

$$a^\dagger = \frac{1}{\sqrt{2}}\left(\xi - \frac{d}{d\xi}\right) \tag{2.150}$$

$$a = \frac{1}{\sqrt{2}}\left(\xi + \frac{d}{d\xi}\right) \tag{2.151}$$

The dagger (†) denotes the Hermitian conjugate. The lowering operator a acts on a harmonic oscillator eigenfunction ψ_ℓ and produces the eigenfunction with quantum number with $\ell - 1$. It lowers the value of the integer ℓ by one. Similarly, the raising operator $a^\dagger$ operates on an eigenfunction ψ_ℓ and raises the value of ℓ by one. In regular and Dirac notation

$$a\psi_\ell(\xi) = \sqrt{\ell}\psi_{\ell-1}(\xi), \quad a|\ell\rangle = \sqrt{\ell}|\ell-1\rangle \tag{2.152}$$
$$a^\dagger\psi_\ell(\xi) = \sqrt{\ell+1}\psi_{\ell+1}(\xi), \quad a^\dagger|\ell\rangle = \sqrt{\ell+1}|\ell+1\rangle$$

These results are the basis for all operator algebras.

What happens if the lowering operator acts on the ground state ($\ell = 0$)?

$$a\psi_0 = ? \tag{2.153}$$

Equation (2.152) gives that $a|0\rangle = \sqrt{0}| - 1\rangle$. The square root of zero is zero, and there is no eigenfunction with $\ell = -1$. The correct answer is found from the definition

$$a|0\rangle = \frac{N_0}{\sqrt{2}}\left(\xi + \frac{d}{d\xi}\right)e^{-\xi^2/2} = \frac{N_0}{\sqrt{2}}(\xi - \xi)e^{-\xi^2/2} = 0 \tag{2.154}$$

The result is zero. One cannot lower the ground state.

Further results are obtained by iterating the above operations. The operator $(a^\dagger)^n$ raises n times the quantum number ℓ, to the value of $\ell + n$:

$$(a^\dagger)^n|\ell\rangle = (a^\dagger)^{n-1}\sqrt{\ell+1}|\ell+1\rangle \tag{2.155}$$

$$= [(n+\ell)(n+\ell-1)\cdots(l+2)(l+1)]^{1/2}|n+\ell\rangle \tag{2.156}$$

$$= \sqrt{\frac{(n+\ell)!}{\ell!}}|n+\ell\rangle \tag{2.157}$$

In (2.157) for the raising operator, set $\ell = 0$ and find

$$\frac{(a^\dagger)^n}{\sqrt{n!}}|0\rangle = |n\rangle \tag{2.158}$$

The eigenfunction of quantum number n is obtained by operating n-times by the raising operator on the ground state $|0\rangle$.

Similar steps with the lowering operator give

$$a^n|\ell\rangle = [\ell(\ell-1)(\ell-2)\cdots(\ell-n-1)]^{1/2}|\ell-n\rangle \tag{2.159}$$

$$= \begin{cases} 0 & \ell < n \\ \sqrt{\frac{\ell!}{(\ell-n)!}}|\ell-n\rangle & \ell \geq n \end{cases} \tag{2.160}$$

The operator language is an elegant way to express the eigenfunctions.

The operators can be expressed in terms of the x- and p-notation. Recall that $\xi = x/x_0$ and $\hbar d/d\xi = ix_0 p$; then

$$a = \frac{1}{x_0\sqrt{2}}\left(x + \frac{i}{m\omega}p\right)$$

$$a^\dagger = \frac{1}{x_0\sqrt{2}}\left(x - \frac{i}{m\omega}p\right) \tag{2.161}$$

Since x and p are Hermitian, then $a^\dagger$ is the Hermitian conjugate of a and vice versa. Add and subtract these identities and find

$$x = \frac{x_0}{\sqrt{2}}(a + a^\dagger)$$

$$p = \frac{\hbar}{ix_0\sqrt{2}}(a - a^\dagger) \tag{2.162}$$

The displacement and momentum operators are expressed by raising and lowering operators.

The raising and lowering operators do not commute. The operator $aa^\dagger$ on a function $f(\xi)$ gives a different result than $a^\dagger a$ acting on the same function. The difference of these two operations is just the function $f(\xi)$:

$$(aa^\dagger - a^\dagger a)f(\xi) = f(\xi) \tag{2.163}$$

This result is proven using the definitions of the operators:

$$aa^\dagger f(\xi) = \frac{1}{2}\left(\xi + \frac{d}{d\xi}\right)\left(\xi - \frac{d}{d\xi}\right)f(\xi) = \frac{1}{2}\left[(\xi^2+1)f - \frac{d^2f}{d\xi^2}\right]$$

$$a^\dagger af(\xi) = \frac{1}{2}\left(\xi - \frac{d}{d\xi}\right)\left(\xi + \frac{d}{d\xi}\right)f(\xi) = \frac{1}{2}\left[(\xi^2-1)f - \frac{d^2f}{d\xi^2}\right]$$

$$(aa^\dagger - a^\dagger a)f(\xi) = f(\xi) \tag{2.164}$$

Subtracting the first two lines gives the third line. This relationship is valid for any function $f(\xi)$, and it is usually written without the function:

$$aa^\dagger - a^\dagger a = 1 \tag{2.165}$$

$$[a, a^\dagger] = aa^\dagger - a^\dagger a = 1 \tag{2.166}$$

The bracket notation is a shorthand for the commutation relation. Later it is written for other operators, such as

$$[A, B] = AB - BA \tag{2.167}$$

An example from section 1.4 is $[x, p] = i\hbar$. Another proof of $[a, a^\dagger] = 1$ is obtained using the x and p definitions of the operators in (2.161):

$$[a, a^\dagger] = \frac{1}{2x_0^2}\left[x + \frac{ip}{m\omega}, x - \frac{ip}{m\omega}\right] \tag{2.168}$$

$$= \frac{i}{2m\omega x_0^2}\{[p, x] - [x, p]\} = 1 \tag{2.169}$$

The Hamiltonian H is also elegantly expressed in terms of raising and lowering operators. The first step is to write

$$x^2 = \frac{x_0^2}{2}(aa + aa^\dagger + a^\dagger a + a^\dagger a^\dagger) \tag{2.170}$$

$$p^2 = -\frac{\hbar^2}{2x_0^2}(aa - aa^\dagger - a^\dagger a + a^\dagger a^\dagger)$$

These operator relations are used in the Hamiltonian. For the potential energy term, the prefactor is $Kx^2/2 \rightarrow Kx_0^2/4 = \hbar\omega/4$. for the kinetic energy term, the prefactor is $p^2/2m \rightarrow -\hbar^2/4mx_0^2 = -\hbar\omega/4$. Using the expressions in (2.170) gives

$$H = \frac{\hbar\omega}{4}[-(aa - aa^\dagger - a^\dagger a + a^\dagger a^\dagger) + (aa + aa^\dagger + a^\dagger + a^\dagger a^\dagger)]$$

$$= \frac{\hbar\omega}{2}(aa^\dagger + a^\dagger a) \tag{2.171}$$

All terms canceled that were aa or $a^\dagger a^\dagger$. The above result for H is correct, but is usually written differently. From the commutation relation,

$$aa^\dagger = a^\dagger a + 1 \tag{2.172}$$

$$H = \hbar\omega\left(a^\dagger a + \frac{1}{2}\right) \tag{2.173}$$

The combination of $a^\dagger a$ operating on an eigenstate gives

$$a^{\dagger}a|n\rangle = a^{\dagger}\sqrt{n}|n-1\rangle = n|n\rangle \tag{2.174}$$

$$H|n\rangle = \hbar\omega\left(n+\frac{1}{2}\right)|n\rangle = E_n|n\rangle \tag{2.175}$$

Thus, $H|n\rangle = E_n|n\rangle$, and the eigenvalue is again found to be $E_n = \hbar\omega(n+\frac{1}{2})$. Even for $n=0$ the eigenvalaue is $E_0 = \hbar\omega/2$, which is called the *zero-point energy*. There is always some motion, even in the lowest energy state. This feature is in accord with the uncertainty principle that the momentum is not zero when the particle is localized to a region.

The last exercise in this section, using operators, is the evaluation of an important integral. It is the matrix element of exp(iqx) between two eigenstates:

$$M_{n\ell}(q) = \int_{-\infty}^{\infty} dx e^{iqx}\phi_n(x)\phi_\ell(x) \tag{2.176}$$

The phrase "matrix element" is the evaluation of an integral that contains any function plus two eigenstates. It may be evaluated in several different ways. With luck, it might be found in a table of integrals in the section on Hermite polynomials. Another approach is using generating functions. Here we give a third approach invented by Feynman.

The first step is to rewrite this expression in terms of operators, using the bra and ket notation. In the exponent, the operator x is written using eqn. (2.162):

$$M_{n\ell}(\lambda) = \langle n|e^{i\lambda(a+a^{\dagger})}|\ell\rangle \tag{2.177}$$

$$\lambda = \frac{qx_0}{\sqrt{2}} \tag{2.178}$$

The matrix element depends on λ. If $\lambda=0$ then $M_{n\ell}(0) = \delta_{n\ell}$.The next step is to separate the exponential into two factors that separately contain exp($i\lambda a$) and exp($i\lambda a^{\dagger}$). These factors contain operators that do not commute, so the separation must be done carefully. Richard Feynman and Roy Glauber independently proved the same theorem in 1951. Feynman called it the *disentangling* of operators. FEYNMAN-GLAUBER THEOREM *If A and B are operators, such that* $[A, B] = C$, *and* $[A, C] = 0$, $[B, C] = 0$, *then*

$$e^{A+B} = e^{A}e^{B}e^{-C/2} \tag{2.179}$$

Usually the theorem is applied to cases where C is a constant, which automatically commutes with any operator. In our case, select

$$A = i\lambda a^{\dagger}, \quad B = i\lambda a, \quad C = (i\lambda)^2[a^{\dagger}, a] = \lambda^2 \tag{2.180}$$

The last relation used $[a^{\dagger}, a] = -1$. With the help of Feynman's theorem, the matrix element is now

$$M_{n\ell} = e^{-\lambda^2/2}\langle n|e^{i\lambda a^{\dagger}}e^{i\lambda a}|\ell\rangle \tag{2.181}$$

The next step is to examine the two factors in the matrix element. Expand the exponents in a power series and use the result in (2.160) for $a^n|\ell\rangle$ to evaluate each term in the series:

$$e^{i\lambda a}|\ell\rangle = \sum_{j=0}^{\infty} \frac{(i\lambda)^j}{j!} a^j|\ell\rangle \tag{2.182}$$

$$= \sum_{j=0}^{\ell} \frac{(i\lambda)^j}{j!} \sqrt{\frac{\ell!}{(\ell-j)!}}|\ell-j\rangle \tag{2.183}$$

The series is ended after $j = \ell$ terms, since the lowering operator gives zero when operating on the state $|0\rangle$.

The left-hand side of the matrix element (2.181) is given by a similar expression. Take the Hermitian conjugate of eqn. (2.183) and replace ℓ by n, and j by α:

$$\langle n|e^{i\lambda a^\dagger} = \sum_{\alpha=0}^{n} \frac{(i\lambda)^\alpha}{\alpha!} \sqrt{\frac{n!}{(n-\alpha)!}}\langle n-\alpha| \tag{2.184}$$

The above two results are combined to obtain a double series:

$$M_{n\ell}(\lambda) = e^{-\lambda^2/2} \sum_{\alpha,j} \frac{(i\lambda)^{\alpha+j}}{\alpha! j!} \sqrt{\frac{n!\ell!}{(n-\alpha)!(\ell-j)!}}\langle n-\alpha|\ell-j\rangle \tag{2.185}$$

The orthogonality relation (2.115) requires the factor $\langle n-\alpha|\ell-j\rangle$ to be zero unless $n-\alpha = \ell-j$. At this point it is required to specify whether n or ℓ is the larger integer. Make an arbitrary choice of $n \geq \ell$. Then use the orthogonality relation to eliminate the summation over $\alpha = n-\ell+j$:

$$M_{n\ell}(\lambda) = e^{-\lambda^2/2}\sqrt{n!\ell!}(i\lambda)^{n-\ell} \sum_{j=0}^{\ell} \frac{(i\lambda)^{2j}}{j!(\ell-j)!(n-\ell+j)!} \tag{2.186}$$

The series is the same one that defines the associated Laguerre polynomial:

$$L_\ell^{n-\ell}(\lambda^2) = n! \sum_{j=0}^{\ell} \frac{(-\lambda^2)^j}{j!(\ell-j)!(n-\ell+j)!} \tag{2.187}$$

$$M_{n\ell}(\lambda) = \sqrt{\frac{\ell!}{n!}}(i\lambda)^{n-\ell} e^{-\lambda^2/2} L_\ell^{n-\ell}(\lambda^2) \tag{2.188}$$

This formula is the final result when $n \geq \ell$. Harmonic oscillator eigenfunctions are found for quantized vibrations of atoms, for electrons in magnetic fields, and for electrons in parabolic potentials. The above matrix element is needed in the discussion of scattering in such systems. The matrix element is introduced here as an exercise in the manipulation of operators, and as a historic example from Feynman's paper on the disentangling of operators.

2.5 Exponential Potential

A simple exponential potential

$$V(x) = Ae^{-\alpha x} \tag{2.189}$$

can be solved exactly in one dimension. This potential is more realistic than a square well, since many potentials in nature are approximately an exponential. Two cases are

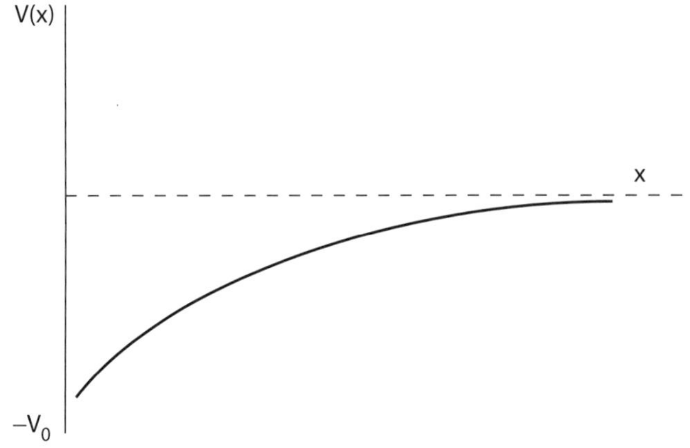

FIGURE 2.9. Attractive exponential potential.

considered: the bound states of an attractive potential, and the scattering states for a repulsive potential. A third case, the continuum states of the attractive potential, is treated in the problem assignment.

2.5.1 *Bound State*

For the attractive potential, the prefactor $A = -V_0$ is negative. The exponent has the factor $\alpha = 2/a$, where a is some characteristic length associated with the range of the potential. It is unphysical to let the potential go to large negative values when $x \to -\infty$, so the potential must be terminated at some point. In the present example, an arbitrary choice is made to terminate it at the origin with an infinite repulsive potential. The potential for the present example is

$$V(x) = \begin{cases} \infty & x < 0 \\ -V_0 e^{-2x/a} & x > 0 \end{cases} \tag{2.190}$$

The constant V_0 is positive so the potential is negative for $x > 0$. This potential is shown in figure 2.9. The infinite potential at the origin again forces the eigenfunctions to have the property $\phi_n(x=0) = 0$.

Schrödinger's equation is solved by changing the variable to

$$y = e^{-x/a} \tag{2.191}$$

Since $0 < x < \infty$, then $1 > y > 0$. Schrödinger's equation in the variable x must be changed to an equation in y. The derivative terms are

$$\frac{d}{dx} = \frac{dy}{dx}\frac{d}{dy} = -\frac{y}{a}\frac{d}{dy} \tag{2.192}$$

$$\frac{d^2}{dx^2} = \frac{1}{a^2}\left[y^2\frac{d^2}{dy^2} + y\frac{d}{dy}\right] \tag{2.193}$$

$$H = -E_a\left[y^2\frac{d^2}{dy^2} + y\frac{d}{dy} + \frac{V_0}{E_a}y^2\right] \tag{2.194}$$

$$E_a = \frac{\hbar^2}{2ma^2} \tag{2.195}$$

Define the potential strength g and the dimensionless eigenvalue β:

$$g^2 = \frac{V_0}{E_a} = \frac{2mV_0a^2}{\hbar^2} \tag{2.196}$$

$$\beta^2 = -\frac{E}{E_a} \tag{2.197}$$

$$0 = \left[y^2\frac{d^2}{dy^2} + y\frac{d}{dy} + g^2y^2 - \beta^2\right]\psi(y) \tag{2.198}$$

For bound states the eigenvalue $E < 0$ so that $\beta^2 > 0$ and β is assumed to be positive. The same potential strength was used earlier in a problem for the square well.

The differential equation in (2.198) is recognized as Bessel's equation, whose solutions are Bessel functions $J_\beta(gy)$. The most general solution, when β is not an integer, is

$$\psi(y) = C_1J_\beta(gy) + C_2J_{-\beta}(gy) \tag{2.199}$$

If β happens to be an integer, the independent solutions are the Bessel function and the Neumann function.

Now it is time to apply the boundary conditions. First consider what happens at large positive values of $x >> a$. At these values y becomes very small. For small values of their argument, the Bessel functions become

$$\lim_{z\to 0} J_\nu(z) = \frac{(z/2)^\nu}{\Gamma(1+\nu)} \tag{2.200}$$

$$\psi(x) = \frac{C_1}{\Gamma(1+\beta)}(g/2)^\beta e^{-\beta x/a} + \frac{C_2}{\Gamma(1-\beta)}(2/g)^\beta e^{\beta x/a} \tag{2.201}$$

The second term contains the factor of $\exp(\beta x/a)$ that diverges as $x >> a$. Since $\psi(x)$ must vanish at large values of x, as required for a bound state, this divergent term must be eliminated by setting $C_2 = 0$. The eigenfunction in eqn. (2.199) is now

$$\psi(x) = C_1J_\beta(ge^{-x/a}) \tag{2.202}$$

The other boundary condition is applied at the point $x = 0$ where the eigenfunction must vanish due to the infinite potential. This second boundary condition requires that

$$0 = J_\beta(g) \tag{2.203}$$

In this expression, the potential strength g is fixed, and the index β is varied until a zero is found for the Bessel function. For a given value of g, there is a valid eigenvalue for every different value of β that satisfies this equation. This situation is similar to the bound states of the square well, which is discussed in section 2.1.

There is a critical value of coupling strength, called g_c, such that no bound states exist for $g < g_c$. The threshold bound state at $g = g_c$ has a zero binding energy ($\beta = 0$). The value

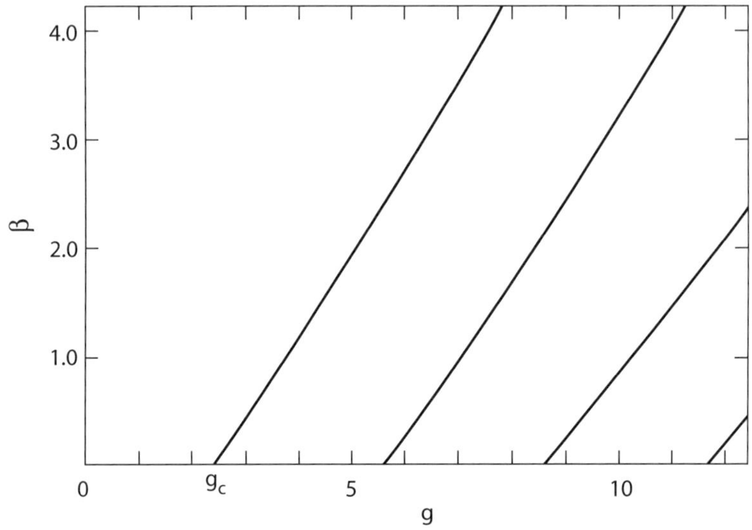

FIGURE 2.10. Bound-state eigenvalues $E = -\beta^2\hbar^2/2ma^2$ of the exponential potential as a function of potential strength g.

of g_c is found from the first zero of the Bessel function $J_0(g_c) = 0$ of zero index. This criterion gives $g_c = 2.4048....$

By using a table of the zeros of Bessel functions, one can produce a graph of the allowed values of β as a function of g. This graph is shown in figure 2.10. The solid lines are the allowed bound states. The lines are almost straight. For every value of g there are only a small number of allowed bound states and there are none for $g < g_c$.

2.5.2 Continuum State

To provide a bit of variety, and to contrast with the last subsection, Schrödinger's equation is now solved for a repulsive exponential potential. The potential function is

$$V(x) = V_0 e^{-2x/a} \tag{2.204}$$

The prefactor V_0 is positive, as is the range a. This potential diverges as x goes to negative infinity, thereby forcing $\psi(x)$ to be zero. It is unnecessary to add additional regions of infinite potential. The above potential function is assumed for all values of x.

The potential energy has no regions with negative values, so there are no bound states. All solutions are continuum states with $E > 0$. Schrödinger's equation is

$$0 = \left[-\frac{\hbar^2}{2m}\frac{d^2}{dx^2} + V_0 e^{-2x/a} - E \right] \psi(k, x) \tag{2.205}$$

Again use the variable transformation $y = \exp(-x/a)$, and the coupling strength has the same definition:

$$g^2 = \frac{2ma^2 V_0}{\hbar^2}, \quad k^2 = \frac{2mE}{\hbar^2}, \quad K = ka \tag{2.206}$$

Also introduce a dimensionless wave vector K. Schrödinger's equation is again changed to the variable y:

$$0=\left[y^2\frac{d^2}{dy^2}+y\frac{d}{dy}-g^2y^2+K^2\right]\psi(k,\,y) \tag{2.207}$$

The last two terms changed sign compared to (2.198). One change is due to the change of sign of V_0, so g^2y^2 changed sign. The other is the difference between bound and continuum states, so $-\beta^2$ became $+K^2$.

Equation (2.207) is still a form of Bessel function. The change of sign of g^2y^2 causes the solution to be Bessel functions of imaginary argument, which are $I_\nu(z)$ and $K_\nu(z)$. If ν is not an integer, one can use $I_{\pm\nu}$ instead of K_ν. The sign change on the energy term means that the index becomes imaginary: $\nu=\pm iK$. So the most general solution to (2.207) is

$$\psi(k,x)=B_1 I_{iK}(gy)+B_2 I_{-iK}(gy) \tag{2.208}$$

Now apply the boundary conditions to determine the constant coefficients $B_{1,2}$. First take the limit $x\to-\infty$ so that $y\to+\infty$. The potential energy diverges in this region, so the eigenfunction must vanish. The asymptotic form for the Bessel function of large argument is

$$\lim_{z\to\infty} I_\nu(z)=\frac{e^z}{\sqrt{2\pi z}}\left[1+O\left(\frac{1}{z}\right)\right] \tag{2.209}$$

The first term in the asymptotic expansion is independent of ν and is the same for both terms in (2.208). The eigenfunction is made to vanish by the condition that $B_1+B_2=0$. It is convenient to redefine $B_1=iC=-B_2$, and the eigenfunction is

$$\psi(k,x)=iC[I_{iK}(ge^{-x/a})-I_{-iK}(ge^{-x/a})] \tag{2.210}$$

The prefactor in front is inserted to make $\psi(k,\,x)$ a real function, since the square bracket is obviously imaginary: it is a function minus its complex conjugate. The constant C is for normalization, and is defined in a later section. The above result is the final formula for the continuum eigenfunction.

An important property of this eigenfunction is its behavior as $x\to+\infty$, which is $y<<1$. The imaginary Bessel functions are expanded for small argument:

$$\lim_{z\to 0} I_\nu(z)=\frac{(z/2)^\nu}{\Gamma(1+\nu)} \tag{2.211}$$

$$\lim_{x\to\infty}\psi(k,\,x)=iC\left[\frac{(g/2)^{iK}}{\Gamma(1+iK)}e^{-iKx/a}-\frac{(g/2)^{-iK}}{\Gamma(1-iK)}e^{iKx/a}\right] \tag{2.212}$$

This expression is a bit formidable at first sight. It simplifies when one notices that most of the terms are constant phase factors. For example, the first term has the factors

$$\frac{(g/2)^{iK}}{\Gamma(1+iK)}=\frac{e^{-i\delta}}{|\Gamma(1+iK)|} \tag{2.213}$$

$$\delta(k)=-K\ln(g/2)-\frac{i}{2}\ln\left[\frac{\Gamma(1+iK)}{\Gamma(1-iK)}\right] \tag{2.214}$$

The quantity $\delta(k)$ is called the *phase shift*. The eigenfunction at large argument can be written in terms of the phase shift:

$$\lim_{x\to\infty} \psi(x) = \frac{2C}{|\Gamma(1+iK)|}\sin(kx+\delta) \tag{2.215}$$

Phase shifts were introduced in section 2.1 in the discussion of continuum states of the square well. There it is noted that when the potential $V(x)$ is a contant, the solution to Schrödinter's equation can be written in terms of $\sin(kx)$, $\cos(kx)$, or $\sin(kx+\delta)$. The same reasoning applies to the present case of the exponential potential, since $V(x)$ goes to zero at large values of positive x. In this region one can always write the solution in the form of a constant times $\sin(kx+\delta)$. Our eigenfunction (2.210) behaves correctly in this limit. The derivation provides the exact formula for the phase shift $\delta(k)$.

The quantum mechanical problem that is solved above is the behavior of a frictionless object that is slid along the ground toward a smooth hill. The "hill" is the increasing potential for negative values of x. Far from the hill ($x>>a$), where the potential is negligible, the wave function is a combination of $\sin(kx)$ or $\cos(kx)$. Alternatively, it can be written as the summation of two exponentials:

$$\psi(x) = Ie^{-ikx}+Re^{ikx} \tag{2.216}$$

The two constant coefficients denote the incoming amplitude (I) and the reflected amplitude (R). The term $I\exp(-ikx)$ denotes a wave traveling to the left, toward the hill. The term $R\exp(ikx)$ denotes the wave that has been reflected from the hill. One sends the object with a certain kinetic energy that is determined by the wave vector k. The intensity I determines how many particles are sent toward the hill, and R gives the number that return. The particle current is

$$j(x) = \frac{\hbar}{2im}\left[\psi^*\frac{d\psi}{dx}-\psi\frac{d\psi^*}{dx}\right] \tag{2.217}$$

$$= -v_k[|I|^2-|R|^2] \tag{2.218}$$

where the velocity is $v_k = \hbar k/m$. The current $j(x)$ does not depend on x since all of the cross terms cancel to zero. The two terms represent the incoming current $(-v_k|I|^2)$ and reflected current $(v_k|R|^2)$. It is assumed that all of the particles sent toward the hill come back. The hill neither creates nor destroys particles, so $|R|^2 = |I|^2$. Therefore, the current $j(x) = 0$.

Since $|R| = |I|$ they have the same magnitude and can differ only by a phase factor:

$$R = -Ie^{2i\delta} \tag{2.219}$$

$$\psi(k,x) = I(e^{-ikx}-e^{ikx+2i\delta}) \tag{2.220}$$

$$= -2iIe^{i\delta}\sin(kx+\delta) \tag{2.221}$$

The wave function is $\sin(kx+\delta)$ times a constant prefactor. This form is required whenever the potential reflects all of the wave amplitude that is directed toward it. The potential can neither absorb nor emit particles, nor can a particle tunnel through it and come out the other side. Also note that a wave function of the form $\psi(k, x) = C' \sin(kx + \delta)$ has a zero value of the current.

It is useful to consider an experiment whose objective is to determine the shape of the hill. The hill is hidden in clouds and is not directly observable. A physicist decides to determine the shape of the hill by sliding particles at the frictionless hill with different velocities v_k and awaiting their return. In classical physics, the measurement is how long $t(v_k)$ it takes for the particle to return as a function of velocity. In quantum mechanics, the only measurable quantity is the phase shift $\delta(k)$. Wave packets are discussed in a later chapter. Using wave packets, we can also approximately $(\Delta t \Delta E \geq \hbar)$ measure the time the object returns.

2.6 Delta-Function Potential

All of the solutions to Schrödinger's equation encountered so far in these notes, have made the following two assumptions regarding the eigenfunctions or wave functions:

1. $\psi(x)$ is continuous everywhere.
2. $d\psi/dx$ is continuous everywhere.

These two assumptions are valid as long as several conditions are met. The first is that Schrödinger's equation is a second-order differential equation, which is always true in nonrelativistic physics. The second assumption is conditional on the form of $V(x)$: the potential has to be continuous except for occasional discontinuities. The wave function and its first derivative are continuous as long as these two conditions are satisfied. For example, the potential for the square well has a discontinuity in $V(x)$ at $x = a$, yet ψ and $d\psi/dx$ are both continuous at $x = a$ and elsewhere.

However, when $V(x)$ contains a potential term with a delta-function $\delta(x - x_0)$, then the derivative $d\psi/dx$ is not continuous at $x = x_0$. The wave function is continuous, so it has a cusp at the point of the delta-function. This feature is derived by writing Schrödinger's equation as

$$\frac{d^2\psi(x)}{dx^2} = -\frac{2m}{\hbar^2}[E - V(x)]\psi(x) \tag{2.222}$$

Integrate this equation, term by term, over the interval from $x_0 - \varepsilon$ to $x_0 + \varepsilon$. Here ε is small, and eventually we take $\varepsilon \to 0$:

$$\begin{aligned}\int_{x_0-\varepsilon}^{x_0+\varepsilon} dx \frac{d^2\psi}{dx^2} &= \left(\frac{d\psi}{dx}\right)_{x_0+\varepsilon} - \left(\frac{d\psi}{dx}\right)_{x_0-\varepsilon} \\ &= -\frac{2m}{\hbar^2} \int_{x_0-\varepsilon}^{x_0+\varepsilon} dx[E - V(x)]\psi(x)\end{aligned}$$

Examine the integral on the right. There are two possible outcomes when trying to evaluate this integral.

- If $V(x)$ is a smooth function of x or has only finite discontinuities, then one can use the mean value theorem over the small interval of the integral:

$$\left(\frac{d\psi}{dx}\right)_{x_0+\varepsilon} - \left(\frac{d\psi}{dx}\right)_{x_0-\varepsilon} = -\frac{2m(2\varepsilon)}{\hbar^2}[E-V(x_0)]\psi(x_0)$$

  In the limit that $\varepsilon \to 0$, the right-hand side vanishes. Then we have shown that the derivative is continuous at the point x_0.

- Let there be a delta-function potential at the point x_0. It has the property that

$$V(x) = \lambda\delta(x-x_0) \tag{2.223}$$

$$\int_{x_0-\varepsilon}^{x_0+\varepsilon} dx\delta(x-x_0) = 1 \tag{2.224}$$

$$\left(\frac{d\psi}{dx}\right)_{x_0+\varepsilon} - \left(\frac{d\psi}{dx}\right)_{x_0-\varepsilon} = \frac{2m\lambda}{\hbar^2}\psi(x_0) \tag{2.225}$$

  Equation (2.225) is the fundamental relationship for the change of slope of a wave function at the point of a delta-function potential. The change of slope is proportional to the amplitude of the delta-function λ as well as the wave function $\psi(x_0)$ at this point.

The slope of the wave function is continuous at all points except those where the potential energy has a delta-function.

As an example, find the bound states of the potential:

$$V(x)=V_0-W\delta(x) \tag{2.226}$$

The constant $W > 0$ is the strength of the delta-function, which is located at the origin. There is also a constant V_0 everywhere. At large distances the potential is V_0. So bound states have $E< V_0$ and continuum states have $E> V_0$.

The way to treat the delta-function is to first solve Schrödinger's equation at the points away from the singularity. For $x\neq 0$ one has the equation

$$\frac{d^2}{dx^2}\psi(x) = \frac{2m}{\hbar^2}(V_0-E)\psi \equiv \alpha^2\psi(x) \tag{2.227}$$

$$\alpha^2 = \frac{2m}{\hbar^2}(V_0-E) \tag{2.228}$$

$$\psi(x) = Ae^{-\alpha x} + Be^{\alpha x} \tag{2.229}$$

Bound-state wave functions must vanish as $|x| \to \infty$:

- For $x > 0$ then $B=0$ in this region: for $x > 0$ the solution has the form of $\exp(-\alpha x)$.
- For $x < 0$ then $A=0$ in this region: for $x<0$ the solution has the form of $\exp(\alpha x)$.
- Since the wave function is continuous everywhere, the solution must have the form

$$\psi(x) = Ae^{-\alpha|x|} \tag{2.230}$$

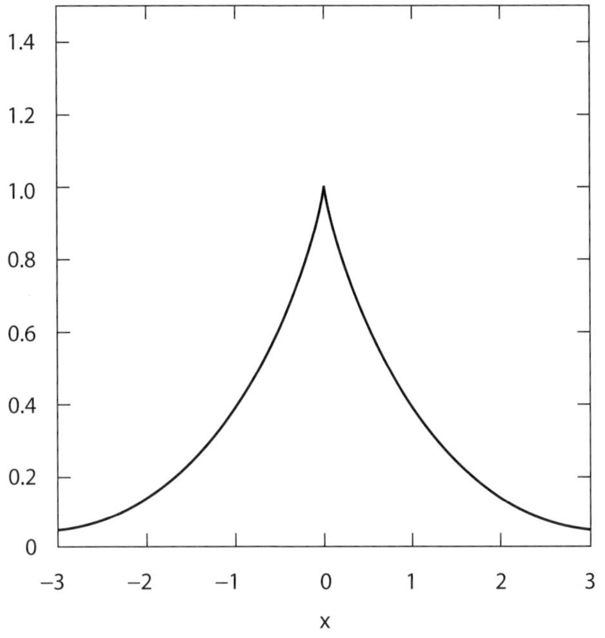

FIGURE 2.11. Bound-state eigenfunction for delta-function potential.

The only possible eigenfunction has a cusp at the origin, figure 2.11. A cusp is possible only with a delta-function potential. Use eqn. (2.225) to find the change of slope at $x=0$, where now the amplitude of the delta-function is W:

$$\left(\frac{d\psi}{dx}\right)_{+\varepsilon} = -\alpha A, \quad \left(\frac{d\psi}{dx}\right)_{-\varepsilon} = \alpha A \tag{2.231}$$

$$-2\alpha A = -\frac{2mW}{\hbar^2} A \tag{2.232}$$

$$\alpha = \frac{mW}{\hbar^2} \tag{2.233}$$

$$E = V_0 - \frac{\hbar^2 \alpha^2}{2m} = V_0 - E_B \tag{2.234}$$

$$E_B = \frac{mW^2}{2\hbar^2} \tag{2.235}$$

There is only one bound state, whose eigenvalue is given above. The parameter W has the units of joule-meter, and the eigenvalue has the units of joules.

The normalization of the eigenfunction is easy for this problem. For bound states in one dimension, the prefactor A is determined by the condition that $|\psi(x)|^2$ is one when integrated over all space. Since the integral is symmetric, one needs to do only the positive half:

$$1 = \int_{-\infty}^{\infty} dx |\psi(x)|^2 = 2A^2 \int_0^{\infty} dx e^{-2\alpha x} = \frac{A^2}{\alpha} \tag{2.236}$$

$$A = \sqrt{\alpha}, \quad \psi(x) = \sqrt{\alpha} e^{-\alpha|x|} \tag{2.237}$$

The eigenfunction for the bound state has been completely determined.

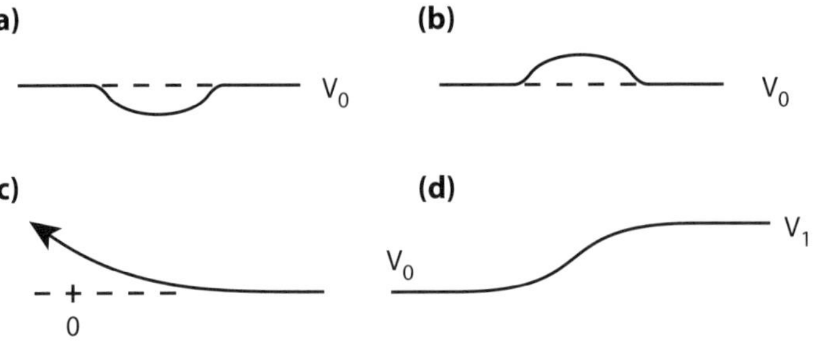

FIGURE 2.12. Potential functions in one dimension.

2.7 Number of Solutions

It is important to realize the number of independent solutions to Schrödinger's equation that are permitted in each range of energy. This section gives some examples of the shapes of potential functions and the number and kind of solutions permitted in each case. Some of these examples are from the exact solutions of the prior sections.

1. *Constant potential.* If the potential has the constant value V_0 throughout all of space, no solution exists for $E < V_0$. Two independent continuum solutions exist for all values of $E > V_0$:

$$k^2 = \frac{2m}{\hbar^2}(E - V_0)$$
$$\psi(x) = Ae^{ikx} + Be^{-ikx} = (A + B)\cos(kx) + i(A - B)\sin(kx) \tag{2.238}$$

   The two solutions can be expressed either as $\exp(\pm ikx)$ or as $\cos(kx)$, $\sin(kx)$.
2. *Constant potential plus attractive region.* In the example shown in figure 2.12a, there is a constant potential everywhere except in a local region where the poential lowers to a value $V_1 < V_0$. For $E > V_0$ there are still two continuum solutions for each value of energy E. The continuum states extend throughout all values of x. They can be given in terms of plane waves except in the region of the potential minimum. For the energy region $V_0 > E > V_1$, bound states may exist. They are usually of finite number and have discrete energy states. If the potential is symmetric, as drawn in the figure, the bound and continuum states have either even or odd parity. The eigenfunctions $\psi_n(x)$ for bound states become small in value away from the potential minimum.
3. *Constant potenial plus local maximum.* As shown in figure 2.12b, the potential has a constant value V_0 except for a local region with a maximum of height V_1. In this case there are no bound states. There are two continuum states for each value of $E > V_0$.
4. *Half-space.* The term "half-space" is used to describe potentials that diverge to positive infinity at some point. The wave function is required to vanish at that point. Figure 2.12c shows a gradual divergence of $V(x)$. This case was solved for the repulsive exponential potential. The condition that the eigenfunction vanish is stringent: both real and imaginary parts must vanish. This requirement eliminates one solution, so that there is only one

solution for each value of $E > V(+\infty)$. There can also be discrete bound states if the potential has an attractive region where $V(x) < V_0$. This possibility is not shown in the figure.

5. *Potential step.* Here there is one constant value V_1 as $x \to \infty$, and a different value V_0 as $x \to -\infty$. Figure 2.12d shows an example where $V_0 < V_1$. The step is shown changing gradually, but it could be a sudden step. For $E < V_0$ there are no solutions. For $V_0 < E < V_1$ there is only one solution. For $E > V_1$ there are two independent solutions.

2.8 Normalization

An eigenfunction is normalized when it has the correct constant coefficient in front. The coefficient may be a function of energy but depend neither on position nor on spin. Eigenfunctions are used in a variety of applications, and it is important to always have the proper normalization. Previously, in this chapter, the normalization has been treated cavalierly, in that it has been found for easy examples but not for others. Now the topic is treated systematically. The general method is derived, and then illustrated by examples. There are several different cases.

2.8.1 *Bound States*

One dimension is treated first. For bound states, the eigenvalues occur only in discrete amounts. The indices n or m is used to denote eigenvalues E_n or E_m and eigenfunctions $\psi_n(x)$ or $\psi_m(x)$. Then the normalization condition is

$$\delta_{nm} = \int_{-\infty}^{\infty} dx \psi_n^*(x) \psi_m(x) \tag{2.239}$$

The Kronecker delta-function has the discrete values of zero ($n \neq m$) or one ($n = m$). The integral vanishes if $n \neq m$. This orthogonality is a general property of second-order differential equations, which is useful in quantum mechanics. The case that $n = m$ pro- provides the normalization coefficient:

$$1 = \int_{-\infty}^{\infty} dx |\psi_n(x)|^2 \tag{2.240}$$

The factor in the integrand is written as ψ_n^2 when the eigenfunctions are real. Bound states in one dimension can always be made real. When the eigenfunctions are complex the factor of $|\psi_n(x)|^2$ is correct.

The integral in eqn. (2.240) was evaluated for the bound state of the delta-function in (2.237). Another bound-state eigenfunction is given in (2.202) for the exponential potential. Its coefficient C_1 is found from the integral

$$1 = C_1^2 \int_0^{\infty} dx J_\beta(g e^{-x/a})^2 \tag{2.241}$$

In this example the integral is over only positive values of x, since $\psi = 0$ for $x < 0$. The above integral is hard to evaluate analytically and is usually done on the computer, where it is easy.

Another example of the bound state is the square well in section 2.1. The following integral is obtained for the eigenfunction in (2.29):

$$1 = C_1^2 \int_0^a dx \sin^2(px) + C_3^2 \int_a^\infty dx e^{-2\alpha x} \tag{2.242}$$

$$1 = \frac{C_1^2}{2}\left[a - \frac{\sin(2pa)}{2p}\right] + \frac{C_3^2}{2\alpha} e^{-2\alpha a} \tag{2.243}$$

Next use eqn. (2.30) to relate C_3 to C_1 and derive an equation in which the only unknown is C_1:

$$1 = \frac{C_1^2}{2}\left[a - \frac{\sin(2pa)}{2p} + \frac{\sin^2(pa)}{\alpha}\right] \tag{2.244}$$

This equation provides the value of C_1. It depends on the eigenvalue through p and α. There is a different value of C_1 for each bound state.

Three-dimensional solutions to Schrödinger's equation are discussed in chapter 5. Each bound state is characterized by three different quantum numbers, one for each dimension. Call these three numbers ℓ, m, n. The normalization of the eigenfunction is given by the three-dimensional integral:

$$1 = \int dxdydz |\psi_{\ell mn}(x, y, z)|^2 \tag{2.245}$$

This integral is evaluated in chapter 5 for several examples.

2.8.2 Box Normalization

The eigenfunction is normalized for a particle in a box. This problem is solved at the beginning of section 2.1. A particle confined to a box $0 < x < L$, with potential walls of infinite height, has the eigenfunction

$$\psi_n(x) = \sqrt{\frac{2}{L}} \sin\left(\frac{n\pi x}{L}\right) \tag{2.246}$$

This case is an example of a bound state, and could have been included in the prior subsection.

The above example is a box with closed ends. An important variation is the periodic box. There it is assumed that the motion of the particle is around the perimeter of a circle of circumference L. When a particle has gone a distance L it is back to where it started. Eigenfunctions have the feature that $\psi(x + nL) = \psi(x)$, where n is any integer. The potential is assumed to be a constant V_0. There are no bound states, but only continuum states with $E > V_0$. The wave vector is

$$k = \sqrt{\frac{2m}{\hbar^2}(E - V_0)} \tag{2.247}$$

the eigenfunctions are plane waves. Although it is possible to use sines and cosines, it is easier to use

$$\psi(x) = Ae^{ikx} + Be^{-ikx} \tag{2.248}$$

The two terms with $\exp(\pm ikx)$ are independent and are treated separately. For each one the constraint $\psi(x+nL) = \psi(x)$ is valid. This contraint requires that

$$e^{ikL} = 1 \tag{2.249}$$

$$k_m = \frac{2\pi m}{L} \tag{2.250}$$

The wave vector k must be an integer multiple of $2\pi/L$. The integer m can be positive or negative: negative values give negative wave vectors, and the particle is going the other way around the circle. Both terms in eqn. (2.248) are not needed. This result is different than the particle in the straight box $0 < x < L$, where $k_{\mathrm{m}} = \pi m/L$ but only positive values of m are allowed.

The general form of the eigenfunction is

$$\psi_m(x) = \frac{e^{ik_m x}}{\sqrt{L}} \tag{2.251}$$

$$\delta_{nm} = \int_0^L dx \psi_n^*(x)\psi_m(x) = \int_0^L \frac{dx}{L} e^{ix(k_m - k_n)} \tag{2.252}$$

Here the normalization constant is $A = 1/\sqrt{L}$. The eigenvalue is

$$E_n = V_0 + \frac{\hbar^2 k_n^2}{2m} = V_0 + n^2 \frac{(2\pi\hbar)^2}{2mL^2} \tag{2.253}$$

The periodic box is a popular model in some branches of physics.

Periodic boundary conditions are also applied to plane waves in two and three dimensions. It is hard to visualize the topology of a two- or three-dimensional object that is connected with periodic boundary conditions. This difficulty is usually ignored, and this boundary condition is quite popular. Let the box have dimensions (L_{x}, L_{y}, L_{z}) in three dimensions. The eigenfunction is

$$\psi_{\ell mn}(\mathbf{r}) = \frac{\exp[i(k_x x + k_y y + k_z z)]}{\sqrt{L_x L_y L_z}} \tag{2.254}$$

$$k_x = \frac{2\pi\ell}{L_x}, \quad k_y = \frac{2\pi m}{L_y}, \quad k_z = \frac{2\pi n}{L_z} \tag{2.255}$$

A common shorthand is the notation

$$\psi(\mathbf{k}, \mathbf{r}) = \frac{\exp(i\mathbf{k}\cdot\mathbf{r})}{\sqrt{\Omega}} \tag{2.256}$$

where $\Omega = L_x L_y L_z$ is the volume of the box.

2.8.3 Delta-Function Normalization

Continuum states are defined as those for which the eigenvalue E is continuous. For E to be continuous, the system must stretch to infinity in at least one direction. Any box of finite length has discrete values of energy. In one dimension, assume that $V(x) \to V_0$ as x gets large in either direction. Continuum states are those with $E > V_0$.

First consider the case of a potential function that is constant everywhere, $V(x) = V_0$. For $E > V_0$ define the wave vector k by

$$E = V_0 + \frac{\hbar^2 k^2}{2m} \tag{2.257}$$

The plane wave eigenfunctions are of the form

$$\psi(k, x) = Ae^{ikx} \tag{2.258}$$

$$\int_{-\infty}^{\infty} dx \psi^*(k', x)\psi(k, x) = A^2 \int_{-\infty}^{\infty} dx e^{ix(k-k')} \tag{2.259}$$

$$= 2\pi A^2 \delta(k-k') \tag{2.260}$$

The integral is one of the definitions of a delta-function:

$$2\pi\delta(k-k') = \int_{-\infty}^{\infty} dx e^{ix(k-k')} \tag{2.261}$$

Equation (2.260) is the basis of delta-function normalization. Since k is a continuous variable, we cannot use a Kronecker delta when two functions have different values of wave vector. Instead, we have to use the continuous delta-function $\delta(k-k')$.

There are two different conventions in physics:

- Set $A = 1$ and the right-hand side of (2.260) is $2\pi\delta(k-k')$. This convention is the easiest and is the one we use.
- Set $A = 1/\sqrt{2\pi}$ and the right-hand side of (2.260) is $\delta(k-k')$, without any other factors. This convention makes a neat right-hand side, but one has to write every plane wave eigenfunction being divided by $\sqrt{2\pi}$.

The above result is immediately extended to higher dimensions. In three dimensions,

$$\int d^3r \psi^*(\mathbf{k}', \mathbf{r})\psi(\mathbf{k}, \mathbf{r}) = (2\pi)^3 \delta^3(\mathbf{k}-\mathbf{k}') \tag{2.262}$$

The normalization is given by a three-dimensional delta-function.

Examine the delta-function integral in (2.261). Expand the exponent in $\exp[ix(k-k')] = \cos[x(k-k')] + i\sin[x(k-k')]$. The sine has an argument that is odd in x, so its integral vanishes. The integral is

$$2\pi\delta(k-k') = \int_{-\infty}^{\infty} dx \cos[x(k-k')] \tag{2.263}$$

$$2\pi\delta(k-k') = 2\int_{0}^{\infty} dx \cos[x(k-k')] \tag{2.264}$$

Since the cosine function is even in x, the integral needs to be done only over the half-space. The last integral is useful for delta-function normalization for half-space problems.

The second example for delta-function normalization is the half-space problem. Imagine that there is a barrier to the left, such as the exponential potential discussed earlier. Far to the right, the eigenfunction is

$$\lim_{x\to\infty} \psi(k, x) = C(k) \sin(kx + \delta) \tag{2.265}$$

Since k is a continuous wave vector, the eigenfunction is normalized according to delta-function normalization:

$$\int_{-\infty}^{\infty} dx \psi(k', x)\psi(k, x) = 2\pi\delta(k-k') \tag{2.266}$$

Note that there is no complex conjugate (ψ^*) in the above integral, since for the half-space problems the eigenfunctions are always real. The normalization in (2.266) requires that $C(k) = 2$ in (2.265). A crude explanation is to write the right-hand side of eqn. (2.265) as

$$C \sin(kx+\delta) = \frac{C}{2i}\left(e^{ikx+i\delta} - e^{-ikx-i\delta}\right) \tag{2.267}$$

The eigenfunction can be viewed as an incoming and reflected plane wave. Each should have unit amplitude so $C = 2$: the factor of i is kept to keep $\psi(k, x)$ real. Note that the constant C does not depend on energy or wave vector.

A better derivation of $C = 2$ is provided here. The derivation is done for the repulsive exponential potential $V(x) = V_0 \exp(-2\alpha x)$ that diverges for negative values of x. We assume the existence of a point x_0, such that for $x > x_0$ the eigenfunction is accurately given by the asymptotic form in (2.265). This assumption is not very restrictive since x_0 may be taken as far to the right as needed. The normalization integral is written as

$$2\pi\delta(k-k') = \int_{-\infty}^{x_0} dx \psi(k, x)\psi(k', x) + C^2 \int_{x_0}^{\infty} dx \sin(kx+\delta)\sin(k'x+\delta') \tag{2.268}$$

where $\delta(k) \equiv \delta$, $\delta(k') \equiv \delta'$. Many terms are generated during the evaluation of these integrals. The delta-function is given by a divergent integral. Here we neglect every integral that is not divergent. We neglect the first term on the right of the above equation (2.268). Since the potential function diverges at negative values of x, the eigenfunction $\psi(k,x)$ goes to zero, and the first integral gives a finite number. Only the second integral diverges at $k = k'$. Use the trigometric identity

$$\sin(A)\sin(B) = \frac{1}{2}[\cos(A-B) - \cos(A+B)] \tag{2.269}$$

to write the second integral as

$$2\pi\delta(k-k') = \frac{C^2}{2}\int_{x_0}^{\infty} dx\{\cos[x(k-k')+\delta-\delta'] - \cos[x(k+k')+\delta+\delta']\} + \text{finite terms} \tag{2.270}$$

Since k and k' are both positive, then $(k + k')$ is never zero, and the second integral is not divergent. Add it to the "finite terms." The first integral diverges if $k = k'$, and this term provides the singular contribution in delta-function normalization. Note that if $k = k'$ then $\delta = \delta'$ and the factor $\delta - \delta'$ vanishes. Write

$$\int_{x_0}^{\infty} dx\{\cos[x(k-k') + \delta - \delta'] = \pi\delta(k-k') - \int_0^{x_0} dx\cos[x(k-k') + \delta - \delta'] \tag{2.271}$$

$$2\pi\delta(k-k') = \frac{C^2\pi}{2}\delta(k-k') + \text{finite terms.} \tag{2.272}$$

The integral over $0 < x < x_0$ is part of the "finite terms." All of the finite terms must add up to zero. Proving this feature is rather hard. However, clearly the coefficient of the delta-function $\delta(k-k')$ requires that $C^2 = 4$, $C = 2$.

As an example of this procedure, the continuum eigenfunction is normalized for the potential in section 2.5. The eigenfunction is given in (2.210) in terms of a coefficient C. The asymptotic form of this wave function is shown in eqn. (2.215). Since the correct normalization on $\sin(kx + \delta)$ is 2, then

$$2 = \frac{2C}{|\Gamma(1+iK)|} \tag{2.273}$$

$$C = |\Gamma(1+iK)| \tag{2.274}$$

$$\psi(k, x) = i|\Gamma(1+iK)|[I_{iK}(ge^{-x/a}) - I_{-iK}(ge^{-x/a})] \tag{2.275}$$

The factor of i in front is a constant phase that is included to keep $\psi(k,x)$ a real function. Constant phase factors are not observable and can be added or discarded to eigenfunctions according to one's whim. Similarly, the absolute magnitude symbol on $\Gamma(1+iK)$ is unnecessary since its phase factor is also unobservable.

2.8.4 The Limit of Infinite Volume

Often it is useful to change from one type of normalization to another. For example, it is usually easier to manipulate formulas while using box normalization. Then, at the end of the calculation, it is usually necessary to sum over states. At this point we let the volume Ω become infinity, so that we get continuum states, and the wave vector summations become continuous integrals. The step of going from box to delta-function normalization is now explained.

In one dimension, the eigenfunction with periodic boundary conditions is

$$\psi(k, x) = \frac{e^{ikx}}{\sqrt{L}}, \quad k = \frac{2\pi n}{L} \tag{2.276}$$

$$\int_0^L dx\psi^*(k', x)\psi(k, x) = \delta_{kk'} \equiv \delta_{nn'} \tag{2.277}$$

The Kronecker delta is used since the states are discrete. Now change this result to delta-function normalization. First, change the variable $y = x - L/2$ so that the integral is $-L/2 < y < L/2$:

$$e^{i(k-k')L/2} \int_{-L/2}^{L/2} \frac{dy}{L} e^{iy(k-k')} = \delta_{kk'} \tag{2.278}$$

Now take the limit that $L \to \infty$. This limit is used so that the values of k and k' are unchanged: the integers n,n' increase proportional to L. The integral is zero unless $k = k'$ so the phase factor in front is zero. Multiply the entire equation by L:

$$\lim_{L\to\infty} \int_{-L/2}^{L/2} dy e^{iy(k-k')} = \lim_{L\to\infty} [L\delta_{kk'}] = 2\pi\delta(k-k') \tag{2.279}$$

The last identity gives the behavior of the Kronecker delta-function in the limit of an infinite box. The same limit can be shown another way. Start with the definition of the delta-function:

$$f(k) = \sum_{k'} f(k')\delta_{kk'} \tag{2.280}$$

$$f(k) = \int dk' f(k')\delta(k-k') \tag{2.281}$$

These two expressions give equivalent identities for the box and delta-function normalization. The first equation should morph to the second as $L \to \infty$. Since $k' = 2\pi n'/L$, increasing values of k' have n' increasing by one integer. Then write $\Delta n' = 1$, $\Delta k' = 2\pi\Delta n'/L$. Rewrite the first equation as

$$f(k) = \sum_{n'} \Delta n' f(k')\delta_{kk'} = \sum_{n'} \Delta k' f(k') \left[\frac{L}{2\pi}\delta_{kk'}\right] \tag{2.282}$$

Now take the limit as $L \to \infty$. Then $\Delta k' \to dk'$ is a continuous variable, and the delta-function changes according to eqn. (2.279):

$$f(k) = \int dk' f(k')\delta(k-k') \tag{2.283}$$

These series of steps change (2.280) into (2.281).

The same procedure works in three dimensions:

$$\lim_{\Omega\to\infty} [\Omega\delta_{\mathbf{k},\mathbf{k}'}] = (2\pi)^3\delta^3(\mathbf{k}-\mathbf{k}') \tag{2.284}$$

In three dimensions the wave vectors are $k_x = 2\pi\ell/L_x$, $k_y = 2\pi m/L_y$, $k_z = 2\pi n/L_z$ and the three-dimensional Kronecker delta is

$$\delta_{\mathbf{k},\mathbf{k}'} \equiv \delta_{\ell\ell'}\delta_{mm'}\delta_{nn'} \tag{2.285}$$

The Kronecker delta is unity only if all three components are the same.

2.9 Wave Packets

Most experiments are done with pulsed beams. Particles are generated or accelerated in groups. Then it is useful to think of the experiment in terms of wave packets. The packet is a local region in space that contains one or more particles. This localization of the particle is always limited by the uncertainty principle $\Delta x \Delta p \geq \hbar$.

The usual plane wave has a solution in one dimension of

$$\phi(k, x, t) = \exp\left[i\left(kx - \frac{\varepsilon(k)t}{\hbar}\right)\right], \quad \varepsilon(k) = \frac{\hbar^2 k^2}{2m} = \hbar\omega_k \tag{2.286}$$

The above applies to nonrelativisitic particles with mass. We wish to localize this particle into a packet that is confined to a limited region of space. The uncertainty principle $\Delta x \Delta k > 1$ reminds us that we will pay a price. The localization in space will introduce a spread in the values of the wave vector. We use a Gaussian function to localized the particle, since it makes all integrals easy. Consider the wave packet

$$\Psi(k_0, x, t) = A \int_{-\infty}^{\infty} dk e^{-(k-k_0)^2/2\Delta^2} e^{ikx - i\omega_k t} \tag{2.287}$$

The constant A is normalization and is defined below. The constant Δ determines the width of the packet in real space and in wave-vector space.

First do the integral. The exponent contains only linear and quadratic powers of k, so it is of standard Gaussian form:

$$\begin{aligned}\Psi(k_0, x, t) &= A \int_{-\infty}^{\infty} dk \exp\left\{-k^2\left(\frac{1}{2\Delta^2} + \frac{it\hbar}{2m}\right) + k\left(ix + \frac{k_0}{\Delta^2}\right) - \frac{k_0^2}{2\Delta^2}\right\} \\ &= A\frac{\sqrt{2\pi}\Delta}{\sqrt{1+itu}} \exp\left[\frac{ixk_0 - \frac{x^2\Delta^2}{2} - it\omega_0}{1+itu}\right] \\ &= A\frac{\sqrt{2\pi}\Delta}{\sqrt{1+itu}} \exp\left[\frac{-\Delta^2(x - v_0 t)^2 + i(k_0 x - \omega_0 t) + itux^2\Delta^2}{2[1+(tu)^2]}\right]\end{aligned} \tag{2.288}$$

$$u = \frac{\hbar\Delta^2}{m}, \quad E_0 = \frac{\hbar^2 k_0^2}{2m} = \hbar\omega_0, \quad v_0 = \frac{\hbar k_0}{m} \tag{2.289}$$

The velocity of the packet is v_0. The expression for the packet function is complicated. A more transparent result is obtained for the particle density:

$$\rho(k_0, x, t) = |\Psi|^2 = \frac{2\pi A^2 \Delta^2}{\sqrt{1+(tu)^2}} \exp\left[-\frac{\Delta^2(x - v_0 t)^2}{1+(tu)^2}\right] \tag{2.290}$$

The density has a Gaussian shape, and moves in the positive x-direction with a velocity v_0. The constant Δ has the dimensions of wave vector. It controls the width of the Gaussian packet: in real space the width is $O(1/\Delta)$.

The normalization constant A is determined by assuming there is only one particle in the packet, so that

$$1 = \int_{-\infty}^{\infty} dx \rho(k_0, x, t) = 2\pi^{3/2} \Delta A^2 \tag{2.291}$$

$$A = \frac{1}{(2\Delta)^{1/2} \pi^{3/4}} \tag{2.292}$$

The packet describes a Gaussian envelope moving in the x-direction with a velocity v_0. The packet function oscillates with the usual plane wave factor of $\exp[i(k_0 x - \omega_0 t)]$. The spreading of the packet is due to the factor of $[1 + (tu)^2]$ that appears various places. The packet is least spread out at $t = 0$. Quite often one can neglect this factor. For example, in the laboratory the packet has to travel a distance L, which takes a time $t = L/v_0$. Then the factor $tu = L\Delta^2/k_0$, which is usually much less than one. Neglecting this factor gives a simple expression for the packet:

$$\Psi(k_0, x, t) = e^{i(k_0 x - t\omega_0)} G(x - v_0 t) \tag{2.293}$$

$$G(x) = \sqrt{\frac{\Delta}{\sqrt{\pi}}} \exp\left[-\frac{1}{2}\Delta^2 x^2\right] \tag{2.294}$$

This simple form will be often used for the shape of the packet function.

Next calculate the kinetic energy, and also the particle current. Both require the derivative of the packet function with respect to x:

$$\frac{\partial}{\partial x} \Psi(k_0, x, t) \approx \Psi(k_0, x, t)[ik_0 - \Delta^2(x - v_0 t)] \tag{2.295}$$

$$E_{KE} = \frac{\hbar^2}{2m} \int dx \left| \frac{\partial \Psi}{\partial x} \right|^2 = \frac{\hbar^2}{2m} \left[k_0^2 + \frac{\Delta^2}{2} \right]$$

$$J = \frac{\hbar}{2mi} \left[\Psi(k_0, x, t)^* \frac{\partial \Psi}{\partial x} - \Psi(k_0, x, t) \frac{\partial \Psi^*}{\partial} \right] = v_0 \rho(k_0, x, t) \tag{2.296}$$

The packet moves with a velocity v_0, which determines the value of the current density. The kinetic energy has two terms. One is E_0, as expected, while the other depends on Δ^2 and is the kinetic energy of the spread in wave vectors.

Homework

1. Derive the numerical value in eV for the bound-state energy of an electron in the one-dimensional square-well potential:

$$V(x) = \begin{cases} -V_0 & \text{for } |x| < b \\ 0 & \text{for } |x| > b \end{cases} \tag{2.297}$$

where $b = 1.0$ Å and $V_0 = 1.0$ eV. What is the critical value of coupling strength g_c for this potential?

2. Make a graph similar to figure 2.8 of the solutions of eqn. (2.37) over the range of $0 < g < 10$. Show that new bound states start at $g_{cn} = \pi\,(n + \frac{1}{2})$.

3. A particle in a one-dimensional square well of infinite sides is confined to the interval $0 < x < L$. The eigenfunctions are given in eqn. (2.19). Prove the relation

$$\sum_{n=1}^{\infty} \phi_n(x)\phi_n(x') = C\delta(x-x') \tag{2.298}$$

by evaluating the summation, and determine the constant C.

4. For the potential shown in figure 2.1, prove eq. (2.51) by doing the integral. ψ_B and ψ_k are the bound and continuum eigenfunctions.

5. Consider in one dimension the solutions to Schrödinger's equation for the half-space problem ($V_0 > 0$):

$$V(x) = \begin{cases} V = \infty & x \leqslant 0 \\ V_0 & 0 < x < a \\ V = 0 & x > a \end{cases} \tag{2.299}$$

Find an expression for the phase shift $\delta(k)$ for the two cases (a) $0 < E < V_0$ and (b) $V_0 < E$.

6. Consider a particle of energy $E > V_0 > 0$ approaching a potential step from the left. The Hamiltonians in the two regions are

$$H = \begin{cases} \dfrac{p^2}{2m_L} & x < 0 \\ \dfrac{p^2}{2m_R} + V_0 & x > 0 \end{cases} \tag{2.300}$$

where $m_L \neq m_R$.

Find the amplitude of the transmitted (T) and reflected wave (R), by matching at $x = 0$ the amplitude of the wave function and the derivative

$$\frac{1}{m_L}\left(\frac{d\phi}{dx}\right)_{x=0^-} = \frac{1}{m_R}\left(\frac{d\phi}{dx}\right)_{x=0^+} \tag{2.301}$$

Show that this choice of matching conserves the current of the particles.

7. A particle of mass m moves in a one-dimensional square-well potential with walls of infinite height: the particle is constrained in the region $-L/2 < x < L/2$.

   a. What are the eigenvalues and eigenfunctions of the lowest two states in energy?
   b. What is the expectation value of the energy for a state that is an equal mixture of these two states?
   c. For the state in (b), what is the probability, as a function of time, that the particle is in the right-hand side of the well?

8. Derive an equation for the eigenvalues of the potential $V(x) = F|x|$, where $F > 0$ for values $-\infty < x < \infty$.

9. Derive an equation for the bound states of the potential

$$V(x) = -V_0 e^{-2|x|/a} \tag{2.302}$$

for all $-\infty < x < \infty$ and $V_0 > 0$.

10. Find the transmission coefficient of a continuum wave going from left to right for the potential in the previous problem.

11. Find the exact eigenvalues for the potential

$$V(x) = \begin{cases} \infty & x < 0 \\ \frac{1}{2}Kx^2 & x > 0 \end{cases} \tag{2.303}$$

Hint: With a little thought, the answer may be found by doing no derivation.

12. Consider the potential in one dimension:

$$V(x) = g\frac{\hbar^2}{2mx^2} \tag{2.304}$$

where $g > 0$ is a dimensionless parameter and solve for $x > 0$.

a. Find the solutions to Schrödinger equation.
b. Derive the phase shift by considering the form of the eigenfunction at large positive x.

Hint: Try a solution of the form $\sqrt{x}\, J_\nu(kx)$ and use the fact that

$$\lim_{z\to\infty} J_\nu(z) = \sqrt{\frac{2}{\pi z}} \cos\left[z - \frac{\pi}{2}\left(\nu + \frac{1}{2}\right)\right] \tag{2.305}$$

13. Evaluate the following integral for the harmonic oscillator using generating functions:

$$M_{n\ell} = \int_{-\infty}^{\infty} dx \phi_n(x) x \phi_\ell(x) \tag{2.306}$$

14. Evaluate the following integral for the harmonic oscillator using generating functions:

$$M_{n\ell}(q) = \int_{-\infty}^{\infty} dx \phi_n(x) \phi_\ell(x) e^{iqx} \tag{2.307}$$

15. For the harmonic oscillator, evaluate the following (s is constant):

a. $[a, H]$
b. $[a^\dagger, H]$

c. $e^{sH} a e^{-sH}$
d. $e^{sH} a^\dagger e^{-sH}$

16. The *squeezed state* is the operator (λ is a constant)

$$S(\lambda) = N(\lambda) e^{\lambda a^\dagger} \tag{2.308}$$

This is also called a *coherent state*.

a. Find the normalization constant $N(\lambda)$ such that $\langle 0|S^\dagger S|0\rangle = 1$.
b. Evaluate the commutator $[a, S]$.
c. Evaluate the commutator $[a^\dagger, S]$.
d. Show that $S|0\rangle$ is an eigenstate of the lowering operator a and find its eigenvalue.

17. Find the transmission and reflection coefficients, from left to right, of a particle scattering off the potential $V(x) = V_0 \Theta(x) + \lambda\delta(x)$, $\lambda > 0$ in one dimension, where $E > V_0$. Find the transmission and reflection coefficients from right to left and verify the relations found from time reversal.

18. Consider in one dimension the bound states of a particle in the pair of delta-function potentials ($W > 0$):

$$V(x) = -W[\delta(x+a) + \delta(x-a)] \tag{2.309}$$

Derive the eigenvalue equation for all possible bound states.

19. Consider the problem of a delta-function potential outside of an infinite barrier at the origin ($a > 0$):

$$V(x) = \begin{cases} \infty & x \leq 0 \\ W\delta(x-a) & x > 0 \end{cases} \tag{2.310}$$

a. Find an analytical expression for the phase shift as a function of k.
b. Plot this on a piece of graph paper for the range $0 < ka < 2\pi$ when $g \equiv 2maW/\hbar^2 = 1$.

20. Consider the one-dimensional Schrödinger equation with a delta-function potential $V(x) = W\delta(x)$. For each value of energy $E > 0$ construct two wave functions that are orthogonal to each other and normalized according to delta-function normalization.

21. Find the formula for the normalization coefficient C_1 for the bound-state wave function in (2.29). Then find the numerical value for the bound state shown in figure 2.2.

22. A one-dimensional Hamiltonian has a ground-state eigenfunction of

$$\psi_0(x) = \frac{A}{\cosh(x/a)} \tag{2.311}$$

where (a, A) are constants. Assume that $V(x) \to 0$ as $|x|/a >> 1$.

a. What is A?
b. What is the exact eigenvalue?
c. What is the exact potential $V(x)$?

23. Derive a wave packet for photons ($\omega = ck$) using (a) a Gaussian packet, and (b) a Lorentzian packet.

3 Approximate Methods

The previous chapter has exact solutions to Schrödinger's equation in one dimension for some simple potentials. There are only a few potentials $V(x)$ for which Schrödinger's equation is exactly solvable. Many other potentials need to be solved besides those with exact solutions. This chapter presents two approximate methods for solving Schrödinger's equation: WKBJ and variational. They were both important in the precomputer days of quantum mechanics, when many problems could be solved only approximately. Now numerical methods usually permit accurate solutions for most problems. However, the present methods are often useful even in the computer age. Many computer solutions are iterative, where one needs to start the process with an approximate answer. The two approximate methods are often useful inputs for these kinds of computer programs. A very modern topic is the path integral method [1] for solving problems in quantum mechanics, and they are based on WKBJ.

When I began teaching quantum mechanics, I also included in this chapter a section on numerical methods of solving Schrödinger's equation. I no longer include this material for two reasons: (1) my codes use FORTRAN, which is no longer popular with younger scientists; (2) most universities have entire courses devoted to numerical methods in physics. Readers are encouraged to learn numerical methods, since they are very important in the present age of science.

3.1 WKBJ

The WKBJ method is named after Wentzel, Kramers, Brillouin, and Jeffreys. British authors sometimes put Jeffreys first (JWKB) and German authors often omit his name (WKB). There were other important contributors, whose names are customarily omitted from the alphabet soup: Green, Liouville, and Rayleigh. Our treatment is brief; further information is in the references [2, 3].

In one dimension, Schrödinger's equation for a single particle is

$$0=\left[-\frac{\hbar^2}{2m}\frac{d^2}{dx^2}+V(x)-E\right]\psi(x) \tag{3.1}$$

The eigenfunction is assumed to be an exponential function of a phase factor $\sigma(x)$:

$$\psi(x)=\exp\left[\frac{i}{\hbar}\sigma(x)\right] \tag{3.2}$$

This step involves no approximation. If $\sigma(x)$ is found exactly, then so is $\psi(x)$. The function $\sigma(x)$ need not be real, and its imaginary part determines the amplitude of $\psi(x)$.

A differential equation is derived for $\sigma(x)$. Let prime denote the derivative with respect to x:

$$\psi'(x)=\frac{i}{\hbar}\sigma'(x)\psi(x) \tag{3.3}$$

$$\psi''(x)=\left[\frac{i}{\hbar}\sigma''-\left(\frac{\sigma'}{\hbar}\right)^2\right]\psi(x) \tag{3.4}$$

The second derivative in (3.4) is inserted into eqn. (3.1). Then multiply the result by $2m$, which produces the differential equation

$$0=(\sigma')^2-i\hbar\sigma''+2m[V(x)-E] \tag{3.5}$$

The equation is nonlinear. It is usually easier to solve a linear equation than one that is nonlinear. So the present approach is not useful if Schrödinger's equation can be solved exactly. Usually it is not useful to turn a linear equation into one that is nonlinear.

The WKBJ method is based on the observation that quantum mechanics reduces to classical mechanics in the limit that $\hbar\to 0$. Classical mechanics is usually a good approximation to the motion of a particle. The main idea behind WKBJ is to treat $\hbar$ as a "small" parameter, and to expand the eigenfunction in a series in this parameter. Experience has shown that the natural function to expand is not $\psi(x)$, but the phase factor $\sigma(x)$:

$$\sigma(x)=\sigma_0(x)+\frac{\hbar}{i}\sigma_1(x)+\left(\frac{\hbar}{i}\right)^2\sigma_2(x)+\cdots \tag{3.6}$$

The individual terms $\sigma_n(x)$ are each assumed to be independent of $\hbar$.

In most perturbation expansions, the small expansion parameter is dimensionless. Planck's constant has the dimensions of joule-second, so one is expanding in a dimensional parameter. Mathematically, this does not make sense, but it seems to work in WKBJ.

Insert the series (3.6) into eqn. (3.5) and then regroup the series so that all terms are together that have the same power of $(\hbar)^n$:

$$0=[\sigma_0'-i\hbar\sigma_1'-\hbar^2\sigma_2'+\cdots]^2-i\hbar[\sigma_0''-i\hbar\sigma_1''+\cdots]+2m(V-E)$$

$$\begin{aligned}0&=\left[(\sigma_0')^2+2m(V-E)\right]-i\hbar[2\sigma_0'\sigma_1'+\sigma_0'']\\&\quad-\hbar^2[(\sigma_1')^2+2\sigma_0'\sigma_2'+\sigma_1'']+O(\hbar^3)\end{aligned} \tag{3.7}$$

The last equation is a series in which each term has the form $(\hbar)^n C_n(x)$. The next step is subtle. We assume that each coefficient $C_n(x) = 0$ separately. The summation of all of the terms is zero. Each term $(\hbar)^n C_n$ has all terms with the same power of $\hbar$. Therefore, this term cannot be canceled by any other term, and must vanish by itself. Setting each term $C_n = 0$ for increasing values of n gives a series of equations:

$$0 = (\sigma_0')^2 + 2m[V(x) - E]$$
$$0 = 2\sigma_0'\sigma_1' + \sigma_0''$$
$$0 = (\sigma_1')^2 + 2\sigma_0'\sigma_2' + \sigma_1'' \tag{3.8}$$

Only three equations are listed here, since that is one more than will be used. The standard WKBJ approximation to the eigenfunction has only the two terms σ_0 and σ_1. Usually σ_2 and higher terms are neglected. Keeping only σ_0 and σ_1 gives an approximate eigenfunction:

$$\psi(x) = \exp\left[\frac{i}{\hbar}\sigma_0(x) + \sigma_1(x) + O(\hbar)\right] \tag{3.9}$$

These equations are now solved. The first one is

$$\sigma_0' = \pm p(x) \tag{3.10}$$

$$p(x) = \sqrt{2m[E - V(x)]} \tag{3.11}$$

$$\sigma_0 = \pm\int^x dx' p(x') + \text{constant} \tag{3.12}$$

There are two solutions, so that (3.9) has two possible terms. Schrödinger's equation is a second-order differential equation, which has two independent solutions. In WKBJ they are given by these two values of σ_0. The symbol p is used since the quantity is actually the classical momentum.

The second equation from (3.8) is easy to solve, once σ_0 is known:

$$\sigma_1' = -\frac{\sigma_0''}{2\sigma_0'} = -\frac{1}{2}\frac{d}{dx}\ln[\sigma_0'] \tag{3.13}$$

$$\sigma_1 = -\frac{1}{2}\ln[p(x)] + \text{constant} \tag{3.14}$$

$$\psi(x) = \frac{A_1}{\sqrt{p(x)}}e^{i\sigma_0(x)/\hbar} + \frac{A_2}{\sqrt{p(x)}}e^{-i\sigma_0(x)/\hbar} \tag{3.15}$$

The last line gives the general form of the eigenfunction in the WKBJ method. The prefactors have the constant coefficients $A_{1,2}$, which are fixed by the boundary conditions. They arise in the derivation from the constants of integration in eqns. (3.12) and (3.14).

The phase factor of σ_1 causes the prefactor of $1/\sqrt{p}$ in front of each term. In classical physics the energy at each point is $E = p^2/2m + V(x)$. The classical momentum is $p(x) = \sqrt{2m[E - V(x)]}$. The probability of finding a particle at a point x is proportional to $|\psi|^2 \sim 1/p(x)$. Fast particles (large p) travel through a region quickly and have a small probability of being there. Slow particles (small p) move slowly through a region.

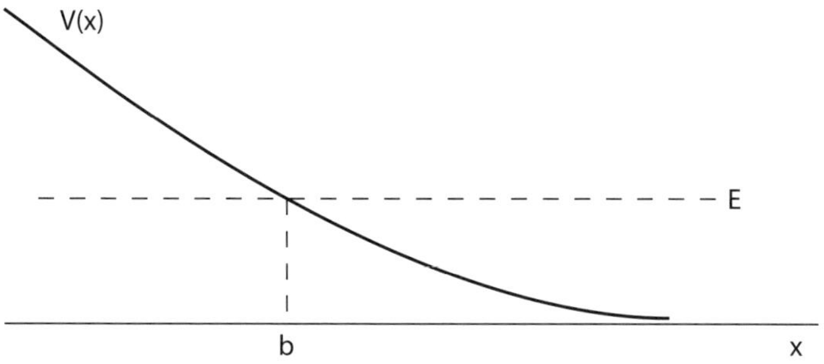

FIGURE 3.1. The turning point $x = b$ is at the point where $E = V(b)$.

Consider the half-space problem shown in figure 3.1. The potential function $V(x)$ is assumed to approach infinity as $x \rightarrow -\infty$. The energy E is marked as the horizontal dashed line. The *turning point* is defined as the value of position $x = b$, where $E = V(b)$. Turning points have a key role in the WKBJ eigenfunctions, since there the classical momentum vanishes: $p(b) = 0$. The turning point b is where all phase integrals start, such as

$$\sigma_0(x) = \int_b^x dx' p(x') \tag{3.16}$$

First consider the eigenfunction for $x < b$. In this region $V(x) > E$ and $p(x)$ is an imaginary number. Write it as $p = i\alpha(x)$:

$$\alpha(x) = \sqrt{2m[V(x) - E]} \tag{3.17}$$

$$\psi(x) = \frac{A_1}{\sqrt{\alpha(x)}} e^{-\frac{1}{\hbar}\int_x^b dx'\alpha(x')} + \frac{A_2}{\sqrt{\alpha(x)}} e^{\frac{1}{\hbar}\int_x^b dx'\alpha(x')} \tag{3.18}$$

Note that the limits of integration have been reversed, since $x < b$. As x moves further to the left, away from the turning point, the first exponent becomes increasingly negative and this term decays to zero. The second term becomes increasingly positive and grows without bound. Usually this behavior is not acceptable, so we eliminate this term by setting $A_2 = 0$. The eigenfunction in (3.18) has only the first term in most cases:

$$\psi(x) = \frac{A_1}{\sqrt{\alpha(x)}} \exp\left\{-\frac{1}{\hbar}\int_x^b dx'\alpha(x')\right\} \tag{3.19}$$

Equation (3.19) is the proper eigenfunction for the potential function $V(x)$ shown in figure 3.1.

Next consider the eigenfunction for $x > b$. Here both $p(x)$ and $\sigma_0(x)$ are real functions, and the exponential factors are phases. The eigenfunction can be written in terms of sines and cosines of the phase integral. In analogy with our past treatment of half-space problems, the eigenfunction is written as the sine of a phase integral plus a constant phase angle β:

$$\psi(x) = \frac{D}{\sqrt{p(x)}} \sin\left[\frac{1}{\hbar}\int_b^x dx' p(x') + \beta\right] \tag{3.20}$$

The amplitude D and phase angle β are determined by matching this eigenfunction onto the evanescent wave given in eqn. (3.19). These two expressions are the same eigenfunction, which is being described in two different regions of space. The method for matching two eigenfunctions is to equate them, and their derivative, at the common point $x = b$. This simple procedure is impossible because both eigenfunctions diverge at the match point since the prefactor has $p \to 0$. In fact, near the turning point, the WKBJ eigenfunction is a poor approximation to the actual eigenfunction.

Matching the two functions in (3.19) and (3.20) gives

$$A_1 = \frac{D}{2}, \quad \beta = \frac{\pi}{4} \tag{3.21}$$

These relations are derived in the books by Headings [2] and the Fromans [3].

The coefficient D is determined by examining the eigenfunction in the limit of $x \to \infty$. There the momentum $p(x)$ approaches the limit $\hbar k$:

$$k^2 = \frac{2m}{\hbar^2}[E - V(\infty)] \tag{3.22}$$

$$\lim_{x\to\infty} \psi(x) = \frac{D}{\sqrt{\hbar k}} \sin\left[\frac{1}{\hbar}\int_b^x dx' p(x') + \frac{\pi}{4}\right] \tag{3.23}$$

Delta-function normalization is used for this continuous eigenfunction, as described in chapter 2. Then the prefactor is just two, so

$$D = 2\sqrt{\hbar k}, \quad A_1 = \sqrt{\hbar k} \tag{3.24}$$

The WKBJ eigenfunction is now completely specified.

The asymptotic expression in eqn. (3.23) can be used to obtain the WKBJ expression for the phase shift $\delta(k)$. Recall that the phase shift is defined as

$$\lim_{x\to\infty} \psi(x) = 2\sin(kx + \delta) \tag{3.25}$$

$$\delta_{\text{WKBJ}}(k) = \frac{\pi}{4} + \lim_{x\to\infty}\left[\frac{1}{\hbar}\int_b^x dx' p(x') - kx\right] \tag{3.26}$$

The second expression follows from (3.23). Equation (3.26) will be used often to derive phase shifts from potential functions.

In deriving (3.26), it is assumed that the right-hand side is independent of x. This assumption is valid for potential functions $V(x)$ that go to zero rapidly with increasing values of x. There are some cases, such as a Coulomb potential $V = g/x$, for which the right-hand side of eqn. (3.26) is not independent of x. In these cases the eigenfunction does not have the asymptotic form given in (3.25), and the phase shift is not defined. There is nothing wrong with the WKBJ approximation in these situations; it is only that one cannot use the idea of a phase shift. The asymptotic form of the eigenfunction can be described by a phase shift only if the right-hand side of eqn. (3.26) is independent of x.

The WKBJ phase shift is found for the potential function

$$V(x) = V_0 e^{-2x/a} \tag{3.27}$$

The turning point is found by $E = V(b)$ or

$$b = \frac{a}{2}\ln\left(\frac{V_0}{E}\right) \tag{3.28}$$

The phase integral is

$$\Theta(x) = \frac{\sqrt{2m}}{\hbar}\int_b^x dx'[E - V_0 e^{-2x'/a}]^{1/2} \tag{3.29}$$

Change the variable of integration to

$$z = e^{-2x'/a}, \quad dx' = -\frac{a}{2}\frac{dz}{z} \tag{3.30}$$

$$\Theta(x) = -\frac{a\sqrt{2m}}{2\hbar}\int_{E/V_0}^{\exp(-2x/a)} \frac{dz}{z}\sqrt{E - V_0 z} \tag{3.31}$$

The indefinite integral can be done exactly:

$$\int \frac{dz}{z}\sqrt{E_0 - V_0 z} = 2\sqrt{E - V_0 z} + \sqrt{E}\ln\left[\frac{\sqrt{E} - \sqrt{E - V_0 z}}{\sqrt{E} + \sqrt{E - V_0 z}}\right] \tag{3.32}$$

$$\Theta(x) = -ka\left\{\sqrt{1 - \frac{V(x)}{E}} + \frac{1}{2}\ln\left[\frac{\sqrt{E} - \sqrt{E - V(x)}}{\sqrt{E} + \sqrt{E - V(x)}}\right]\right\}$$

$$\delta_{\text{WKBJ}}(k) = \frac{\pi}{4} + \lim_{x\to\infty}[\Theta(x) - kx] \tag{3.33}$$

To find the WKBJ phase shift, it is necessary to evaluate $\Theta(x)$ as $x \to \infty$. Then $V(x)$ becomes very small, and $V(x)/E$ is a small number. Use this ratio as an expansion parameter:

$$\lim_{x\to\infty}\sqrt{E - V(x)} = \sqrt{E} - \frac{V}{2\sqrt{E}} + O(V^2) \tag{3.34}$$

$$\lim_{x\to\infty}\Theta(x) = -ka\left\{1 + \frac{1}{2}\ln\left[\frac{V(x)}{4E}\right] + O\left(\frac{V}{E}\right)\right\} \tag{3.35}$$

The argument of the logarithm contains the factor of

$$\ln[V(x)] = \ln[V_0 e^{-2x/a}] = -\frac{2x}{a} + \ln[V_0] \tag{3.36}$$

$$\Theta(x) = kx - ka\left\{1 + \frac{1}{2}\ln\left[\frac{V_0}{4E}\right] + O\left(\frac{V}{E}\right)\right\} \tag{3.37}$$

Recall that $K = ka$, $g^2 = 2mV_0a^2/\hbar^2$. The WKBJ and exact phase shifts are

$$\delta_{\text{WKBJ}}(k) = \frac{\pi}{4} - K\left[1 + \ln\left(\frac{g}{2K}\right)\right] \tag{3.38}$$

$$\delta(k) = -K\ln\left(\frac{g}{2}\right) - \frac{i}{2}\ln\left[\frac{\Gamma(1 + iK)}{\Gamma(1 - iK)}\right] \tag{3.39}$$

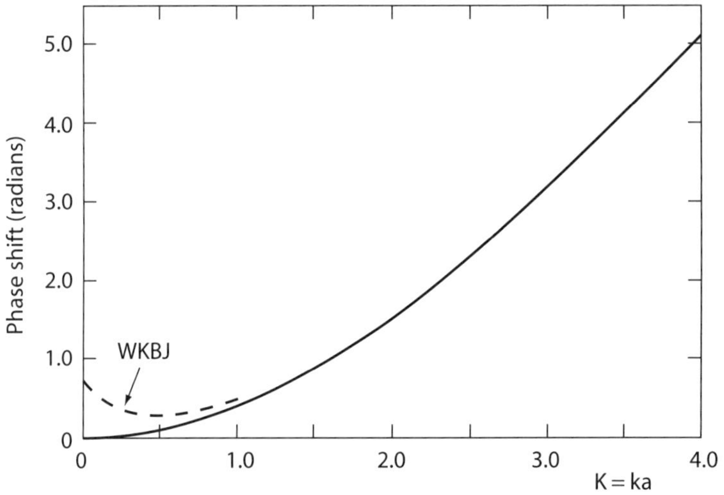

FIGURE 3.2. The solid line is the exact phase shift for the repulsive exponential potential as a function of $K = ka$ when $g = 1.0$. The dashed line is the WKBJ result.

The latter formula is from the previous chapter. Both terms contain the factor of $-K\ln(g/2)$ but otherwise do not look much alike.

Figure 3.2 shows a plot of these two phase-shift formulas for $g = 1.0$. The solid line is the exact result, and the dashed line is the WKBJ result at low energy. At high energy the two lines merge and only one is shown. The WKBJ result is accurate at large values of $K = ka$. It is inaccurate at low values of K since it goes to $\pi/4$ rather than zero. This result is typical of WKBJ, since this theory is always more accurate at large values of particle energy. The classical approximation is better in this case. The eigenfunction is also more accurate for larger energies. This feature is in accord with our intuition, in which energetic particles should behave in a more classical fashion. The WKBJ approximation is an excellent method of describing the motion of particles except those states with the lowest energy.

Phase shifts are studied more thoroughly in chapter 10 on scattering theory. Phase shifts are important in physics since they completely determine the elastic scattering of a particle from a potential or from another particle. An important feature is the behavior of the phase shift $\delta(k)$ as the wave vector k goes to zero. *Levinson's theorem* states that $\delta(k = 0) = m\pi$, where m is the number of bound states of the potential. There are no bound states for the repulsive exponential potential, which is why the phase shifts simply go to zero. Levinson's theorem further indicates the feature that Coulomb potentials do not have phase shifts, since they have an infinite number of bound states.

3.2 Bound States by WKBJ

The prior section discussed the calculation of continuum eigenfunctions and phase shifts using the WKBJ approximation. This section shows how to use WKBJ to find eigenvalues

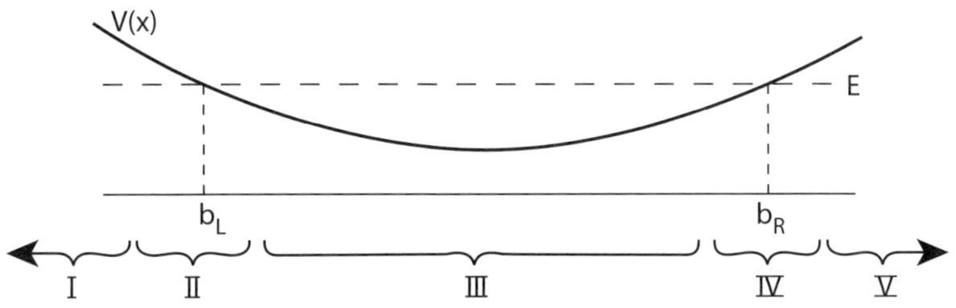

FIGURE 3.3. The five regions of an eigenfunction for a bound state.

and eigenvectors of bound states in one dimension. The similar problem for three dimensions is treated in chapter 5.

Figure 3.3 shows a potential $V(x)$ that has a single smooth minimum. Multiple minima usually causes multiple bound states. The case of periodic minima is very important in solids. Only one minimum is treated here.

For one minimum, an eigenstate of energy E requires two turning points. As shown in figure 3.3, they are called the left b_L and right b_R turning points. Both obey $E = V(b_j)$ for $j = (L, R)$. The construction of the eigenfunction requires space to be divided into regions: here there are five regions. Regions II and IV are near the two turning points. Regions I and V are in the two evanescent regions, where the eigenfunction decays away from the turning point.

Region III is between the turning points. Here $E > V(x)$ and the eigenfunction has sinusoidal character. Careful construction of the eigenfunction in this region is the key to obtaining the eigenvalue equation. As one starts to write down the eigenfunction in region III, it is apparent there are two obvious choices: the phase integral can be started at either turning point:

$$\psi_L(x) = \frac{C_L}{\sqrt{p(x)}} \sin\left[\Theta_L(x) + \frac{\pi}{4}\right] \tag{3.40}$$

$$\psi_R(x) = \frac{C_R}{\sqrt{p(x)}} \sin\left[\Theta_R(x) + \frac{\pi}{4}\right] \tag{3.41}$$

$$\Theta_L(x) = \frac{1}{\hbar}\int_{b_L}^{x} dx' p(x') \tag{3.42}$$

$$\Theta_R(x) = \frac{1}{\hbar}\int_{x}^{b_R} dx' p(x') \tag{3.43}$$

Is the proper eigenfunction $\psi_L(x)$ or $\psi_R(x)$? The answer is that they should be the same eigenfunction, so it should not make any difference. The equality $\psi_L = \psi_R$ is obeyed only for certain values of energy E. These values are the discrete eigenvalues E_n. Bound states exist only for these energies, since only for these values is the eigenfunction matched correctly at both turning points.

Define $\phi_L(x)$ and $\phi_R(x)$ as the two phases factors in eqns. (3.40)–(3.41). Their sum is the total phase ϕ_T:

$$\phi_R = \Theta_R + \frac{\pi}{4}, \quad \phi_L = \Theta_L + \frac{\pi}{4} \tag{3.44}$$

$$\phi_T = \phi_L + \phi_R = \frac{\pi}{2} + \frac{1}{\hbar}\int_{b_L}^{b_R} dx' p(x') \tag{3.45}$$

The total phase is independent of x.

The eigenvalue condition for the existence of bound states is obtained from the constraint that $\psi_L(x) = \psi_R(x)$. In ψ_L, write $\phi_L = \phi_T - \phi_R$, and then use the trigometric identity for the sine of the difference of two angles:

$$\psi_L(x) = \frac{C_L}{\sqrt{p(x)}} \sin[\phi_T - \phi_R] \tag{3.46}$$

$$= \frac{C_L}{\sqrt{p(x)}} [\sin(\phi_T)\cos(\phi_R) - \sin(\phi_R)\cos(\phi_T)] \tag{3.47}$$

$$= \psi_R = \frac{C_R}{\sqrt{p(x)}} \sin(\phi_R) \tag{3.48}$$

The last two expressions must be equal. The extra term in (3.47) with $\cos(\phi_R)$ must be eliminated. Since $\phi_{R,L}(x)$ depends on position, and ϕ_T does not, we can eliminate terms only by setting the value of ϕ_T. In this case, it is

$$\phi_T = m\pi, \qquad m = 1, 2, 3, \ldots \tag{3.49}$$

Then $\sin(\phi_T) = 0$, $\cos(\phi_T) = (-1)^m$, and $\psi_L = \psi_R$ for all values of x provided:

$$C_R = -(-1)^m C_L \tag{3.50}$$

The two requirements (3.49, 3.50) are sufficient to make $\psi_L(x) = \psi_R(x)$ for all values of x. The same eigenfunction is obtained regardless of whether the phase integral is started from the left or right turning point.

The eigenvalue condition (3.49) is usually manipulated by bringing the $\pi/2$ term acrossed the equal sign and multiplying by $\hbar$. The integer m is replaced by $m = n + 1$, where n is any positive integer including zero:

$$\int_{b_L}^{b_R} dx p(x) = \pi\hbar\left(n + \frac{1}{2}\right) \tag{3.51}$$

Equation (3.51) is the final form for the eigenvalue condition in WKBJ, for the case that the turning points are smooth functions of $V(x)$.

Students should recognize the above expression as a form of the old Bohr-Sommerfeld quantization conditions from the early days of quantum mechanics:

$$\oint dx p(x) = 2\pi\hbar\left(n + \frac{1}{2}\right) \tag{3.52}$$

The circle in the integral symbol means to take the integral through a complete oscillation from one turning point to the other and back to the first. The Bohr-Sommerfeld equation (3.52) is just twice the WKBJ expression (3.51). In the early history of quantum mechanics, the Bohr-Sommerfeld condition was proposed as the exact equation for bound states. It is not exact, but instead is just the WKBJ expression. However, as shown below, this expression gives the exact eigenvalue equation for several simple potential functions $V(x)$. That is why it was possible to mistake this formula for the exact eigenvalue equation. However, for other potential functions $V(x)$ it gives an approximate but accurate eigenvalue equation. Examples of exact and approximate eigenvalue equations are given below.

3.2.1 *Harmonic Oscillator*

The potential function for the harmonic oscillator in one dimension is $V = Kx^2/2$. It is inserted into the eigenvalue equation (3.51):

$$\int dx[m(2E_n - Kx^2)]^{1/2} = \pi\hbar\left(n + \frac{1}{2}\right) \tag{3.53}$$

The turning points are always easy to find, since they are the values of x where the argument of the square root is zero. For the harmonic oscillator they are at

$$b_L = -x_0, \quad b_R = x_0, \; x_0^2 = 2E_n/K \tag{3.54}$$

$$\pi\hbar\left(n + \frac{1}{2}\right) = \sqrt{mK}\int_{-x_0}^{x_0} dx\sqrt{x_0^2 - x^2} \tag{3.55}$$

The indefinite integral has an analytic expression. But in the present case, set $x = x_0 \sin(\alpha)$, $dx = d\alpha x_0 \cos(\alpha)$, and the integral is

$$\pi\hbar\left(n + \frac{1}{2}\right) = \sqrt{mK}x_0^2\int_{-\pi/2}^{\pi/2} d\alpha \cos^2(\alpha) = \frac{\pi}{2}\sqrt{mK}\left(\frac{2E_n}{K}\right) \tag{3.56}$$

$$E_n = \hbar\sqrt{\frac{K}{m}}\left(n + \frac{1}{2}\right) \tag{3.57}$$

The above expression for E_n is the exact eigenvalue derived in chapter 2. The WKBJ approximation gives the exact eigenvalue for the harmonic oscillator. The poor quality of the WKBJ eigenfunction at turning points does not affect the quality of the eigenvalue.

3.2.2 *Morse Potential*

The Morse potential is defined as

$$V(x) = A[e^{-2(x-x_0)/a} - 2e^{-(x-x_0)/a}] \tag{3.58}$$

It is shown in figure 3.4. The potential has a single minimum at $x = x_0$ of depth A. The eigenvalue spectrum contains both bound and continuum states. This potential is a fair approximation to the real potential between neutral atoms or molecules. There is an attractive well and a repulsive region. Schrödinger's equation for this potential can be solved exactly, for both bound and continuum states, using confluent hypergeometric functions. This technique is discussed in chapter 5.

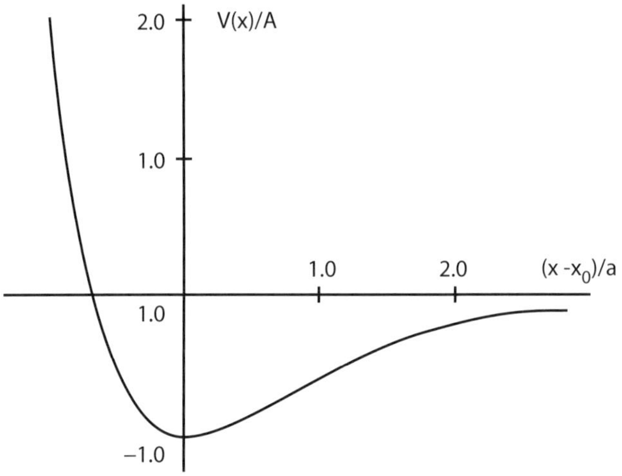

FIGURE 3.4. The Morse potential.

In the Morse potential, the trick to solving Schrödinger's equation exactly, or by WKBJ, is to make the variable change $y = \exp[-(x - x_0)/a]$. In WKBJ the integral becomes

$$x = x_0 - a\ln(y), \quad dx = -a\frac{dy}{y} \tag{3.59}$$

$$\pi\hbar\left(n+\frac{1}{2}\right) = \sqrt{2ma}\int\frac{dy}{y}\left[E_n - A(y^2-2y)\right]^{1/2} \tag{3.60}$$

The turning points are found by setting to zero the argument of the square root in $p(x)$:

$$y^2 - 2y - E/A = 0, \quad y_{1,2} = 1 \pm \sqrt{1 + E_n/A} \tag{3.61}$$

$$\pi\hbar\left(n+\frac{1}{2}\right) = \sqrt{2mAa}\int_{y_2}^{y_1}\frac{dy}{y}\left[(y_1 - y)(y - y_2)\right]^{1/2} \tag{3.62}$$

The definite integral is found from standard tables

$$\int_{y_2}^{y_1}\frac{dy}{y}\left[(y_1 - y)(y - y_2)\right]^{1/2} = \frac{\pi}{2}\left[\sqrt{y_1} - \sqrt{y_2}\,\right]^2 \tag{3.63}$$

$$= \pi\left[1 - \sqrt{-E_n/A}\right] \tag{3.64}$$

$$E_n = -A\left[1 - \frac{\hbar(n+\frac{1}{2})}{\sqrt{2ma^2A}}\right]^2 \tag{3.65}$$

Bound states exist only if the expression inside of the square bracket is positive. Bound states do not exist for large values of the integer n. Even for $n = 0$, there is a minimum value of A_c for which no bound states exist when $A < A_c$. Define the potential strength for the Morse potential as

$$g^2 = \frac{2ma^2A}{\hbar^2} \tag{3.66}$$

and bound states exist only if $g > \frac{1}{2}$. The eigenvalue equation found using WKBJ is exact.

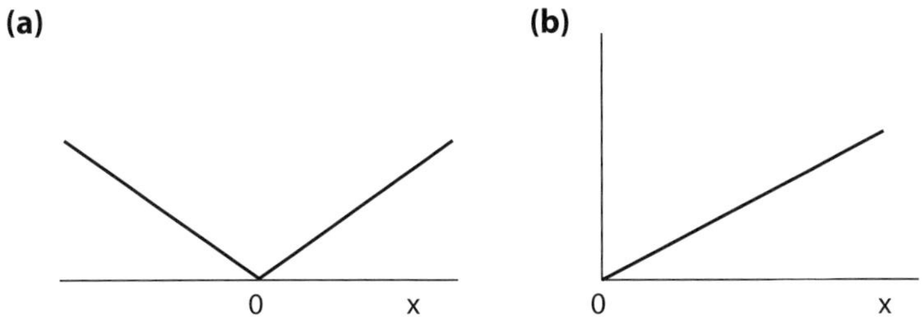

FIGURE 3.5. (a) Symmetric ramp. (b) Half-space ramp.

3.2.3 Symmetric Ramp

Next consider the potential function $V(x)=F|x|$, where $F>0$. The potential function is shown in figure 3.5. The eigenvalue equation in WKBJ is

$$\pi\hbar\left(n+\frac{1}{2}\right)=\sqrt{2m}\int dx\sqrt{E_n-F|x|} \tag{3.67}$$

The expression $|x|$ is best handled by recognizing that the integral for positive and negative values of x gives the same contribution. So write the integral as

$$\pi\hbar\left(n+\frac{1}{2}\right)=2\sqrt{2m}\int_0^{E_n/F}dx\sqrt{E_n-Fx} \tag{3.68}$$

$$=2\sqrt{2m}\left[-\frac{2}{3F}(E_n-Fx)^{3/2}\right]_0^{E/F} \tag{3.69}$$

$$=\frac{4\sqrt{2m}}{3F}E_n^{3/2} \tag{3.70}$$

The eigenvalue equation from WKBJ is

$$E_n=E_F\left[\frac{3\pi}{4}\left(n+\frac{1}{2}\right)\right]^{2/3} \tag{3.71}$$

$$E_F=\left(\frac{\hbar^2F^2}{2m}\right)^{1/3} \tag{3.72}$$

All of the dimensional factors go into E_F, which has the units of energy. The WKBJ solution is not exact.

The potential $V(x)=F|x|$ is symmetric $[V(-x)=V(x)]$. The eigenfunctions can be separated into even and odd parity. In WKBJ, the solutions with even values of integer n have even parity $[V(-x)=V(x)]$, while those with odd values of n have odd parity $[\psi(-x)=-\psi(x)]$. The exact solution to Schrödinger's equation for this potential function are given in terms of Airy functions. For $x>0$,

$$0=\left[-\frac{\hbar^2}{2m}\frac{d^2}{dx^2}+Fx-E\right]\psi(x) \tag{3.73}$$

$$\psi(x)=CAi(\xi) \tag{3.74}$$

$$\xi=(x-E/F)/x_0,\ x_0^3=\frac{\hbar^2}{2mF} \tag{3.75}$$

Table 3.1 Antisymmetric Eigenvalues E_n/E_F for the Exact and WKBJ Solutions

n	Exact	WKBJ
1	2.3381	2.3202
3	4.0879	4.0818
5	5.5206	5.5171
7	6.7867	6.7844

- The symmetric solutions have zero slope at the matching point $x = 0$, so the eigenvalue condition is

$$0 = Ai'\left(-\frac{E_n}{E_F}\right) \tag{3.76}$$

The prime denotes derivative.

- The antisymmetric solutions have zero value at the matching point $x = 0$, so the eigenvalue condition is

$$0 = Ai\left(-\frac{E_n}{E_F}\right) \tag{3.77}$$

The eigenvalues can be found from a table of Airy functions.

Table 3.1 compares the eigenvalues for the antisymmetric solutions between the exact and WKBJ methods. The WKBJ is accurate but not exact. The accuracy increases with larger values of n. This trend is in accord with the idea that WKBJ is more accurate at larger energies, where the particle motion is more classical. Similar results are found for the symmetric solutions.

3.2.4 Discontinuous Potentials

Chapter 2 discussed a number of potentials that had discontinuities, such as a particle in a box. The treatment of the prior section was valid when the potential function $V(x)$ had a finite derivative at the turning points. The slope $F = dV/dx$ at $x = b_j$ is assumed finite and well behaved. This subsection considers the opposite case where F is infinite due to discontinuities in the potential function.

The changes from the prior case are discussed using the prior equations:

$$\phi_R = \Theta_R + \frac{\pi}{4}, \quad \phi_L = \Theta_L + \frac{\pi}{4} \tag{3.78}$$

$$\phi_T = \phi_L + \phi_R = \frac{\pi}{2} + \frac{1}{\hbar}\int_{b_L}^{b_R} dx' p(x') = \pi(n+1) \tag{3.79}$$

The phase integral contains the factor of $\pi/4$. This factor comes from the integration through the turning point, in the case that the turning point has a finite slope. When the turning point has a discontinuity, the factor of $\pi/4$ is absent.

The first case is a particle in a box $0<x<L$ of infinite walls. The two turning points both have infinite walls, and the factor of $\pi/4$ is absent from both. The above equations get changed to

$$\phi_R = \Theta_R, \ \phi_L = \Theta_L \tag{3.80}$$

$$\phi_T = \phi_L + \phi_R = \frac{1}{\hbar}\int_{b_L}^{b_R} dx' p(x') = \pi(n+1) \tag{3.81}$$

Inside the box the momentum $p = \sqrt{2mE_n}$ is independent of x. The above phase integral is

$$L\sqrt{2mE_n} = \pi\hbar(n+1) \tag{3.82}$$

$$E_n = (n+1)^2 \frac{\hbar^2\pi^2}{2mL^2}, \quad n = 0, 1, 2, \ldots \tag{3.83}$$

This solution is another example where WKBJ gives the exact eigenvalue. The eigenfunction is also exact. Getting the exact result from WKBJ requires knowing when to use, or when not to use, the phase constant $\pi/4$.

Another example is the half-space linear potential. Consider the potential function

$$V(x) = \infty, \quad \text{if} \quad x \le 0 \tag{3.84}$$

$$= Fx, \quad \text{if} \quad x > 0 \tag{3.85}$$

The phase integral on the left starts at a discontinuous potential ($b_L = 0$), so the constant $\pi/4$ is absent. The phase integral on the right goes through a finite slope at the turning point ($b_R = E/F$), so the factor of $\pi/4$ is present

$$\phi_R = \Theta_R + \frac{\pi}{4}, \quad \phi_L = \Theta_L \tag{3.86}$$

$$\phi_T = \phi_L + \phi_R = \frac{\pi}{4} + \frac{1}{\hbar}\int_{b_L}^{b_R} dx' p(x') = \pi(n+1) \tag{3.87}$$

The phase integral is similar to eqn. (3.69):

$$\sqrt{2m}\int_0^{E/F} dx\sqrt{E - Fx} = \pi\hbar\left(n + \frac{3}{4}\right) \tag{3.88}$$

$$\frac{2\sqrt{2m}}{3F}[E_n]^{3/2} = \pi\hbar\left(n + \frac{3}{4}\right) \tag{3.89}$$

$$E_n = E_F\left[\frac{3\pi}{2}\left(n + \frac{3}{4}\right)\right]^{2/3} \tag{3.90}$$

The eigenvalue equation is similar to that found earlier in (3.71) for the symmetric potential. The symmetric potential has even- and odd-parity solutions. The odd-parity solutions had odd values of n in (3.71) and their eigenfunction vanished at the origin [$\psi(x=0) = 0$]. If we set $n \to 2n+1$ in (3.71) to generate the odd solutions, then we get exactly (3.90). The half-space potential has the same eigenfunctions and eigenvalues as the odd-parity solutions to the symmetric potential.

3.3 Electron Tunneling

WKBJ is often used to calculate the tunneling rate of particles through potential barriers. Of course, it is better to use exact eigenfunctions for the evanescent waves, but they are often unavailable. WKBJ is usually an accurate method of obtaining the tunneling rates.

A potential barrier has $V(x) > E$. For this case the momentum is an imaginary variable. The amplitude of the eigenfunction in the barrier has the form

$$\psi(x) = T_0 \exp[-\alpha(x)] \tag{3.91}$$

$$\alpha(x) = \frac{\sqrt{2m}}{\hbar}\int_0^x dx' \sqrt{V(x') - E} \tag{3.92}$$

The prefactor is usually taken to be a constant T_0, with no momentum term. The barrier is considered to be over the interval $0 < x < L$, and the phase integral is started at $x = 0$. There is never a phase constant such as $\pi/4$ for the evanescent wave. The tunneling probability is the absolute magnitude squared of the eigenfunction at $x = L$:

$$P = |\psi(L)|^2 = |T_0|^2 \exp[-2\alpha(L)] \tag{3.93}$$

The first case is a simple repulsive square well of height $V_0 > 0$ and width L. The tunneling probability is

$$\alpha(L) = \gamma L, \quad \gamma = \frac{\sqrt{2m}}{\hbar}\sqrt{V_0 - E} \tag{3.94}$$

$$P = |\psi(L)|^2 = |T_0|^2 \exp[-2\gamma L] \tag{3.95}$$

The same result is found when solving the exact eigenfunction.

The next case is called *Fowler-Nordheim* tunneling. The usual experimental geometry is to put a positive voltage on the surface of a metal to assist electrons to exit the surface. Figure 3.6 shows the surface region. The shaded region on the left shows the occupied electron states $E < \mu$, where μ is the chemical potential. The electrons are confined to the metal by a step potential that has a work function $e\phi$ from the chemical potential. The external potential is represented by an electric field E that makes $F = eE$. The potential function is $V(x) = e\phi - Fx$, where its zero is defined as the chemical potential. The tunneling exponent is

$$\alpha(x) = \frac{\sqrt{2m}}{\hbar}\int_0^x dx' \sqrt{e\phi - Fx' - E} \tag{3.96}$$

The electron exits the triangular barrier at the point $x' = L = (e\phi - E)/F$. The integral is similar to those for the linear potential

$$\alpha(L) = \frac{2}{3}\frac{\sqrt{2m}}{\hbar F}[e\phi - E]^{3/2} \tag{3.97}$$

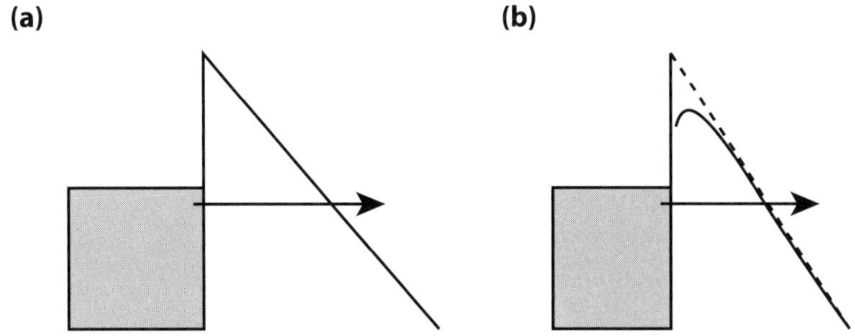

FIGURE 3.6. Fowler-Nordheim tunneling. (a) Electrons must tunnel through a triangular barrier. (b) The addition of the image potential reduces the barrier height.

$$P = |T_0|^2 \exp[-2\alpha(L)] \tag{3.98}$$

The interesting feature is that the tunneling rate has an exponent that is inversely proportional to the applied field F. At large values of the electric field, the exponent becomes small and the tunneling is very rapid. The factor $e\phi - E$ must be positive or there is no need to tunnel.

When an electron is outside of the surface of a perfect conductor, it has an image potential $-e^2/4x$. A better theory of Fowler-Nordheim tunneling includes this image potential:

$$V(x) = e\phi - Fx - \frac{e^2}{4x} \tag{3.99}$$

This potential is shown in figure 3.6b. There are now two turning points $b_{L,R}$ that are both positive. They are found as the points where $V(b) = E$:

$$0 = e\phi - Fb - E - \frac{e^2}{4b} \tag{3.100}$$

$$b_j = \xi \pm \sqrt{\xi^2 - e^2/4F}, \quad \xi = \frac{e\phi - E}{2F} \tag{3.101}$$

$$\alpha = \frac{\sqrt{2mF}}{\hbar^2} \int_{b_L}^{b_R} \frac{dx'}{\sqrt{x'}} \sqrt{(b_R - x')(x' - b_L)} \tag{3.102}$$

$b_R = \xi + \sqrt{\cdots}$ and $b_L = \xi - \sqrt{\cdots}$. The factor of $\xi^2 - e^2/4F$ must be positive. In figure 3.6b, this constraint means that E is less than the top of the potential barrier.

The above integral is expressed in terms of complete elliptic integrals:

$$\alpha = \frac{2\sqrt{2mFb_R}}{3\hbar^2}[2\xi E(p) - 2b_L K(p)], \quad p^2 = \frac{b_R - b_L}{b_R} \tag{3.103}$$

This expression is usually evaluated on the computer. The image correction to Fowler-Nordheim tunneling is most important at large values of field F.

3.4 Variational Theory

The variational method is useful for finding the eigenfunction and eigenvalue of the lowest bound state of a Hamiltonian. The lowest bound state is called the *ground state*.

The method cannot be used for continuum states nor for most excited states. The ground state is the single most important state in a quantum mechanical system, so the variational method is important. The variational method can also be used to find the ground state of a collection of particles, and this method is discussed in later chapters.

Assume there is a Hamiltonian H that describes any number of particles in any number of dimensions. It has a set of eigenfunctions ψ_n and eigenvalues ε_n that obey the relationship

$$H\psi_n = \varepsilon_n \psi_n \tag{3.104}$$

Presumedly they are too hard to find exactly, or else there is no need to use approximate methods such as variational theory. The ground-state eigenvalue is labeled with $n=0$, (ε_0, ψ_0). For a system of N-particles, the eigenfunction $\psi_n(\mathbf{r}_1, \mathbf{r}_2, \ldots, \mathbf{r}_N)$ depends on $3N$ space variables. There are also spin variables. They are important, and are discussed in chapter 4.

The variational method is based on the following theorem:

Theorem: *If $\phi(\mathbf{r}_1, \mathbf{r}_2, \ldots, \mathbf{r}_N)$ is any function of the 3N space coordinates, then*

$$\frac{\int \Pi_j^N d^3r_j \phi^*(\mathbf{r}_1, \ldots, \mathbf{r}_N) H \phi(\mathbf{r}_1, \ldots, \mathbf{r}_N)}{\int \Pi_j^N d^3r_j |\phi(\mathbf{r}_1, \ldots, \mathbf{r}_N)|^2} \geq \varepsilon_0 \tag{3.105}$$

Proof: The eigenfunctions ψ_n form a complete set of states and are assumed to be orthogonal. The arbitrary function ϕ can be expanded in terms of these functions:

$$\phi = \sum_n a_n \psi_n \tag{3.106}$$

$$a_n = \int \Pi_j^N d^3r_j \psi_n^* \phi \tag{3.107}$$

This expansion can be used to evaluate the two integrals in eqn. (3.105). We also use eqn. (3.104):

$$H\phi = \sum_n a_n \varepsilon_n \psi_n \tag{3.108}$$

$$\int \phi^* H \phi = \sum_n |a_n|^2 \varepsilon_n \tag{3.109}$$

$$\int \phi^* \phi = \sum_n |a_n|^2 \tag{3.110}$$

The expression (3.105) is manipulated by multiplying by the left-hand-side denominator, and then moving the right-hand side across the inequality sign:

$$\sum_n |a_n|^2 \varepsilon_n \geq \varepsilon_0 \sum_n |a_n|^2 \tag{3.111}$$

$$\sum_n |a_n|^2 (\varepsilon_n - \varepsilon_0) \geq 0 \tag{3.112}$$

The last equation is correct. The quantity $|a_n|^2$ is positive or zero, and $(\varepsilon - \varepsilon_0)$ is positive or zero: recall that ε_0 is the minimum value of ε_n. Therefore, the left-hand side of the last equation must be positive or zero. Inverting the steps in the proof, we have shown that eqn. (3.105) is valid. The equality holds if $\phi=\psi_0$. In that case one has guessed the exact ground-state eigenfunction. Otherwise, the left-hand side of (3.105) is larger than ε_0.

The variational method is based on the theorem in eqn. (3.105). The left-hand side is a *functional* of ϕ:

$$\mathcal{E}(\phi)=\frac{\int \Pi_j^N d^3r_j \phi^*(\mathbf{r}_1, \ldots, \mathbf{r}_N) H \phi(\mathbf{r}_1, \ldots, \mathbf{r}_N)}{\int \Pi_j^N d^3r_j |\phi(\mathbf{r}_1, \ldots, \mathbf{r}_N)|^2} \tag{3.113}$$

The notation $\mathcal{E}(\phi)$ means the value of $\mathcal{E}$ depends on the form of ϕ. The best eigenvalue is the lowest value of $\mathcal{E}(\phi)$. "Lowest value" means smallest positive value or largest negative value. The theorem states that $\mathcal{E}(\phi)$ is always greater than or equal to the exact ground-state eigenvalue ε_0. Making $\mathcal{E}(\phi)$ as small as possible is the closest one can come to the correct ε_0.

The function ϕ is called the *trial eigenfunction*. The variational procedure is a minimization procedure. One constructs a trial function ϕ that depends on m parameters: $a_1, a_2, \ldots, a_m$. The integrals are all evaluated. The resulting energy functional $\mathcal{E}(\phi)$ depends on the parameters: $a_1, a_2, \ldots, a_m$. They are independently varied to find the minimum energy:

$$0=\frac{\mathcal{E}}{a_1} \tag{3.114}$$

$$0=\frac{\mathcal{E}}{a_2} \tag{3.115}$$

$$0=\vdots \tag{3.116}$$

$$0=\frac{\mathcal{E}}{a_m} \tag{3.117}$$

This set of m equations is sufficient to determine the optimal values of the m parameters a_m. The functional $\mathcal{E}(\phi)$ has its minimum value at these parameters, and one has found the lowest eigenvalue for this trial eigenfunction.

As an example, consider in one dimension the nonrelativistic Hamiltonian for a single particle:

$$H=-\frac{\hbar^2}{2m}\frac{d^2}{dx^2}+V(x) \tag{3.118}$$

The potential is assumed to have attractive regions, such that bound states exist. The first step is to choose a trial eigenfunction in terms of some parameters. Two of many possible choices are

$$\phi(x)=Ae^{-ax^2}(1+bx+cx^2) \tag{3.119}$$

$$\phi(x)=\frac{A}{\cosh[\alpha(x-x_0)]} \tag{3.120}$$

The first expression has three variational parameters: a, b, c. The prefactor A is not a variational parameter since it cancels out between numerator and denominator in eqn.

(3.105). In this case the energy depends on $\mathcal{E}(a, b, c)$. The second trial wave function has two variational parameters α, x_0.

The variational calculation proceeds in several steps. The first one is to choose the form of the trial eigenfunction. Some standard choices in one dimension are polynomials multiplied by exponentials or by Gaussians. In three dimensions, they are polynomials in r multiplied by exponential or Gaussian functions of r. Increasing the number of variational parameters increases the accuracy of the result, but also increases the algebraic complexity. Most of our examples have only one variational parameter, since with two or more one has to minimize on the computer.

3.4.1 *Half-Space Potential*

The first example is the half-space linear potential $[V(x) = Fx]$ that was solved in a previous section, both exactly and by WKBJ. The variational result will be compared to these earlier results.

The first step is to choose the form for the trial eigenfunction. It is important to choose one that has the right shape in x-space. For example, since $V(x)$ diverges at the point $x = 0$, the eigenfunction must vanish at $x = 0$. The trial eigenfunction must be multiplied by a polynomial x^s, where $s > 0$. This polynomial ensures that $\phi(x = 0) = 0$. Secondly, the eigenfunction should not have any cusps except at points where the potential function has delta-functions. Thirdly, for the linear potential, it is expected that the eigenfunction must vanish as $x \to \infty$. With these constraints, consider three possible choices:

$$\phi = Ax^s e^{-\alpha x} \tag{3.121}$$

$$\phi = Ax^s e^{-(\alpha x)^2} \tag{3.122}$$

$$\phi = Ax^s e^{-(\alpha x)^{3/2}} \tag{3.123}$$

Each trial eigenfunction has two parameters: s and α.

The first trial eigenfunction is an exponential, the second is a Gaussian, and the third has an exponent with $x^{3/2}$. This latter guess is from the asymptotic limit of the Airy function. To make the calculation as simple as possible, choose $s = 1$ and the first choice above, with a simple exponential.

A variational calculation always has at least three integrals to perform: the kinetic energy KE, the potential energy PE, and the normalization I:

$$\mathcal{E}(\phi) = \frac{\text{KE} + \text{PE}}{\text{I}} \tag{3.124}$$

$$\text{KE} = -\frac{\hbar^2}{2m}\int dx\, \phi^* \frac{d^2}{dx^2}\phi = \frac{\hbar^2}{2m}\int dx \left|\frac{d\phi}{dx}\right|^2 \tag{3.125}$$

$$\text{PE} = \int dx\ V(x)|\phi|^2 \tag{3.126}$$

$$\text{I} = \int dx\ |\phi|^2 \tag{3.127}$$

Two possible forms are given for the kinetic energy KE. The second is derived from the first by an integration by parts. We prefer the second form, since one only has to take a single derivative of the trial eigenfunction. The kinetic energy is always a positive number. This is guaranteed by the second form, but is also true using the first expression. Below, when we give trial eignfunctions ϕ, we also give $d\phi/dx$ for the kinetic energy.

The three integrals for the half-space linear potential $[V = Fx]$ are

$$\phi = Axe^{-\alpha x}, \quad \frac{d\phi}{dx} = A(1-\alpha x)e^{-\alpha x} \tag{3.128}$$

$$\mathrm{I} = A^2\int_0^\infty dx x^2 e^{-2\alpha x} = 2!\frac{A^2}{(2\alpha)^3} \tag{3.129}$$

$$\mathrm{KE} = \frac{\hbar^2 A^2}{2m}\int_0^\infty dx(1-\alpha x)^2 e^{-2\alpha x} = \frac{\hbar^2 A^2}{2m}\frac{1}{4\alpha} = \frac{\hbar^2\alpha^2}{2m}\mathrm{I} \tag{3.130}$$

$$\mathrm{PE} = A^2 F\int_0^\infty dx x^3 e^{-2\alpha x} = \frac{3!FA^2}{(2\alpha)^4} = \frac{3F}{2\alpha}\mathrm{I} \tag{3.131}$$

The energy functional is

$$\mathcal{E}(\alpha) = \frac{\hbar^2\alpha^2}{2m} + \frac{3F}{2\alpha} \tag{3.132}$$

The variational parameter is α. It is varied to find its value α_0 at which $E(\alpha_0)$ has its smallest value:

$$0 = \frac{d\mathcal{E}}{d\alpha} = \frac{\hbar^2\alpha_0}{m} - \frac{3F}{2\alpha_0^2} \tag{3.133}$$

The value of α_0 is found by solving the above equation:

$$\alpha_0^3 = \frac{3Fm}{2\hbar^2}, \quad \alpha_0 = \left(\frac{3Fm}{2\hbar^2}\right)^{1/3} \tag{3.134}$$

$$\mathcal{E}(\alpha_0) = \frac{\hbar^2\alpha_0^2}{2m} + \frac{3F}{2\alpha_0} = \left(\frac{\hbar^2 F^2}{2m}\right)^{1/3}\left[\left(\frac{3}{4}\right)^{2/3} + \frac{3}{2}\left(\frac{4}{3}\right)^{1/3}\right] \tag{3.135}$$

$$= \frac{3^{5/3}}{2^{4/3}}E_F = 2.476E_F \tag{3.136}$$

The result depends on the dimensional parameter E_F defined in section 3.2 for the linear potential.

The variational result has a minimum eigenvalue $\mathcal{E}(\alpha_0) = 2.476E_F$. This result is about 6% higher than the exact result $2.338E_F$ given in table 3.1. The one-parameter variational result has produced a result that is reasonably accurate, but not as precise as the WKBJ value for the same potential. The trial eigenfunction is

$$\psi(x) = \frac{Ax}{\sqrt{I}}\exp[-\alpha_0 x] \tag{3.137}$$

The prefactor has the correct normalization.

3.4.2 *Harmonic Oscillator in One Dimension*

The second example is the harmonic oscillator potential in one dimension $V(x) = Kx^2/2$. The trial eigenfunction is

$$\phi(x) = A \exp\left[-\frac{1}{2}\alpha^2 x^2\right] \tag{3.138}$$

$$\frac{d\phi}{dx} = -A\alpha^2 x \exp\left[-\frac{1}{2}\alpha^2 x^2\right] \tag{3.139}$$

The exact ground-state eigenfunction is a Gaussian, so the above trial eigenfunction will yield the exact answer. This choice satisfies several criteria. First, it is a symmetric function in x, which is expected from a symmetric potential $[V(-x) = V(x)]$. It also decays to zero at large values of x. Note that we could also try $\exp[-\alpha|x|]$, but that has a cusp at the origin. A cusp is not wanted except for potentials with delta-functions. Another quite suitable trial eigenfunction is $\phi = A/\cosh(\alpha x)$, but that is harder to integrate.

Using the Gaussian trial eigenfunction, the three integrals are

$$\mathrm{I} = A^2 \int_{-\infty}^{\infty} dx\, e^{-\alpha^2 x^2} = \frac{\sqrt{\pi}}{\alpha} A^2 \tag{3.140}$$

$$\mathrm{KE} = \frac{\hbar^2 \alpha^4}{2m} A^2 \int_{-\infty}^{\infty} dx\, x^2 e^{-\alpha^2 x^2} = \frac{\hbar^2 \alpha \sqrt{\pi}}{4m} A^2 = \frac{\hbar^2 \alpha^2}{4m} \mathrm{I} \tag{3.141}$$

$$\mathrm{PE} = \frac{K}{2} A^2 \int_{-\infty}^{\infty} dx\, x^2 e^{-\alpha^2 x^2} = \frac{\sqrt{\pi} K}{4m\alpha^3} A^2 = \frac{K}{4\alpha^2} \mathrm{I} \tag{3.142}$$

The energy functional is

$$\mathcal{E}(\alpha) = \frac{\hbar^2 \alpha^2}{4m} + \frac{K}{4\alpha^2} \tag{3.143}$$

The parameter α has the dimensions of inverse length. Both terms in the above expression have the units of energy. It is always good to check the dimensions.

The minimum value of energy is found by varying the parameter α to find the value α_0 at the minimum of energy:

$$0 = \frac{d\mathcal{E}}{d\alpha} = \frac{\hbar^2 \alpha_0}{2m} - \frac{K}{2\alpha_0^3} \tag{3.144}$$

$$\alpha_0^4 = \frac{mK}{\hbar^2} \tag{3.145}$$

$$\mathcal{E}(\alpha_0) = \frac{\hbar^2}{2m}\left(\frac{mK}{\hbar^2}\right)^{1/2} + \frac{K}{4}\left(\frac{\hbar^2}{mK}\right)^{1/2} = \frac{1}{2}\hbar\sqrt{\frac{K}{m}} = \frac{1}{2}\hbar\omega$$

$$\psi_0(x) = \sqrt{\frac{\alpha_0}{\sqrt{\pi}}} \exp\left[-\frac{1}{2}\alpha_0^2 x^2\right] \tag{3.146}$$

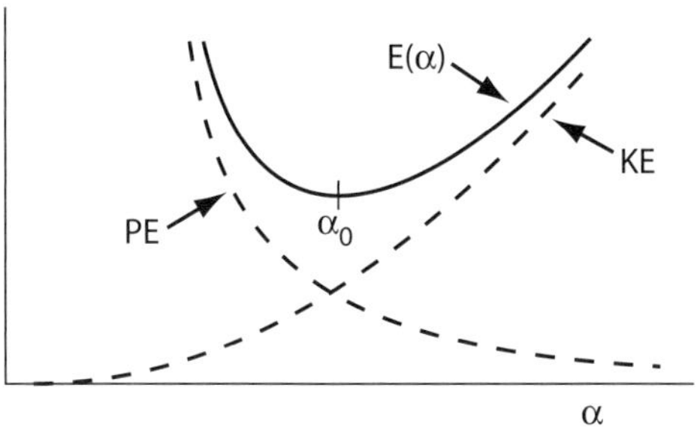

FIGURE 3.7. Plotted as a function of α are the kinetic energy (KE), the potential energy (PE), and the total energy E. The minimum in the total energy is at α_0.

The above result is recognized as the exact eigenvalue and exact eigenfunction of the $n=0$ state of the harmonic oscillator.

The variational method illustrates quite well the competition between kinetic and potential energies. For large values of α, the eigenfunction has a narrow spread in x-space; the kinetic energy is quite high and the potential energy is quite low. For small values of α, the eigenfunction is quite spreadout in x-space; the kinetic energy is quite low but the potential energy is quite high. The minimization procedure finds just the right balance between kinetic and potential energy.

Figure 3.7 shows a graph of the kinetic and potential energies as a function of α. Also shown is the total energy $E(\alpha) = \mathrm{KE} + \mathrm{PE}$. The figure illustrates the trade-off between kinetic and potential energies in the minimization of the total energy. This procedure is an example of the uncertainty principle $\Delta x \Delta p \geq \hbar$, where a confinement in x causes a spread in momentum, and vice versa.

The variational procedure can be used to find the minimum energy of an excited state as long as the excited state has a different quantum number than the ground state, such as different parity or different angular momentum. As an example, consider the following trial eigenfunction for the one-dimensional harmonic oscillator:

$$\phi = Ax \exp\left[-\frac{1}{2}\alpha^2 x^2\right] \tag{3.147}$$

This eigenfunction has odd parity $[\phi(-x) = -\phi(x)]$, and is orthogonal to the ground state: the latter has even parity. The variational calculation can be done with this trial eigenfunction, and one finds the eigenvalue $\mathcal{E} = 3\hbar\omega/2$. This eigenvalue is exact for the $n=1$ state of the harmonic oscillator. The variational procedure worked for this one excited state since it has a different parity than the ground state. In three-dimensional calculations, the variational method can be used to find the lowest eigenvalue for states of different angular momentum. This procedure is used in finding atomic eigenfunctions. Some variational calculations in three dimensions are discussed in chapter 5.

References

1. L. S. Schulman, *Techniques and Applications of Path Integration* (Wiley, New York, 1981)
2. J. Headings, *An Introduction to Phase Integral Methods* (Wiley, New York, 1962)
3. N. Froman and P.O. Froman, *JWKB Approximation: Contribution to the Theory* (Interscience, New York, 1965)

Homework

1. Use WKBJ to calculate the phase shift of a particle in the one-dimensional potential $V(x)=0$, $x>0$ and $V(x)=|Fx|$, $x<0$.

2. Use WKBJ to calculate the phase shift of a particle in the one-dimensional potential $V(x)=0$, $x>0$ and $V(x)=Kx^2/2$, $x<0$.

3. Use WKBJ to derive an equation for the phase shift $\delta(k)$ of an electron in the one-dimensional potential ($V_0>0$, $a>0$):

$$V(x)=\begin{cases}\infty & x\leq 0\\ -V_0(1-x/a) & 0<x<a\\ 0 & a<x\end{cases} \tag{3.148}$$

4. The phase $\phi(x)=\Theta(x)+\beta$ has $\beta=\pi/4$ for potentials with smooth turning points, and $\beta=0$ for an infinite step potential. Examine the value of $\beta(k)$ for a finite step potential. Solve exactly, as in chapter 2, the eigenfunctions for the potential:

$$V(x)=\begin{cases}V_0 & x<0\\ 0 & x>0\end{cases} \tag{3.149}$$

for $0<E<V_0$. For $x>0$ the eigenfunction has the form $\sin(kx+\delta)$. In this case $\delta(k)=\beta(k)$. For what value of E/V_0 is $\beta=\pi/4$?

5. Use WKBJ to find the eigenvalue of bound states in the one-dimensional potential:

$$V(x)=0 \quad \text{for} \quad |x|<a \tag{3.150}$$

$$=\frac{K}{2}(|x|-a)^2 \quad \text{for} \quad |x|>a \tag{3.151}$$

6. Use WKBJ to find the bound-state energies E_n in the symmetric quartic potential $V(x)=Kx^4$. *Hint:*

$$\int_0^1 dy\sqrt{1-y^4}=\frac{\Gamma(\frac{1}{4})\Gamma(\frac{3}{2})}{4\Gamma(\frac{7}{4})}=0.8740 \tag{3.152}$$

7. Use the trial eigenfunction $\phi = Ax\,\exp[-(\alpha x)^p]$ to find the lowest eigenvalue for the half-space linear potential for $p=\frac{3}{2}$ and $p=2$. Which of the three exponents $p=(1, \frac{3}{2}, 2)$ gives the lowest eigenvalue?

8. Use the variational method to find the ground-state energy of the following potential $(-\infty < x < \infty)$:

$$V(x) = -V_0 \exp[-(x/a)^2] \tag{3.153}$$

$$g = \frac{2ma^2V_0}{\hbar^2} = \sqrt{8} \tag{3.154}$$

Choose your own trial wave function, and obtain a numerical answer for the binding energy as a fraction of V_0.

9. Use the following trial function $\phi(x) = Ax\,\exp(-\alpha x)$ to find the ground-state energy variationally for the half-space potential:

$$V(x) = \begin{cases} \infty & x<0 \\ -\dfrac{e^2}{x} & 0<x \end{cases} \tag{3.155}$$

where e and m are the charge and mass of an electron.

10. Show that the $\phi(x)$ used in problem 9 is an exact eigenstate of the Hamiltonian, and find its eigenvalue.

11. For the potential $V(x)$ in problem 9, find all of the bound-state energies by WKBJ.

12. For the potential $V(x)$ in problem 9, use WKBJ to derive an expression for the continuum eigenfunction.

13. Use the variational method to find the eigenvalue of the harmonic oscillator with the trial eigenfunction $\phi = Ax\,\exp[-\alpha^2x^2/2]$.

14. Use the variational method to solve for the ground-state eigenvalue of the quartic potential $[V(x) = Kx^4]$ using a Gaussian trial wavefunction.

15. At the interface between a metal and a semiconductor, a Schottky barrier is formed due to a depletion region. The potential function is

$$V(x) = A(x_0 - x)^2 \quad \text{for} \quad 0<x<x_0 \tag{3.156}$$

$$= 0 \quad \text{for } x>x_0, x<0 \tag{3.157}$$

where $x=0$ is the interface; the metal is at $x<0$ and the semiconductor at $x>0$. Use WKBJ to calculate the tunneling exponent $2\int dx\,\alpha(x)$ of an electron with energy $E, > Ax_0^2 > E > 0]$ going from the metal to the semiconductor.

16. A potential $V(x)$ is an inverted parabola:

$$V(x) = \begin{cases} K(a^2 - x^2) & |x| < a \\ 0 & |x| > a \end{cases} \tag{3.158}$$

Calculate the WKBJ tunneling exponent $\alpha = \int k(x)dx$ for values of $0 < E < Ka^2$.

4 Spin and Angular Momentum

Angular momentum is an important entity in quantum mechanics. The two major contributors to the angular momentum are the spin ($\vec{s}$) of the particle and the orbital angular momentum ($\vec{\ell}$) from the rotational motion. In classical mechanics, the total angular momentum $\vec{j} = \vec{\ell} + \vec{s}$ is obtained by the vector addition of the two component vectors. In quantum mechanics, all three forms of angular momentum ($\vec{s}, \vec{\ell}, \vec{j}$) are individually quantized and only selected values are allowed. Then the problem of adding angular momentum $\vec{j} = \vec{\ell} + \vec{s}$ is more complicated, since both constituents and final value are quantized.

In systems of more than one particle, the problem of combining all of the spins and all of the orbital motions becomes important in constructing the many-particle eigenfunction. The many-particle aspects are deferred until chapter 9. The present chapter is concerned with understanding the properties of only one or two separate components of angular momentum.

The eigenfunctions and eigenvalues for both spin and orbital angular momentum are derived in this chapter. The discussion is based on the Heisenberg approach to quantum mechanics. At the start, some operators are defined and their commutation relations established. Then the eigenfunction and eigenvalues are derived by the manipulation of operators. The advantage of this method is its generality: the final formulas are valid for all values of angular momentum. They apply equally well to spin or orbital motion.

4.1 Operators, Eigenvalues, and Eigenfunctions

In this section the symbol for orbital angular momentum is $\vec{M} = \mathbf{r} \times \mathbf{p}$:

$$M_x = yp_z - zp_y \tag{4.1}$$

$$M_y = zp_x - xp_z \tag{4.2}$$

$$M_z = xp_y - yp_x \tag{4.3}$$

$$M^2 = M_x^2 + M_y^2 + M_z^2 \tag{4.4}$$

4.1.1 Commutation Relations

The operators M_j do not commute, since r_α does not commute with p_β when $\alpha = \beta$:

$$[r_\alpha, p_\beta] = i\hbar\delta_{\alpha\beta} \tag{4.5}$$

The first example is

$$[M_x, M_y] = [yp_z - zp_y, zp_x - xp_z] \tag{4.6}$$
$$= yp_x[p_z, z] + xp_y[z, p_z] \tag{4.7}$$
$$= i\hbar(-yp_x + xp_y) = i\hbar M_z \tag{4.8}$$
$$[M_y, M_z] = i\hbar M_x \tag{4.9}$$
$$[M_z, M_x] = i\hbar M_y \tag{4.10}$$

The last two commutators are derived in the same fashion as the first one. The three commutation relations (4.8)–(4.10), along with the definition of M^2 in (4.4), provide the starting point for the derivation of the eigenvalues and eigenfunctions. Although these equations are derived for orbital angular momentum, the components of spin angular momentum obey the same commutation relations. In fact, the components of any kind of angular momentum obey these relationships. The results of this section are quite general. This generality is the advantage of using the Heisenberg method of solving the operator equations.

The first step in the derivation is to derive some more commutation relations. These additional relations are found from the three starting ones in eqns. (4.8)–(4.10). The first is to show that M^2 commutes with any component M_j. It is sufficient to prove it for one of them, say for M_z:

$$[M^2, M_z] = [M_x^2 + M_y^2 + M_z^2, M_z] = ? \tag{4.11}$$

Since M_z commutes, with itself, it also commutes with M_z^2. One of the three components above gives zero:

$$[M_z^2, M_z] = 0 \tag{4.12}$$

Below we consider the case of M_x^2 and M_y^2. In each case, we add and subtract a term such as $M_xM_zM_x$ to evaluate the commutators:

$$[M_x^2, M_z] = M_x^2M_z - M_zM_x^2 \tag{4.13}$$
$$= M_x(M_xM_z - M_zM_x) + (M_xM_z - M_zM_x)M_x$$
$$= -i\hbar(M_xM_y + M_yM_x) \tag{4.14}$$
$$[M_y^2, M_z] = M_y^2M_z - M_zM_y^2 \tag{4.15}$$
$$= M_y(M_yM_z - M_zM_y) + (M_yM_z - M_zM_y)M_y$$
$$= i\hbar(M_yM_x + M_xM_y) \tag{4.16}$$
$$[M_x^2 + M_y^2, M_z] = 0 \tag{4.17}$$

It has been shown that $[M^2, M_z] = 0$. M_x and M_y also commute with M^2:

$$[M^2, M_x] = 0 = [M^2, M_y] \tag{4.18}$$

Since M^2 commutes with M_z, the eigenstates can be required to be simultaneous eigenvectors of both operators. Each will have different eigenvalues. Each eigenvalue is assigned a symbol: j with M^2 and m with M_z. The eigenstates are denoted by $|j, m\rangle$. The dimensions of angular momentum are those of Planck's constant $\hbar$, so the eigenvalues include this symbol. The dimensionless quantities m and f_j are also used:

$$M_z|j, m\rangle = \hbar m|j, m\rangle \tag{4.19}$$

$$M^2|j, m\rangle = \hbar^2 f_j|j, m\rangle \tag{4.20}$$

These definitions do not in any way restrict the possible values of m and j. At the moment they are just symbols.

It would be just as easy to choose the pair of operators (M^2, M_x) or (M^2, M_y). The choice of (M^2, $M_{\hat{z}}$) is conventional but arbitrary. Since (M_x, M_y, M_z) do not commute with each other, only one of them can be paired with M^2. The state $|j, m\rangle$ is not an eigenstate of M_x or M_y. The choice of M_z makes the z-axis the basis for the quantization of angular momentum. Any other direction can be chosen with equal correctness.

4.1.2. *Raising and Lowering Operators*

Define two more operators, called the *raising operator* ($L^\dagger$) and the *lowering operator* (L):

$$L^\dagger = M_x + iM_y \tag{4.21}$$

$$L = M_x - iM_y \tag{4.22}$$

The dagger symbol (†) denotes the Hermitian conjugate. Since the angular momentum components are Hermitian, the raising and lowering operators are mutual Hermitian conjugates. Similar raising and lowering operators were introduced in section 2.4 for the harmonic oscillator. Here they have the same meaning, since they raise or lower the z-component of angular momentum.

Next explore some of the operator relationships. Since M_x and M_y commute with M^2, then so do the raising and lowering operators:

$$[M^2, L] = 0 \tag{4.23}$$

$$[M^2, L^\dagger] = 0 \tag{4.24}$$

However, they do not commute with M_z or with each other:

$$\begin{aligned} [L, L^\dagger] &= [M_x - iM_y, M_x + iM_y] \\ &= -i[M_y, M_x] + i[M_x, M_y] = -2\hbar M_z \end{aligned} \tag{4.25}$$

$$[M_z, L] = [M_z, M_x - iM_y] = i\hbar(M_y + iM_x) = -\hbar L \tag{4.26}$$

$$[M_z, L^\dagger] = \hbar L^\dagger \tag{4.27}$$

The operator M^2 is expressed in terms of the raising and lowering operators. First use eqns. (4.21)–(4.22) to get

$$M_x = \frac{1}{2}(L + L^\dagger), \quad M_y = \frac{i}{2}(L - L^\dagger) \tag{4.28}$$

$$M^2 = \frac{1}{4}(L + L^\dagger)^2 + \frac{i^2}{4}(L - L^\dagger)^2 + M_z^2 \tag{4.29}$$

$$= \frac{1}{2}(L^\dagger L + LL^\dagger) + M_z^2 \tag{4.30}$$

Since L and $L^\dagger$ do not commute, the order in which they are written is important. These commutation relations and definitions complete the first stage of the derivation.

4.1.3 Eigenfunctions and Eigenvalues

The next step is to solve the operator relations to determine the properties of the eigenfunctions. It is convenient to use the Dirac notation with bras and kets. A ket is any eigenstate on the right ($|j, m\rangle$); a bra is the eigenstate on the left ($\langle j, m|$). The bra is the Hermitian conjugate of the ket. They are assumed to be orthogonal:

$$\langle j, m | j', m' \rangle = \delta_{jj'} \delta_{mm'} \tag{4.31}$$

The expectation value of an operator is discussed and defined in chapters 1 and 2. For any function F of angular momentum, its expectation value is obtained by sandwiching the operator between a ket on the right and a bra on the left:

$$\langle F \rangle = \langle j, m | F | j, m \rangle \tag{4.32}$$

A matrix element is a similar expression, but the bra and ket may not be the same state:

$$\langle F \rangle (jj', mm') = \langle j', m' | F | j, m \rangle \tag{4.33}$$

Quite often these expectation values and matrix elements can be found using operator techniques. In the next section the state $|j, m\rangle$ is given an explicit representation in terms of matrices. Then one could do a matrix multiplication to get the expectation values.

The present derivation proceeds by taking matrix elements of commutation relations. Many properties can be derived by this procedure. The first one uses eqn. (4.24):

$$0 = \langle j', m' | [M^2, L^\dagger] | j, m \rangle = \langle j', m' | M^2 L^\dagger - L^\dagger M^2 | j, m \rangle \tag{4.34}$$

Since the commutator is zero, its matrix element equals zero. Use eqn. (4.20) to evaluate $M^2 |j, m\rangle = \hbar^2 f_j |j, m\rangle$ or its Hermitian conjugate $\langle j', m' | M^2 = \hbar^2 f_j' \langle j', m'|$:

$$0 = \hbar^2 \langle j', m' | (f_{j'} L^\dagger - L^\dagger f_j) | j, m \rangle \tag{4.35}$$

$$= \hbar^2 (f_{j'} - f_j) \langle j', m' | L^\dagger | j, m \rangle \tag{4.36}$$

Note that we took $\langle j', m' | M^2 L^\dagger | j, m \rangle$ and operated M^2 to the left on the bra. It could not be taken to the right since the operator $L^\dagger$ is in the way.

Equation (4.36) is zero. If $j = j'$ it vanishes because $f_j = f_{j'}$. If $j \neq j'$ it must vanish because the matrix element of $L^\dagger$ is zero

$$\langle j', m'|L^{\dagger}|j, m\rangle = \delta_{jj'} L^{\dagger}_{j;m'm} \tag{4.37}$$

The matrix element is zero unless $j = j'$. If $j = j'$ it can be nonzero and is represented by the unknown c-number $L^{\dagger}_{j;m'm}$ Further information regarding this matrix element is found by evaluating the commutator in (4.27) when $j = j'$:

$$\langle j, m'|[M_z, L^{\dagger}]|j, m\rangle = \hbar\langle j, m'|L^{\dagger}|j, m\rangle \tag{4.38}$$

$$\langle j, m'|M_z L^{\dagger} - L^{\dagger} M_z|j, m\rangle = \hbar\langle j, m'|L^{\dagger}|j, m\rangle \tag{4.39}$$

On the left side of the equal sign, evaluate $M_z|j, m\rangle = \hbar m|j, m\rangle$ and $\langle j, m'|M_z = \hbar m'\langle j, m'|$. Also shift the term on the right of the equal sign to the left:

$$\hbar\langle j, m'|m' L^{\dagger} - L^{\dagger} m|j, m\rangle = \hbar\langle j, m'|L^{\dagger}|j, m\rangle \tag{4.40}$$

$$\hbar(m' - m - 1)L^{\dagger}_{j;\, m'm} = 0 \tag{4.41}$$

The expression on the left vanishes. If $m' = m + 1$, the factor in parentheses vanishes. If $m' \neq m + 1$, the matrix element vanishes. If $m' = m + 1$, the matrix element does not vanish, but is still unknown:

$$L^{\dagger}_{j;m'm} = \delta_{m' = m+1} \hbar q_j(m) \tag{4.42}$$

The function $q_j(m)$ is still unknown. The matrix element on the left side of eqn. (4.37) has been shown to vanish unless $j' = j$, $m' = m + 1$. The raising operator $L^{\dagger}$ raises the value of m.

Similar relationships may be established for the lowering operator L. It is possible to derive the result by going through the same steps. However, a shortcut is possible by recognizing that L is just the Hermitian conjugate of $L^{\dagger}$. The matrix element of L is just the Hermitian conjugate of the matrix element for $L^{\dagger}$:

$$\{\langle j', m'|L^{\dagger}|j, m\rangle = \delta_{jj'}\delta_{m',\, m+1}\hbar q_j(m)\}^{\dagger} \tag{4.43}$$

$$\langle j, m|L|j', m'\rangle = \delta_{jj'}\delta_{m',\, m+1}\hbar q_j(m)^{*} \tag{4.44}$$

Now m' is on the right (in the ket), while m is on the left (in the bra). Since $m' = m + 1$, the operator L lowers the value of the quantum number m.

Both matrix elements depend on the function $q_j(m)$. It is found by examining the matrix element of another commutator (4.25):

$$\langle j, m|[L, L^{\dagger}]j, m\rangle = -2\hbar\langle j, m|M_z|j, m\rangle \tag{4.45}$$

$$\langle j, m|LL^{\dagger} - L^{\dagger}L|j, m\rangle = -2\hbar^2 m \tag{4.46}$$

This expression is not simple to evaluate, since $|j, m\rangle$ is an eigenstate of neither L nor $L^{\dagger}$. It is necessary to insert a complete set of states between L and $L^{\dagger}$. This step employs the completeness relation in terms of the identity operator I:

$$I = \sum_{j', m'} |j', m'\rangle\langle j', m'| \tag{4.47}$$

This identity is inserted in the two terms on the left in eqn. (4.46). Inserting the identity does not change the value:

$$\sum_{j',m'}[\langle j,m|L|j',m'\rangle\langle j',m'|L^{\dagger}|j,m\rangle$$
$$-\langle j,m|L^{\dagger}|j',m'\rangle\langle j',m'|L|j,m\rangle]=-2\hbar^2 m \tag{4.48}$$

The only nonzero matrix elements have $j'=j$. The matrix elements of $L^{\dagger}$ raise the value of m, while those for L lower it. The only nonzero term in the summation is

$$\langle j,m|L|j,m+1\rangle\langle j,m+1|L^{\dagger}|j,m\rangle \tag{4.49}$$
$$-\langle j,m|L^{\dagger}|j,m-1\rangle\langle j,m-1|L|j,m\rangle=-2\hbar^2 m$$
$$\hbar^2\{|q_j(m)|^2-|q_j(m-1)|^2\}=-2\hbar^2 m \tag{4.50}$$

Cancel $\hbar^2$ from each side and find

$$|q_j(m)|^2-|q_j(m-1)|^2=-2m \tag{4.51}$$

This equation is solved to find $|q_j(m)|^2$ as a function of m. Expand this function in a power series in m. The coefficients a_n of this series could be functions of j. In fact, some are and some are not:

$$|q_j(m)|^2=a_0+a_1m+a_2m^2+a_3m^3+\cdots \tag{4.52}$$
$$|q_j(m-1)|^2=a_0+a_1(m-1)+a_2(m-1)^2+a_3(m-1)^3+\cdots \tag{4.53}$$
$$|q_j(m)|^2-|q_j(m-1)|^2=a_1+a_2(2m-1)+a_3(3m^2-3m+1)+\cdots \tag{4.54}$$
$$-2m=a_1+a_2(2m-1)+a_3(3m^2-3m+1)+\cdots \tag{4.55}$$

The solution to eqn. (4.55) is

$$a_1=a_2=-1 \tag{4.56}$$
$$a_n=0 \quad \text{if } n\geq 3 \tag{4.57}$$

The higher coefficients a_3, a_4, etc. are zero since they multiply higher powers of m and there are no such terms on the left of the equal sign. No information is obtained regarding the first term a_0, and it is nonzero. So far the solution has the form

$$|q_j(m)|^2=a_0-m(m+1) \tag{4.58}$$

An important feature is that $|q_j(m)|^2$ cannot be negative, since it is the absolute magnitude squared of a function. This gives the condition $a_0\geq m(m+1)$.

The derivation is nearly completed. All the information is in hand to present the final arguments. So far it has been proved that when $L^{\dagger}$ operates upon an eigenstate $|j,\, m\rangle$, it produces the eigenstate $|j,\, m+1\rangle$ multiplied by a coefficient $\hbar q_j(m)$:

$$L^{\dagger}|j,m\rangle=\hbar q_j(m)|j,m+1\rangle \tag{4.59}$$
$$L|j,m\rangle=\hbar q_j(m-1)^*|j,m-1\rangle \tag{4.60}$$

These two relationships were proven above. When $L^\dagger$ operates on $|j, m\rangle$ it creates a state $|j, m'\rangle$ where $m' = m+1$. Similarly, the operation of L on $|j, m\rangle$ creates a pure state of $|j, m-1\rangle$.

Consider the effect of operating on an eigenstate $|j, m\rangle$ by n-successive raising operators. This operation is denoted as $(L^\dagger)^n$. Each operation raises m by unity. After raising n times, one ends up with the state $|j, m+n\rangle$:

$$(L^\dagger)^2|j,m\rangle = \hbar q_j(m)L^\dagger|j,m+1\rangle = \hbar^2 q_j(m)q_j(m+1)|j,m+2\rangle$$

$$(L^\dagger)^n|j,m\rangle = \hbar^n q_j(m)q_j(m+1)\cdots q_j(m+n-1)|j,m+n\rangle \tag{4.61}$$

The final expression contains the product of n-factors of the form $q_j(m+\ell)$, where ℓ is an integer between zero and $n-1$. The value of m can be raised further by operating more times by $L^\dagger$.

Repeated operations with the raising operator eventually creates problems. The value of $(m+n)$ becomes large enough that the expression

$$|q_j(m+n)|^2 = a_0 - (m+n)(m+n+1) \tag{4.62}$$

becomes negative on the right-hand side. No matter what value is chosen for a_0, we can find a value of n large enough to make this expression negative. This feature violates the constraint that it must always be a positive number. This constraint was derived from the commutation relations and is a fundamental feature of angular momentum.

The problem is resolved by selecting the value of a_0 to equal

$$a_0 = N(N+1), \quad N = m + n_u \tag{4.63}$$

$$|q_j(m+n_u)|^2 = a_0(N)[1-1] = 0 \tag{4.64}$$

In eqn. (4.61), if one of the $q_j(m+n_u) = 0$, then further operations by the raising operator continue to give zero. This choice of a_0 means that $|j, N\rangle$ has the largest value of N in the system. One cannot raise the value of m past the value of $m = N$. Attempts to raise this state by operating by the raising operator simply give zero.

Repeat this argument for the lowering operator. Each operation by L lowers the value of m by one. Repeated lowering will make the m value negative:

$$(L)^{n_\ell}|j,m\rangle = \hbar^{n_\ell} q_j(m-1)^* q_j(m-2)^* \cdots q_j(m-n_\ell)^*|j,m-n_\ell\rangle \tag{4.65}$$

Repeated operations by L have the same problem as repeated operations by $L^\dagger$. When $m - n_\ell$ has large negative values, then

$$|q_j(m-n_\ell)|^2 = a_0 - (m-n_\ell)(m-n_\ell+1) \tag{4.66}$$

$$= a_o - (n_\ell - m)(n_\ell - m - 1) \tag{4.67}$$

where $n_\ell - m - 1$ is a positive number or zero. The right-hand side is negative for large values of $n_\ell - m$. Again it is necessary to find some way to truncate the series of operations. The series is truncated [see (4.63)] if $N = n_\ell - m - 1$, where $N = m + n_u$ is the value used to truncate the upper sequence. Combining these two definitions of N gives

$$N = n_\ell - m - 1 = m + n_u \tag{4.68}$$

$$m = \frac{1}{2}(n_\ell - 1 - n_u) \tag{4.69}$$

The same value for N can be used to truncate the upper and lower sequences provided m satisfies eqn. (4.69). Since n_ℓ and n_u are both integers, then m is either an integer or a half-integer. Two examples are

1. If $m=2$ and $n_u=1$ then $N=3$ and $a_0=(3)(4)=12$. A value of $n_\ell=6$ makes $n_\ell - m - 1 = 3 = N$.
2. If $m=-\frac{3}{2}$ and $n_u=4$ then $N=\frac{5}{2}$ and $a_0=(\frac{5}{2})(\frac{7}{2})=\frac{35}{4}$. Then $n_\ell=2$ and $n_\ell-1-m=\frac{5}{2}=N$.

The fact that m can only be an integer or a half-integer is of fundamental importance. In quantum mechanics, it is found that all values of angular momentum come in integer or half-integer values. This fundamental law of nature is a direct consequence of the commutation relations for angular momentum operators. Note that we started the derivation using classical angular momentum operators $\vec{M}=\mathbf{r}\times\mathbf{p}$, which occur only with integer values. However, the operator algebra also permits a solution with half-integers. Many fundamental particles such as electrons, protons, neutrons, and neutrinos have spin-$\frac{1}{2}$.

The quantity N is the largest allowed value of m since $a_0=N(N+1)$. The largest negative value of m is $-N$. Trying to lower this value gives

$$L|j,-N\rangle = \hbar q_j(-N-1)|j,-N-1\rangle \tag{4.70}$$

$$|q_j(-N-1)|^2 = a_0-(-N-1)(-N) = a_0 - N(N+1) = 0$$

The number of allowed values of $-N \le m \le N$ is $2N+1$.

The final step in the derivation is to determine the meaning of the quantum number j. It enters the eigenvalue of M^2. We are now ready to determine this quantity using eqn. (4.30). The terms in this equation give

$$LL^\dagger|j,m\rangle = \hbar q_j(m)L|j,m+1\rangle = \hbar^2|q_j(m)|^2|j,m\rangle \tag{4.71}$$

$$L^\dagger L|j,m\rangle = \hbar q_j(m-1)^* L^\dagger|j,m-1\rangle = \hbar^2|q_j(m-1)|^2|j,m\rangle \tag{4.72}$$

$$M_z^2|j,m\rangle = \hbar^2 m^2|j,m\rangle \tag{4.73}$$

Now evaluate $M^2|j,\, m\rangle$:

$$M^2|j,m\rangle = \frac{\hbar^2}{2}\{|q_j(m)|^2 + |q_j(m-1)|^2 + 2m^2\}|j,m\rangle \tag{4.74}$$

$$= \frac{\hbar^2}{2}\{a_0 - m(m+1) + a_0 - (m-1)m + 2m^2\}|j,m\rangle$$

$$M^2|j,m\rangle = \hbar^2 a_0|j,m\rangle \tag{4.75}$$

The eigenvalue of M^2 is just $\hbar^2 a_0$. The result was supposed to be $\hbar^2 f_j$, so that $f_j = a_0$. Recall that $a_0 = N(N+1)$, where N is the largest value of m. So j is identified as $j = N$, since that is the only variable in a_0.

The various results are now collected:

$$M_z|j, m\rangle = \hbar m|j, m\rangle \tag{4.76}$$

$$M^2|j, m\rangle = \hbar^2 j(j+1)|j, m\rangle \tag{4.77}$$

$$L^\dagger|j, m\rangle = \hbar\sqrt{j(j+1) - m(m+1)}|j, m+1\rangle \tag{4.78}$$

$$L|j, m\rangle = \hbar\sqrt{j(j+1) - m(m-1)}|j, m-1\rangle \tag{4.79}$$

These four equations are the basic, important relationships for angular momentum. Recall that m and j can both be integers or both be half-integers. The values of m change in integer steps over the range $-j \le m \le j$, for a total of $(2j+1)$ values.

It is interesting to compare these results to the operators of the harmonic oscillator. The latter are discussed in section 2.4. Angular momentum starts with three operators (M_x, M_y, M_z), while the harmonic oscillator has two (x, p). Raising and lowering operators were constructed for each case. The commutation relations for the two systems had important differences. For the harmonic oscillator $[a, a^\dagger] = 1$, the commutator is a c-number. For the angular momentum, the commutator is another operator: $[L, L^\dagger] = 2\hbar M_z$. The end result is also different. For the harmonic oscillator, the eigenstates $|n\rangle$ permitted all values of positive n, even up to very large values. For the angular momentum, the value of j goes to arbitrarily large positive integers or half-integers but m is bounded by j.

4.2 Representations

The relationships between operators and eigenfunctions in (4.76)–(4.79) are the important equations of angular momentum. One can get along very well in physics by using these relationships as they are written. There is really no need to write down a specific form for the eigenstates $|j, m\rangle$.

A representation is a set of specific functions that describe the operators and eigenfunctions. There are two different kinds of representations that are used most commonly. One is a matrix form, where the operators are square matrices of dimension $(2j+1)$ and the eigenfunctions are vectors of the same length. The matrix representation can be applied to an integer or half-integer value of j. Integer values of j also have a representation in terms of polynomials in (x, y, z), while the operators have the derivative form shown in eqns. (4.1)–(4.3). Examples of both types of representations are given below.

$j = \frac{1}{2}$.

The usual way of representing $j = \frac{1}{2}$ is using the Pauli spin matrices σ_j. The angular momentum operator is $M_j = (\hbar/2)\sigma_j$. The Pauli matrices are

$$\sigma_x = \begin{pmatrix} 0 & 1 \\ 1 & 0 \end{pmatrix}, \quad \sigma_y = \begin{pmatrix} 0 & -i \\ i & 0 \end{pmatrix} \tag{4.80}$$

$$\sigma_z = \begin{pmatrix} 1 & 0 \\ 0 & -1 \end{pmatrix}, \quad I = \begin{pmatrix} 1 & 0 \\ 0 & 1 \end{pmatrix} \tag{4.81}$$

The identity operator is I. The raising and lowering operators are

$$L^{\dagger} = M_x + iM_y = \hbar \begin{pmatrix} 0 & 1 \\ 0 & 0 \end{pmatrix} \tag{4.82}$$

$$L = M_x - iM_y = \hbar \begin{pmatrix} 0 & 0 \\ 1 & 0 \end{pmatrix} \tag{4.83}$$

$$M_z = \frac{\hbar}{2} \begin{pmatrix} 1 & 0 \\ 0 & -1 \end{pmatrix} \tag{4.84}$$

$$M_x^2 = M_y^2 = M_z^2 = \frac{\hbar^2}{4} \begin{pmatrix} 1 & 0 \\ 0 & 1 \end{pmatrix} \tag{4.85}$$

$$M^2 = \frac{3\hbar^2}{4} I \tag{4.86}$$

The eigenstates are column vectors for kets and row vectors for bras. The elements are zeros or ones:

$$\left|\frac{1}{2}, \frac{1}{2}\right\rangle = \begin{pmatrix} 1 \\ 0 \end{pmatrix}, \quad \left|\frac{1}{2}, -\frac{1}{2}\right\rangle = \begin{pmatrix} 0 \\ 1 \end{pmatrix} \tag{4.87}$$

$$\left\langle\frac{1}{2}, \frac{1}{2}\right| = (1, 0), \quad \left\langle\frac{1}{2}, -\frac{1}{2}\right| = (0, 1) \tag{4.88}$$

The two states $|\frac{1}{2}, m\rangle$ are eigenstates of both M_z and M^2, as required by the original assumption in the derivation:

$$M_z \begin{pmatrix} 1 \\ 0 \end{pmatrix} = \frac{\hbar}{2} \begin{pmatrix} 1 \\ 0 \end{pmatrix} \tag{4.89}$$

$$M_z \begin{pmatrix} 0 \\ 1 \end{pmatrix} = -\frac{\hbar}{2} \begin{pmatrix} 0 \\ 1 \end{pmatrix} \tag{4.90}$$

The factors $q_j(m)$ are all zero or one:

$$L^{\dagger} \begin{pmatrix} 0 \\ 1 \end{pmatrix} = \hbar \begin{pmatrix} 1 \\ 0 \end{pmatrix}, \quad L^{\dagger} \begin{pmatrix} 1 \\ 0 \end{pmatrix} = 0 \tag{4.91}$$

$$L \begin{pmatrix} 1 \\ 0 \end{pmatrix} = \hbar \begin{pmatrix} 0 \\ 1 \end{pmatrix}, \quad L \begin{pmatrix} 0 \\ 1 \end{pmatrix} = 0 \tag{4.92}$$

A common application for the $j = \frac{1}{2}$ representation is in describing the spins of fermions such as electrons or nucleons. The eigenstate with $m = \frac{1}{2}$ is usually called "spin-up," while $m = -\frac{1}{2}$ is called "spin-down." The operators are often represented by the symbol s rather than by M or L. (s_x, s_y, s_z) are the three spin components, while $s^{(+)}$ and $s^{(-)}$ are the raising and lowering operators. They are given by Pauli matrices $s_j = (\hbar/2)\sigma_j$.

$\underline{j = 1.}$

Matrices. The $j = 1$ angular momentum state can be represented either by matrices or by polynomials. The matrix form is usually used when describing the spin of a particle or a composite particle. The polynomial representation is used for the orbital motion of the particle. Both representations are given here, beginning with the matrix one.

A matrix representation is constructed by writing down all of the eigenfunctions as column vectors. They have zeros in all spots except for 1 at one spot. The position of the "1" is determined by the value of m. For $j=1$ the three eigenvectors are

$$|1,1\rangle = \begin{pmatrix} 1 \\ 0 \\ 0 \end{pmatrix}, \quad |1,0\rangle = \begin{pmatrix} 0 \\ 1 \\ 0 \end{pmatrix}, \quad |1,-1\rangle = \begin{pmatrix} 0 \\ 0 \\ 1 \end{pmatrix} \tag{4.93}$$

The next step is to construct the raising and lowering matrices $L^\dagger$ and L. They are mutual Hermitian conjugates, so constructing one is sufficient. Since $L^\dagger$ raises m by unity, it only has elements one spot above the diagonal. Similarly, the matrix for L has nonzero elements only one spot below the diagonal. These elements are $\hbar q_j(m)$. For $j=1$ then

$$q_j(m) = \sqrt{j(j+1) - m(m+1)} \tag{4.94}$$

$$q_1(1) = 0 \tag{4.95}$$

$$q_1(0) = \sqrt{2} \tag{4.96}$$

$$q_1(-1) = \sqrt{2} \tag{4.97}$$

These values immediately give the raising and lowering matrices:

$$L^\dagger = \hbar\sqrt{2}\begin{pmatrix} 0 & 1 & 0 \\ 0 & 0 & 1 \\ 0 & 0 & 0 \end{pmatrix} \tag{4.98}$$

$$L = \hbar\sqrt{2}\begin{pmatrix} 0 & 0 & 0 \\ 1 & 0 & 0 \\ 0 & 1 & 0 \end{pmatrix} \tag{4.99}$$

The next step is to construct M_x, M_y, M_z. The matrix M_z is diagonal with elements $\hbar m$. The matrices M_x, M_y are found from the raising and lowering matrices:

$$M_x = \frac{1}{2}(L^\dagger + L) = \frac{\hbar}{\sqrt{2}}\begin{pmatrix} 0 & 1 & 0 \\ 1 & 0 & 1 \\ 0 & 1 & 0 \end{pmatrix} \tag{4.100}$$

$$M_y = \frac{i}{2}(L - L^\dagger) = \frac{i\hbar}{\sqrt{2}}\begin{pmatrix} 0 & -1 & 0 \\ 1 & 0 & -1 \\ 0 & 1 & 0 \end{pmatrix} \tag{4.101}$$

$$M_z = \hbar\begin{pmatrix} 1 & 0 & 0 \\ 0 & 0 & 0 \\ 0 & 0 & -1 \end{pmatrix} \tag{4.102}$$

The magnitude matrix is $M^2 = 2\hbar^2 I$, where I is always the identity matrix. The value of 2 comes from $j(j+1)$ when $j=1$.

r-Space. For integer values of j, it is easy to construct an r-space representation. The eigenfunctions are polynomials of (x, y, z). The operators are always the same, regardless of the integer value of j. They are given at the beginning of this chapter:

$$M_x = yp_z - zp_y = -i\hbar\left(y\frac{\partial}{\partial z} - z\frac{\partial}{\partial y}\right) \tag{4.103}$$

$$M_y = zp_x - xp_z = -i\hbar\left(z\frac{\partial}{\partial x} - x\frac{\partial}{\partial z}\right) \tag{4.104}$$

$$M_z = xp_y - yp_x = -i\hbar\left(x\frac{\partial}{\partial y} - y\frac{\partial}{\partial x}\right) \tag{4.105}$$

The symbol L_z is also used for M_z. The raising and lowering operators are

$$L = M_x - iM_y = -i[z(p_x - ip_y) - (x - iy)p_z] \tag{4.106}$$

$$= -\hbar\left[z\left(\frac{\partial}{\partial x} - i\frac{\partial}{\partial y}\right) - (x - iy)\frac{\partial}{\partial z}\right] \tag{4.107}$$

$$L^\dagger = i[z(p_x + ip_y) - (x + iy)p_z] \tag{4.108}$$

$$= \hbar\left[z\left(\frac{\partial}{\partial x} + i\frac{\partial}{\partial y}\right) - (x + iy)\frac{\partial}{\partial z}\right] \tag{4.109}$$

In spherical coordinates, $x = r\sin(\theta)\cos(\phi)$, $y = r\sin(\theta)\sin(\phi)$, $z = r\cos(\theta)$. Then operators are

$$L_x = i\hbar\left[\sin(\phi)\frac{\partial}{\partial\theta} + \frac{\cos(\phi)}{\tan(\theta)}\frac{\partial}{\partial\phi}\right] \tag{4.110}$$

$$L_y = i\hbar\left[-\cos(\phi)\frac{\partial}{\partial\theta} + \frac{\sin(\phi)}{\tan(\theta)}\frac{\partial}{\partial\phi}\right] \tag{4.111}$$

$$L_z = \frac{\hbar}{i}\frac{\partial}{\partial\phi} \tag{4.112}$$

$$L^\dagger = \hbar e^{i\phi}\left[\frac{\partial}{\partial\theta} + \cot(\theta)\frac{\partial}{\partial\phi}\right] \tag{4.113}$$

$$L = \hbar e^{-i\phi}\left[-\frac{\partial}{\partial\theta} + \cot(\theta)\frac{\partial}{\partial\phi}\right] \tag{4.114}$$

The spherical harmonic functions $Y_\ell^m(\theta, \phi)$ are introduced in chapter 5:

$$Y_\ell^m(\theta, \phi) = C_{\ell m} P_\ell^{|m|}(\theta) e^{im\phi} \tag{4.115}$$

$$C_{\ell m} = (-1)^m \sqrt{\frac{2\ell + 1}{4\pi}\frac{(\ell - |m|)!}{(\ell + |m|)!}} \tag{4.116}$$

where $P_\ell^{|m|}$ are associated Legendre polynomials. The angular momentum functions $(j = \ell)$ are just r^ℓ times the spherical harmonics with only part of the factor of $C_{\ell m}$:

$$|\ell, m\rangle = D_{\ell m} r^\ell P_\ell^{|m|}(\theta) e^{im\phi} \tag{4.117}$$

$$D_{\ell m} = (-1)^m \sqrt{\frac{(\ell - |m|)!}{(\ell + |m|)!}} \tag{4.118}$$

The normalization is found from the integration over 4π solid angle:

$$\int_0^{2\pi} d\phi \int_0^{\pi} \sin(\theta) d\theta \langle \ell, m | \ell, m \rangle = \frac{4\pi}{2\ell+1} r^{2\ell} \tag{4.119}$$

Some examples are

$$|0,0\rangle = 1 \tag{4.120}$$

$$|1,1\rangle = -\sqrt{\frac{1}{2}}(x+iy) = -\sqrt{\frac{1}{2}} r \sin(\theta) e^{i\phi} \tag{4.121}$$

$$|1,0\rangle = z = r\cos(\theta) \tag{4.122}$$

$$|1,-1\rangle = \sqrt{\frac{1}{2}}(x-iy) = \sqrt{\frac{1}{2}} r \sin(\theta) e^{-i\phi} \tag{4.123}$$

The last three equations are the eigenstates for $j=1$.

If one knows the Legendre polynomial $P_\ell(\theta)$ then one can construct the rest of the basis set using raising and lowering operators (4.107)–(4.109). An example is for $\ell = 2$:

$$P_2(\theta) = \frac{1}{2}[3\cos^2(\theta) - 1] \tag{4.124}$$

$$|2,0\rangle = \frac{1}{2}[3z^2 - r^2] = \frac{1}{2}[2z^2 - x^2 - y^2] \tag{4.125}$$

Using the raising operator gives, with $q_2(0) = \sqrt{6}$:

$$L^\dagger|2,0\rangle = \hbar\sqrt{6}|2,1\rangle = -\hbar 3z(x+iy) \tag{4.126}$$

$$|2,1\rangle = -\sqrt{\frac{3}{2}} z(x+iy) \tag{4.127}$$

The first equality in eqn. (4.126) comes from eqn. (4.59), and the second comes by using the operator in (4.109). In the second line, we solve for $|2, 1\rangle$. Note that this procedure gives the correct normalization of the eigenfunction. Raising again with $q_2(1) = 2$ gives

$$L^\dagger|2,1\rangle = \hbar 2|2,2\rangle = \hbar\sqrt{\frac{3}{2}}(x+iy)^2 \tag{4.128}$$

$$|2,2\rangle = \sqrt{\frac{3}{8}}(x+iy)^2 \tag{4.129}$$

The application of the lowering operator to (4.125) gives

$$|2,-1\rangle = \sqrt{\frac{3}{2}} z(x-iy) \tag{4.130}$$

$$|2,-2\rangle = \sqrt{\frac{3}{8}}(x-iy)^2 \tag{4.131}$$

This completes the discussion of the r-space representation.

The operator for M^2 is written out in (x, y, z) coordinates:

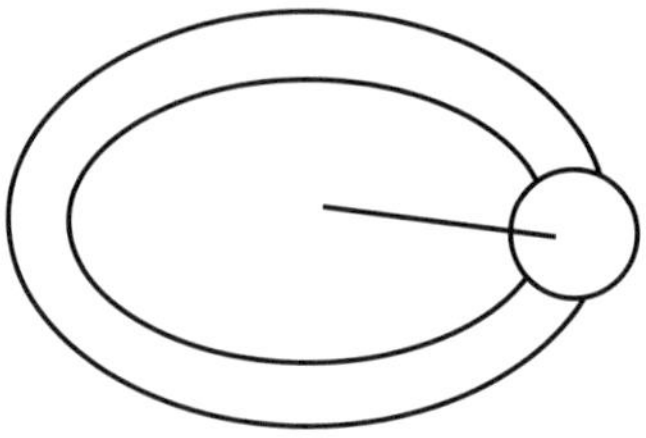

FIGURE 4.1. Bead on wire ring.

$$M^2 = -\hbar^2 \left[(x^2+y^2+z^2)\left(\frac{\partial^2}{\partial x^2}+\frac{\partial^2}{\partial y^2}+\frac{\partial^2}{\partial z^2}\right) - \left(x\frac{\partial}{\partial x}+y\frac{\partial}{\partial y}+z\frac{\partial}{\partial z}\right) - \left(x\frac{\partial}{\partial x}+y\frac{\partial}{\partial y}+z\frac{\partial}{\partial z}\right)^2 \right] \quad (4.132)$$

This complicated expression is greatly simplified when writing it in spherical coordinates: $x = r\sin(\theta)\cos(\phi)$, $y = r\sin(\theta)\sin(\phi)$, $z = r\cos(\theta)$. Then it is $r^2\nabla^2$ minus the terms involving radial derivatives:

$$M^2 = -\hbar^2[r^2\nabla^2 - (\vec{r}\cdot\vec{\nabla}) - (\vec{r}\cdot\vec{\nabla})^2] \quad (4.133)$$

$$M^2 = -\hbar^2\left[\frac{1}{\sin(\theta)}\frac{\partial}{\partial\theta}\left(\sin(\theta)\frac{\partial}{\partial\theta}\right) + \frac{1}{\sin^2(\theta)}\frac{\partial^2}{\partial\phi^2}\right] \quad (4.134)$$

This operator gives an eigenvalue of $\hbar^2\ell(\ell+1)$ when operating on the spherical harmonics:

$$M^2 Y_\ell^m(\theta,\phi) = \hbar^2\ell(\ell+1)Y_\ell^m(\theta,\phi) \quad (4.135)$$

This result is important when discussing three-dimensional solutions to Schrödinger's equation in the next chapter.

4.3 Rigid Rotations

The quantum mechanics of rigid rotors is quite simple. One solves Schrödinger's equation while fixing the radial variable r. As an example, consider the problem of a circle of wire with a radius R. Let a hollow bead of mass M_b slide around the wire. The angle θ denotes the position of the bead along the wire, as shown in figure 4.1. It is the only variable that can change. The Hamiltonian contains a kinetic energy term and a potential energy term. The potential energy confines the bead to the wire. If the ring is horizontal then gravity is not an issue. Then the potential energy does not depend on the angle θ, and V can be treated as a constant V_0. The Hamiltonian is

$$H = -\frac{\hbar^2}{2M_bR^2}\frac{d^2}{d\theta^2} + V_0 \quad (4.136)$$

$$H\phi_n(\theta) = E_n\phi_n(\theta) \quad (4.137)$$

$$\phi_n(\theta) = \frac{e^{in\theta}}{\sqrt{2\pi}} \quad (4.138)$$

$$E_n = n^2 \frac{\hbar^2}{2M_b R^2} + V_0 = n^2 E_1 + V_0 \tag{4.139}$$

The kinetic energy term contains only the derivative with respect to angle. The eigenfunction must have the form $\exp(in\theta)$. The constant n must be an integer to have $\phi_n(\theta + 2\pi) = \phi_n(\theta)$. Note that for the bead the moment of inertia is $I = M_b R^2$. The angular momentum operator is

$$L = \frac{\hbar}{i}\frac{d}{d\theta}, \quad L\phi_n = \hbar n \phi_n \tag{4.140}$$

The angular momentum is quantized.

A solid object can be characterized by the three moments of inertia about the three axes: (I_x, I_y, I_z). The kinetic energy term is then

$$\hat{T} = \frac{1}{2}\left[\frac{M_x^2}{I_x} + \frac{M_y^2}{I_y} + \frac{M_z^2}{I_z}\right] \tag{4.141}$$

where M_j^2 are the kinetic energy terms about the axis r_j. A special case is the rotation of an object around a fixed axis: say the z-direction. The only variable is the rotation angle θ. All we need to know about the object is its moment of inertia I_z, and the analysis is identical to that of the bead:

$$\phi_n(\theta) = \frac{e^{in\theta}}{\sqrt{2\pi}} \tag{4.142}$$

$$E_n = n^2 \frac{\hbar^2}{2I_z} = n^2 E_1 \tag{4.143}$$

Another special case is the rotation of a spherical top, which is a solid with $I_x = I_y = I_z \equiv I$. The kinetic energy term is

$$\hat{T} = \frac{M^2}{2I} \tag{4.144}$$

where M^2 was introduced in section 4.2. The eigenstates are the spherical harmonics $Y_\ell^m(\theta, \phi)$, and

$$E_\ell = \ell(\ell+1)\frac{\hbar^2}{2I} \tag{4.145}$$

Another special case is a diatomic molecule of identical atoms, such as O_2 or N_2. As shown in figure 4.2, $I_x = I_y \equiv I_\perp \neq I_z$. In fact, for rotations around the axis connecting the two atoms, $I_z \ll I_\perp$. Most of the mass is in the nuclei. For (I_x, I_y), the value of is $I_x = 2M(d/2)^2$, where d is the separation of the two nuclei. However, $I_z \sim Mb^2$, where b is the radius of the nuclei, which is about 10^{-5} of d. Then the kinetic energy term can be written as

$$\hat{T} = \frac{1}{2}\left[\frac{M^2}{I_\perp} + \left(\frac{1}{I_z} - \frac{1}{I_\perp}\right)M_z^2\right] \tag{4.146}$$

$$\hat{T} Y_\ell^m(\theta, \phi) = \frac{\hbar^2}{2}\left[\frac{\ell(\ell+1)}{I_\perp} + m^2\left(\frac{1}{I_z} - \frac{1}{I_\perp}\right)\right] Y_\ell^m(\theta, \phi) \tag{4.147}$$

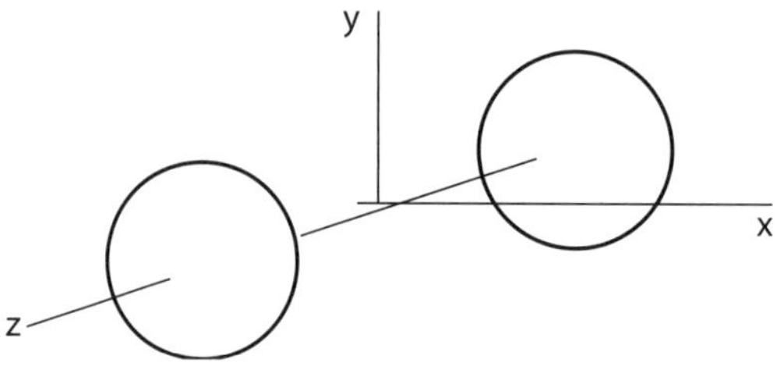

FIGURE 4.2. The three axes of a diatomic molecule.

Since $I_z \ll I_\perp$, then $1/I_z \gg 1/I_\perp$. At ordinary temperatures it is energetically unfavorable to have $m \neq 0$, so the diatomic molecules has eigenvalues and eigenfunctions:

$$\hat{T} Y_\ell^0(\theta, \phi) = \frac{\hbar^2 \ell(\ell+1)}{2I_\perp} Y_\ell^0(\theta, \phi) \tag{4.148}$$

The difference between this case, and the spherical top, is the degeneracy of each energy level. For the spherical top it is $(2\ell+1)$, while for the diatomic molecule it is one.

4.4 The Addition of Angular Momentum

Many problems in quantum mechanics require the addition of two or more angular momentum variables. One example is for electron states in atoms, where the spin and orbital motion are separate angular momentum components. In classical mechanics, angular momentum is a vector, and the addition of angular momentum is just the addition of vectors. In quantum mechanics, each angular momentum variable has the type of quantized eigenstates described in the prior two sections. When two angular momenta $\vec{j}_1$ and $\vec{j}_2$ are to be combined to get a third one $\vec{J}$, this operation can be represented as vector addition:

$$\vec{J} = \vec{j}_1 + \vec{j}_2 \tag{4.149}$$

The process of "adding" these angular momenta is different than the adding of vectors. The two separate momenta have quantized eigenstates: $|j_1, m_1\rangle$, $|j_2, m_2\rangle$. Their combination has its own set of quantized eigenstates $|J, M\rangle$. For the addition of angular momentum, it is assumed that the states of $|J, M\rangle$ are found from combinations of the component states $|j_1, m_1\rangle$, $|j_2, m_2\rangle$. The process of vector addition in quantum mechanics means that the eigenstates of $|J, M\rangle$ are linear combinations of the products of the eigenstates of the constituents. This relationship is written as

$$|J, M\rangle = \sum_{m_1 m_2} (j_1 j_2 m_1 m_2 | JM) |j_1, m_1\rangle |j_2, m_2\rangle \tag{4.150}$$

The factor $(j_1 j_2 m_1 m_2 | JM)$ is a constant coefficient. Unfortunately, it must contain six symbols, since it depends on six parameters. There are many other common ways of writing the same coefficient, with the same six symbols:

$$C_{j_1 j_2}(JM; m_1 m_2), \quad C^{JM}_{j_1 m_1, j_2 m_2} \tag{4.151}$$

They are called *Clebsch-Gordon coefficients.* The process of adding angular momentum in quantum mechanics is to find these coefficients. The standard approach in most textbooks is to prove many theorems involving properties of these coefficients. This approach tends to obscure the fact that they are actually simple numbers such as $1/\sqrt{2}$ or $1/\sqrt{3}$. The present approach to these coefficients is to derive them for a few simple but important cases. The simple cases are all that is needed for most applications. The forbidding formulas are mentioned only at the end.

One rule is worth mentioning first: $M = m_1 + m_2$. The m-component of the total is the summation of the m-components of the two constituents. The summation in (4.150) is not over (m_1, m_2) separately, but is constrained by $m_2 = M - m_1$.

The first example is the vector addition of two systems, each with spin-$\frac{1}{2}$. This addition is denoted as $(\frac{1}{2}) \times (\frac{1}{2})$. To simplify the notation, α_j denotes particle j as spin-up, and β_j denotes it as spin-down. Explicit representations for (α, β) are given in the prior section.

The first step is to construct the eigenstate with the maximum component of M for the combined system. The maximum value of M is achieved when the spins of the two components are both aligned in the up direction. For $(\frac{1}{2}) \times (\frac{1}{2})$ the maximum combination is $M = 1$, which is obtained by adding $m_1 = \frac{1}{2}$ to $m_2 = \frac{1}{2}$. This eigenstate is

$$|1, 1\rangle = \alpha_1 \alpha_2 \tag{4.152}$$

Since $M = 1$, then J must be an integer that is at least one. In fact, the maximum value of $J = j_1 + j_2$. For the present case that is $J = 1$. The normalization coefficient on the right side of the above equation is one. Then the normalization is

$$\langle 1, 1|1, 1\rangle = \langle \alpha_1|\alpha_1\rangle \langle \alpha_2|\alpha_2\rangle = 1 \tag{4.153}$$

The state of minimum M is constructed similarly. It has $M = -1$ and is composed of adding the two spin-down components:

$$|1, -1\rangle = \beta_1 \beta_2 \tag{4.154}$$

Equations (4.152)–(4.154) give two of the three states $|1, M\rangle$.

In the discussion of the present example, the Pauli exclusion principle is being ignored. It is assumed that the two angular momenta $j_1 = \frac{1}{2}$ and $j_2 = \frac{1}{2}$ are from two distinguishable particles. For distinguishable particles, such as an electron and a neutron, there is no need to worry about the exclusion principle as to whether two identical particles occupy the same state. For identical particles, such as two electrons, one has to keep an eye out for several other restrictions when adding angular momenta. The Pauli exclusion principle is that no two identical fermions can occupy the same eigenstate at the same time. If they are in different orbital states, then there is no restriction on whether the two fermions can be in the same spin state. If two identical fermions are in the same orbital state, they cannot be also in the same spin state. In this case the exclusion principle excludes the occupation of the $|1, 1\rangle$ and $|1, -1\rangle$ states. Another restriction is that the total wave function of the two fermion systems must be antisymmetric under the exchange of the coordinates of the two fermions. Clearly, neither $|1, 1\rangle$ nor $|1, -1\rangle$ have this property. Neither does the state with $J=1$, $M=0$. The exclusion principle rules out all three states with $|1, M\rangle$. The related topics of wave function symmetry, Pauli exclusion principle, and permitted spin arrangements are treated at length in chapter 9. The present discussion is only to alert the reader that the addition of angular momentum for identical particles must be done in agreement with other symmetry rules such as provided by the exclusion principle.

Now return to finding the states of $(\frac{1}{2})\times(\frac{1}{2})$. Apply the lowering operator to both sides of eqn. (4.152):

$$L|1,1\rangle = L(\alpha_1\alpha_2) = (L\alpha_1)\alpha_2 + \alpha_1(L\alpha_2) \tag{4.155}$$

The chain rule has been used on the right. Since the raising and lowering operators contain derivatives, one can use the chain rule when operating on a product of functions. The evaluation on the left is $L|1,1\rangle = \hbar\sqrt{2}|1,0\rangle$. On the right it is $L\alpha_j = \hbar\beta_j$. The final expression is

$$\hbar\sqrt{2}|1,0\rangle = \hbar(\beta_1\alpha_2 + \alpha_1\beta_2) \tag{4.156}$$

$$|1,0\rangle = \frac{1}{\sqrt{2}}(\beta_1\alpha_2 + \alpha_1\beta_2) \tag{4.157}$$

The same result is obtained by the operation of

$$L^{\dagger}|1,-1\rangle = L^{\dagger}(\beta_1\beta_2) \tag{4.158}$$

The operation $L|1, 0\rangle$ gives the correct value of $|1, -1\rangle$.

The four orthogonal functions α_1, β_1, α_2, β_2 can be combined into four linearly independent product functions with different indices:

$$\alpha_1\alpha_2, \alpha_1\beta_2, \beta_1\alpha_2, \beta_1\beta_2$$

The combined system should have four states. So far three have been identified: $|1, M\rangle$, with $M=1, 0, -1$. The degeneracy of the $J=1$ state is $(2J+1)=3$, so all of them have been found. The remaining product state has $(2J'+1)=1$ so $J'=0$. The state with $J'=0$ must have $M'=0$. It is composed of product states that have $m_1+m_2=0$, such as $\alpha_1\beta_2$ and $\alpha_2\,\beta_1$. $|0, 0\rangle$ must also be orthogonal to other states such as $|1, 0\rangle$. Therefore, it can only be

$$|0,0\rangle = \pm\frac{1}{\sqrt{2}}(\alpha_1\beta_2 - \beta_1\alpha_2) \tag{4.159}$$

The sign in front is arbitrary.

The normalization uses the feature that

$$\langle\alpha_i|\beta_j\rangle = 0, \quad \langle\alpha_i|\alpha_j\rangle = \delta_{ij}, \quad \langle\beta_i|\beta_j\rangle = \delta_{ij} \tag{4.160}$$

Therefore,

Table 4.1 Vector Addition of $(\frac{1}{2})(\frac{1}{2})$ gives State $|J=1, M\rangle\, J=|0, 0\rangle$

	$\|1, 1\rangle$	$\|1, 0\rangle$	$\|0, 0\rangle$	$\|1, -1\rangle$
$\alpha_1\alpha_2$	1			
$\alpha_1\beta_2$		$\frac{1}{\sqrt{2}}$	$-\frac{1}{\sqrt{2}}$	
$\beta_1\alpha_2$		$\frac{1}{\sqrt{2}}$	$\frac{1}{\sqrt{2}}$	
$\beta_1\beta_2$				1

$$\langle 0,0|0,0\rangle = \frac{1}{2}\langle(\alpha_1\beta_2-\beta_1\alpha_2)(\alpha_1\beta_2-\beta_1\alpha_2)\rangle \tag{4.161}$$

$$= \frac{1}{2}[\langle\alpha_1|\alpha_1\rangle\langle\beta_2|\beta_2\rangle + \langle\alpha_2|\alpha_2\rangle\langle\beta_1|\beta_1\rangle] = 1 \tag{4.162}$$

Similarly,

$$\langle 0,0|1,0\rangle = \frac{1}{2}\langle(\alpha_1\beta_2-\beta_1\alpha_2)(\alpha_1\beta_2+\beta_1\alpha_2)\rangle \tag{4.163}$$

$$= \frac{1}{2}[\langle\alpha_1|\alpha_1\rangle\langle\beta_2|\beta_2\rangle - \langle\alpha_2|\alpha_2\rangle\langle\beta_1|\beta_1\rangle] \tag{4.164}$$

$$= \frac{1}{2}[1-1] = 0 \tag{4.165}$$

These results are summarized in table 4.1. Blank values are zero. The values of 0, 1, and $\pm 1/\sqrt{2}$ are the Clebsch-Gordon coefficients for the present example of $(\frac{1}{2})\times(\frac{1}{2})$. The four-by-four array has 16 possible values. Ten are zero, two are one, and four are $\pm 1/\sqrt{2}$.

Such tables are a convenient way to represent the results. One can read them down as a column:

$$|1,0\rangle = \frac{1}{\sqrt{2}}(\beta_1\alpha_2 + \alpha_1\beta_2) \tag{4.166}$$

$$|0,0\rangle = \frac{1}{\sqrt{2}}(\beta_1\alpha_2 - \alpha_1\beta_2) \tag{4.167}$$

One can also read them across horizontal rows:

$$\alpha_1\beta_2 = \frac{1}{\sqrt{2}}[|1,0\rangle - |0,0\rangle] \tag{4.168}$$

$$\beta_1\alpha_2 = \frac{1}{\sqrt{2}}[|1,0\rangle + |0,0\rangle] \tag{4.169}$$

There are several rules for constructing tables of Clebsch-Gordon coefficients. The first is that for any value of M the terms in the series (4.150) have $M = m_1 + m_2$. This feature makes most of the Clebsch-Gordon coefficients be zero.

The second rule concerns the allowed values of J. The maximum J is $j_1 + j_2$ and the minimum J is $|j_1 - j_2|$. Values of J occur for all integer steps between the minimum and maximum value. For example, the multiplication $(\frac{3}{2})\times(1)$ has values of $J=\frac{5}{2}, \frac{3}{2}, \frac{1}{2}$. For $(4)\times(2)$ the values of $J=2, 3, 4, 5, 6$. One, and only one, value of J is allowed for each permitted value. For example, in $(4)\times(2)$ a value such as $J=3$ can occur only once. This restriction does not apply for the multiplication of three or more angular momenta, but does apply for two.

These rules give exactly the correct number of new states. One starts with $N=(2j_1+1)(2j_2+1)$ states in the two component systems. It is easy to check that the combined system has

$$\sum_{|j_1-j_2|}^{j_1+j_2}(2J+1)=(2j_1+1)(2j_2+1) \tag{4.170}$$

The Clebsch-Gordon coefficients obey some orthogonality relationships. For example, in table 4.1, treat each row of numbers as a column vector. All of these four column vectors are orthogonal. Similarly, treat each row as a vector and the four row vectors are orthogonal. This orthogonality is a feature of all such tables of Clebsch-Gordon coefficients. The formal statement is

$$\delta_{JJ'}\delta_{MM'}=\sum_{m_1}(j_1j_2m_1, M-m_1|JM)(j_1j_2m_1, M'-m_1|J'M')$$

$$\delta_{m_1m_1'}\delta_{m_2m_2'}=\sum_{JM}(j_1j_2m_1m_2|JM)(j_1j_2m_1'm_2'|JM) \tag{4.171}$$

These are some of the forbidding formulas mentioned at the beginning of the section.

Homework

1. The harmonic oscillator has

$$H=\hbar\omega\left(a^\dagger a+\frac{1}{2}\right) \tag{4.172}$$

$$[a, a^\dagger]=1, \quad [a^\dagger, a^\dagger]=0, \quad [a, a]=0 \tag{4.173}$$

Starting from just these operator relations, derive the eigenvalue spectrum of the harmonic oscillator. Also find how the raising and lowering operators affect the eigenstates.

2. Prove the following result, where $\vec{\sigma}=(\sigma_x, \sigma_y, \sigma_z)$ are Pauli spin matrices, and $\vec{A}=(A_x, A_y, A_z)$ and $\vec{B}=(B_x, B_y, B_z)$ are ordinary vectors:

$$(\vec{\sigma}\cdot\vec{A})(\vec{\sigma}\cdot\vec{B})=\vec{A}\cdot\vec{B}I+i\vec{\sigma}\cdot(\vec{A}\times\vec{B}) \tag{4.174}$$

where I is the identity matrix.

3. Construct a matrix representation for the $j=\frac{3}{2}$ angular momentum state.

4. Construct an r-space representation for the eigenstates with $j = 3$.

5. Calculate the moment of inertia $I_\perp$ for a diatomic molecule when it is rotating on an axis perpendicular to the line between the two nuclei. Then calculate the value of $E_\perp = \hbar^2/2I_\perp$ in temperature units $(T_1 = E_\perp/k_B)$ for the three diatomic molecules: H_2, N_2, O_2. The values for the nuclear separation can be found in the *Handbook of Chemistry and Physics.*

6. A molecule consists of four atoms in a perfect square of side a and atomic mass M. What are the eigenvalues of rigid rotation in three dimensions?

7. Derive a table of Clebsch-Gordon coefficients for $(\frac{1}{2})\times(1)$ and also $(\frac{1}{2})\times(\frac{3}{2})$.

8. Derive a table of Clebsch-Gordon coefficients for $(1)\times(1)$. List only states with $m \geq 0$.

9. Write down the angular momentum states obtained by combining three spin-$\frac{1}{2}$ states. Treat the particles as distinguishable.

5 Two and Three Dimensions

Most quantum mechanical problems are three dimensional. Nuclei, atoms, solids, and stars are systems that must be solved in three dimensions. Two-dimensional solutions are required when a particle is restricted to move on a surface. Examples are when electrons move on the surface of liquid helium or in a semiconductor quantum well. Many of the techniques discussed in prior chapters are now applied to two and three dimensions. Some exact solutions are provided, as well as approximate methods such as WKBJ and variational. Spin and angular momentum are both utilized.

5.1 Plane Waves in Three Dimensions

The most important solution to Schrödinger's equation is where the potential vanishes everywhere. This case is called by a variety of names: plane wave, noninteracting particles, and free particles.

Schrödinger's equation, when $V = 0$, has only the kinetic energy term and the eigenvalue. It is written below in Cartesian (x, y, z), polar (ρ, θ, z), and spherical (r, θ, ϕ) coordinates:

$$-\frac{\hbar^2\nabla^2}{2m}\psi(\mathbf{r}) = E\psi(\mathbf{r}) \tag{5.1}$$

$$\nabla^2\psi = \begin{cases} \left(\frac{\partial^2}{\partial x^2} + \frac{\partial^2}{\partial y^2} + \frac{\partial^2}{\partial z^2}\right)\psi \\ \frac{1}{\rho}\frac{\partial}{\partial\rho}\left(\rho\frac{\partial\psi}{\partial\rho}\right) + \frac{1}{\rho^2}\frac{\partial^2\psi}{\partial\theta^2} + \frac{\partial^2\psi}{\partial z^2} \\ \frac{1}{r^2}\frac{\partial}{\partial r}\left(r^2\frac{\partial\psi}{\partial r}\right) + \frac{1}{r^2\sin(\theta)}\frac{\partial}{\partial\theta}\left(\sin(\theta)\frac{\partial\psi}{\partial\theta}\right) + \frac{1}{r^2\sin^2(\theta)}\frac{\partial^2\psi}{\partial\phi^2} \end{cases} \tag{5.2}$$

The choice of coordinate system usually depends on the symmetry of the potential term $V(\mathbf{r})$, and whether it is given in terms of Cartesian, polar, or spherical coordinates. When

$V = 0$ one can use any coordinate system. Cartesian coordinates are similar to solving the one-dimensional case. Spherical coordinates are used here: $x = r\sin(\theta)\cos(\phi)$, $y = r\sin(\theta)\,\sin(\phi)$, $z = r\cos(\theta)$.

In spherical coordinates, Schrödinger's equation (5.1) is separable. The solution can be written as a product of three functions: one of r, one of θ, and one of ϕ:

$$\psi(\mathbf{r}) = R(r)f(\theta)g(\phi) \tag{5.3}$$

$$g(\phi) = e^{im\phi} \tag{5.4}$$

where m is an integer. The integer is required by the condition that $g(\phi + 2\pi) = g(\phi)$. The other angular function is expressed as an associated Legendre polynomial:

$$f(\theta) = P_l^{|m|}(\theta) \tag{5.5}$$

The product of the two angular functions plus the proper normalization is called a *spherical harmonic*:

$$Y_\ell^m(\theta, \phi) = N_{\ell m} e^{im\phi} P_\ell^{|m|}(\theta) \tag{5.6}$$

$$N_{\ell m} = \varepsilon \sqrt{\frac{2\ell + 1}{4\pi} \frac{(\ell - |m|)!}{(\ell + |m|)!}} \tag{5.7}$$

where $\varepsilon = 1$ if $m < 0$, and $\varepsilon = (-1)^m$ if $m \geq 0$. The most frequently used spherical harmonics are those with small values of (ℓ, m). Note that ℓ is angular momentum. It is an integer and $-\ell \leq m \leq \ell$:

$$Y_0^0 = \sqrt{\frac{1}{4\pi}} \qquad Y_2^0 = \sqrt{\frac{5}{16\pi}}[3\cos^2(\theta) - 1] \tag{5.8}$$

$$Y_1^0 = \sqrt{\frac{3}{4\pi}}\cos(\theta) \qquad Y_2^{\pm 1} = \mp\sqrt{\frac{15}{8\pi}}\sin(\theta)\cos(\theta)e^{\pm i\phi} \tag{5.9}$$

$$Y_1^{\pm 1} = \mp\sqrt{\frac{3}{8\pi}}\sin(\theta)e^{\pm i\phi} \qquad Y_2^{\pm 2} = \sqrt{\frac{15}{32\pi}}\sin^2(\theta)e^{\pm 2i\phi} \tag{5.10}$$

Multiplying these spherical harmonics by $\sqrt{4\pi/(2\ell+1)}r^\ell$ gives the eigenstates $|\ell, m\rangle$ of the angular momentum operator for integer values of $j = \ell$. The properties of these functions are discussed in chapter 4.

The normalization of the spherical harmonics is given by the orthogonality relation:

$$\int_0^{2\pi} d\phi \int_0^{\pi} d\theta \sin(\theta) Y_\ell^{m*}(\theta, \phi) Y_{\ell'}^{m'}(\theta, \phi) = \delta_{\ell\ell'}\delta_{mm'} \tag{5.11}$$

They obey an eigenvalue equation that contains the angular part of the operator ∇^2 in eqn. (5.2):

$$\left[\frac{1}{\sin(\theta)}\frac{\partial}{\partial\theta}\left(\sin(\theta)\frac{\partial}{\partial\theta}\right) + \frac{1}{\sin^2(\theta)}\frac{\partial^2}{\partial\phi^2}\right] Y_\ell^m = -\ell(\ell+1)Y_\ell^m \tag{5.12}$$

The eigenfunction in (5.2) is written with the angular functions given by the spherical harmonic:

$$\psi(\mathbf{r}) = R(r) Y_\ell^m(\theta, \phi) \tag{5.13}$$

$$\nabla^2[RY_\ell^m] = \left[\frac{1}{r^2}\frac{\partial}{\partial r}\left(r^2\frac{\partial R}{\partial r}\right) - \frac{\ell(\ell+1)R}{r^2}\right] Y_\ell^m \tag{5.14}$$

The last equation gives the result for $\nabla^2[RY_\ell^m]$. The spherical harmonic can be factored out of Schrödinger's equation, which leaves an equation for the unknown radial function $R(r)$:

$$E = \frac{\hbar^2 k^2}{2m} \tag{5.15}$$

$$0 = \frac{1}{r^2}\frac{\partial}{\partial r}\left(r^2\frac{\partial R}{\partial r}\right) - \frac{\ell(\ell+1)R}{r^2} + k^2 R \tag{5.16}$$

This differential equation is related to Bessel's equation of half-integer index. They are usually combined with a prefactor and called a *spherical Bessel function* [1]:

$$j_\ell(z) = \sqrt{\frac{\pi}{2z}} J_{\ell+1/2}(z) \tag{5.17}$$

$$y_\ell(z) = \sqrt{\frac{\pi}{2z}} Y_{\ell+1/2}(z) \tag{5.18}$$

where $J_\nu(z)$, $Y_\nu(z)$ are the usual Bessel and Neumann functions. The latter is sometimes denoted as $N_\nu(z)$. It should not be confused with the spherical harmonic. Note that the spherical Bessel functions are written with lowercase symbols.

The differential equation (5.16) has the general solution for the radial function:

$$R_\ell(kr) = C_1 j_\ell(kr) + C_2 y_\ell(kr) \tag{5.19}$$

where $C_{1,2}$ are constant coefficients to be determined by the boundary and initial conditions. The radial function has the quantum numbers (ℓ, k). The functions $y_\ell(z)$ diverge at the origin $z = 0$ and this solution is never permitted for vanishing value of r. When $V = 0$ for all values of r, then $y_\ell(kr)$ must be eliminated by setting $C_2 = 0$. However, there are many cases where the solution $y_\ell(kr)$ is needed, such as the states outside of a spherical box. The function $y_\ell(kr)$ is always used when solutions are away from zero.

There are many problems where $V = 0$ in some regions of space, while $V \neq 0$ elsewhere. The two plane wave solutions are used whenever $V =$ constant: if $V = V_0$, then $k^2 = 2m(E - V_0)/\hbar^2$. Below are given some properties of spherical Bessel functions:

$$j_0(z) = \frac{\sin(z)}{z} \qquad y_0(z) = -\frac{\cos(z)}{z} \tag{5.20}$$

$$\begin{aligned} j_1(z) &= \frac{\sin(z)}{z^2} - \frac{\cos(z)}{z} & y_1 &= -\frac{\cos(z)}{z^2} - \frac{\sin(z)}{z} \\ j_2(z) &= \left(\frac{3}{z^3} - \frac{1}{z}\right)\sin(z) - 3\frac{\cos(z)}{z^2} & y_2(z) &= -\left(\frac{3}{z^3} - \frac{1}{z}\right)\cos(z) - 3\frac{\sin(z)}{z^2} \end{aligned} \tag{5.21}$$

The behaviors in the limit of small argument, and then for large argument, are

$$\lim_{z\to 0} j_\ell(z) = \frac{z^\ell}{1 \cdot 3 \cdot 5 \cdots (2\ell+1)} \tag{5.22}$$

$$\lim_{z\to 0} y_\ell(z) = -\frac{1\cdot 3\cdot 5\cdots(2\ell-1)}{z^{\ell+1}} \tag{5.23}$$

$$\lim_{z\to\infty} j_\ell(z) = \frac{1}{z}\sin\left(z - \frac{\pi\ell}{2}\right) \tag{5.24}$$

$$\lim_{z\to\infty} y_\ell(z) = -\frac{1}{z}\cos\left(z - \frac{\pi\ell}{2}\right) \tag{5.25}$$

The two spherical Hankel functions are defined:

$$h_\ell^{(1)}(z) = j_\ell(z) + iy_\ell(z) \tag{5.26}$$

$$h_\ell^{(2)}(z) = j_\ell(z) - iy_\ell(z) \tag{5.27}$$

Using the above expressions for the spherical Bessel functions, it is easy to show that

$$h_0^{(1)} = -\frac{i}{z}e^{iz} \tag{5.28}$$

$$h_1^{(1)} = -e^{iz}\left[\frac{1}{z} + \frac{i}{z^2}\right] \tag{5.29}$$

The first spherical Hankel function is always $\exp(iz)$ times a polynomial in $(1/z)$. The phase factor indicates an outgoing spherical wave. The second spherical Hankel function is the complex conjugate of the first one. It has a phase of $\exp(-iz)$ for an incoming spherical wave. These properties of spherical Bessel functions are sufficient information for most applications.

When $V = 0$ or a constant, only continuum solutions are allowed for $E > V_0$. If $V_0 = 0$, then E can have any positive value or zero. The most general form of the solution, which is regular at the origin ($r = 0$), is

$$\psi(\mathbf{k}, \mathbf{r}) = \sum_{\ell=0}^{\infty}\sum_{m=-\ell}^{\ell} C_{\ell m} j_\ell(kr) Y_\ell^m(\theta, \phi) \tag{5.30}$$

$$E(k) = \frac{\hbar^2 k^2}{2m} \tag{5.31}$$

The coefficients $C_{\ell m}$ depend on the boundary and initial conditions. There is an infinite number of terms in the series, and each has the same eigenvalue $E(k)$. Each solution is distinguished by having different values for the two quantum numbers (ℓ, m).

At this point the reader may be confused. The solution (5.30) appears to be different than the usual one, which is the plane wave solution

$$\psi(\mathbf{k}, \mathbf{r}) = \exp(i\mathbf{k}\cdot\mathbf{r}) \tag{5.32}$$

What is the relationship between the two solutions (5.30) and (5.32)? The plane wave solution (5.32) can be expanded in a power series in angular momentum:

$$e^{i\mathbf{k}\cdot\mathbf{r}} = \sum_\ell (2\ell+1) i^\ell j_\ell(kr) P_\ell(\theta') \tag{5.33}$$

This expansion is the generating function for spherical Bessel functions.

The angle θ' is between the vectors **k** and **r**. The angles (θ, ϕ) are those that **r** makes with respect to the (x, y, z) coordinate system. Let (θ'', ϕ'') be the angles that **k** makes with respect to the same coordinate system. According to the law of cosines, these angles are related by

$$\cos(\theta') = \cos(\theta)\cos(\theta'') + \sin(\theta)\sin(\theta'')\cos(\phi - \phi'') \tag{5.34}$$

The law of cosines is derived by writing $\vec{r} = r[\sin(\theta)\cos(\phi), \sin(\theta)\sin(\phi), \cos(\theta)]$ and $\vec{k} = k[\sin(\theta'')\cos(\phi''), \sin(\theta'')\sin(\phi''), \cos(\theta'')]$ in (x, y, z) coordinates, and then

$$\vec{k}\cdot\vec{r} = kr\cos(\theta') = k_x x + k_y y + k_z z \tag{5.35}$$

$$= kr[\sin(\theta)\sin(\theta'')(\cos(\phi)\cos(\phi'') + \sin(\phi)\sin(\phi'')) + \cos(\theta)\cos(\theta'')] \tag{5.36}$$

$$= kr[\sin(\theta)\sin(\theta'')\cos(\phi - \phi'') + \cos(\theta)\cos(\theta'')] \tag{5.37}$$

which completes the proof.

The addition theorem for angular momentum is

$$(2\ell + 1)P_\ell(\theta') = 4\pi \sum_{m=-\ell}^{\ell} Y_\ell^m(\theta, \phi) Y_\ell^{m*}(\theta'', \phi'') \tag{5.38}$$

The law of cosines is the $\ell = 1$ case of the addition theorem:

$$3\cos(\theta') = 3[\cos(\theta)\cos(\theta'') + \sin(\theta)\sin(\theta'')\cos(\phi - \phi'')] \tag{5.39}$$

$$= 4\pi\frac{3}{4\pi}\left\{\cos(\theta)\cos(\theta'') + \frac{1}{2}\sin(\theta)\sin(\theta'')\left[e^{i(\phi-\phi'')} + e^{-i(\phi-\phi'')}\right]\right\}$$

The last bracket is $2\cos(\theta - \theta'')$ and the two expressions are identical. The expansion in eqn. (5.33) becomes

$$e^{i\mathbf{k}\cdot\mathbf{r}} = 4\pi\sum_{\ell m} i^\ell Y_\ell^m(\theta'', \phi'')^* j_\ell(kr) Y_\ell^m(\theta, \phi) \tag{5.40}$$

$$C_{\ell m} = 4\pi i^\ell Y_\ell^m(\theta'', \phi'')^* \tag{5.41}$$

The expansion has exactly the form of eqn. (5.30), with the coefficient $C_{\ell m}$ shown above. The plane wave solution (5.32) is obtained by a particular choice for the coefficients $C_{\ell m}$. Equation (5.30) is the most general solution, while (5.32) is a particular solution.

5.2 Plane Waves in Two Dimensions

Plane wave solutions are also important in two-dimensional problems. Usually the variables are Cartesian (x, y) or polar (ρ, θ) coordinates. In this dimension the Laplacian is

$$\nabla^2\psi = \begin{cases} \left(\dfrac{\partial^2}{\partial x^2} + \dfrac{\partial^2}{\partial y^2}\right)\psi \\ \dfrac{1}{\rho}\dfrac{\partial}{\partial\rho}\left(\rho\dfrac{\partial\psi}{\partial\rho}\right) + \dfrac{1}{\rho^2}\dfrac{\partial^2\psi}{\partial\theta^2} \end{cases} \tag{5.42}$$

In polar coordinates, the general solution for angular momentum n is

$$\psi_n = R(\rho)e^{in\Theta} \tag{5.43}$$

The angular function must have the form shown, since increasing θ by 2π radians does not change the eigenfunction. This constraint requires n to be an integer, which could be positive or negative. The equation for the radial function is

$$\left(\frac{d^2}{d\rho^2} + \frac{1}{\rho}\frac{d}{d\rho} - \frac{n^2}{\rho^2} + k^2\right)R(\rho) = 0 \tag{5.44}$$

The solutions to this differential equation are Bessel functions $J_n(k\rho)$, $Y_n(k\rho)$. The most general solution is

$$\psi(k, \rho, \theta) = \sum_n [D_n(k)J_n(k\rho) + F_n(k)Y_n(k\rho)]e^{in\Theta} \tag{5.45}$$

$$E(k) = \frac{\hbar^2 k^2}{2m} \tag{5.46}$$

where $D_n(k)$ and $F_n(k)$ are normalization coefficients. Note that the coefficients D_n, F_n could be complex. One could also write the angular term using $\cos(n\theta)$, $\sin(n\theta)$. The Bessel $J_n(z)$ and Neumann $Y_n(z)$ functions are the radial solution. The Neumann function $Y_n(z)$ diverges at the origin ($z = 0$), so $F_n = 0$ if the plane wave solution goes to the origin.

Again we can relate the two-dimensional expansion of the plane wave to the above solution:

$$e^{i\mathbf{k}\cdot\mathbf{r}} = e^{ik\rho\cos(\theta-\theta')} = \sum_{n=-\infty}^{\infty} i^n J_n(k\rho)e^{in(\theta-\theta')} \tag{5.47}$$

Here θ is the angle that $\mathbf{r}$ makes with the x-axis, and θ' is the angle that $\mathbf{k}$ makes with the same axis. The above series is the generating function for the Bessel function.

The behaviors in the limit of small argument, and then for large argument, are

$$\lim_{z\to 0} J_n(z) = \frac{(z/2)^n}{n!} \tag{5.48}$$

$$\lim_{z\to 0} Y_0(z) \sim \frac{2}{\pi}\ln\left(\frac{z}{2}\right) \tag{5.49}$$

$$\lim_{z\to 0} Y_n(z) \sim \frac{(n-1)!}{\pi}\left(\frac{2}{z}\right)^n, \quad n>0 \tag{5.50}$$

$$\lim_{z\to\infty} J_n(z) = \sqrt{\frac{2}{\pi z}}\cos\left(z - \frac{\pi n}{2} - \frac{\pi}{4}\right) \tag{5.51}$$

$$\lim_{z\to\infty} Y_n(z) = \sqrt{\frac{2}{\pi z}}\sin\left(z - \frac{\pi n}{2} - \frac{\pi}{4}\right) \tag{5.52}$$

The Hankel functions are

$$H_n^{(1)}(z) = J_n(z) + iY_n(z) \tag{5.53}$$

$$H_n^{(2)}(z) = J_n(z) - iY_n(z) \tag{5.54}$$

The first Hankel function is an outgoing wave in two dimensions, while the second one is an incoming wave. These properties are sufficient information for most applications.

5.3 Central Potentials

Central potentials have the feature that the potential function $V(r)$ depends on the magnitude of the vector $\mathbf{r}$, but not on its direction. These potentials in three dimensions are also called *spherically symmetric*. Central potentials occur often in atomic and nuclear physics. The eigenvalue equation is

$$0 = (H-E)\psi = \left[-\frac{\hbar^2}{2m}\nabla^2 + V(r) - E\right]\psi(\mathbf{r}) \tag{5.55}$$

This equation will be solved in two and three dimensions.

5.3.1 Central Potentials in 3D

Central potentials in three dimensions have the feature that the eigenfunctions can be separated into a function of $R(r)$ times the angular functions. The latter is a spherical harmonic:

$$\psi(\mathbf{r}) = R(r) Y_\ell^m(\theta, \phi) \tag{5.56}$$

This separation is exact for central potentials and makes their solution particularly simple. However, if the potential function depends also on angle, such as $V(r, \theta)$, then such separation is not possible. In eqn. (5.56) only the radial function $R(r)$ is unknown. Its equation is derived from (5.55) to be, in spherical coordinates,

$$0 = \frac{1}{r^2}\frac{d}{dr}\left(r^2\frac{dR}{dr}\right) - \left[\frac{\ell(\ell+1)}{r^2} + \frac{2m}{\hbar^2}[V(r)-E]\right]R(r) \tag{5.57}$$

This important differential equation is the basis for many atomic and nuclear calculations.

The above equation is rarely written in this form. Instead, it is modified because of the equivalence of the following two ways of writing the double-derivative term:

$$\frac{1}{r^2}\frac{d}{dr}\left(r^2\frac{dR}{dr}\right) = \frac{d^2R}{dr^2} + \frac{2}{r}\frac{dR}{dr} \tag{5.58}$$

$$\frac{1}{r}\frac{d^2(rR)}{dr^2} = \frac{d^2R}{dr^2} + \frac{2}{r}\frac{dR}{dr} \tag{5.59}$$

$$\frac{1}{r^2}\frac{d}{dr}\left(r^2\frac{dR}{dr}\right) = \frac{1}{r}\frac{d^2(rR)}{dr^2} \tag{5.60}$$

The second expression for the double derivative is substituted into eqn. (5.57), in place of the first. Then it is natural to define the unknown *radial function* as

$$\chi(r) = rR(r) \tag{5.61}$$

$$0 = \frac{d^2\chi(r)}{dr^2} - \left[\frac{\ell(\ell+1)}{r^2} + \frac{2m}{\hbar^2}[V(r)-E]\right]\chi(r) \tag{5.62}$$

The last equation is found by multiplying every term in (5.57) by r. The final equation solves for the radial function. After solving for $\chi(r)$, the radial part of the eigenfunction is found from $R(r) = \chi(r)/r$. Equation (5.62) has a simple form and seems to be a one-dimensional problem with an effective potential:

$$V_\ell(r) = V(r) + \frac{\hbar^2}{2mr^2}\ell(\ell+1) \tag{5.63}$$

$$0 = \left[\frac{d^2}{dr^2} - \frac{2m}{\hbar^2}[V_\ell(r) - E]\right]\chi(r) \tag{5.64}$$

The second term in the potential provides the centrifugal barrier.

Many of our solutions in one dimension can be carried over to three dimensions, particularly when $\ell = 0$. The close connection between the wave equation in one and three dimensions means that waves behave similarly in these two cases. Schrödinger's equation in two dimensions cannot be cast as a type of one-dimensional problem, and waves in two dimensions have different properties than those in one or three dimensions.

Eigenfunctions are generally well behaved at the origin ($r = 0$). Even for divergent potentials, such as the Coulomb $V(r) = C/r$, $R(0)$ is a finite number. That means that $\chi(r) = rR(r)$ goes to zero at $r = 0$. Generally eqn. (5.62) is solved with one boundary condition that $\chi(r = 0) = 0$. In one dimension, the eigenfunction was sometimes made to vanish at the origin by inventing potentials that diverged at that point. In three dimensions, the potential $V(r)$ may not diverge, but $\chi(r = 0) = 0$.

In one dimension, radial eigenfunctions at large values of distance could be expressed in terms of phase shifts. The same is true of three dimensions. Two cases must be distinguished. The first is where there is a net charge $Z_i e$ on the atom causing $V(r)$. Then an electron has a long-ranged Coulomb potential that goes as $Z_i e^2/r$ at large values of r. This potential is sufficiently long-ranged that a particle is never out of its influence, and the usual phase shifts cannot be defined. The eigenfunctions for Coulomb potentials have the form of Whittaker functions. They are discussed in a later section.

Any potential that goes to zero as $O(1/r^2)$ at large r, or even faster, permits the asymptotic form of the continuum eigenfunction to be defined in terms of phase shifts $\delta_\ell(k)$. The phase shifts depend on the angular momentum and the wave vector k. If the scattering potential has a magnetic moment, it could also depend on the azimuthal quantum number m. If the potential $V(r)$ vanishes everywhere, the eigenfunctions are plane wave states $R(r) = j_\ell(kr)$. The asymptotic form is given in eqn. (5.24). The phase shift is defined as the change in the phase of the spherical Bessel function:

$$(V = 0) \quad \lim_{r\to\infty} R(r) \to \frac{D}{kr}\sin\left(kr - \frac{\pi\ell}{2}\right) \tag{5.65}$$

$$(V \neq 0) \quad \lim_{r\to\infty} R(r) \to \frac{D}{kr}\sin\left[kr - \frac{\pi\ell}{2} + \delta_\ell(k)\right] \tag{5.66}$$

The phase shift is defined only in regions where the potential is a constant or zero. The notation ($V \neq 0$) means that the potential is nonzero in some other region of space. Usually we solve for the function $\chi(r)$. Its asymptotic limits are

$$(V=0) \quad \lim_{r\to\infty} \chi(r) \to \sin(kr - \pi\ell/2) \tag{5.67}$$

$$(V\neq 0) \quad \lim_{r\to\infty} \chi(r) \to \sin[kr - \pi\ell/2 + \delta_\ell(k)] \tag{5.68}$$

The only difference between one and three dimensions is the extra factor of $\pi\ell/2$ in the three-dimensional phase.

Quite often the central potential goes to zero very rapidly with r. Such behavior is found for exponential potentials $V \sim \exp(-r/a)$ or power laws $V \sim 1/r^n$ for $n > 2$. Then for intermediate values of r, the potential $V(r)$ is small, while the centrifugal potential $\hbar^2\ell(\ell+1)/2mr^2$ is large. For these intermediate values of r, one is effectively solving the plane wave equation with a centrifugal potential. Here the most general solution is (5.19)

$$R(r) = C_1\left[j_\ell(kr) + \frac{C_2}{C_1} y_\ell(kr)\right] \tag{5.69}$$

The ratio C_2/C_1 can range in value from $-\infty$ to $+\infty$. The phase shift is introduced by setting this ratio equal to $-\tan(\delta_\ell)$:

$$R(r) = C_1[j_\ell(kr) - \tan(\delta_\ell) y_\ell(kr)] \tag{5.70}$$

It is permissible to use $y_\ell(kr)$ since the above formula is being applied only at large values of r. For smaller values of r, where $V(r) \neq 0$, the eigenfunction cannot be expressed in terms of plane waves.

This method of defining the phase shift is consistent with the earlier definition (5.66). This equivalence is shown by taking the limit of (5.70) as $kr >> 1$ using (5.24):

$$\lim_{kr>>1} R(r) \to \frac{C_1}{kr}\left[\sin\left(kr - \frac{\pi\ell}{2}\right) + \tan(\delta_\ell)\cos\left(kr - \frac{\pi\ell}{2}\right)\right] \tag{5.71}$$

$$\to \frac{C_1}{kr\cos(\delta_\ell)}\left[\cos(\delta_\ell)\sin\left(kr - \frac{\pi\ell}{2}\right) + \sin(\delta_\ell)\cos\left(kr - \frac{\pi\ell}{2}\right)\right]$$

$$\to \frac{C_1}{kr\cos(\delta_\ell)}\sin\left(kr - \frac{\pi\ell}{2} + \delta_\ell\right) \tag{5.72}$$

The two terms are combined by using the trigonometric formula for the sine of the summation of two angles. The asymptotic form of (5.70) is exactly the earlier definition (5.66). The factor of $\cos(\delta_\ell)$ in the prefactor gets absorbed into the normalization constant.

As an example, solve for the continuum states of the three-dimensional attractive square well shown in figure 5.1a:

$$V(r) = \begin{cases} -V_0, & 0 < r < a \\ 0, & a < r \end{cases} \tag{5.73}$$

The eigenfunction in the two regions of space is as follows:

- $0 < r < a$ has $p^2 = 2m(E + V_0)/\hbar^2$

  $$R_\ell(kr) = A j_\ell(pr) \tag{5.74}$$

  There is no term $y_\ell(pr)$ since $R(r)$ cannot diverge at the origin.

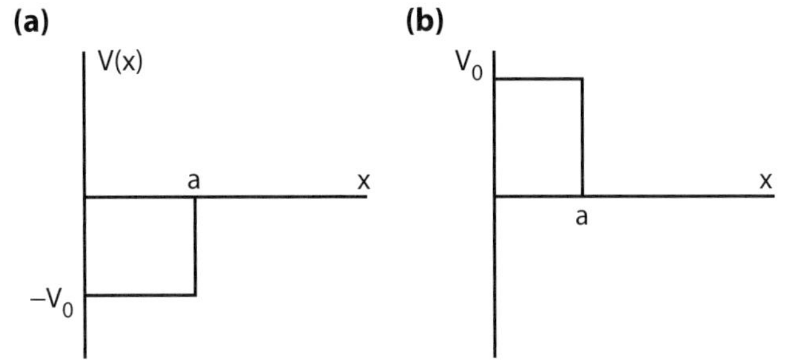

FIGURE 5.1 Square well in three dimensions: (a) attractive, (b) repulsive.

- $a < r$ has $k^2 = 2mE/\hbar^2$

$$R(r) = B[\, j_\ell(kr) - \tan(\delta_\ell) y_\ell(kr)] \tag{5.75}$$

Now there is a term $y_\ell(kr)$ since this solution is valid only away from the origin. The second-order equation has two unknown constants, which are (B, δ_ℓ).

Match these two solutions at $r = a$ by equating the eigenfunction and its first derivative:

$$Aj_\ell(pa) = B[\, j_\ell(ka) - \tan(\delta_\ell) y_\ell(ka)] \tag{5.76}$$

$$Apj'_\ell(pa) = Bk[\, j'_\ell(ka) - \tan(\delta_\ell) y'_\ell(ka)] \tag{5.77}$$

where a prime denotes derivative. Divide these two equations: divide the left side by the left side, and the right side by the right side. This step eliminates the constants (A, B). One is left with an equation in which the only unknown is the tangent of the phase shift:

$$p\frac{j'_\ell(pa)}{j_\ell(pa)} = k\frac{j'_\ell(ka) - \tan(\delta) y'_\ell(ka)}{j_\ell(ka) - \tan(\delta) y_\ell(ka)} \tag{5.78}$$

$$\tan(\delta_\ell) = \frac{pj'_\ell(pa) j_\ell(ka) - kj_\ell(pa) j'_\ell(ka)}{pj'_\ell(pa) y_\ell(ka) - kj_\ell(pa) y'_\ell(ka)} \tag{5.79}$$

It is an interesting exercise to evaluate this expression for $\ell = 0$ and to show that it reduces to

$$\frac{\tan(ka + \delta_0)}{k} = \frac{\tan(pa)}{p} \tag{5.80}$$

$$\delta_0(k) = -ka + \tan^{-1}\left[\frac{k}{p}\tan(pa)\right] \tag{5.81}$$

The Hamiltonian of the three-dimensional square well will occur often in later chapters, as it is used to discuss scattering theory and other topics.

A related problem is the repulsive square well so that $V(r) = +V_0$ for $0 < r < a$ shown in figure 5.1 b. Let $E < V_0$ so the solution inside of the square well is a decaying exponential rather than sinusodial. For $0 < r < a$ define a constant:

$$\alpha^2 = \frac{2m}{\hbar^2}(V_0 - E) \tag{5.82}$$

$$R(r) = Ai_\ell(\alpha r) + Bk_\ell(\alpha r) \tag{5.83}$$

$$i_\ell(z) = \sqrt{\frac{\pi}{2z}} I_{\ell+1/2}(z),\ k_\ell(z) = \sqrt{\frac{\pi}{2z}} I_{-\ell-1/2}(z) \tag{5.84}$$

$I_n(z)$ and $K_n(z)$ are Bessel functions of imaginary argument. The functions $i_\ell(z)$, $k_\ell(z)$, are their spherical form. They are equivalent to the one-dimensional functions $\sinh(\alpha x)$, $\cosh(\alpha x)$. Their properties are described in ref. [1]. The function $k_\ell(z)$ diverges at the origin, so this term is discarded by setting $B = 0$. So the eigenfunction for this problem is

$$0 < r < a \quad R(r) = Ai_\ell(\alpha r) \tag{5.85}$$

$$a < r \quad R(r) = C[j_\ell(kr) - \tan(\delta_\ell) y_\ell(kr)] \tag{5.86}$$

The final solution is found by matching these two forms at $r = a$.

5.3.2 Central Potential in 2D

The derivation is quite similar in two dimensions. The angular functions have the form $\exp(in\theta)$, where n is an integer. The differential equation for the radial function is

$$\left\{ \frac{d^2}{d\rho^2} + \frac{1}{\rho}\frac{d}{d\rho} - \frac{n^2}{\rho^2} - \frac{2m}{\hbar^2}[V(\rho) - E] \right\} R(\rho) = 0 \tag{5.87}$$

Outside of the potential region, where $V \to 0$, the most general form for the radial function for continuum states is

$$R_n(\rho) = A[J_n(k\rho) - \tan(\delta_n) Y_n(k\rho)] \tag{5.88}$$

where $\tan[\delta_n(k)]$ is the phase shift for scattering from that potential. Using the asymptotic forms for the Bessel functions, then at large argument

$$\lim_{k\rho >> 1} R_n(\rho) = \frac{A}{\cos(\delta_n)} \sqrt{\frac{2}{\pi k\rho}} [\cos(\delta_n)\cos(\Phi) - \sin(\delta_n)\sin(\Phi)] \tag{5.89}$$

$$= \frac{A}{\cos(\delta_n)} \sqrt{\frac{2}{\pi k\rho}} \cos(\Phi + \delta_n), \quad \Phi = k\rho - \frac{\pi}{4}(2n+1) \tag{5.90}$$

A simple example of a problem that can be solved exactly has a repulsive potential energy $\sim 1/\rho^2$ with a dimensionless coupling constant g:

$$V(\rho) = \frac{\hbar^2 g}{2m\rho^2} \tag{5.91}$$

$$0 = \left[\frac{d^2}{d\rho^2} + \frac{1}{\rho}\frac{d}{d\rho} - \frac{n^2 + g}{\rho^2} + k^2 \right] R(\rho) \tag{5.92}$$

$$R_n(\rho) = J_{\pm\nu}(k\rho), \quad \nu = \sqrt{n^2 + g} \tag{5.93}$$

In three dimensions, we found it useful to define the function $\chi(r) = rR(r)$. The similar function in two dimensions is $\chi(\rho) = \sqrt{\rho} R(\rho)$. The differential equation for $\chi(\rho)$ is

$$\left\{ \frac{d^2}{d\rho^2} - \frac{n^2 - 1/4}{\rho^2} - \frac{2m}{\hbar^2}[V(\rho) - E] \right\} \chi_n(\rho) = 0 \tag{5.94}$$

This has the form of an equation in one dimension with an effective potential

$$V_n(\rho) = V(\rho) + \frac{\hbar^2}{2m}\frac{n^2-1/4}{\rho^2} \tag{5.95}$$

This formula is sometimes useful.

5.4 Coulomb Potentials

Coulomb potentials of the form $V(r) = -Ze^2/r$ in cgs units, or $-Ze^2/4\pi\varepsilon_0 r$ in SI units, occur often in physics. One example is the hydrogen atom, where $Z=1$ and e is the unit of electrical charge. Other atoms have $Z > 1$. On another energy scale, the gravitational potential is $V = -Gm_1m_2/r$.

For Coulomb potentials, Z can be either positive or negative, so the potentials are attractive or repulsive. Only continuum solutions are allowed for repulsive potentials. Attractive potentials have both bound and continuum eigenstates. There are an infinite number of bound states for the attractive Coulomb potential. This case is a notable exception to the rule of thumb that the number of bound states of a potential is usually finite.

5.4.1 *Bound States*

Our first example is the bound states of a simple Coulomb potential $V(r) = -Ze^2/r$, $Z > 0$. Usually Z is an integer. The particle is assumed to be an electron, and the source of the potential is an atomic nucleus. The nucleus is assumed fixed at the origin. It is convenient to rewrite Schrödinger's equation in dimensionless form. The unit of length is the Bohr radius a_0 and the unit of energy is the Rydberg E_{Ry}:

$$a_0 = \frac{\hbar^2}{me^2} = 0.0529 \text{ nm} \tag{5.96}$$

$$E_{Ry} = \frac{\hbar^2}{2ma_0^2} = \frac{e^2}{2a_0} = 13.60 \text{ eV} \tag{5.97}$$

Since this problem has a central potential, the steps of the last section are followed to derive a differential equation for the radial function $\chi(r) = rR(r)$. This equation is written below twice: once using the r-variable, and again in dimensionless form using $\rho = r/a_0$ and $\varepsilon = -E/E_{Ry}$

$$0 = \left[\frac{d^2}{dr^2} - \frac{\ell(\ell+1)}{r^2} + \frac{2mZe^2}{\hbar^2 r} + \frac{2mE}{\hbar^2}\right]\chi(r) \tag{5.98}$$

$$0 = \left[\frac{d^2}{d\rho^2} - \frac{\ell(\ell+1)}{\rho^2} + \frac{2Z}{\rho} - \varepsilon\right]\chi(\rho) \tag{5.99}$$

At this point, most books just provide the eigenvalues and eigenfunctions. We will show the derivation. Our approach takes a bit longer, but the technique can be applied to other problems.

First examine the differential equation at large values of ρ, where the terms in ρ^{-1} and ρ^{-2} are neglected:

$$0 = \left[\frac{d^2}{d\rho^2} - \varepsilon\right]\chi(\rho) \tag{5.100}$$

The solution to this differential equation is $\chi = A\exp(\pm\sqrt{\varepsilon}\rho)$. The choice $A\exp(+\sqrt{\varepsilon}\rho)$ is discarded since the bound-state eigenfunction must vanish at large values of ρ. Similarly, the differential equation at small values of ρ has the dominant term for $\ell \neq 0$:

$$0 = \left[\frac{d^2}{d\rho^2} - \frac{\ell(\ell+1)}{\rho^2}\right]\chi(\rho) \tag{5.101}$$

The solution to this differential equation is $\chi = \rho^{-\ell}$, or $\rho^{\ell+1}$. The choice $\rho^{-\ell}$ is discarded since the radial function must vanish as $\rho \to 0$. Try a general solution that incorporates these two asymptotic limits. Define an unknown function $G(\rho)$ as

$$\chi(\rho) = \rho^{\ell+1} G(\rho) \exp[-\sqrt{\varepsilon}\rho] \tag{5.102}$$

This step involves no approximation. An exact solution for $G(\rho)$ provides an exact solution for $\chi(\rho)$. Generally, one finds that $G(\rho)$ is a polynomial in ρ with a few terms. The factors of $\rho^{\ell+1}$ and $\exp[-\sqrt{\varepsilon}\rho]$ provide the correct asymptotic limits for small or large values of ρ.

The ansatz (5.102) is inserted into the differential eqn. (5.99). Taking the second derivatives produces a differential equation for $G(\rho)$:

$$\frac{d\chi}{d\rho} = e^{-\sqrt{\varepsilon}\rho}\left[(\ell+1)\rho^{\ell} G - \sqrt{\varepsilon}\rho^{\ell+1} G + \rho^{\ell+1}\frac{dG}{d\rho}\right] \tag{5.103}$$

$$\begin{aligned}\frac{d^2\chi}{d\rho^2} = \rho^{\ell+1} e^{-\sqrt{\varepsilon}\rho}\bigg[\frac{d^2G}{d\rho^2} + 2\frac{dG}{d\rho}\left(\frac{\ell+1}{\rho} - \sqrt{\varepsilon}\right)\\ + \left(\frac{\ell(\ell+1)}{\rho^2} + \varepsilon - 2\sqrt{\varepsilon}\frac{(\ell+1)}{\rho}\right) G\bigg]\end{aligned} \tag{5.104}$$

The second derivative is used in eqn. (5.99). The factor of $\rho^{\ell+1}\exp(-\sqrt{\varepsilon}\rho)$ is canceled from each term. There remains only a differential equation for $G(\rho)$. The terms in $\varepsilon + \ell(\ell+1)/\rho^2$ are canceled:

$$0 = \frac{d^2G}{d\rho^2} + 2\left(\frac{\ell+1}{\rho} - \sqrt{\varepsilon}\right)\frac{dG}{d\rho} + \frac{2}{\rho}[Z - \sqrt{\varepsilon}(\ell+1)]G \tag{5.105}$$

This equation is multiplied by ρ and divided by $2\sqrt{\varepsilon}$:

$$0 = \frac{\rho}{2\sqrt{\varepsilon}}\frac{d^2G}{d\rho^2} + \left(\frac{\ell+1}{\sqrt{\varepsilon}} - \rho\right)\frac{dG}{d\rho} + \left(\frac{Z}{\sqrt{\varepsilon}} - (\ell+1)\right)G \tag{5.106}$$

Equation (5.106) is a standard differential equation, which is satisfied by the *confluent hypergeometric function*. All eigenfunctions of the Coulomb potential involve these functions, so a discussion is necessary.

5.4.2 Confluent Hypergeometric Functions

The standard notation for the two confluent hypergeometric functions are $F(a, b, z)$ and $U(a, b, z)$. Each are solutions of the differential equation:

$$0 = z\frac{d^2w}{dz^2} + (b-z)\frac{dw}{dz} - aw \tag{5.107}$$

$$w = C_1 F(a, b, z) + C_2 U(a, b, z) \tag{5.108}$$

The function $F(a, b, z)$ is unity at the origin ($z = 0$), and is defined by a power series in z:

$$F(a, b, z) = 1 + \frac{a}{b}\frac{z}{1!} + \frac{a(a+1)}{b(b+1)}\frac{z^2}{2!} + \frac{a(a+1)(a+2)}{b(b+1)(b+2)}\frac{z^3}{3!} + \cdots \tag{5.109}$$

The function $U(a, b, z)$ diverges at the origin and cannot be defined as a power series in z. It can be defined as a power series in inverse powers of z:

$$U(a, b, z) = \frac{1}{z^a}\sum_{n=0}^{\infty}\frac{(-1)^n (a)_n (1+a-b)_n}{n! z^n} \tag{5.110}$$

$$(a)_n = a(a+1)(a+2)\cdots(a+n-1) \tag{5.111}$$

Probably the most convenient way to evaluate it is numerically, using an integral definition such as

$$U(a, b, z) = \frac{1}{\Gamma(a)}\int_0^{\infty} dt e^{-zt} t^{a-1}(1+t)^{b-a-1} \tag{5.112}$$

where $\Gamma(a)$ is the gamma-function. The asymptotic behavior at large z is

$$\lim_{z\to\infty} F(a, b, z) \to \frac{\Gamma(b)}{\Gamma(a)} e^z z^{a-b}\left[1 + O\left(\frac{1}{z}\right)\right] \tag{5.113}$$

$$\lim_{z\to\infty} U(a, b, z) \to \frac{1}{z^a}\left[1 + O\left(\frac{1}{z}\right)\right] \tag{5.114}$$

The asymptotic expression for F is valid as long as a is not a negative integer or zero. If $a = -n$, the series in (5.109) has only $(n+1)$ terms. The asymptotic expression is given by the last term, which is proportional to z^n. These properties are the ones needed for the treatment of Coulomb potentials.

5.4.3 Hydrogen Eigenfunctions

The differential equation (5.106) for $G(\rho)$ is identical to the one for the confluent hypergeometric function in (5.107), where the parameters are

$$a = \ell + 1 - \frac{Z}{\sqrt{\varepsilon}} \tag{5.115}$$

$$b = 2(\ell + 1) \tag{5.116}$$

$$z = 2\sqrt{\varepsilon}\rho \tag{5.117}$$

Do not confuse capital Z, the charge on the nucleus, with lower case z. The general solution to eqn. (5.106) is

$$G(\rho) = C_1 F(\ell+1-Z/\sqrt{\varepsilon},\ 2\ell+2,\ 2\sqrt{\varepsilon}\rho) \\ + C_2 U(\ell+1-Z/\sqrt{\varepsilon},\ 2\ell+2,\ 2\sqrt{\varepsilon}\rho) \tag{5.118}$$

The eigenfunction must be finite at the origin, so set $C_2 = 0$ to eliminate the solution U. The radial function and the radial eigenfunction are

$$\chi(\rho) = C_1 \rho^{\ell+1} e^{-\sqrt{\varepsilon}\rho} F(\ell+1-Z/\sqrt{\varepsilon},\ 2\ell+2,\ 2\sqrt{\varepsilon}\rho) \tag{5.119}$$

$$R(\rho) = C_1 \rho^{\ell} e^{-\sqrt{\varepsilon}\rho} F(\ell+1-Z/\sqrt{\varepsilon},\ 2\ell+2,\ 2\sqrt{\varepsilon}\rho) \tag{5.120}$$

The next step is to determine the eigenvalues. The eigenfunction is well behaved in the limit of small ρ. Examine the solution at large values of ρ. The asymptotic expansion of the confluent hypergeometric function is given in (5.113), so

$$\lim_{\rho\to\infty} \chi(\rho) \to C_1 \rho^{-Z/\sqrt{\varepsilon}} e^{\sqrt{\varepsilon}\rho} \frac{\Gamma(2\ell+2)}{\Gamma(\ell+1-Z/\sqrt{\varepsilon})(2\sqrt{\varepsilon})^{\ell+1+Z/\sqrt{\varepsilon}}}$$

This expression diverges at large values of ρ because of the exponential factor $\exp(\sqrt{\varepsilon}\rho)$. The radial function in (5.119) has an exponent of $\exp(-\sqrt{\varepsilon}\rho)$. The confluent hypergeometric function contributes $\exp(2\sqrt{\varepsilon}\rho)$, which leads to the divergence.

This behavior is not acceptable for bound states. Some way must be found to curtail the divergent behavior at large ρ. The salvation comes by requiring that a in $F(a, b, z)$ be a negative integer or zero. This choice truncates the series for F after n terms, and its asymptotic limit is a polynomial ρ^n. In that case the eigenfunction goes to zero when $\rho\to\infty$. The requirement that a be a negative integer or zero is

$$-n_r = a = \ell + 1 - \frac{Z}{\sqrt{\varepsilon}} \tag{5.121}$$

This expression is solved for the eigenvalue:

$$\varepsilon = \frac{Z^2}{n^2}, \quad n = n_r + \ell + 1 \tag{5.122}$$

$$E_n = -\frac{Z^2}{n^2} E_{\mathrm{Ry}} \tag{5.123}$$

where n_r is the *radial quantum number* and n is the *principal quantum number*. The angular momentum is ℓ. The eigenvalues for the hydrogen atom are obtained by setting $Z = 1$.

The radial eigenfunction for the Coulomb bound states is written in terms of the confluent hypergeometric function:

$$R(\rho) = C_1 \rho^{\ell} e^{-Z\rho/n} F(-n_r,\ 2\ell+2,\ 2Z\rho/n) \tag{5.124}$$

This expression is correct, although it must be normalized with the proper choice of C_1. Usually this eigenfunction is written in a different notation. When $a=-n_r$ the confluent hypergeometric function is identical to an associated Laguerre polynomial:

$$F\left(-n_r,\, 2\ell+2,\, \frac{2Z\rho}{n}\right)=\frac{n_r!}{(2\ell+2)_{n_r}}L_{n_r}^{(2\ell+1)}\left(\frac{2Z\rho}{n}\right) \tag{5.125}$$

$$(2\ell+2)_n=(2\ell+2)(2\ell+3)\cdots(2\ell+1+n) \tag{5.126}$$

$$R(\rho)=N\rho^{\ell}e^{-Z\rho/n}L_{n_r}^{(2\ell+1)}(2Z\rho/n) \tag{5.127}$$

The prefactor N is a normalization coefficient that depends on the quantum numbers $(\ell,\, n_r)$:

$$1=\int_0^{\infty}\rho^2 d\rho R(\rho)^2 \tag{5.128}$$

Bound states in three dimensions have three quantum numbers. For the hydrogen atom they are $\ell,\, m,\, n_r$. They are associated with the three variables $\theta,\, \phi,\, r$. Atomic states are usually described in a notation where the letters s, p, d, f are given instead of $\ell=0, 1, 2, 3$. A subscript on the letter denotes the value of m. For example, $4d_z$ means $n=4$, $\ell=2$, $m=0$; and $3p_1$ means $n=3$, $\ell=1$, $m=1$. The subscript z is used instead of $m=0$.

Some normalized radial functions for $R_{n\ell}$ are given below. Instead of angular momentum ℓ, we use the alternative notation of $s(\ell=0)$, $p(\ell=1)$, $d(\ell=2)$:

$$R_{1s}(\rho)=2Z^{3/2}e^{-Z\rho} \tag{5.129}$$

$$R_{2s}(\rho)=\left(\frac{Z}{2}\right)^{3/2}(2-Z\rho)e^{-Z\rho/2} \tag{5.130}$$

$$R_{3s}(\rho)=2\left(\frac{Z}{3}\right)^{3/2}e^{-Z\rho/3}\left[1-\frac{2}{3}Z\rho+\frac{2}{27}(Z\rho)^2\right] \tag{5.131}$$

$$R_{2p}(\rho)=\frac{Z^{5/2}}{\sqrt{4!}}\rho e^{-Z\rho/2} \tag{5.132}$$

$$R_{3p}(\rho)=\frac{(2Z)^{5/2}}{3^{7/2}}\rho e^{-Z\rho/3}\left(1-\frac{Z\rho}{6}\right) \tag{5.133}$$

$$R_{3d}(\rho)=\frac{Z^{7/2}}{81}\sqrt{\frac{8}{15}}\rho^2 e^{-Z\rho/3} \tag{5.134}$$

Earlier chapters discussed expectation values of operators. Some results for hydrogen eigenfunctions ($Z=1$) are, in Dirac notation,

$$\langle n\ell|r|n\ell\rangle=\frac{a_0}{2}[3n^2-\ell(\ell+1)] \tag{5.135}$$

$$\langle n\ell|r^2|n\ell\rangle=\frac{a_0^2}{2}n^2[5n^2+1-3\ell(\ell+1)] \tag{5.136}$$

$$\langle n\ell|\frac{1}{r}|n\ell\rangle=\frac{1}{n^2a_0} \tag{5.137}$$

$$\langle n\ell|\frac{1}{r^2}|n\ell\rangle=\frac{2}{(2\ell+1)n^3a_0^2} \tag{5.138}$$

Matrix elements are integrals of operators with two different eigenfunctions. An important example is

$$\langle 2p_z|z|1s\rangle = \int d^3 r \phi_{2p_z}(\mathbf{r}) z \phi_{1s}(r) \tag{5.139}$$

The notation $2p_z$ means $n=2$, $\ell=1$, $m=0$, and $1s$ is $n=1$, $\ell=0$. The two eigenfunctions, including radial and angular parts, are

$$\phi_{1s} = \frac{R_{1s}(r)}{\sqrt{4\pi a_0^3}} = \frac{1}{\sqrt{\pi a_0^3}} e^{-r/a_0} \tag{5.140}$$

$$\phi_{2p_z} = R_{2p}(\rho) Y_1^0(\theta, \phi) = \frac{r\cos(\theta)}{\sqrt{32\pi a_0^5}} e^{-r/2a_0} \tag{5.141}$$

Now do the integral, with $z = r\cos(\theta)$ and $\rho = r/a_0$:

$$\langle 2p_z|z|1s\rangle = \frac{a_0}{\pi\sqrt{32}} \int_0^\infty d\rho \rho^4 e^{-3\rho/2} \int_0^\pi d\theta \sin(\theta)\cos^2(\theta) \int_0^{2\pi} d\phi$$

The $d\phi$ integral gives 2π, the $d\theta$ integral gives $\frac{2}{3}$, and the $d\rho$ integral gives $4!\left(\frac{2}{3}\right)^5$. The final answer is

$$\langle 2p_z|z|1s\rangle = a_0 \frac{2^{15/2}}{3^5} = 0.745 a_0 \tag{5.142}$$

Other matrix elements are given in the problem sets.

Both nucleons and ions are characterized by having a strong potential in the core region at small values of r. Outside of the core region, the potential is purely Coulombic. The eigenvalues of the radial equation are not given by the integers n_r, since they are influenced by the core region. In this case, in the Coulombic region a is no longer an integer. The function $F(a, b, z)$ cannot be used in this case, since it diverges at large values of ρ. Discard this function. Outside the core region, when a is not an integer, use the other confluent hypergeometric function $U(a, b, z)$. Its asymptotic expansion in (5.113) converges properly. The radial function outside of the inner core region is expressed in terms of an effective quantum number $n^* = Z/\sqrt{\varepsilon}$ that is generally not an integer:

$$\chi_{n^*\ell}(\rho) = C_2 \rho^{\ell+1} e^{-Z\rho/n^*} U(\ell+1-n^*, 2\ell+2, 2Z\rho/n^*) \tag{5.143}$$

So both forms of the confluent hypergeometric function are useful for describing bound states of the Coulombic potential. The function F is used when the $1/r$ potential extends to the origin. The function U is needed when the Coulombic potential is outside of an ion core region of an atom.

Whittaker functions are these two forms for the radial function of the Coulombic potential. They are defined using $\xi = 2\rho\sqrt{\varepsilon}$, $K = Z/\sqrt{\varepsilon} \equiv n^*$:

$$M_{K,\ell+1/2}(\xi) = \xi^{\ell+1} e^{-\xi/2} F(\ell+1-K, 2\ell+2, \xi) \tag{5.144}$$

$$W_{K,\ell+1/2}(\xi) = \xi^{\ell+1} e^{-\xi/2} U(\ell+1-K, 2\ell+2, \xi) \tag{5.145}$$

This completes the discussion of bound states of the Coulombic potential.

5.4.4 Continuum States

Continuum states have a positive energy. The only change in the analysis is that the reduced energy $\varepsilon = -E/E_{Ry}$ is negative, so $\sqrt{\varepsilon}$ is imaginary. Define a dimensionless wave vector $k = \pm i\sqrt{\varepsilon}$, $(k^2 = E/E_{Ry})$. Only the radial eigenfunction is affected by this change. The parameter $a = \ell + 1 \pm iZ/k$ is complex. The two independent solutions for the radial function are

$$\chi(\rho) = \rho^{\ell+1}[C_1 e^{ik\rho} F(\ell+1-iZ/k,\ 2\ell+2,\ -2ik\rho) \\ + C_2 e^{-ik\rho} F(\ell+1+iZ/k,\ 2\ell+2,\ 2ik\rho)] \tag{5.146}$$

The constants $C_{1,2}$ are determined by the boundary conditions. The method of choosing them is now described.

The radial function is examined at large values of $k\rho$. Use the asymptotic forms for $F(a, b, z)$ in (5.113). One term in the above equation is

$$\lim_{k\rho>>1} F(\ell+1-iZ/k,\ 2\ell+2,\ -2ik\rho) \rightarrow \frac{\Gamma(2\ell+2)e^{-2ik\rho}}{\Gamma(\ell+1-iZ/k)}(-2ik\rho)^{-\ell-1-iZ/k} \\ \rightarrow \frac{\Gamma(2\ell+2)e^{-\pi Z/2k}}{|\Gamma(\ell+1-iZ/k)|}\exp\left[-i[2k\rho - \frac{\pi}{2}(\ell+1) + \eta_\ell + \frac{Z}{k}\ln(2k\rho)]\right] \tag{5.147}$$

The phase factor η_ℓ is

$$\Gamma(\ell+1-iZ/k) = |\Gamma(\ell+1-iZ/k)|e^{i\eta_\ell} \tag{5.148}$$

The phase factor in (5.147) contains the term $(Z/k)\ln(2k\rho)$. It is this term that prevents the radial function from having the simple asymptotic expression of the form $\sin(k\rho + \delta_\ell - \pi\ell/2)$. The term with $\ln(2k\rho)$ in the asymptotic phase in (5.147) shows that a particle is never completely away from the reach of a Coulomb potential. We can still include a phase shift in the eigenfunction, but it is a change of phase beyond all of the terms in (5.147).

The solution of Schrödinger's equation outside of a core is discussed in section 2.6. One pair of independent solutions is the incoming and the outgoing wave. They had the same intensity, the same $|C_j|^2$, since the core potential neither creates nor destroys particles. The inward and outward currents have to be equal. The same argument is applied here. The first term in (5.146) has the asymptotic form of an incoming wave, while the second term has the asymptotic form of an outgoing wave. The magnitude of their amplitudes must be the same if they describe the eigenfunction of a continuum state outside of a central core region. If the core region causes a phase shift $\delta_\ell(k)$, the difference in amplitude between the outgoing and incoming waves is

$$C_2 = C_1 \exp[2i\delta_\ell] \tag{5.149}$$

The asymptotic form for the radial function in (5.146) is

$$\lim_{k\rho>>1} \chi(\rho) = \frac{2iC_1 e^{i\delta_\ell} e^{-\pi Z/2k}\Gamma(2\ell+2)}{(2k)^{\ell+1}|\Gamma(\ell+1-iZ/k)|}\sin(\Theta) \tag{5.150}$$

$$\Theta = k\rho - \frac{\pi\ell}{2} + \frac{Z}{k}\ln(2k\rho) + \eta_\ell + \delta_\ell \tag{5.151}$$

The phase shift δ_ℓ is zero if the Coulomb potential extends to the origin. The phase shift occurs typically in an electron scattering from an ion. Outside of the ion radius, the potential is purely Coulomb's law, and the eigenfunctions are Whittaker functions. Inside of the ion core, they have a very different form.

The remaining example of eigenfunctions of the Coulomb potential is when it is repulsive ($V = Ze^2/r$). Only continuum states exist. The repulsive potential has no bound states. The continuum eigenfunction is identical to the one above, except change the sign of Z. Continuum eigenfunctions are also represented as confluent hypergeometric functions or as Whittaker functions.

5.5 WKBJ

5.5.1 Three Dimensions

The WKBJ solution in three dimensions is very similar to the that for one dimension. There are two major changes. The first is that WKBJ finds the radial function $\chi(r)$, since this quantity has the simple form for the second derivative. The other major change is using the potential function in (5.63) that contains the centrifugal barrier. The centrifugal term has to be changed slightly to be suitable for WKBJ. The effective potential for WKBJ is

$$V_\ell = V(r) + \frac{\hbar^2}{2mr^2}\left(\ell + \frac{1}{2}\right)^2 \tag{5.152}$$

There is a factor of $(\ell + \frac{1}{2})^2 = \ell(\ell+1) + \frac{1}{4}$. Where did the extra factor of $\frac{1}{4}$ come from? The centrifugal potential usually has $\ell(\ell+1)$.

As described in chapter 3, the WKBJ eigenfunction has different forms for different regions. In three dimensions, it is the radial function $\chi(r)$ that has these various forms. When the energy $E < V_\ell$, the radial function is evanescent:

$$\alpha(r) = \left\{\frac{\hbar^2}{r^2}\left(\ell + \frac{1}{2}\right)^2 + 2m[V(r) - E]\right\}^{1/2} \tag{5.153}$$

$$\chi(r) = \frac{C_1}{2\sqrt{\alpha(r)}}\exp\left[-\int_r^b \frac{dr'}{\hbar}\alpha(r')\right] + \frac{C_2}{2\sqrt{\alpha(r)}}\exp\left[\int_r^b \frac{dr'}{\hbar}\alpha(r')\right]$$

As $r \to 0$ the radial function must vanish as $r^{\ell+1}$. The factor of $(\ell + \frac{1}{2})^2$ gives this correctly. Set $C_2 = 0$ and consider the first exponent. The largest term in $\alpha(r')$ is the centrifugal term:

$$\alpha(r') \sim \frac{\hbar(\ell + \frac{1}{2})}{r'} \tag{5.154}$$

$$-\int_r^b \frac{dr'}{\hbar}\alpha(r') = \left(\ell + \frac{1}{2}\right)\ln\left(\frac{r}{b}\right) \tag{5.155}$$

$$\chi(r) = C_1 \frac{\sqrt{r}}{2\sqrt{\hbar(\ell+\frac{1}{2})}} \exp\left[\left(\ell+\frac{1}{2}\right)\ln\left(\frac{r}{b}\right)\right] \tag{5.156}$$

$$= r^{\ell+1} \frac{C_1}{2\sqrt{\hbar(\ell+\frac{1}{2})} \quad b^{\ell+1/2}} \tag{5.157}$$

The prefactor $1/\sqrt{\alpha(r)}$ is always going to give $\sqrt{r}$ in the radial function at small r. The other power $r^{\ell+1/2}$ can only come from the phase integral in the exponent. It gives the correct dependence only if the centrifugal term has the factor $(\ell+\frac{1}{2})^2$. This form for the centrifugal potential should be used only for WKBJ calculations. For all other calculations, use the form $\ell(\ell+1)$.

The centrifugal potential is a term of $O(\hbar^2)$. In the derivation of WKBJ in section 3.1, the phase factor $\sigma(r)$ in the exponent was expanded in a power series in $O(\hbar^n)$. Only the first two terms of this series were retained, and terms of $O(\hbar^2)$ were neglected. The centrifugal potential is the order of the neglected terms, so perhaps it should be neglected also. However, it makes an important contribution to the result, and good answers are obtained only by including it.

When $E > V_\ell$ the radial function in WKBJ has the sinusodial form

$$p(r) = \left\{2m[V(r)-E] - \frac{\hbar^2}{r^2}\left(\ell+\frac{1}{2}\right)^2\right\}^{1/2} \tag{5.158}$$

$$\chi(r) = \frac{C_1}{\sqrt{p(r)}} \sin\left[\frac{1}{\hbar}\int_b^r dr' p(r') + \frac{\pi}{4}\right] \tag{5.159}$$

The phase constant $\pi/4$ is used whenever the turning point b has a smooth potential. For an abrupt potential, the factor of $\pi/4$ is absent. The WKBJ formula for the phase shift is

$$\delta(k) = \frac{\pi}{4} + \lim_{r\to\infty}[\Theta(r) - kr] \tag{5.160}$$

The Bohr-Sommerfeld equation for bound states is still

$$\oint dr\, p(r) = 2\pi\hbar\left(n_r + \frac{1}{2}\right) \tag{5.161}$$

These formulas are very similar to those of one dimension. Wave properties in one and three dimensions are similar. The only new wrinkle in three dimension is the addition of the centrifugal potential (5.152).

5.5.2 3D Hydrogen Atom

WKBJ is used to calculate the eigenvalues of the hydrogen atom in three dimensions. The potential is $V(r) = -e^2/r$. Use eqn. (5.161) to find the negative values of E that satisfy

$$\int dr\left[2m\left(E+\frac{e^2}{r}\right) - \frac{\hbar^2}{r^2}\left(\ell+\frac{1}{2}\right)^2\right]^{1/2} = \pi\hbar\left(n_r+\frac{1}{2}\right) \tag{5.162}$$

Divide the equation by $\hbar$, which makes both sides dimensionless. Change variables to atomic units, where $\rho = r/a_0$ and $E = -\alpha^2 E_{\rm Ry}$. Take the factor of ρ out of the bracket:

$$\int \frac{d\rho}{\rho}\left[-\alpha^2\rho^2 + 2\rho - \left(\ell+\frac{1}{2}\right)^2\right]^{1/2} = \pi\left(n_r + \frac{1}{2}\right) \tag{5.163}$$

The factor within the square root is a quadratic expression, which has two roots $\rho_{1,2}$. They are the two turning points, since they are the points where the classical momentum vanishes:

$$\rho_{1,2} = \frac{1}{\alpha^2}\left[1 \pm \sqrt{1-\alpha^2\left(\ell+\frac{1}{2}\right)^2}\right] \tag{5.164}$$

$$\pi\left(n_r + \frac{1}{2}\right) = \alpha\int_{\rho_1}^{\rho_2}\frac{d\rho}{\rho}\sqrt{(\rho_2-\rho)(\rho-\rho_1)} \tag{5.165}$$

The integral is given in chapter 3:

$$\int_{\rho_1}^{\rho_2}\frac{d\rho}{\rho}\sqrt{(\rho_2-\rho)(\rho-\rho_1)} = \frac{\pi}{2}\left[\rho_1+\rho_2-2\sqrt{\rho_1\rho_2}\right] \tag{5.166}$$

$$\pi\left(n_r+\frac{1}{2}\right) = \pi\left[\frac{1}{\alpha} - \left(\ell+\frac{1}{2}\right)\right] \tag{5.167}$$

$$\frac{1}{\alpha} = n_r + \ell + 1 \equiv n \tag{5.168}$$

$$E_n = -\frac{E_{\rm Ry}}{n^2} \tag{5.169}$$

The WKBJ method produces the exact eigenvalue for this classic problem. Again, n_r is the radial quantum number and n is the principal quantum number. The angular momentum is ℓ. The WKBJ form for the centrifugal barrier with $(\ell+\frac{1}{2})^2$ is important for getting the right form for these quantum numbers.

5.5.3 *Two Dimensions*

The WKBJ approximation in two dimensions starts with Schrödinger's equation for central potentials written in polar coordinates:

$$\psi(\mathbf{r}) = R(r)e^{i\ell\theta} \tag{5.170}$$

$$0 = \left[\hbar^2\left(\frac{d^2}{dr^2} + \frac{1}{r}\frac{d}{dr}\right) + p^2(r)\right]R(r) \tag{5.171}$$

$$p^2(r) = 2m[E - V(r)] - \frac{\hbar^2\ell^2}{r^2}, \quad V_\ell = V(r) + \frac{\hbar^2\ell^2}{2mr^2} \tag{5.172}$$

In two dimensions we use the correct centrifugal potential—it does not need the type of change required for three dimensions. This feature will be shown after we derive the formula.

The 2D (two-dimensional) WKBJ formula is found using the same steps for one and three dimensions. The radial part of the eigenfunction is defined as a complex phase factor. The latter obeys the differential equation found from eqn. (5.171):

$$R(r) = \exp\left[\frac{i}{\hbar}\sigma(r)\right] \tag{5.173}$$

$$\frac{dR}{dr} = \frac{i}{\hbar}\frac{d\sigma(r)}{dr}R \tag{5.174}$$

$$\frac{d^2R}{dr^2} = \left[-\frac{1}{\hbar^2}\left(\frac{d\sigma}{dr}\right)^2 + \frac{i}{\hbar}\frac{d^2\sigma}{dr^2}\right]R \tag{5.175}$$

$$0 = -\left(\frac{d\sigma}{dr}\right)^2 + i\hbar\left[\frac{d^2\sigma}{dr^2} + \frac{1}{r}\frac{d\sigma}{dr}\right] + p^2(r) \tag{5.176}$$

Equation (5.176) is derived by inserting the derivatives into (5.171). The function $\sigma(r)$ is expanded in a perturbation expansion in powers of Planck's constant:

$$\sigma(r) = \sigma_0 + \frac{\hbar}{i}\sigma_1(r) + \cdots \tag{5.177}$$

This expansion is inserted in eqn. (5.176) and terms are collected with the same powers of $\hbar$. One generates a series of equations:

$$0 = -\left(\frac{d\sigma_0}{dr}\right)^2 + p^2(r) \tag{5.178}$$

$$0 = 2\frac{d\sigma_0}{dr}\frac{d\sigma_1}{dr} + \frac{d^2\sigma_0}{dr^2} + \frac{1}{r}\frac{d\sigma_0}{dr} \tag{5.179}$$

These equations are simple to solve:

$$\frac{d\sigma_0}{dr} = \pm p(r), \quad \sigma_0(r) = \pm\int^r dr' p(r') \tag{5.180}$$

$$2\frac{d\sigma_1}{dr} = -\left[\frac{d}{dr}\ln\left(\frac{d\sigma_0}{dr}\right) + \frac{1}{r}\right] \tag{5.181}$$

$$\sigma_1 = -\frac{1}{2}\ln\left[rp(r)\right] \tag{5.182}$$

$$R(r) = \frac{C_1}{\sqrt{rp(r)}}\sin\left[\frac{1}{\hbar}\int_b^r dr' p(r') + \frac{\pi}{4}\right] \tag{5.183}$$

The phase factor has the same constant $\pi/4$ that is used when the turning point has a finite value of dV/dr. The above formula is used when the momentum $p(r)$ is real.

The plane wave solution in two dimensions is given exactly when $V = 0$ in terms of Bessel functions $J_\ell(kr)$. Their properties at small and large values of (kr) are given in eqns. (5.48, 5.51):

$$\lim_{z\to 0} J_\ell(z) = \frac{(z/2)^\ell}{\ell!} \tag{5.184}$$

$$\lim_{z\to\infty} J_\ell(z) = \sqrt{\frac{2}{\pi z}}\cos\left(z - \frac{\pi\ell}{2} - \frac{\pi}{4}\right) \tag{5.185}$$

Note that the prefactor at large $z = kr$ has $1/\sqrt{z}$. The WKBJ solution in (5.183) has the same dependence.

Consider the behavior as $z \to 0$. The exact solution for $V = 0$ has $R \sim z^{\ell}$. The evanescent form of WKBJ is

$$\alpha(r) = \left[\frac{\hbar^2\ell^2}{r^2} - 2mE\right]^{1/2} \approx \frac{\hbar\ell}{r} \tag{5.186}$$

$$R(r) = \frac{C_1}{2\sqrt{r\alpha(r)}} \exp\left[-\frac{1}{\hbar}\int_r^b dr'\alpha(r')\right] \tag{5.187}$$

$$= \frac{C_1}{2\sqrt{\hbar\ell}} \exp\left[\ell \ln\left(\frac{r}{b}\right)\right] = r^{\ell}\frac{C_1}{2b^{\ell}\sqrt{\hbar\ell}} \tag{5.188}$$

The correct behavior $(\sim r^{\ell})$ is obtained when using the form for the momentum in (5.172). The centrifugal term is $\hbar^2\ell^2/r^2$. This completes the formal derivation of WKBJ in two-dimensional central potentials.

The exact solution for the hydrogen atom in two dimensions is given as a homework problem. The solution employs confluent hypergeometric functions. It is interesting to find the WKBJ prediction for the eigenvalues. The eigenvalue equation in 2D is the same as in 1D or 3D:

$$\pi\hbar\left(n_r + \frac{1}{2}\right) = \int dr\, p(r) \tag{5.189}$$

where now we use the 2D form for $p(r)$ in eqn. (5.172). In fact, the integral is exactly the same as the one done above for 3D. The only change is to replace $\left(\ell + \frac{1}{2}\right)$ by ℓ. From eqn. (5.167) we find

$$\pi\left(n_r + \frac{1}{2}\right) = \pi\left[\frac{1}{\alpha} - \ell\right] \tag{5.190}$$

$$\frac{1}{\alpha} = n_r + \ell + \frac{1}{2} \equiv n - \frac{1}{2} \tag{5.191}$$

$$E_n = -\frac{E_{R_y}}{\left(n - \frac{1}{2}\right)^2} \tag{5.192}$$

where $n = 1, 2, \ldots$. The lowest eigenvalue is $E_0 = -4E_R$. The binding energy of the hydrogen atom in two dimensions is four times that of three dimensions. The WKBJ prediction gives the exact result. In 2D the Rydberg series is based on half-integers rather than whole integers.

5.6 Hydrogen-like Atoms

There are many examples of atoms or ions where a single electron is trapped in bound states outside of a closed shell of other electrons. A simple example is lithium, which has three electrons bound to a nucleus of charge three. Two electrons are bound in the lowest state, the 1*s* orbital. They fill this shell and it is "closed": no more electrons can be bound in it. This filling is a result of the Pauli exclusion principle, which is discussed in detail in chapter 9.

The third electron in lithium is bound in a higher energy state. The quantum numbers of electrons in atoms are (n, ℓ, m) even when the states are not hydrogenic. The largest binding energy for the third electron in lithium is in the $2s$ state. The electron outside of the closed shell is easily excited to other states that lie higher in energy. This outer electron spends part of the time during its orbit being far from the nucleus and far outside of the inner shell of $1s$ electrons. At these far distances the effective potential is $-e^2/r$. The ion core has a charge of $|e|$: the nucleus has $3|e|$ and the two $1s$ electrons give $-2|e|$. The net charge $(e<0)$ of the ion is $-e$, so the potential energy on the outer electron is $-e^2/r$. When the outer electron is far from the ion core, it has the same potential function as the hydrogen atom. In this region, the eigenfunction is a Whittaker function. During its orbit, the outer electron spends part of its time inside the ion core, and here its potential is decidedly not hydrogenic.

5.6.1 Quantum Defect

Do the energy levels of the outer electron obey a Rydberg series? They do, but it has a modified form, due to the core potential. An *effective quantum number* n^* is defined according to

$$E_n = -\frac{Z_i^2 E_{Ry}}{(n^*)^2}, \quad n^* = Z_i\sqrt{-\frac{E_{Ry}}{E_n}} \tag{5.193}$$

where Z_i is the charge on the ion; for lithium it is one. The energy levels $E_n < 0$ for the various atomic states have been measured using optical probes, and are tabulated in Moore's volume [2]. These values can be used to calculate the effective quantum number n^* for these states. The results are shown in table 5.1. The Rydberg energy is $E_{Ry} = 13.60$ eV. The effective quantum numbers n^* are not integers, but they increase in value by nearly integer steps. To a good approximation, they can be written as

$$n^* = n - C_\ell(n) \tag{5.194}$$

where $C_\ell(n)$ is defined as the *quantum defect*. For lithium, the quantum defects at large n are $C_s = 0.40$, $C_p = 0.05$. The values are smaller for p-states $(\ell = 1)$ since the electron orbit is mostly outside of the ion core. For higher angular momentum, the quantum defect is nearly zero.

Sodium (Na) is below lithium in the periodic table. It has an atomic number of $Z = 11$. Ten electrons are in the ion core $(1s)^2(2s)^2(2p)^6$ and one electron is in an outer shell $(Z_i = 1)$. The lowest eigenstate for this outer electron is denoted as $(3s)$. Its quantum defects are $Cs = 1.35$, $Cp = 0.86$. The energy levels for the p-state have a small energy splitting due to the spin–orbit interaction. We used the series for the $p_{1/2}$ state $\left(j = \frac{1}{2}, \ell = 1, s = \frac{1}{2}\right)$. Spin–orbit interaction is discussed fully in chapter 6.

The alkali series of atoms are lithium (Li), sodium (Na), potassium (K), rubidium (Rb), cesium (Cs), and francium (Fr). Each has a single electron outside of a closed shell that forms an ion core. Each of the alkali atoms has a Rydberg-like series for its outer electron. For each alkali atom, one can make a table of values similar to table 5.1.

Table 5.1 Atomic Energy Levels E_n in Lithium and Sodium and their Effective Quantum Number n^*.

State	$-E_n$(eV)	n^*	State	$-E_n$(eV)	n^*
Li			Li		
$2s$	5.390	1.588	$2p$	3.519	1.97
$3s$	2.018	2.596	$3p$	1.556	2.96
$4s$	1.051	3.598	$4p$	0.872	3.95
$5s$	0.643	4.600	$5p$	0.555	4.95
$6s$	0.434	5.600	$6p$	0.384	5.95
Na			Na		
$3s$	5.138	1.626	$3p(\frac{1}{2})$	3.037	2.116
$4s$	1.947	2.643	$4p(\frac{1}{2})$	1.386	3.133
$5s$	1.022	3.647	$5p(\frac{1}{2})$	0.794	4.138
$6s$	0.629	4.649	$6p(\frac{1}{2})$	0.515	5.140

The alkaline earth atoms are the next column in the periodic table: beryllium (Be), magnesium (Mg), calcium (Ca), strontium (Sr), and barium (Ba). For this series of atoms, the neutral atom has two electrons outside of the closed shell. If the atom is singly ionized by the removal of one outer electron the remaining ion has a single electron outside of a closed shell with $Z_i = 2$. The energy levels of the single electron outside of the ion core also has a Rydberg-like behavior that can be described by a quantum defect.

Trivalent ions with closed shells, such as Al^{3+} and Ga^{3+}, also bind an additional electron with a Rydberg-type series. These different cases are the topic of this section.

5.6.2 WKBJ Derivation

WKBJ theory is used to explain the origin of the quantum defect $C_\ell(n)$. The eigenvalue equation for bound states is derived for a potential that is Coulombic outside of a closed shell. The core region makes an added contribution to the phase, which is the explanation of the quantum defect.

Figure 5.2 shows the effective potential $V_\ell(r)$ from (5.152) outside of a core region:

$$V_\ell(r) = -\frac{Z_i e^2}{r} + \frac{\hbar^2}{2mr^2}\left(\ell + \tfrac{1}{2}\right)^2 \tag{5.195}$$

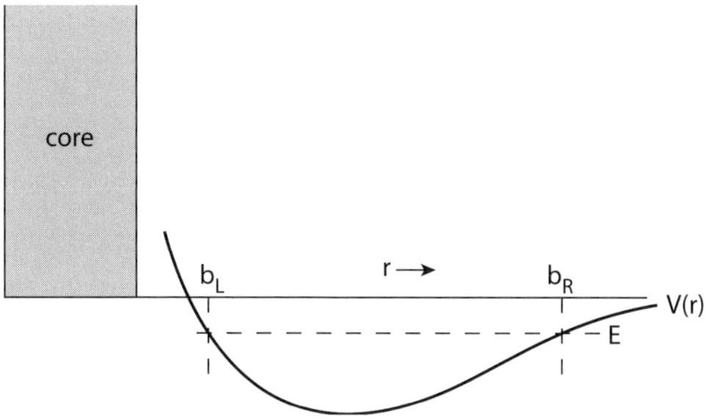

FIGURE 5.2 Effective potential of an electron outside of an ion core.

There is an attractive Coulomb interaction with the ion core $(-Z_i e^2/r)$ and a repulsive centrifugal potential. The centrifugal potential causes the upturn in the potential function at small values of r. For a bound state of energy E, there are two turning points. They are drawn as both being outside of the core region, although that may not be true in all cases.

Recall the reasoning of section 3.3 for finding bound states by WKBJ. The left ϕ_L and right ϕ_R phase integrals are defined and their summation must be an integer multiple of π:

$$\phi_L = \frac{\pi}{4} + \frac{1}{\hbar}\int_{b_L}^{r} dr' p(r') \tag{5.196}$$

$$\phi_R = \frac{\pi}{4} + \frac{1}{\hbar}\int_{r}^{b_R} dr' p(r') \tag{5.197}$$

$$\phi_L + \phi_R = \pi(n_r + 1) \tag{5.198}$$

The phase factor of $\pi/4$ in ϕ_L comes from matching the radial function for $r > b_L$ onto the evanescent wave when $r < b_L$. The eigenfunction for $r < b_L$ is changed by the core region and so is its phase constant. Add an additional phase constant πC_ℓ and rewrite (5.196) as

$$\phi_L = \frac{\pi}{4} + \pi C_\ell(n) + \frac{1}{\hbar}\int_{b_L}^{r} dr' p(r') \tag{5.199}$$

$$\pi(n_r + 1) = \pi\left[\tfrac{1}{2} + C_\ell\right] + \frac{1}{\hbar}\int_{b_L}^{b_R} drp(r) \tag{5.200}$$

The integral on the right can be evaluated as long as the interval $b_L < r < b_R$ is outside of the core region. The integral is exactly the same one done for the hydrogen atom in three dimensions. Using (5.166)–(5.169) with the quantum defect added gives

$$\left[\frac{1}{\alpha} - \left(\ell + \tfrac{1}{2}\right)\right] = n_r + \tfrac{1}{2} - C_\ell \tag{5.201}$$

$$\frac{1}{\alpha}=n_r+\ell+1-C_\ell=n-C_\ell \tag{5.202}$$

$$E_n=-\frac{Z_i^2 E_{Ry}}{(n-C_\ell)^2} \tag{5.203}$$

The WKBJ method produces the exact eigenvalue for this example. The principal quantum number $n = n_r + \ell + 1$ is modified by the quantum defect C_ℓ. The quantum defect enters as a change in the phase of the eigenfunction due to the core region. The phase changes since the potential in eqn. (5.195) does not apply to all regions of space, but is changed in the core region.

Quite often it is useful to know the eigenfunction for the electron outside of the core region. The most accurate function is a Whittaker function with the effective quantum number $n^* = Z_i/\sqrt{\varepsilon}$:

$$\chi(\rho)=C_1 W_{n^*,\,\ell+1/2}(2Z\rho/n^*) \tag{5.204}$$

This simple expression is quite useful.

5.6.3 *Expectation Values*

The effective quantum numbers are used to estimate expectation values of various functions of r. When using hydrogen eigenfunctions, these expectation values depend on the quantum numbers (n, ℓ). For hydrogen-like atoms, assume they have the same functional form with n^* substituted for n. If a_0 is the Bohr radius and $Z_i = 1$, we get

$$\langle n^*,\ell|r^2|n^*,\ell\rangle=\frac{a_0^2}{2}(n^*)^2[5n^{*2}+1-3\ell(\ell+1)] \tag{5.205}$$

$$\langle n^*,\ell|r|n^*,\ell\rangle=\frac{a_0}{2}[3n^{*2}-\ell(\ell+1)] \tag{5.206}$$

$$\langle n^*,\ell\left|\frac{1}{r}\right|n^*,\ell\rangle=\frac{1}{a_0(n^*)^2} \tag{5.207}$$

These formulas are approximate but usually quite accurate. The accuracy is better for $\langle r^2\rangle$ since that weighs large r, where the electron is outside of the core. Here the accuracy is four significant digits. The results are least accurate for $\langle 1/r\rangle$ since then the electron is closer to the ion cores. The effective quantum numbers can also be used to estimate matrix elements between two states of the outer electron. Bates and Damgaard [3] coined the phrase "Coulomb approximation" to describe the use of effective quantum numbers in the calculation of matrix elements. They show that the method is quite accurate when compared to either experiment or rigorous calculations.

5.7 Variational Theory

The variational theory for two and three dimensions is similar to the theory for one dimension, as discussed chapter 3. One still picks a trial eigenfunction $\phi(r)$, but now it could depend on two or three variables. The three integrals for normalization, kinetic energy, and potential energy are done in two or three dimensions. For a single particle interacting with a potential $V(r)$ in three dimensions:

$$I = \int d^3 r |\phi(\mathbf{r})|^2 \tag{5.208}$$

$$\mathrm{KE} = \frac{\hbar^2}{2m} \int d^3 r |\vec{\nabla}\phi|^2 \tag{5.209}$$

$$\mathrm{PE} = \int d^3 r V(r) |\phi(r)|^2 \tag{5.210}$$

$$E(\phi) = \frac{\mathrm{KE} + \mathrm{PE}}{\mathrm{I}} \tag{5.211}$$

The trial eigenfunction $\phi(r)$ is expressed in terms of parameters a, b, c, which can be varied to find the minimum energy. In one dimension, $\phi(x) = A \exp(-\alpha|x|)$ has a cusp at the origin, which makes it an unsuitable trial function except for delta-function potentials. In two or three dimensions, the trial function $\phi(r) = A \exp(-\alpha r)$ is quite suitable, since it is not defined for $r < 0$ and has no cusp.

5.7.1 Hydrogen Atom: $n = 1$

The first example is for the ground state of the hydrogen atom in three dimensions. The Hamiltonian is

$$H = -\frac{\hbar^2 \nabla^2}{2m} - \frac{e^2}{r} \tag{5.212}$$

The ground state has zero angular momentum. Its trial function has no angular dependence and is only a function of the radial distance r. Possible trial functions are

$$\phi_1(r) = A \exp[-\alpha r] \tag{5.213}$$

$$\phi_2(r) = A \exp\left[-\frac{1}{2}\alpha^2 r^2\right] \tag{5.214}$$

$$\vec{\nabla}\phi_2 = -\alpha^2 r \phi_2 \tag{5.215}$$

The obvious choice for a trial function is (5.213). The exact eigenfunction is an exponential and this trial function gives the exact eigenvalue, where $\alpha = 1/a_0$.

To provide some variety, we do the variational calculation with the Gaussian (5.214). Since it is not the exact form of the eigenfunction, we should find a binding energy less than one Rydberg. The three integrals are

$$\mathrm{I} = 4\pi \mathrm{A}^2 \int_0^\infty \mathrm{d}r r^2 \mathrm{e}^{-(\alpha r)^2} = \frac{\pi^{3/2} \mathrm{A}^2}{\alpha^3} \tag{5.216}$$

$$\mathrm{KE} = 4\pi \frac{\hbar^2 A^2}{2m} \alpha^4 \int_0^\infty r^4 dr e^{-(\alpha r)^2} = \frac{3\pi^{3/2} \hbar^2 A^2}{4 m \alpha} \tag{5.217}$$

$$\mathrm{PE} = -4\pi e^2 A^2 \int_0^\infty dr r e^{-(\alpha r)^2} = -\frac{2\pi e^2 A^2}{\alpha^2} \tag{5.218}$$

$$E(\alpha) = \frac{3}{4} \frac{\hbar^2 \alpha^2}{m} - \frac{2 e^2 \alpha}{\sqrt{\pi}} \tag{5.219}$$

Find the minimum value of $E(\alpha)$ by varying α:

$$0 = \frac{dE}{d\alpha} = \frac{3\hbar^2\alpha_0}{2m} - \frac{2e^2}{\sqrt{\pi}} \tag{5.220}$$

$$\alpha_0 = \frac{4}{3\sqrt{\pi}}\left(\frac{e^2 m}{\hbar^2}\right) = \frac{4}{3\sqrt{\pi}a_0} \tag{5.221}$$

The optimal value α_0 of the variational parameter is proportional to $1/a_0$, where a_0 is the Bohr radius. The energy $E(\alpha_0)$ is the minimum value for a Gaussian trial function. Recall that $E_{Ry} = e^2/2a_0 = \hbar^2/2ma_0^2$:

$$E(\alpha_0) = E_{Ry}\left[\frac{3}{2}\left(\frac{4}{3\sqrt{\pi}}\right)^2 - \frac{4}{\sqrt{\pi}}\left(\frac{4}{3\sqrt{\pi}}\right)\right] \tag{5.222}$$

$$= -\frac{8}{3\pi}E_{Ry} = -0.849E_{Ry} \tag{5.223}$$

The exact eigenvalue is $-E_{Ry}$. The variational calculation found a lower binding energy by 15%. Obviously, the Gaussian trial function is a poor approximation to the hydrogen ground-state eigenfunction.

5.7.2 Hydrogen Atom: $\ell = 1$

Angular momentum is a good quantum number if the potential $V(r)$ is spherically symmetric and nonrelativistic. In this case, any eigenfunction can be written as a product of an angular function and a radial function. The angular function is a spherical harmonic:

$$\psi(\mathbf{r}) = R_{n\ell}(r)Y_\ell^m(\theta, \phi) \tag{5.224}$$

The variational method can be used to find the lowest eigenvalue for each separate value of angular momentum ℓ. This procedure is illustrated by finding the lowest energy state with $\ell = 1$ for the hydrogen atom. The trial eigenfunction must have the correct angular dependence. For $\ell = 1$, $m = 0$, it has a prefactor of $r\cos(\theta) = z$:

$$\phi(r) = Aze^{-\alpha r} \tag{5.225}$$

$$\vec{\nabla}\phi = Ae^{-\alpha r}\left[\hat{z} - \frac{\alpha z}{r}\vec{r}\right] \tag{5.226}$$

$$|\vec{\nabla}\phi|^2 = A^2e^{-2\alpha r}\left[1 - \alpha r(2 - \alpha r)\cos^2(\theta)\right] \tag{5.227}$$

Now evaluate the three integrals. Since $z^2 = r^2\cos^2(\theta)$, the angular integral often has the form

$$\int_0^\pi d\theta \sin(\theta)\cos^2(\theta)\int_0^{2\pi} d\phi = \frac{4\pi}{3} \tag{5.228}$$

This factor is found in all three integrals:

$$\mathrm{I}=\frac{4\pi A^2}{3}\int_0^\infty drr^4e^{-2\alpha r}=\frac{\pi A^2}{\alpha^5} \tag{5.229}$$

$$\mathrm{PE}=-\frac{4\pi A^2e^2}{3}\int_0^\infty drr^3e^{-2\alpha r}=-\frac{\pi e^2A^2}{2\alpha^4} \tag{5.230}$$

$$\mathrm{KE}=\frac{4\pi A^2}{2m}\int_0^\infty drr^2e^{-2\alpha r}\left[1-\frac{2\alpha r}{3}+\frac{\alpha^2r^2}{3}\right] \tag{5.231}$$

$$=\frac{\pi\hbar^2A^2}{2m\alpha^3} \tag{5.232}$$

$$E(\alpha)=\frac{\hbar^2\alpha^2}{2m}-\frac{e^2\alpha}{2} \tag{5.233}$$

Minimizing $E(\alpha)$ with respect to α gives the optimal value of $\alpha_0=1/(2a_0)$. The eigenvalue is

$$E(\alpha_0)=-\frac{E_{\mathrm{Ry}}}{4} \tag{5.234}$$

The variational result gives the exact eigenvalue for the $n=2$ state of the hydrogen atom. In this case $n=2$ was found using ($\ell=1$, $n_r=0$, $m=0$). The exact result was obtained since the trial function (5.225) is identical to the exact eigenfunction.

The variational method can be applied to eigenstates besides the ground state. It is necessary that the state have an angular momentum that is different than the ground state. We could not use variational theory to solve for the $n=2$ state that had ($\ell=0$, $n_r=1$, $m=0$), since it gives $-E_{\mathrm{Ry}}$. The angular momentum ℓ must be different in each case. Since these states are called excited states, the variational method can be used to find some, but not all, of the excited states.

5.7.3 *Helium Atom*

The ground state of the helium atom is accurately described by a simple, one-parameter, variational calculation. The atom has two electrons, and the variational calculation is done for the two-electron system. The nucleus is regarded as fixed. Most of the modern employment of the variational technique is in solving many-particle wave functions. The helium atom, with just two electrons, provides a simple example of a many-particle system. Some important concepts, such as relaxation energy, are introduced and calculated.

The two electrons are both in the (1*s*) orbital state. They have antiparallel spins that are arranged into a spin-singlet state:

$$|0,\,0\rangle=\frac{1}{\sqrt{2}}[\alpha_1\beta_2-\alpha_2\beta_1] \tag{5.235}$$

Many-electron wave functions are discussed in chapter 9. Here we give a brief introduction, sufficient to solve the present problem.

The two-electron wave function must change sign when the particle labels are interchanged. The spin singlet has this property, since exchanging the (1, 2) subscripts in (5.235) changes the sign. Two particle wave functions can be written as the product of a spin term and an orbital term:

$$\psi(\mathbf{r}_1, m_1; \mathbf{r}_2, m_2) = \phi(\mathbf{r}_1, \mathbf{r}_2)\chi(m_1, m_2) \tag{5.236}$$

$$\psi(\mathbf{r}_2, m_2; \mathbf{r}_1, m_1) = -\psi(\mathbf{r}_1, m_1; \mathbf{r}_2, m_2) \tag{5.237}$$

where $m_{1,2}$ are spin labels. The second line give the antisymmetric requirement. Since the spin part is antisymmetric $[\chi(m_2, m_1) = -\chi(m_1, m_2)]$, the orbital part must be symmetric: $[\phi(\mathbf{r}_2, \mathbf{r}_1) = \phi(\mathbf{r}_1, \mathbf{r}_2)]$. The simple way to make it symmetric, with both electrons in the same orbital state, is to use

$$\phi(\mathbf{r}_1, \mathbf{r}_2) = \mathbf{A}\exp[-\alpha(\mathbf{r}_1 + \mathbf{r}_2)/\mathbf{a}_0] \tag{5.238}$$

$$\vec{\nabla}_1 \phi(r_1, r_2) = -\frac{\alpha}{a_0}\hat{r}_1 \phi(r_1, r_2) \tag{5.239}$$

The (1s) orbital has no angular dependence, so the radial part of the two-particle wave function is the product of two exponentials. The variational parameter is α. It is dimensionless, since the exponent has been normalized by the Bohr radius a_0. The same variational parameter must be used for both electrons to keep the orbital part symmetric. Also shown is $(\vec{\nabla}_1 \phi)$, and a similar results holds for $(\vec{\nabla}_2 \phi)$. They are needed in the evaluation of the kinetic energy operator.

The Hamiltonian for the two-electron system is

$$H = -\frac{\hbar^2}{2m}(\nabla_1^2 + \nabla_2^2) - 2e^2\left(\frac{1}{r_1} + \frac{1}{r_2}\right) + \frac{e^2}{|\mathbf{r}_1 - \mathbf{r}_2|} \tag{5.240}$$

The first two terms are the kinetic energy. The next two terms are the potential energy for the interaction between the electrons and the nucleus: the nucleus is an alpha-particle of charge two. The last term is the repulsive interaction between the two electrons. The last term is the most interesting, and the hardest to calculate. Electron–electron interactions are what make many-electron systems difficult to describe.

Without the electron–electron interaction, each electron interacts only with the nucleus. The Hamiltonian would be

$$H = H_1 + H_2 \tag{5.241}$$

$$H_j = -\frac{\hbar^2 \nabla_j^2}{2m} - \frac{2e^2}{r_j} \tag{5.242}$$

$$E_j = -\frac{4E_{Ry}}{n^2}, \quad E_g = -8E_{Ry} \tag{5.243}$$

Each electron has the standard solution for a Coulomb potential of charge $Z = 2$. The ground state of the two-electron system is E_g. Without electron–electron interactions, each electron behaves independently of the others.

Electron–electron interactions make the atomic problem into a true three-body problem: a nucleus and two electrons. There are no known exact solutions for this problem. Only approximate solutions are available. The variational method is probably the most popular method of solving the ground state of two electrons in the helium atom. For atoms with more electrons, other methods, such as density functional theory, are successful. They are discussed in chapter 9.

The variational calculation for helium has four integrals. The electron–electron interaction is evaluated separately as the fourth integral. The spin part is properly normalized $\langle 0, 0 \mid 0, 0\rangle = 1$ and requires no more discussion. The normalization integral is

$$\mathrm{I} = \int d^3 r_1 \int d^3 r_2 |\phi(r_1, r_2)|^2 = A^2 \left[4\pi \int_0^\infty r^2 dr e^{-2\alpha r/a_0} \right]^2 \tag{5.244}$$

$$= A^2 \left(\frac{\pi a_0^3}{\alpha^3} \right)^2 \tag{5.245}$$

The kinetic energy integral is

$$\mathrm{KE} = \frac{\hbar^2}{2m} \int d^3 r_1 \int d^3 r_2 [|\vec{\nabla}_1 \phi|^2 + |\vec{\nabla}_2 \phi|^2] \tag{5.246}$$

$$= \frac{\hbar^2 \alpha^2}{2m a_0^2} \int d^3 r_1 \int d^3 r_2 [|\phi|^2 + |\phi|^2] = 2\alpha^2 E_{\mathrm{Ry}} \ \mathrm{I} \tag{5.247}$$

There are two terms that provide equal contributions. For a system of N-electrons, the KE integral has N-terms. All terms are identical. One needs to evaluate only one term and multiply the result by N. Each term must be identical due to the required symmetry of the wave function ϕ $(\mathbf{r}_1, \mathbf{r}_2, \mathbf{r}_3, \ldots, \mathbf{r}_N)$. If it changes sign by interchanging variables $(\mathbf{r}_i, \mathbf{r}_j)$, then $|\vec{\nabla}_i \phi|^2$ must give the same result as $|\vec{\nabla}_j \phi|^2$. This argument can be extended to all different pairs of variables, so that all must give the same result.

There is a similar factorization for the potential energy between the electrons and the nucleus:

$$\mathrm{PE} = -2e^2 \int d^3 r_1 \int d^3 r_2 |\phi(r_1, r_2)|^2 \left(\frac{1}{r_1} + \frac{1}{r_2} \right) \tag{5.248}$$

Since $\phi(\mathbf{r}_2, \mathbf{r}_1) = \phi(\mathbf{r}_1, \mathbf{r}_2)$, the two terms $1/r_i$ must give the same result:

$$\mathrm{PE} = -4e^2 \int \frac{d^3 r_1}{r_1} \int d^3 r_2 |\phi(r_1, r_2)|^2 \tag{5.249}$$

$$= -4e^2 A^2 \frac{\pi a_0^3}{\alpha^3} \left[4\pi \int_0^\infty r_1 dr_1 e^{-2\alpha r_1/a_0} \right] \tag{5.250}$$

$$= -\frac{4e^2 \alpha}{a_0} \mathrm{I} = -8\alpha E_{\mathrm{R_y}} \mathrm{I} \tag{5.251}$$

The integral over $d^3 r_2$ just gives the normalization integral. Without electron–electron interactions, we find that $\alpha = 2$: the kinetic energy is $8E_{\mathrm{Ry}}$, the potential energy is $-16E_{\mathrm{Ry}}$, and the total energy is $-8E_{\mathrm{Ry}}$. This result agrees with (5.243).

The last integral is for the electron–electron interactions. Here the angular integrals must be evaluated carefully and are written separately from the radial integrals:

$$\mathrm{EE} = e^2 \int d^3 r_1 \int d^3 r_2 \frac{|\phi(r_1, r_2)|^2}{|\mathbf{r}_1 - \mathbf{r}_2|} \tag{5.252}$$

$$= e^2 A^2 \int_0^\infty r_1^2 dr_1 e^{-2\alpha r_1/a_0} \int_0^\infty r_2^2 dr_2 e^{-2\alpha r_2/a_0} \Theta(r_1, r_2)$$

$$\Theta(r_1, r_2) = \int d\Omega_1 \int d\Omega_2 \frac{1}{|\mathbf{r}_1 - \mathbf{r}_2|} \tag{5.253}$$

The unit of solid angle in spherical coordinates is $d\Omega_i = d\phi_i \sin(\theta_i)\, d\theta_i$. The last equation gives the angular part of the integrand. Its evaluation is simple when using one coordinate system. We can call any direction $\hat{z}$. For doing the integral $d\Omega_1$, call $\mathbf{r}_2 = r_2\hat{z}$. Then the denominator has the factor of $|\mathbf{r}_1 - \mathbf{r}_2| = \sqrt{r_1^2 + r_2^2 - 2r_1r_2\cos(\theta_1)}$, where $\mathbf{r}_1 \cdot \mathbf{r}_2 = r_1r_2\cos(\theta_1)$. With this choice of coordinate system, the integrand has no dependence on ϕ_1, ϕ_2,θ_2 and these integrals are easy. The integral over $d\theta_1$ can also be done exactly:

$$\Theta(r_1, r_2) = 8\pi^2 \int_0^\pi \frac{d\theta_1 \sin(\theta_1)}{\sqrt{r_1^2 + r_2^2 - 2r_1r_2\cos(\theta_1)}} \tag{5.254}$$

$$= 8\pi^2 \left[\frac{\sqrt{r_1^2 + r_2^2 - 2r_1r_2\cos(\theta_1)}}{r_1r_2} \right]_0^\pi \tag{5.255}$$

$$= \frac{8\pi^2}{r_1r_2}[r_1 + r_2 - |r_1 - r_2|] = \frac{(4\pi)^2}{r_>} \tag{5.256}$$

The notation $r_>$ denotes the maximum of (r_1, r_2). Put the largest one in the denominator. This result comes from $|r_1 - r_2|$, which equals $(r_1 - r_2)$ if $r_1 > r_2$ and equals $r_2 - r_1$ if $r_1 < r_2$.

This result for $\Theta(r_1, r_2)$ is inserted into the integral for EE. The integral dr_2 is divided into two segments, depending whether $r_2 < r_1$, $(r_> = r_1)$ or $r_2 > r_1$, $(r_> = r_2)$:

$$\mathrm{EE} = (4\pi eA)^2 \int_0^\infty dr_1 r_1^2 e^{-2\alpha r_1/a_0} \tag{5.257}$$

$$\times \left[\frac{1}{r_1} \int_0^{r_1} dr_2 r_2^2 e^{-2\alpha r_2/a_0} + \int_{r_1}^\infty dr_2 r_2 e^{-2\alpha r_2/a_0} \right] \tag{5.258}$$

Some labor is saved by showing that the two terms in backets are equal, so only one needs to be evaluated. In the second term, interchange dummy variables r_1 and r_2, and then interchange the order of the two integrals. These steps make the second term identical to the first:

$$\mathrm{EE} = 2(4\pi eA)^2 \int_0^\infty dr_1 r_1 e^{-2\alpha r_1/a_0} \int_0^{r_1} dr_2 r_2^2 e^{-2\alpha r_2/a_0} \tag{5.259}$$

The double integral is evaluated in dimensionless units, $x = 2\alpha r_1/a_0$, $y = 2\alpha r_2/a_0$:

$$\mathrm{EE} = (e\pi A)^2 \left(\frac{a_0}{\alpha}\right)^5 \int_0^\infty x dx e^{-x} \int_0^x dy y^2 e^{-y} \tag{5.260}$$

The two integrals in succession are

$$\int_0^x dy y^2 e^{-y} = 2 - e^{-x}(2 + 2x + x^2)$$

$$\int_0^\infty x dx e^{-x}[2 - e^{-x}(2 + 2x + x^2)] = \frac{5}{8} \tag{5.261}$$

$$\mathrm{EE} = \frac{5}{8}\frac{e^2\alpha}{a_0}\mathrm{I} = \frac{5\alpha}{4} E_{\mathrm{Ry}}\mathrm{I} \tag{5.262}$$

The energy functional $E(\alpha)$ is obtained by adding the contributions from kinetic, potential, and electron–electron interactions, and dividing by I:

$$E(\alpha) = \left(2\alpha^2 - 8\alpha + \frac{5}{4}\alpha\right) E_{\text{Ry}} \tag{5.263}$$

$$= \left(2\alpha^2 - \frac{27}{4}\alpha\right) E_{\text{Ry}} \tag{5.264}$$

The $-8\alpha E_{\text{Ry}}$ from the potential energy is combined with the $(\frac{5}{4})\alpha E_{\text{Ry}}$ from the electron–electron interactions to give $-(\frac{27}{4})\alpha E_{\text{Ry}}$. The total energy functional is minimized with respect to α:

$$0 = \frac{dE}{d\alpha} = \left(4\alpha_0 - \frac{27}{4}\right) E_{\text{Ry}} \tag{5.265}$$

$$\alpha_0 = \frac{27}{16} \tag{5.266}$$

$$E(\alpha_0) = -2\left(\frac{27}{16}\right)^2 E_{\text{Ry}} = -5.695 E_{\text{Ry}} \tag{5.267}$$

The experimental answer is -5.808 E_{Ry}. The variational calculation is too high by 0.113 E_{Ry}, which is an error of less than 2%.

The first remark is actually a question to the reader: How is the experimental answer obtained? The experimental values for the ground-state energy are found by a method that is conceptually simple; the experiments themselves are a little harder. The ground-state energy is the summation of the ionization energies of all of the electrons. Measure the value of energy required to remove the first electron, then the second electron, etc. These values are added up and that is the total ground-state energy of the experimental system. For helium there are only two electrons to remove. It is also the classical answer: The total energy is the work needed to remove all of the particles to infinity.

Electron–electron interactions have a big influence on the answer. Without this interaction, the ground-state energy is of two independent electrons. Each electron has energy $-4E_{\text{Ry}}$ in a $Z = 2$ Coulomb potential. The total ground-state energy is $-8E_{\text{Ry}}$. Electron–electron interactions change the result from -8 to -5.8 E_{Ry}. Consider how electron–electron interactions affect the motion of a single electron. Its potential energy comes from the nucleus $-2e^2/r$ and the other electron. As a crude estimate, guess that one electron spends half of its time inside the orbit of the other and half outside. When it is outside, the effective Coulomb potential is reduced to a charge of one. When it is inside the orbit of the other electron, the effective charge is not reduced and stays at two. This crude estimate predicts the Coulomb potential is $-2e^2/r$ half of the time and $-e^2/r$ the other half, for an average Coulomb potential of $-Ze^2/r$, $Z = \frac{3}{2}$. Then the binding energy of a single electron is $-Z^2 E_{\text{Ry}} = -2.25$ E_{Ry}. This value is doubled for two electrons, giving a crude estimate of -4.5 E_{Ry} for helium. Obviously, this simple model is a poor estimate compared to -5.8 E_{Ry}. It does provide some insight as to why the binding energy is reduced from -8 E_{Ry}: The two electrons take turns screening each other from the nucleus. The variational calculation has shown that the best value for the effective charge is not $Z = \frac{3}{2}$, but instead is the variational parameter $Z = \alpha_0 = \frac{27}{16}$.

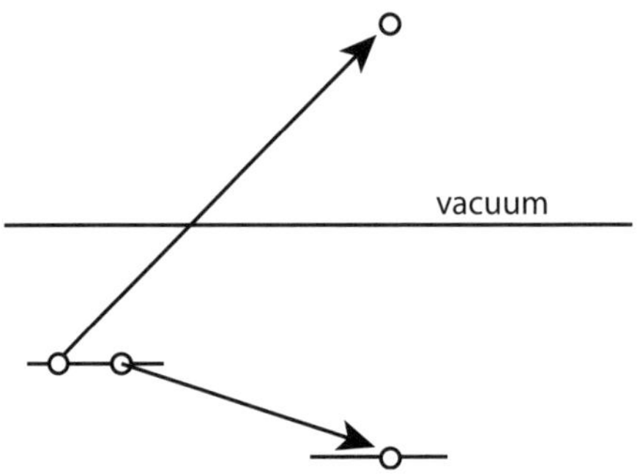

FIGURE 5.3 Removing one electron causes the energy of the second one to relax with a higher binding energy.

The ground-state energy is the summation of the ionization energies of the electrons. What are they for helium? After removing one electron, the second electron is by itself in the Coulomb potential with $Z = 2$. It has a binding energy of $-4\ E_{\rm Ry}$. That is the energy needed to remove it from the nucleus. Since the total binding energy is $-5.81\ E_{\rm Ry}$, the ionization energy to remove the first electron is $5.81 - 4.00 = 1.81\ E_{\rm Ry} = 24.6$ eV. The two electrons have quite different ionization energies. The two electrons are identical and indistinguishable. A more precise statement is that less energy is required to remove the first electron than the second one.

A popular method of illustrating energy states of many-electron systems is to draw a set of energy level diagrams. How do you draw the diagram of the two-electron helium atom? What are the energy levels? These excellent questions have no rigorous answer in a many-electron system. There are two numbers associated with the two-electron system: the total binding energy $-5.81\ E_{\rm Ry}$, and the ionization energy of the first electron $-1.81\ E_{\rm Ry}$. We could draw the energy levels of the two electrons at (i) $-1.81\ E_{\rm Ry}$, (*ii*) $-5.81\ E_{\rm Ry}$, or (iii) $-(5.81/2)\ E_{\rm Ry} = -2.90\ E_{\rm Ry}$. Take your pick.

The most sensible choice is (iii): $-2.9\ E_{\rm Ry}$. Introduce the concept of *relaxation energy*. When one electron is removed, the remaining electron changes its energy from $-2.9\ E_{\rm Ry}$ to $-4.0\ E_{\rm Ry}$. This process is called relaxation, and is illustrated in figure 5.3. The electron that remains bound has its energy changed by $4.0 - 2.9 = 1.1\ E_{\rm Ry}$. The latter is the relaxation energy. Note that the relaxation energy is given to the departing electron: Its ionization energy goes from $-2.9 + 1.1 = 1.8\ E_{\rm Ry}$. So when one electron leaves, the remaining electrons in the atom can increase their binding energy. The total increase is the relaxation energy. It is transferred to the departing electron, which reduces its apparent binding energy.

Koopman's theorem states that the ionization energy equals the negative of the eigenvalue. Because there is relaxation energy in all systems, Koopman's theorem is never true.

Energy levels usually represent eigenvalues of one-electron Hamiltonians. Here the Hamiltonian is for two electrons, and has no one-electron eigenvalues. In chapter 9, we reconsider the many-electron atom and derive an effective and approximate one-electron Hamiltonian for the atom. Of course, this approximate Hamiltonian has eigenvalues.

However, they are a computational method of replacing a complex many-electron problem by an approxmate set of one-electron problems. The eigenvalues of this approximate Hamiltonian have no physical meaning.

The helium atom is quite well described by the one-parameter variational theory. Of course, much better calculations have been done. In 1930 Hylleraas achieved an accuracy of four significant figures, and by 1962 Pekeris found the ground-state energy of helium to ten significant figures.

5.8 Free Particles in a Magnetic Field

An important experimental arrangement has charged particles in a constant magnetic field. Classically, the particles make circular orbits perpendicular to the magnetic field at the cyclotron frequency $\omega_c = eB/mc$. Similar motion is found from the solution using quantum mechanics. The charged particles have plane-wave motion parallel to the field. The motion parallel to the field is well described by the plane-wave theory of earlier sections.

5.8.1 *Gauges*

In the theory of electricity and magnetism, potentials may be either scalar $\phi(\mathbf{r})$ or vector $\mathbf{A}(\mathbf{r})$. The preceding sections used only the scalar potentials $V(\mathbf{r}) \equiv e\phi(\mathbf{r})$. Magnetic fields are introduced into Schrödinger's equation using vector potentials. The technique is quite simple: the particle momentum $\mathbf{p}_i$ is replaced in the kinetic energy by $[\mathbf{p}_i - q_i\mathbf{A}(\mathbf{r}_i)/c]$, where c is the speed of light and q_i is the charge of the particle. In nonrelativistic quantum mechanics, the vector and scalar potentials enter Schrödinger's equation for one particle in the form

$$H = \frac{1}{2m}\left[\mathbf{p} - \frac{q}{c}\mathbf{A}(\mathbf{r})\right]^2 + V(\mathbf{r}) - \vec{\mu}\cdot\vec{B} \tag{5.268}$$

The previous chapters have assumed $\mathbf{A} = 0$. The last term is from the spin of the electron and this Pauli interaction $\vec{\mu}\cdot\vec{B} = \Delta\sigma_z$, where σ_z is a Pauli spin matrix. $\Delta = |\vec{\mu}||\vec{B}|$. The spin interaction is initially neglected, but is included in a later discussion.

This form of the Hamiltonian is derived and justified in chapter 8. A key feature of the derivation is that the momentum is shown to equal

$$\mathbf{p} = m\mathbf{v} + \frac{q}{c}\mathbf{A} \tag{5.269}$$

where $\mathbf{v}$ is the velocity of the particle. The kinetic energy term is $mv^2/2$, which leads to the above form. Also shown in chapter 8 is that the above form depends on a choice of gauge, and here we are using the *Coulomb gauge*.

A vector potential can be used to describe any type of magnetic field $\mathbf{B}$, whether static or oscillatory, homogeneous or inhomogeneous. The magnetic field is always given by $\mathbf{B} = \nabla\times\mathbf{A}$. The present section is concerned with a charged particle in a static, homogeneous, magnetic field. The vector potential must be chosen so that it obeys the following two equations:

$$\nabla \cdot \mathbf{A} = 0 \tag{5.270}$$

$$\nabla \times \mathbf{A} = \hat{z}B \tag{5.271}$$

where B is a constant, and the magnetic field is put in the $\hat{z}$ direction. Equation (5.270) is the Coulomb gauge condition, which is explained in chapter 8. Also specify $\phi(\mathbf{r}, t) = 0$: there is no scalar potential.

Many vector potentials satisfy the above criteria, and there is no unique choice. Three different, valid, vector potentials are

$$\mathbf{A} = B(-y, 0, 0) \tag{5.272}$$

$$\mathbf{A} = B(0, x, 0) \tag{5.273}$$

$$\mathbf{A} = \frac{B}{2}(-y, x, 0) \tag{5.274}$$

Each different vector potential is a choice of gauge condition. Since the eigenvalues do not depend on this choice, they are called *gauge invariant*. The first gauge is chosen, for no particular reason, and then the Hamiltonian for a particle of charge q and mass m in a constant magnetic field B is

$$H_a = \frac{1}{2m}\left[(p_x + \frac{q}{c}yB)^2 + p_y^2 + p_z^2\right] \tag{5.275}$$

The subscript a denotes the choice of gauge. If the particle has a spin, and a magnetic moment $\vec{\mu}$, there is also the Pauli term of $-\vec{\mu} \cdot \mathbf{B}$. The Pauli term is ignored for the present, since the emphasis here is on the orbital motion.

5.8.2 Eigenfunctions and Eigenvalues

The vector potential term only contains the variable y, and only affects motion in the y-direction. The eigenfunctions have plane wave motion in both the x- and z-directions. Both p_x and p_z commute with the Hamiltonian, and can be assigned eigenvalues k_x, k_z. The eigenfunction has the form

$$\psi_a(\mathbf{r}) = \frac{e^{i(k_x x + k_z z)}}{\sqrt{L_x L_z}} f(y) \tag{5.276}$$

where L_x, L_z are the lengths of the volume in these two directions. The unknown function $f(y)$ is determined by operating on the above eigenfunction by the Hamiltonian:

$$H_a \psi_a = E\psi_a \tag{5.277}$$

$$H_a \psi_a = \frac{e^{i(k_x x + k_z z)}}{2m\sqrt{L_x L_z}}\left[\left(\hbar k_x + \frac{qB}{c}y\right)^2 + p_y^2 + \hbar^2 k_z^2\right] f(y) \tag{5.278}$$

Define some symbols such as the z-component of kinetic energy ε_z, the cyclotron frequency ω_c, and the unit of magnetic length ℓ:

$$\varepsilon_z = \frac{\hbar^2 k_z^2}{2m}, \quad \omega_c = \frac{qB}{mc} \tag{5.279}$$

$$\ell^2 = \frac{\hbar}{m\omega_c} = \frac{\hbar c}{qB} \tag{5.280}$$

$$y_0 = -\frac{\hbar c k_x}{qB} = -k_x \ell^2 \tag{5.281}$$

These symbols permit the Hamiltonian to be written in compact form:

$$H_a \psi_a = \frac{e^{i(k_x x + k_z z)}}{\sqrt{L_x L_z}} \left[\varepsilon_z + \frac{p_y^2}{2m} + \frac{m\omega_c^2}{2}(y - y_0)^2 \right] f(y) \tag{5.282}$$

The order of the terms has been inverted. The first term ε_z is known. The last two terms are those of a simple harmonic oscillator of frequency ω_c, where the center of the oscillation is at y_0. Using the harmonic oscillator solution from chapter 2, the eigenvalues and eigenfunctions of (5.282) are

$$E_n(k_z) = \varepsilon_z + \hbar\omega_c \left(n + \frac{1}{2} \right) \tag{5.283}$$

$$\psi_a(\mathbf{r}) = \frac{e^{i(k_x x + k_z z)}}{\sqrt{L_x L_z}} \phi_n(y - y_0) \tag{5.284}$$

where ϕ_n are harmonic oscillator eigenfunctions. The Hamiltonian (5.268) has been solved exactly. Different values of n are called *Landau levels.* The ground state has $n = 0$.

The quantum number k_z gives the kinetic energy in the z-direction. The quantum number k_x has no influence on the eigenvalue. The particle has plane-wave motion along the direction parallel to the magnetic field. It has harmonic oscillator motion in the plane perpendicular to the magnetic field. The quantum number k_x determines the center of the oscillation since $y_0 = -k_x \ell^2$.

Since the eigenvalue does not depend on k_x, the most general eigenfunction with this eigenvalue is a linear combination of functions with different k_x:

$$\psi_a(\mathbf{r}) = \frac{e^{ik_z z}}{\sqrt{L_x L_z}} \sum_{k_x} e^{ik_x x} G(k_x) \phi_n(y + k_x \ell^2) \tag{5.285}$$

We are using box normalization, so the values of $k_x = 2j\pi/L_x$ are discrete. Any function $G(k_x)$ can be used without affecting the eigenvalue.

The above derivation is valid for the vector potential (5.272). Similar results are found using (5.273). The difference is the harmonic oscillator is in the x-direction:

$$H_b = \frac{1}{2m} \left[p_x^2 + \left(p_y - \frac{qB}{c} x \right)^2 + p_z^2 \right] \tag{5.286}$$

$$E_n(k_z) = \varepsilon_z + \hbar\omega_c \left(n + \frac{1}{2} \right) \tag{5.287}$$

$$\psi_b(\mathbf{r}) = \frac{e^{i(k_y y + k_z z)}}{\sqrt{L_x L_z}} \phi_n(x - k_y \ell^2) \tag{5.288}$$

Note that the eigenvalue is the same. What is the relationship between the two eigenfunctions (5.284) and (5.288)? To answer that question, consider the effect of a phase factor on H_a. Define g(**r**) according to

$$\psi_a(\mathbf{r}) = e^{-ixy/\ell^2} g(\mathbf{r}) \tag{5.289}$$

$$H_a\psi_a = \frac{1}{2m}\left[\left(p_x + \frac{q}{c}yB\right)^2 + p_y^2 + p_z^2\right] e^{-ixy/\ell^2} g(\mathbf{r}) \tag{5.290}$$

$$= \frac{e^{-ixy/\ell^2}}{2m}[p_x^2 + (p_y - \frac{qB}{c}x)^2 + p_z^2] g(\mathbf{r}) \tag{5.291}$$

$$= e^{-ixy/\ell^2} H_b g(\mathbf{r}) \tag{5.292}$$

The phase factor changes H_a into H_b. The two eigenfunctions must be related in a similar way:

$$\psi_a(\mathbf{r}) = e^{-ixy/\ell^2} \psi_b(\mathbf{r}) \tag{5.293}$$

This identity is true for a particular choice of $G(k_x)$ in (5.285).

The different choices of vector potential give the same eigenvalues. The high degree of degeneracy for the eigenvalue is related to the center of the classical orbit. The classical picture has the particle going in circular orbits perpendicular to the magnetic field. The center of the orbit may be anyplace in the box without changing the eigenvalue. The arbitrary location of the orbit is the reason the eigenvalue does not depend on $G(k_x)$. Different choices of $G(k_x)$ put the orbit at different locations.

5.8.3 Density of States

The density of states $N(\varepsilon)$ is defined in three dimensions as the number of particles per unit volume, per unit energy. The physical idea: given a box of known volume in a fixed magnetic field, how many particles can be put into the box? How many can be put into the $n = 0$ Landau level, into the $n = 1$ Landau level, etc.?

We continue to use the Hamiltonian H_a and its eigenfunction in (5.284). It has three quantum numbers (n, k_x, k_z). The density of states is

$$N(E) = \frac{1}{L_x L_y L_z} \sum_{nk_xk_z} \delta[E - E_n(k_z)] \tag{5.294}$$

The volume of our box is $\Omega = L_x L_y L_z$. The summations over (k_x, k_z) can be changed to continuous integrals:

$$\frac{1}{L_x}\sum_{k_x} = \int \frac{dk_x}{2\pi}, \quad \frac{1}{L_z}\sum_{k_z} = \int \frac{dk_z}{2\pi} \tag{5.295}$$

Since the eigenvalue depends on k_z, the integral over this variable eliminates the delta-function of energy:

$$\int \frac{dk_z}{2\pi}\delta\left[E-\varepsilon_z-\hbar\omega_c\left(n+\frac{1}{2}\right)\right]=\sqrt{\frac{2m}{\hbar^2}}\frac{1}{2\pi\sqrt{E-\hbar\omega_c(n+\frac{1}{2})}}$$

The result is nonzero only when the argument of the square root is positive.

Since the energy does not depend on the variable k_x, its integral gives the difference of the two limits:

$$\int_{k_\ell}^{k_u} dk_x = k_u - k_\ell \tag{5.296}$$

What determines these limits? They are constrained by the requirement that the particle is in the box. The center of the harmonic oscillator is at y_0, and we require that $0<y_0<L_y$. Neglecting the minus sign ($y_0 = k_x\ell^2$), we get the condition that

$$0<k_x<\frac{L_y}{\ell^2},\quad k_\ell=0,\quad k_u=\frac{L_y}{\ell^2} \tag{5.297}$$

The factor of L_y cancels the same factor in the denominator of (5.294). The final result is

$$N(E)=\frac{1}{(2\pi\ell)^2}\sqrt{\frac{2m}{\hbar^2}}\sum_{n=0}^{n_E}\frac{1}{\sqrt{E-\hbar\omega_c(n+\frac{1}{2})}} \tag{5.298}$$

and n_E is the integer nearest to $E/(\hbar\omega_c)$. The final result has the units of inverse joule·meter^3.

It is useful to take the limit that the magnetic field goes to zero. Then the cyclotron energy $\varepsilon = n\hbar\omega_c$ becomes very small. The summation over n can be changed into a continuous integral. Neglect the small zero point energy:

$$\sum_{n=0}^{n_E}\frac{dn}{\sqrt{E-\hbar\omega_c(n+\frac{1}{2})}}=\frac{1}{\hbar\omega_c}\int_0^E\frac{d\varepsilon}{\sqrt{E-\varepsilon}} \tag{5.299}$$

$$=\frac{2}{\hbar\omega_c}\sqrt{E} \tag{5.300}$$

The prefactor contains the denominator $\ell^2\hbar\omega_c = \hbar^2/m$, so the result is now independent of magnetic field:

$$N(E)=\frac{1}{4\pi^2}\left(\frac{2m}{\hbar^2}\right)^{3/2}\sqrt{E} \tag{5.301}$$

$$=\int\frac{d^3k}{(2\pi)^3}\delta\left(E-\frac{\hbar^2k^2}{2m}\right) \tag{5.302}$$

The final answer is just the density of states of a three-dimensional free particle system.

5.8.4 *Quantum Hall Effect*

The quantum Hall effect (QHE) is experimentally observed at low temperatures in semiconductor superlattices or similar microstructures. Molecular beam epitaxy is used to grow semiconductors with alternate layers of different materials. The electrons are confined to move in one layer. The motion is entirely two dimensional. We shall model this system as if it were an ideal two-dimensional electron gas. The semiconductors have energy bands with quadratic dispersion. The curvature can be described as an effective mass m^*:

$$E_c(k) = E_c(0) + \frac{\hbar^2 k^2}{2m^*} + O(k^4) \tag{5.303}$$

GaAs is often used for this measurement. The effective mass in the conduction band has a value about $m^* = 0.07m_e$, where m_e is the actual mass of the electron. This small value makes the cyclotron frequency $\omega_c = eB/m^*c$ very large compared to most other semiconductors. It is easy to make $\hbar\omega_c/k_BT >> 1$ so the Landau levels are well separated thermodynamically.

In a two-dimensional system, just eliminate the z-variable from the prior discussion. The magnetic field is perpendicular to the xy plane. The Hamiltonian in one gauge is

$$H_2 = \frac{1}{2m}\left[\left(p_x + \frac{q}{c}yB\right)^2 + p_y^2\right] - \vec{\mu}\cdot\vec{B} \tag{5.304}$$

$$E_{nm} = \hbar\omega_c\left(n + \frac{1}{2}\right) - m\Delta, \qquad m = \pm 1 \tag{5.305}$$

$$\psi(x, y) = \frac{e^{ik_x x}}{\sqrt{L_x}}\phi_n(y + k_x\ell^2) \tag{5.306}$$

The density of state in two dimensions is

$$N_2(E) = \frac{1}{L_x L_y}\sum_{nk_x m}\delta[E - E_{nm}], \qquad \sum_{k_x} = \frac{L_x L_y}{2\pi\ell^2} \tag{5.307}$$

$$= \frac{1}{2\pi\ell^2}\sum_{nm}\delta\left[E - \hbar\omega_c\left(n + \frac{1}{2}\right) + m\Delta\right] \tag{5.308}$$

Again the summation over k_x has been converted to $L_xL_y/2\pi\ell^2$. The density of states is a series of delta-functions at energies given by the Landau levels and the spin orientation.

What value of magnetic field is required to put all of the electrons into the lowest Landau level $n = 0$? The two-dimensional density of electrons n_0 (units are inverse area) is given by

$$n_0 = \int dE N_2(E) n_F(E - \mu) \tag{5.309}$$

$$= \frac{1}{2\pi\ell^2}\sum_{nm}\frac{1}{e^{\beta[\hbar\omega_c(n+1/2) - m\Delta - \mu]} + 1} \tag{5.310}$$

$$n_F(E - \mu) = \frac{1}{e^{\beta(E-\mu)} + 1} \tag{5.311}$$

The factor of $n_F(E - \mu)$ is the fermion occupation factor, and $\beta \equiv 1/k_BT$. The Pauli term is $-m\Delta$, where $\Delta = g\mu_0 B$, and μ_0 is the electron magnetic moment. Spin-up and spin-down particles have different energies. It is possible to have all particles with spin-up in one Landau level. If the chemical potential μ is between the first two Landau levels

($\hbar\omega_c/2 < \mu < 3\hbar\omega_c/2$) and the temperature is small ($k_B T << \hbar\omega_c$), then all electrons are in the lowest Landau level of $n = 0$. The density of particles in that case is

$$n_0 = \frac{1}{2\pi\ell^2} \tag{5.312}$$

Some typical experimental numbers are presented. The factor in the denominator is

$$2\pi\ell^2 = \frac{hc}{eB} = \frac{\phi_0}{B} \tag{5.313}$$

The flux quantum is $\phi_0 = hc/e$ in cgs units and $\phi_0 = h/e$ in SI units. It has a numerical value of $\phi_0 = 4.13567\times 10^{-15}$ in units of joules per ampere, which is also tesla square meters. A typical electron density in a quantum well is $n_0 \sim 10^{15}$ per square meters, which requires a magnetic field of $B \sim 4$ tesla.

The integer quantum Hall effect is the observation of plateaus in the Hall voltage at certain values of magnetic field. These plateaus occur when $2\pi\ell^2\, n_0$ has integer values, which occurs when Landau levels are exactly filled. We write eqn. (5.310) as

$$2\pi\ell^2 n_0 = \sum_{nm} \frac{1}{e^{\beta[\hbar\omega_c(n+1/2)-m\Delta-\mu]}+1} \equiv \nu \tag{5.314}$$

where ν is the summation. The summation depends on the chemical potential μ, but at very low temperatures, experiments show a staircase of values as a function of B.

The classical analysis of the Hall effect has the motion of electrons governed by

$$m\frac{d}{dt}\vec{v} = e\left[\vec{E} + \frac{1}{c}\vec{v}\times\vec{B}\right] - \frac{m\vec{v}}{\tau} \tag{5.315}$$

where τ is the lifetime of the electron for scattering from defects and ion vibrations (phonons). The velocity $\vec{v}$ is the average velocity of the distribution of electrons. Assume a constant magnetic field in the $\hat{z}$-direction and current flowing down a bar in the $\hat{x}$ direction:

$$\dot{v}_x = \frac{e}{m}\left[E_x + \frac{v_y B}{c}\right] - \frac{v_x}{\tau} \tag{5.316}$$

$$\dot{v}_y = \frac{e}{m}\left[E_y - \frac{v_x B}{c}\right] - \frac{v_y}{\tau} \tag{5.317}$$

$$\dot{v}_z = \frac{e}{m}E_z - \frac{v_z}{\tau} \tag{5.318}$$

The measurement is dc, so set all time derivatives to zero. The currents can flow only in the $\hat{x}$-direction, so that $v_y = v_z = 0$. There is no electric field in the $\hat{z}$ direction. So the equations are

$$v_x = \frac{e\tau}{m}E_x \tag{5.319}$$

$$E_y = \frac{v_x B}{c} \tag{5.320}$$

The bar has dimensions L_x, L_y, L_z. The current I_x (in amperes) flowing down the bar is

$$I_x = nev_x L_y L_z = n_0 e v_x L_y \tag{5.321}$$

Here the electron density n is the three-dimensional density (number electrons per volume). The areal density is $n_0 = nL_z$ (number electrons per area). The Hall voltage is

$$V_y = E_y L_y = \frac{v_x B}{c} L_y = I_x \frac{B}{n_0 e c} \tag{5.322}$$

The transverse resistance is

$$R_{xy} = \frac{V_y}{I_x} = \frac{B}{n_0 e c} \tag{5.323}$$

The factors on the right are similar to those that go into the integer $\nu = 2\pi\ell^2 n_0$. In fact, the relationship is

$$R_{xy} = \frac{\Omega_0}{\nu}, \quad \Omega_0 = \frac{h}{e^2} = 25,812.8 \text{ ohms} \tag{5.324}$$

We have introduced the quantum of resistance $\Omega_0 = h/e^2$. The integer quantum Hall effect finds this result experimentally. The transverse resistance is given by Ω_0 divided by integers, where each integer means the full occupation of another Landau level.

Numerous experiments have shown the plateaus are due to a complex phenomenon. Semiconductors always have impurities and defects. These imperfections cause local potential fluctuations in the material. They tend to bind the electrons in the quantized orbits. Most of the electrons are actually localized and are unable to move. The electrons that conduct electricity are on the edges of the sample, and are in skipping orbits. See Jain [6] for a discussion of this phenomenon.

5.8.5 Flux Quantization

An interesting phenomenon in superconductors is the quantization of magnetic flux in the center of hollow cylinders. The phenomenon was first reported simultaneously by Deaver and Fairbanks [7] and by Doll and Näbauer [8].

Let $\mathcal{A}$ denote a finite area, which could be a circle or a rectangle. The scalar dS is an infinitesimal part of this area, and the total area is obtained by an integral:

$$\mathcal{A} = \int dS \tag{5.325}$$

Let Φ denote the total magnetic flux through this finite area, which is obtained by integrating the normal component of the magnetic field $\vec{B}$:

$$\Phi = \int dS\hat{n} \cdot \vec{B} = \int dS\hat{n} \cdot \vec{\nabla} \times \vec{A} \tag{5.326}$$

where $\vec{A}$ is the vector potential. Next use Gauss's theorem to change the surface integral into a line integral $\vec{d\ell}$ around the circumference of the area:

$$\Phi = \oint \vec{d\ell} \cdot \vec{A} \tag{5.327}$$

The magnetic flux in a region equals the line integral of the vector potential around the circumference.

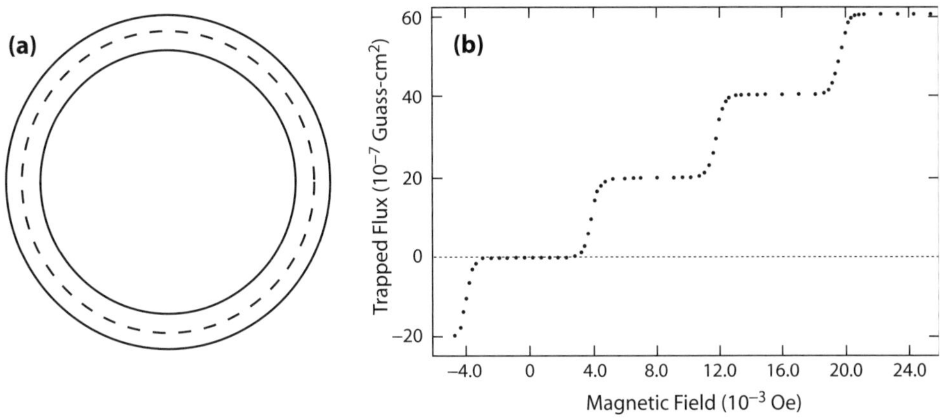

FIGURE 5.4 (a) A hollow cylinder of a superconductor. Dashed line shows path of integration contour. (b) Experimental data of Goodman and Deaver [9].

Flux quantization can be derived using the Bohr-Sommerfeld quantization condition:

$$2\pi\hbar\left(n+\frac{1}{2}\right) = \oint d\vec{\ell}\cdot\vec{p} \tag{5.328}$$

Use the fact that $\vec{p} = m\vec{v} + (q/c)\vec{A}$ to change the above formula to

$$h\left(n+\frac{1}{2}\right) = \oint d\vec{\ell}\cdot\left[m\vec{v}+\frac{q}{c}\vec{A}\right] \tag{5.329}$$

$$= m\oint d\vec{\ell}\cdot\vec{v} + \frac{q}{c}\Phi \tag{5.330}$$

Figure 5.4a shows a hollow cylinder of a superconductor. The magnetic field is perpendicular to the plane of the drawing and goes along the axis of the cylinder. The line integral $\oint d\vec{\ell}$ is taken in a circle inside of the superconductor. It encloses the magnetic flux in the hollow part of the cylinder. The flux does not penetrate into the superconductor. Along this path, the velocity integral $\oint d\vec{\ell}\cdot\vec{v} = 0$. In a superconductor, the electrons are paired, so $q = 2e$, where e is the charge on an electron. The above equation is

$$\Phi = \frac{hc}{2e}\left(n+\frac{1}{2}\right) = \frac{\phi_0}{2}\left(n+\frac{1}{2}\right) \tag{5.331}$$

This equation states that the magnetic flux Φ in the hollow cylinder is multiples of half the flux quantum ϕ_0. As the magnetic field is increased from zero, the flux inside the cylinder is not a linear function of field, but is a series of steps, as in a staircase. The results of Goodman and Deaver [9] are shown in figure 5.4b. The stepwise behavior is evident.

The observation is an example of a macroscopic quantum state. To see the steps, all of the electron states in the cylinder must have a coherent phase. Since the dimensions of the cylinder are millimeters, this phase is coherent over macroscopic dimensions. This coherence is an important property of the superconducting state.

References

1. M. Abramowitz and I.A. Stegun, *Handbook of Mathematical Functions* (many publishers)
2. C.E. Moore, *Atomic Energy Levels* (U.S. Gov. Printing Office, Washington, DC, 1971)
3. D.R. Bates and A. Damgaard, *Philos. Trans. R. Soc.* **A242**, 101 (1949)
4. E.A. Hylleraas, *Z. Phys.* **65**, 209 (1930)
5. C.L. Pekeris, *Phys. Rev.* **126**, 1470 (1962)
6. J.K. Jain, *Composite Fermions* (Cambridge University Press, Cambridge, UK, 2007)
7. B.S. Deaver and W.M. Fairbanks, *Phys. Rev. Lett.* **7**, 43 (1961)
8. R. Doll and M. Näbauer, *Phys. Rev. Lett.* **7**, 51 (1961)
9. W.L. Goodman and B.S. Deaver, *Phys. Rev. Lett.* **24**, 870 (1970)

Homework

1. Verify the addition theorem for $\ell = 2$.

2. Find the exact eigenvalue equation for any value of angular momentum for a particle in a spherical square well:

$$V(r) = \begin{cases} 0 & 0 < r < a \\ \infty & a < r \end{cases} \tag{5.332}$$

   a. What is the specific result for $\ell = 0$?

   b. Look up a table of zeros of spherical Bessel functions (e.g., *Handbook of Mathematical Functions*). Use these data to list, in order, the six lowest eigenvalues considering all possible values of angular momentum.

3. A deuteron is a bound state of a neutron and a proton. Fermi calculated the binding energy assuming the nuclear forces could be approximated by a spherical square well of depth $V(r) = -V_0$ for $r < a$ and $V(r) = 0$ for $r > a$. His values were $V_0 = 36$ MeV and $a = 2.0$ fm. What numerical value did Fermi get for the bound-state energy, in MeV? *Hint*: Use the reduced mass in relative coordinates.

4. Calculate the s-wave ($\ell = 0$) phase shift for a repulsive ($V_0 > 0$) square-well potential:

$$V(r) = \begin{cases} V_0 & 0 < r < a \\ 0 & a < r \end{cases} \tag{5.333}$$

   Assume that $V_0 > E > 0$ and find the limit of $\delta_0(k)$ in the limit that $V_0 \to \infty$.

5. The hard-sphere potential in 3D has $V(r) = \infty$ for $r < a$ and $V = 0$ for $r > a$. This potential forces the radial wave function to vanish at $r = a$, $R(a) = 0$. Derive the formula for the phase shift $\delta_\ell(k)$ for all angular momentum. Use it to derive simple expressions for the phase shift for $\ell = 0$ and $\ell = 1$.

6. In three dimensions, a spherically symmetric potential has the form of a delta-function away from the origin: $V(r) = -\lambda\delta(r-a)$, where λ is a positive constant.

a. Find the eigenvalue equation for an *s*-wave bound state.
b. Show that λ has to have a minimum value for bound states to exist. What is that value?

7. Does an attractive three-dimensional delta-function potential bind a particle in three dimensions? Explain your answer. Use atomic units so that the Hamiltonian is

$$H = -\nabla^2 - \beta\delta^3(\vec{r}) \tag{5.334}$$

where $\beta > 0$ is a constant.

8. Find the exact eigenvalue equation for any value of angular momentum for a particle in a circular square well in two dimensions:

$$V(\rho) = \begin{cases} 0 & 0 < \rho < a \\ \infty & a < \rho \end{cases} \tag{5.335}$$

Look up a table of zeros of Bessel functions (e.g., in the *Handbook of Mathematical Functions*). Use these data to list, in order, the six lowest eigenvalues considering all possible values of angular momentum.

9. Use the series definition of the confluent hypergeometric function $F(a, b, z)$ to show it obeys its differential equation:

$$F(a, b, z) = \sum_{n=0} \frac{z^n}{n!} \frac{(a)_n}{(b)_n}, \qquad (a)_n = a(a+1)\cdots(a+n-1)$$

$$0 = z\frac{d^2F}{dz^2} + (b-z)\frac{dF}{dz} - aF \tag{5.336}$$

10. Use confluent hypergeometric functions to find the exact solution to the Hamiltonian of the hydrogen atom in one dimension. *Hint:* $\phi(x=0) = 0$.

11. Solve the Hamiltonian for the bound states of the hydrogen atom in two dimensions. Find the eigenvalues. *Hint*: Use polar coordinates, and the solution for $R(\rho)$ involves confluent hypergeometric functions.

12. Find the exact eigenvalue and eigenfunction of the *s*-wave ground state of an electron in the potential

$$V(r) = -\frac{e^2}{a+r} \tag{5.337}$$

where a is a positive constant and e is the charge of the electron.

13. Using hydrogen bound-state wave functions, evaluate the following integrals, which are given using Dirac notation:

a. $\langle 1s|2p_z\rangle$ (5.338)

b. $\langle 1s|p_z|2p_z\rangle$ (5.339)

c. $\langle 1s|z|3p_z\rangle$ (5.340)

14. Solve for the exact eigenvalues and eigenstates of the three-dimensional harmonic oscillator using spherical coordinates $V(r) = Kr^2/2$. *Hint*: Set $z = r^2 m\omega/\hbar$ and find the equation for $G(z)$, where

$$\chi(r) = z^{(\ell+1)/2} e^{-z/2} G(z) \tag{5.341}$$

15. For the three-dimensional harmonic oscillator, what is the degeneracy $G(N)$ of each level? That is, how many different states, as a function of N, have the same energy $\hbar\omega\left(N + \frac{3}{2}\right)$?

16. The WKBJ method gave poor results for the hydrogen atom in 1D. In 3D we had to add a potential term

$$\delta V = \frac{\hbar^2}{8mr^2} \tag{5.342}$$

to get a good answer. Try the same thing in 1D. Add the above potential (with x instead of r) to the potential of the 1D hydrogen atom. Show that (a) this gives the proper form of the WKBJ wave function at small x, and (b) it gives the correct eigenvalue using WKBJ.

17. Kratzer's molecular potential has a minimum at $V(r = a) = -D$ and simulates the binding of two atoms:

$$V(r) = D\left[\left(\frac{a}{r}\right)^2 - 2\left(\frac{a}{r}\right)\right] \tag{5.343}$$

Find the exact eigenvalue spectrum for bound states in three dimensions.

18. Use WKBJ to find the eigenvalues of the potential of the previous problem.

19. Use the three-dimensional form of WKBJ to obtain the eigenvalues of the three-dimensional harmonic oscillator. What is the lowest eigenvalue?

20. Construct a table of effective quantum numbers n^* for the ns series of K*I* and Ca*II*. Ca*II* means one electron outside of the closed shell in calcium. The data can be found on the physics library reference shelf in C.E. Moore, *Atomic Energy Levels* (QC453.M58, Vol. 1).

21. Use the variational method to determine the eigenvalue of the ground state of the hydrogen atom in two dimensions. Choose your own trial eigenfunction.

22. Use the variational theory to solve for the ground-state energy of two 1*s* electrons in a Coulomb potential of charge Z. The result for $Z = 2$ should reproduce the helium result in section 5.7.

a. $Z = 1$: Does H^- exist?. What do you predict for its binding energy?
b. $Z = 3$: Compare with the Li^+ ion, whose experimental ionization energies are 75.3 and 121.8 eV for the two electrons.

23. Use the variational method to find the ground-state energy of two particles in one dimension bound to a delta-function potential, and which interact with a delta-function potential:

$$H = -\frac{\hbar^2}{2m}\left[\frac{\partial^2}{\partial x_1^2} + \frac{\partial^2}{\partial x_2^2}\right] - \lambda[\delta(x_1) + \delta(x_2) - \delta(x_1 - x_2)]$$

24. Use the variational method to solve for the ground-state energy of a hydrogen atom in a constant electric field F along the $\hat{z}$-direction. The Hamiltonian is

$$H = -\frac{\hbar^2}{2m}\nabla^2 - \frac{e^2}{r} - eFr\cos(\theta) \tag{5.344}$$

Use a trial wave function of the form

$$\phi(r,\,\theta) = Ae^{-r/a}\left[1 - \lambda\frac{r}{a}\cos(\theta)\right] \tag{5.345}$$

where a is the Bohr radius and λ is the variational parameter.

a. Find the value of λ that minimizes the energy.
b. Express the energy as a function of F.
c. Expand (b) in a power series in F about $F = 0$. If α is the polarizability, the correct answer has the form

$$E(F) = E_{Ry} - \frac{\alpha}{2}F^2 + O(F^4) \tag{5.346}$$

What value for α is predicted by the variational procedure?

25. Consider the Hamiltonian of an electron of charge e bound in a three-dimensional harmonic oscillator potential, which is also subject to a static electric field $\vec{F}$. The potential energy is

$$V(\mathbf{r}) = \frac{K}{2}r^2 - e\vec{F}\cdot\vec{r} \tag{5.347}$$

Find the exact eigenvalue spectrum. What is the polarizability α of a charged particle bound in a harmonic oscillator potential?

26. Find the eigenfunctions and eigenstates of an electron confined to two dimensions that is subject to an electric and magnetic field ($\vec{F} \perp \vec{B}$). The Hamiltonian is

$$H = \frac{1}{2m}\left[p_x^2 + \left(p_y - \frac{eBx}{c}\right)^2\right] - eFx \tag{5.348}$$

27. Consider the Hamiltonian of an electron confined to move only in the (xy)-plane. There is a magnetic field in the z-direction and a quadratic potential in the plane. Show that in the symmetric gauge the Hamiltonian can be written as

$$H = \frac{1}{2m}\left[\left(p_x + \frac{eBy}{2c}\right)^2 + \left(p_y - \frac{eBx}{2c}\right)^2\right] + \frac{K}{2}(x^2 + y^2)$$

Solve this Hamiltonian exactly using polar coordinates. This problem originally solved by V. Fock.

28. Consider the Hamiltonian of an electron confined to move only in the (xy)-plane. There is a magnetic field in the z-direction. Consider the Hamiltonian in the symmetric gauge. Show that the wave function $|n\rangle$

$$H = \frac{1}{2m}\left\{\left[p_x + \frac{eBy}{2c}\right]^2 + \left[p_y - \frac{eBx}{2c}\right]^2\right\} \tag{5.349}$$

$$|n\rangle = A(x - iy)^n \exp[-(x^2 + y^2)/4\ell^2] \tag{5.350}$$

is an exact eigenstate of this Hamiltonian and of the z-component of angular momentum. What is the eigenvalue?

6 Matrix Methods and Perturbation Theory

6.1 H and H_0

When solving any Hamiltonian, the first step is to try to solve it exactly. Finding an exact solution is always a bit of luck. Usually an exact solution is impossible, and then one is faced with a variety of choices. WKBJ, variational, and numerical methods are all available as possible approaches. However, probably the most important approximation is perturbation theory.

The discussion of perturbation theory is divided into two categories. The first is the effect of static perturbations: those that are independent of time. They are treated in this chapter. The second category is perturbations that depend on time, which are discussed in the next two chapters. Matrix methods of solving Hamiltonians are also discussed since they share a common formalism with perturbation theory.

The first step in either the matrix method or perturbation theory is to find another Hamiltonian H_0 that can be solved exactly. The step of choosing H_0 should be done to optimize the following desirable characteristics:

1. Its eigenstates and eigenvalues should resemble those of the exact problem as much as possible.
2. Its eigenfunctions should be simple functions that are easy to generate and integrate.

The first requirement is needed because H_0 is used as the basis for solving the full Hamiltonian H. This process converges faster whenever the solutions to H_0 resemble those of H. For example, if

$$H = \frac{p^2}{2m} + U(\mathbf{r}) \tag{6.1}$$

and $U(\mathbf{r})$ is a potential of short range that has a few bound states, then choose

$$H_0 = \frac{p^2}{2m} + U_0(\mathbf{r}) \tag{6.2}$$

where $U_0(\mathrm{r})$ is also of short range with a few bound states. For example, U_0 could be a spherical square well.

The second criterion, that the eigenfunctions are simple, just makes life easier when it is time to evaluate the integrals for matrix elements. The evaluation of matrix elements is an essential feature of this chapter and we want to make it as painless as possible.

The eigenvalues of H_0 are denoted with a superscript (0), and the eigenstates are ϕ_n. The eigenvalues of H are denoted without the superscript, and its eigenfunctions are ψ_n:

$$H_0\phi_n = E_n^{(0)}\phi_n \tag{6.3}$$

$$H\psi_n = E_n\psi_n \tag{6.4}$$

The goal of the theory is to derive the eigenfunctions ψ_n and eigenvalues E_n of H by starting from the known values ϕ_n and $E_n^{(0)}$ of H_0.

Since H and H_0 are different, define $V(\mathbf{r})$ as their difference:

$$H = H_0 + V \tag{6.5}$$

The notation suggests that V is a type of potential energy. This identification is customary but not essential. In the usual application of perturbation theory, $V = U - U_0$. However, there are rare occasions where the kinetic energy has a term that contributes to V.

6.2 Matrix Methods

The equations for the matrix method are derived in this section. The method is quite general and can be applied to any Hamiltonian: relativisitic, nonrelativisitic, one-particle, many-particle, etc. Every Hamiltonian discussed in this chapter has the feature that it cannot be solved analytically. Otherwise, there is no need to bother with approximate techniques.

The eigenfunctions ϕ_n are a complete set of states, so any function of position can be expanded in these functions. Each exact eigenfunction ψ_n is expanded in the set ϕ_m in terms of coefficients C_{nm} that need to be determined:

$$\psi_n = \sum_m C_{nm}\phi_m \tag{6.6}$$

Operate on the above equation by $H = H_0 + V$. On the left one gets (6.4) while on the right-side of the equal sign use (6.3):

$$H\psi_n = E_n\psi_n = E_n\sum_m C_{nm}\phi_m \tag{6.7}$$

$$H\psi_n = \sum_m C_{nm}[H_0 + V]\phi_m = \sum_m C_{nm}[E_m^{(0)} + V]\phi_m \tag{6.8}$$

Subtract these two equations and find

$$0 = \sum_m C_{nm}[E_m^{(0)} - E_n + V(\mathbf{r})]\phi_{\mathrm{m}} \tag{6.9}$$

Multiply this equation on the left by ϕ_ℓ and then integrate over $d\tau$, which represents all space and spin variables:

$$\sum_m C_{nm}[E_n - E_m^{(0)}] \int d\tau \phi_\ell^* \phi_m = \sum_m C_{nm} \int d\tau \phi_\ell^* V \phi_m \tag{6.10}$$

The states are assumed to be orthogonal and normalized:

$$\int d\tau \phi_\ell^* \phi_m = \delta_{\ell m} \tag{6.11}$$

$$\int d\tau \phi_\ell^* V \phi_m = \langle \ell | V | m \rangle \equiv V_{\ell m} \tag{6.12}$$

Equation (6.10) is simplified to

$$[E_n - E_\ell^{(0)}] C_{n\ell} = \sum_m C_{nm} V_{\ell m} \tag{6.13}$$

Equation (6.13) is the fundamental equation of the matrix method. The known quantities in this expression are the eigenvalues $E_\ell^{(0)}$ and matrix elements $V_{\ell m}$. The latter are just integrals which involve the known functions $V(r)$ and $\phi_n(\mathbf{r})$.

Equation (6.13) can be solved by a variety of techniques. The first is the matrix method. It is formally exact but numerically cumbersome. Equation (6.13) is exactly equivalent to finding the eigenvalues and eigenvectors of a matrix using the method of determinants. The determinantal equation is

$$0 = \det|[E_\ell^{(0)} - \lambda]\delta_{\ell m} + V_{\ell m}| \tag{6.14}$$

It gives a set of eigenvalues λ_n that are just the desired values of E_n. The coefficients C_{nm} are found from the eigenvectors of the matrix for each eigenvalue. The matrix/determinant method is formally exact. The name "matrix element" for $V_{\ell m}$ arises since they are the elements of the Hamiltonian matrix. They obey $V_{\ell m} = V_{m\ell}^*$ since the Hamiltonian is Hermitian. The diagonal elements V_{mm} are real.

The problem with using the matrix method is the matrix has a dimensionality that is infinite. The number of eigenstates ϕ_n is infinite. Usually it includes a summation over all bound and continuum states, and the latter are continuous and infinite in number. Potentials without continuum states, such as the harmonic oscillator, have an infinite number of bound states. So the above determinant has an infinite number of rows and columns.

In practice, an accurate solution can be obtained for the states lowest in energy by using a finite matrix. The matrix is just truncated after N terms, and the determinant is evaluated for $N \times N$. Convergence is tested by increasing the value of N. Modern computers make this calculation easy in many cases.

States are called *degenerate* when they have the same eigenvalue. If a perturbation acts on degenerate states, the matrix method is useful for determining how the degeneracy is altered or lifted by the potential V. Examples are provided for systems where N is small.

6.2.1 2×2

The most important case has $N = 2$, so the determinant is 2×2:

$$0 = \det\begin{vmatrix} E_1 + V_{11} - \lambda & V_{12} \\ V_{21} & E_2 + V_{22} - \lambda \end{vmatrix} \tag{6.15}$$

which has the solution

$$\lambda = \frac{1}{2}\left[E_1 + E_2 + V_{11} + V_{22} \pm \sqrt{(E_1 + V_{11} - E_2 - V_{22})^2 + 4|V_{12}|^2}\right] \tag{6.16}$$

If the perturbation is small $[V_{12} << (E_1 - E_2)]$, the argument of the square root can be expanded to give the approximate result

$$\lambda = \frac{1}{2}\left\{E_1 + E_2 + V_{11} + V_{22} \pm \left[E_1 + V_{11} - E_2 - V_{22} + \frac{2|V_{12}|^2}{E_1 + V_{11} - E_2 - V_{22}}\right]\right\} \tag{6.17}$$

$$\lambda_1 = E_1 + V_{11} + \frac{|V_{12}|^2}{E_1 + V_{11} - E_2 - V_{22}} \tag{6.18}$$

$$\lambda_2 = E_2 + V_{22} - \frac{|V_{12}|^2}{E_1 + V_{11} - E_2 - V_{22}} \tag{6.19}$$

A frequent use of the 2×2 is where two states ϕ_1, ϕ_2 have the same eigenvalue so $E_1 = E_2$, $V_{11} = V_{22}$. Then the eigenvalues in (6.16) are

$$\lambda_\pm = E_1 + V_{11} \pm |V_{12}| \tag{6.20}$$

In this case the change in the eigenvalue is first order in V_{12}, while in (6.17) it is second order $O(|V_{12}|^2)$. The linear dependence on the off-diagonal matrix elements is possible only when the diagonal elements are degenerate. It is shown below, in the section on perturbation theory, that the dependence on $O(|V_{\ell m}|^2)$ is the usual dependence. Only in degenerate perturbation theory is the eigenvalue linearly dependent on the matrix element.

For degenerate states, the above matrix becomes

$$0 = \det\begin{vmatrix} E_1 + V_{11} - \lambda & V_{12} \\ V_{12}^* & E_1 + V_{11} - \lambda \end{vmatrix} \tag{6.21}$$

$$\lambda_\pm = E_1 + V_{11} \pm |V_{12}| \tag{6.22}$$

In this case, $E_2 = E_1$ and $V_{22} = V_{11}$. The two energy levels (λ_+, λ_-) are split by the interaction term $\pm|V_{12}|$.

6.2.2 Coupled Spins

Another example of matrix methods is the coupling of two spins by the Hamiltonian:

$$H = 2A\vec{s}_1 \cdot \vec{s}_2 \tag{6.23}$$

where $\vec{s}_1, \vec{s}_2$ are both spin-$\frac{1}{2}$ operators. The prefactor $2A$ is a constant. There are several ways to diagonalize this Hamiltonian. The matrix method is straightforward, and is done below. There is also a trick method that is worth remembering.

The trick method uses the feature that the total spin is $\vec{S}=\vec{s}_1+\vec{s}_2$. Then the Hamiltonian can be written in simple form:

$$\vec{S}\cdot\vec{S}=(\vec{s}_1+\vec{s}_2)^2=\vec{s}_1\cdot\vec{s}_1+\vec{s}_2\cdot\vec{s}_2+2\vec{s}_1\cdot\vec{s}_2 \tag{6.24}$$

$$H=A[\vec{S}\cdot\vec{S}-\vec{s}_1\cdot\vec{s}_1-\vec{s}_2\cdot\vec{s}_2] \tag{6.25}$$

$$=A\hbar^2[S(S+1)-s_1(s_1+1)-s_2(s_2+1)] \tag{6.26}$$

The last formulas used the results of chapter 4 that $\vec{S}\cdot\vec{S}=\hbar^2 S(S+1)$ for all spins. If s_1 and s_2 are both spin$-\frac{1}{2}$ then the addition of two spin-$\frac{1}{2}$ states gives two possible states, $S=0$ or $S=1$:

$$E_S=A\hbar^2\left[S(S+1)-\frac{3}{2}\right] \tag{6.27}$$

$$E_0=-\frac{3}{2}A\hbar^2 \quad \text{for} \quad S=0 \tag{6.28}$$

$$E_1=\frac{1}{2}A\hbar^2 \quad \text{for} \quad S=1 \tag{6.29}$$

The last two formulas give the two possible eigenstates. $S=0$ is the spin singlet, and there is one of these ($M=0$). $S=1$ is the spin triplet, and there are three of these ($M=-1, 0, 1$). The eigenstates are found using Clebsch-Gordon coefficients, as discussed in chapter 4. The Hamiltonian (6.23) has been solved exactly.

The same Hamiltonian (6.23) can be solved using the matrix method. Sometimes the above trick is not obvious. The matrix method always works and is straightforward. Denote α_j, β_j as spin-up and spin-down for spin $j=1, 2$.

There are four possible states in the coupled system of two spin-$\frac{1}{2}$ particles:

$$\phi_1=\alpha_1\alpha_2, \quad \phi_2=\alpha_1\beta_2 \tag{6.30}$$

$$\phi_3=\beta_1\alpha_2, \quad \phi_4=\beta_1\beta_2 \tag{6.31}$$

The Hamiltonian (6.23) is written in terms of raising and lowering operators:

$$H=A[2s_{1z}s_{2z}+s_1^{(+)}s_2^{(-)}+s_1^{(-)}s_2^{(+)}] \tag{6.32}$$

The operators are

$$s_{jz}=\frac{\hbar}{2}\begin{pmatrix}1 & 0\\ 0 & -1\end{pmatrix}, \quad s^{(+)}=\hbar\begin{pmatrix}0 & 1\\ 0 & 0\end{pmatrix}, \quad s^{(-)}=\hbar\begin{pmatrix}0 & 0\\ 1 & 0\end{pmatrix}$$

$$\alpha=\begin{pmatrix}1\\ 0\end{pmatrix}, \quad \beta=\begin{pmatrix}0\\ 1\end{pmatrix} \tag{6.33}$$

$$s^{(+)}\beta=\hbar\alpha, \quad s^{(-)}\alpha=\hbar\beta, \quad s_z\alpha=\frac{\hbar}{2}\alpha, \quad s_z\beta=-\frac{\hbar}{2}\beta \tag{6.34}$$

The matrix elements are evaluated using the formulas from chapter 4. The 4×4 matrix has 16 elements. Fortunately, most are zero. The diagonal matrix elements are

$$V_{nn} = \langle n|2As_{1z}s_{2z}|n\rangle \tag{6.35}$$

$$V_{11} = \frac{1}{2}A\hbar^2 = V_{44} \tag{6.36}$$

$$V_{22} = -\frac{1}{2}A\hbar^2 = V_{33} \tag{6.37}$$

$$V_{23} = \langle 2|As_1^{(+)}s_2^{(-)}|3\rangle = A\hbar^2 \tag{6.38}$$

The raising and lowering operators give zero contribution to the diagonal elements. The only off-diagonal elements are $V_{23} = V_{32} = A\hbar^2$. The Hamiltonian matrix gives a determinant ($\tilde{A} = A\hbar^2$):

$$0 = \det\begin{vmatrix} \frac{1}{2}\tilde{A} - \lambda & 0 & 0 & 0 \\ 0 & -\frac{1}{2}\tilde{A} - \lambda & \tilde{A} & 0 \\ 0 & \tilde{A} & -\frac{1}{2}\tilde{A} - \lambda & 0 \\ 0 & 0 & 0 & \frac{1}{2}\tilde{A} - \lambda \end{vmatrix} \tag{6.39}$$

The reader should take some time to understand how the above matrix is derived. For example, can you show why only two of the off-diagonal elements are nonzero? The Hamiltonian in (6.32) does not change the value of the magnetic quantum number $M = m_1 + m_2$ when it operates on any state. A combination such as $s_1^{(+)}s_1^{(-)}$ maintains the total value of M, since it lowers one spin and raises another. This operator gives zero when operating on either $\phi = \alpha_1\alpha_2$ or $\phi_4 = \beta_1\beta_2$. Both of these states are the only ones with that value of M, which is $M = 1$ for ϕ_1 and $M = -1$ for ϕ_4. The Hamiltonian operating on either of these states cannot change it to another state, so they have no off-diagonal elements. The nonzero entries are the same as appear in a table of Clebsch-Gordon coefficients.

It is often necessary to evaluate matrix elements in quantum mechanics. Many of them vanish because of similar arguments about the preservation of quantum numbers. It is important to become familiar with such arguments and to use them, to spare the tedium of evaluating matrix elements that are going to vanish for simple reasons. Other quantum numbers that are useful for evaluting matrix elements are parity and angular momentum.

The determinant in (6.39) has Block diagonal form. There are three separate blocks that are diagonalized separately. The first and last block have no matrix elements connecting them to other rows or columns. They are easily solved to get, in both cases,

$$\lambda = \frac{1}{2}\tilde{A} \tag{6.40}$$

These two eigenvalues are two of the three triplet states. The middle block is 2×2 and is

$$0 = \det\begin{vmatrix} -\frac{1}{2}\tilde{A} - \lambda & \tilde{A} \\ \tilde{A} & -\frac{1}{2}\tilde{A} - \lambda \end{vmatrix} \tag{6.41}$$

$$\lambda = -\frac{1}{2}\tilde{A} \pm \tilde{A} = \tilde{A}\left[\frac{1}{2}, -\frac{3}{2}\right] \tag{6.42}$$

There are three eigenvalues with $E_1 = \tilde{A}/2$ and one with $E_0 = -3\tilde{A}/2$. The same result was found by the trick method. The direct diagonalization of the Hamiltonian matrix can always be relied on to give the correct answer.

6.2.3 Tight-Binding Model

The tight-binding model is useful for describing electron eigenstates in solids and molecules. In chemistry it is called the Hückel model, and in solid-state physics it is the tight-binding model.

Here we consider only the simplest version of the model. Assume that the system has a collection of N identical atoms and each has an single electron in an s-state outside of a closed shell. Assume that the separation of adjacent atoms is always at the same distance. Denote the atomic orbital as $\phi(\mathbf{r} - \mathbf{R}_j)$ when the atom is at $\mathbf{R}_j$. Also assume that these orbitals are orthogonal when located on different sites:

$$\int d^3 r \phi(\mathbf{r}-\mathbf{R}_j)\phi(\mathbf{r}-\mathbf{R}_m) = \delta_{jm} \tag{6.43}$$

A method of arranging this orthogonality was invented by Löwdin. A possible wave function, widely used in chemistry, is called LCAO (linear combination of atomic orbitals):

$$\psi(\mathbf{r}) = \sum_{j=1}^{N} a_j \phi(\mathbf{r} - \mathbf{R}_j) \tag{6.44}$$

The coefficients a_j are determined by solving for the eigenfunctions. Schrödinger's equation for stationary states is

$$H\psi = E\psi \tag{6.45}$$

Multiply the above equation from the left by $\int d^3 r \phi^*(\mathbf{r} - \mathbf{R}_m)$ and do the integral on both sides of the equation:

$$Ea_m = \sum_j a_j \langle m|H|j\rangle \tag{6.46}$$

The tight-binding model assumes that the matrix element has a simple structure:

1. If $m = j$, then $\langle m|H|m\rangle = E_0$.
2. If m and j are nearest neighbors, then $\langle m|H|j\rangle = V$.
3. If m and j are distant neighbors, the matrix element is zero.

The Hamiltonian matrix has the dimensions of $N \times N$. It has E_0 for all diagonal elements. The off-diagonal elements are V for nearest neighbors, and zero otherwise.

As an example, consider a linear chain of N equally spaced atoms. Figure 6.1a shows an example, where the atoms are at the corners, and the lines indicate bonds. Equation (6.46) has the structure ($j = 1, 2, \ldots, N$)

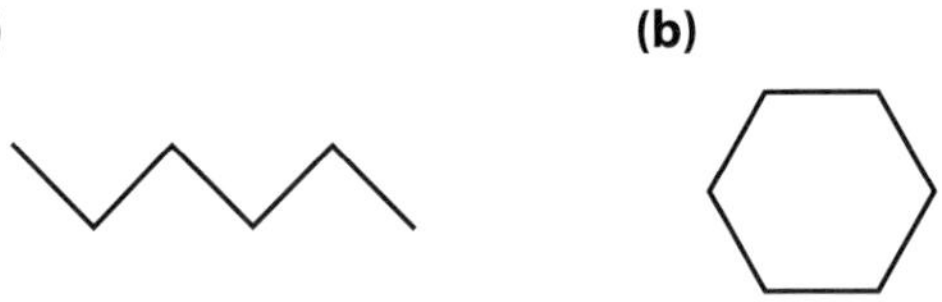

FIGURE 6.1. (a) A linear chain of atoms. (b) The carbon atoms in benzene.

$$Ea_1 = E_0 a_1 + Va_2 \tag{6.47}$$

$$Ea_j = E_0 a_j + V(a_{j+1} + a_{j-1}), j \neq 1, N \tag{6.48}$$

$$Ea_N = E_0 a_N + Va_{N-1} \tag{6.49}$$

The atoms at the two ends ($j=1$, N) have only one neighbor, while the others have two. The Hamiltonian matrix is tridiagonal. It has E_0 for all diagonal elements, and V above and below the diagonal

$$H = \begin{pmatrix} E_0 & V & 0 & \cdots & 0 \\ V & E_0 & V & \cdots & 0 \\ 0 & V & E_0 & V & \cdots \\ \vdots & \vdots & \vdots & \ddots & \vdots \\ 0 & 0 & \cdots & V & E_0 \end{pmatrix} \tag{6.50}$$

Since the Hamiltonian is a matrix, the eigenfunctions are a column vector:

$$\psi = \begin{bmatrix} a_1 \\ a_2 \\ \vdots \\ a_N \end{bmatrix} \tag{6.51}$$

The components of the eigenvector are the coefficients in (6.44). Try a solution of the form

$$a_j = \sin(j\theta + \delta) \tag{6.52}$$

For all except the end atoms, the eigenvalue equation (6.48) gives

$$\begin{aligned}[E - E_0] \sin(j\theta + \delta) &= V\{\sin[(j+1)\theta + \delta] + \sin[(j-1)\theta + \delta]\} \\ &= 2V\cos(\theta)\sin(j\theta + \delta)\end{aligned} \tag{6.53}$$

$$E = E_0 + 2V\cos(\theta) \tag{6.54}$$

The constant δ and the constraints on values of θ are determined by the two equations at the ends. For $j=1$, we have from (6.47)

$$0 = (E_0 - E)\sin(\theta + \delta) + V\sin(2\theta + \delta) \tag{6.55}$$

$$0 = V[-2\cos(\theta)\sin(\theta + \delta) + \sin(2\theta + \delta)] \tag{6.56}$$

The last equation was derived using $E - E_0 = 2V\cos(\theta)$. The first term is manipulated using

$$2\cos(a)\sin(b) = \sin(b+a) + \sin(b-a) \tag{6.57}$$

$$0 = V[-\sin(2\theta+\delta) - \sin(\delta) + \sin(2\theta+\delta)] \tag{6.58}$$

The last equation has the solution that $\delta = 0$. In this case there is no need for the constant phase δ in eqn. (6.52). It is nonzero for other cases, so one should always include it until proven to be zero.

The permissible values of θ are determined by the last eqn. (6.49):

$$0 = (E-E_0)\sin(N\theta) + V\sin[(N-1)\theta] \tag{6.59}$$

$$= V\{-2\cos(\theta)\sin(N\theta) + \sin[(N-1)\theta]\} \tag{6.60}$$

$$= V\{-\sin[(N+1)\theta] - \sin[(N-1)\theta] + \sin[(N-1)\theta]\} \tag{6.61}$$

$$0 = \sin[(N+1)\theta] \tag{6.62}$$

The last equation determines the eigenvalues. Permissibile values of θ are given by $(N+1)\theta = \alpha\pi$, where α is a positive integer:

$$\theta_\alpha = \frac{\pi\alpha}{N+1} \tag{6.63}$$

In practice, the allowed values of $\alpha = 1, 2, \ldots, N-1, N$. Setting $\alpha = (N+1)$ makes $a_j = 0$. For $\alpha > N+1$, the eigenfunctions are the same as the ones for $1 \le \alpha \le N$. The eigenvalues for an electron hopping along the chain are

$$E_\alpha = E_0 + 2V\cos\left[\frac{\pi\alpha}{N+1}\right] \tag{6.64}$$

This completes the discussion of electron motion on a linear chain.

Another geometry has the N atoms arranged in a circle. An example from chemistry is the p_z orbitals of carbon in the benzene molecule ($N = 6$).

The Hamiltonian matrix is now

$$H = \begin{pmatrix} E_0 & V & 0 & \cdots & V \\ V & E_0 & V & \cdots & 0 \\ 0 & V & E_0 & V & \cdots \\ \vdots & \vdots & \vdots & \ddots & \vdots \\ V & 0 & \cdots & V & E_0 \end{pmatrix} \tag{6.65}$$

The matrix has tridiagonal form, except the far corners have elements V to connect the states $j=1$ with $j=N$. They are neighbors in the ring.

This problem is identical to periodic boundary conditions for plane waves. Now take

$$a_j = e^{ij\theta} \tag{6.66}$$

The constraint that $a_{j+N} = a_j$ requires that

$$\theta = \frac{2\pi\alpha}{N}, \quad \alpha = 0, \pm 1, \pm 2, \cdots \tag{6.67}$$

$$E_\alpha = E_0 + 2V\cos\left[\frac{2\pi\alpha}{N}\right] \tag{6.68}$$

One needs n values of α. If N is an even integer, the values of α are $0, \pm 1, \pm 2, \ldots, \pm N/2$, where $\pm N/2$ give the same solution. If N is an odd integer, the values of α are $0, \pm 1, \ldots, \pm(N-1)/2$. Negative values of α have the wave traveling counterclockwise around the ring. Note that this solution is quite different than the linear chain. Further examples are given in the homework assignments.

6.3 The Stark Effect

The Stark effect is the linear splitting of some energy levels in a constant electric field F in atomic hydrogen. The phrase "linear splitting" means that the change in energy levels occurs to the first power of the electric field: $\Delta E \propto F$. This behavior is quite unusual and seems to occur only in hydrogen. The reason is that the hydrogen atom has many different states with the same energy but with different angular momentum. For example, for $n = 2$ there are four degenerate orbital states: one s and three p. For $n = 3$ there is one s, three p, and five d for a total of nine orbital states. As explained above, if the perturbation is

$$U = -e\vec{F}\cdot\vec{r} \tag{6.69}$$

a change in energy proportional to $|U| \propto F$ happens only if there are several degenerate states. Atomic hydrogen is the only atom with this high degree of degeneracy.

In atoms besides hydrogen, states with different angular momentum have different energies. The degeneracy does not occur and changes in energy go as $|U|^2 \propto F^2$. Only atomic hydrogen has the Stark effect.

The hydrogen atom is solved relativistically in chapter 11. The energy levels are changed slightly by relativistic corrections. States with the same principle quantum number no longer all have the same energy. These small corrections are neglected in the present discussion. Here we use the nonrelativistic theory, and states have energy $E_n = -E_{\text{Ry}}/n^2$.

The interaction is given in (6.69), where the constant field is $\vec{F}$. Since the ground state of the hydrogen atom is isotropic, we can choose any direction as $\hat{z}$. Choose it in the direction of $\vec{F}$ so the interaction becomes

$$U = -eFz = -eFr\cos(\theta) \tag{6.70}$$

In this problem, H_0 is the hydrogen atom, whose solutions were given in the previous chapter.

The $n = 1$ state of atomic hydrogen does not have a Stark effect. There is only one orbital state and no orbital degeneracy. In a later section we show that the lowest energy correction for the 1s state goes as $O(F^2)$ for small values of F.

The Stark effect is found in hydrogen for all values of $n \geq 2$. The simplest case is $n = 2$, which is solved here. The case $n = 3$ is given as a problem assignment.

Atomic hydrogen has four orbital states with $n = 2$. They are labeled as $2s$, $2p_z$, $2p_{+1}$, $2p_{-1}$. The $2p$ states have $\ell = 1$: $m = 0$ for $2p_z$ and $m = \pm 1$ for $2p_{\pm 1}$. These four states are used as the basis for the diagonalization of a 4×4 matrix. When $F=0$, these four states have the same energy. The electric field causes nonzero matrix elements between two of these states, and these two have a change in energy proportional to F. The other two states are changed only to $O(F^2)$. The theory reduces to diagonalization of a 2×2 matrix, as was done in (6.15).

The 4×4 matrix has sixteen possible matrix elements to evaluate for $U(\mathbf{r})$ in (6.70). All diagonal elements are zero, because of parity arguments. Only one pair of off-diagonal elements are nonzero:

$$M=\langle 2s|U|2p_z\rangle=-eF\int d^3r\phi_{2s}(\mathbf{r})r\cos(\theta)\phi_{2pz}(\mathbf{r}) \tag{6.71}$$

Recall the eigenfunctions from the prior chapter, where a_0 is the Bohr radius:

$$\phi_{2s}=\frac{1}{\sqrt{8\pi a_0^3}}[1-r/2a_0]e^{-r/2a_0} \tag{6.72}$$

$$\phi_{2pz}=\frac{r\cos(\theta)}{\sqrt{32\pi a_0^5}}e^{-r/2a_0} \tag{6.73}$$

The integral in (6.71) has two parts. There is an angular integral with the integrand of $\cos^2(\theta)$, which gives $4\pi/3$. There is a radial integral

$$\int_0^\infty r^4dre^{-r/a_0}\left(1-\frac{r}{2a_0}\right)=a_0^5\left(4!-\frac{5!}{2}\right)=-36a_0^5 \tag{6.74}$$

Combining both integrals gives

$$M=eF\left(\frac{4\pi}{3}\right)\frac{36a_0^5}{16\pi a_0^4}=3ea_0F \tag{6.75}$$

This matrix element is the only one that is nonzero. Why the others vanish is discussed below.

The matrix to diagonalize is

$$0=\begin{vmatrix} E_2-\lambda & M & 0 & 0 \\ M & E_2-\lambda & 0 & 0 \\ 0 & 0 & E_2-\lambda & 0 \\ 0 & 0 & 0 & E_2-\lambda \end{vmatrix} \tag{6.76}$$

The order of states is $2s$, $2p_z$, $2p_{+1}$, $2p_{-1}$. The matrix has block-diagonal form with a 2×2 part in the upper left corner and two blocks with single elements. The four eigenvalues are

$$\lambda=E_2,\ E_2,\ E_2+M,\ E_2-M \tag{6.77}$$

The four eigenstates are

1. One $\lambda=E_2$ has ϕ_{2p+}.
2. One $\lambda=E_2$ has ϕ_{2p-}.
3. $\lambda=E_2+M$ has eigenfunction

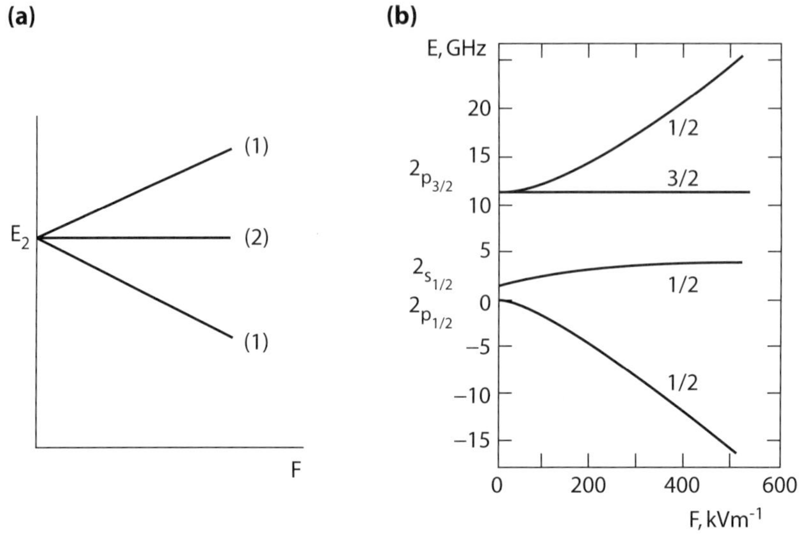

FIGURE 6.2. (a) Energy levels in nonrelativistic theory. (b) Levels including relativistic corrections.

$$\phi_+ = \frac{1}{\sqrt{2}}[\phi_{2s} + \phi_{2pz}] \tag{6.78}$$

4. $\lambda = E_2 - M$ has eigenfunction

$$\phi_- = \frac{1}{\sqrt{2}}[\phi_{2s} - \phi_{2pz}] \tag{6.79}$$

Figure 6.2a shows a graph of the energy levels vs. field F. The numbers in parentheses are the degeneracy of the level. For a nonzero value of F, the fourfold degeneracy of the $n=2$ state is changed to three different energy levels. Two are independent of the field. Two have a linear dependence on the value of field F. Such a linear splitting is possible only when matrix elements exist between degenerate states.

A relativistic solution of the hydrogen atom using the Dirac equation is presented in chapter 11. The relativistic solution shows the four states of the $n=2$ level are not degenerate. These energy levels in an electric field are shown in figure 6.2b. Now there are four levels at nonzero electric field. The p-orbitals ($\ell=1$), combined with the electron spin, produce angular momentum states of $j=\frac{3}{2}$ and $j=\frac{1}{2}$. The $2p_{3/2}$ states are higher in energy than $2p_{1/2}$ and $2s_{1/2}$. This difference is caused by the spin–orbit interaction, which is treated later in this chapter. The Dirac equation predicts that $2s_{1/2}$ and $2p_{1/2}$ are degenerate. However, experimentally, they also have a small splitting called the *Lamb shift*. These energy shifts are small. For $\nu = 10$ GHz, then $h\nu = 4\times10^{-5}$ eV.

Now consider the matrix shown in (6.76). Why are most of the elements zero? Several quantum numbers are used to show that most elements vanish. The first is parity.

The concept of parity is based on whether an eigenfunction or wave function changes sign or does not change sign when the coordinate frame is inverted. Eigenfunctions have a definite parity only if the potential function that generates the eigenfunctions also has it:

$V(-\mathbf{r}) = V(\mathbf{r})$. If the potential lacks parity, so do the eigenfunctions—they are of mixed parity. This potential function is different than the one $U(\mathbf{r})$ below in the matrix elements. For example, in the hydrogen atom $V = -e^2/|\mathbf{r}|$ has even parity, while the perturbation $U(\mathbf{r}) = -e\vec{F} \cdot \vec{r}$ has odd parity.

1. Functions $f_e(\mathbf{r})$ with even parity obey $f_e(\mathbf{r}) = f_e(-\mathbf{r})$. Atomic orbitals with even values of angular momentum have even parity. *s*- and *d*-states have even parity.
2. Functions $f_o(\mathbf{r})$ with odd parity obey $f_o(-\mathbf{r}) = -f_o(\mathbf{r})$. Atomic orbitals with odd values of angular momentum have odd parity. *p*- and *f*- states have odd parity.

The concept of parity is useful for showing that matrix elements sometimes vanish. The argument is that if $M = -M$, then $M = 0$:

$$M_{ij} = \int d^3 r \phi_i(\mathbf{r})^* U(\mathbf{r}) \phi_j(\mathbf{r}) \tag{6.80}$$

Now, make a variable change $\mathbf{r}' = -\mathbf{r}$. Then $d^3r' = d^3r$. Depending on the parity of the different functions, one has $\phi_i(-\mathbf{r})^* = \pm\phi_i(\mathbf{r})^*$, $\phi_j(-\mathbf{r}) = \pm\phi_j(\mathbf{r})$, $U(-\mathbf{r}) = \pm U(\mathbf{r})$. The variable change gives the following:

1. If $M_{ij} = +M_{ij}$, the matrix element may be nonzero.
2. If $M_{ij} = -M_{ij}$, the matrix element is zero.

In the present case, $U(-\mathbf{r}) = -U(\mathbf{r})$ and the potential function has odd parity. If both ϕ_i and ϕ_j have the same parity—both are even or both are odd—the matrix element vanishes. For the Stark effect, matrix elements of $U(\mathbf{r})$ between different *p*-states all vanish. Also, all diagonal matrix elements vanish. This feature eliminates many of the matrix elements in (6.76).

The magnetic quantum number *m* is also important in evaluating matrix elements. In the Stark effect, the interaction $U(\mathbf{r})$ has $m = 0$, as do the states $2s$ and $2p_z$. The value of *m* is one for $2p_{+1}$, and is minus one for $2p_{-1}$. The integral over azimuthal angle gives zero unless the summation of all *m* components is zero:

$$\int_0^{2\pi} d\phi e^{im'\phi} = 2\pi\delta_{m'=0} \tag{6.81}$$

Since *U* has no dependence on *m*, the matrix element is zero unless $m' = m_i - m_j = 0$. This requirement makes the matrix element be zero between $2s$ and $2p_{+1}$, and $2s$ and $2p_{-1}$. Combining parity on one hand and *m* on the other, we have shown that all matrix elements in (6.76) are zero except *M* between $2s$ and $2p_z$. They are shown to vanish without doing any cumbersome integrals.

The Stark effect for $n = 3$ is assigned as a homework problem. There are nine degenerate orbital states, so the Hamiltonian matrix is 9×9. Using parity and $m_i = m_j$ immediately sets to zero most of the matrix elements. There are nonzero matrix elements among the three states with $m = 0$ ($3s$, $3p_z$, $3d_z$), between two with $m = 1$ ($3p_{+1}$, $3d_{+1}$), and between two with $m = -1$ ($3p_{-1}$, $3d_{-1}$). The states $3d_{2+}$ and $3d_{2-}$ have no nonzero matrix elements with other states. The 9×9 matrix has been reduced to one 3×3 block, two 2×2 blocks, and two 1×1 blocks. Considering parity eliminates some matrix elements within these blocks.

6.4 Perturbation Theory

6.4.1 General Formulas

Perturbation theory is another method of solving eqn. (6.13), which is rewritten here:

$$(E_n - E_\ell^{(0)})C_{n\ell} = \sum_m C_{nm} V_{\ell m} \tag{6.82}$$

The method provides results for the eigenvalues E_n and also for the coefficients C_{nm} of the eigenfunction. These quantities are not obtained exactly in perturbation theory. Instead, the idea is to identify a smallness parameter λ, which is taken as dimensionless. All unknown quantities are expanded as a power series in λ^n. The method is most successful when only one or two terms in the series are sufficient for an accurate answer. Rapid convergence is achieved by making λ as small as possible.

A simple estimate of the smallness parameter is the ratio of the matrix elements to energy differences:

$$\lambda \propto \frac{V_{nm}}{E_n^{(0)} - E_m^{(0)}} \tag{6.83}$$

The crudeness of this definition is apparent when it is realized that there are many different states and many different combinations of matrix elements and energy denominators. Perhaps λ is some weighted average of these ratios.

Perturbation theory is derived by expanding all unknown quantities in a power series in the dimensionless parameter ξ:

$$E_n = \sum_{\ell=0}^{\infty} \xi^\ell E_n^{(\ell)} \tag{6.84}$$

$$C_{nm} = \sum_{\ell=0}^{\infty} \xi^\ell C_{nm}^{(\ell)} \tag{6.85}$$

Each successive term is regarded as being smaller than the prior ones. Eventually the parameter ξ is set equal to unity, so the terms $(E_n^{(\ell)}, C_{nm}^{(\ell)})$ are the quantities that are becoming smaller. The nth order term is roughly proportional to λ^n, where λ is the smallness parameter in eqn. (6.83). The parameter ξ is *not* the smallness parameter λ. The parameter ξ is used to keep track of the degree or power of smallness. Equation (6.82) is recast as a power series in ξ.

The potential term on the right in (6.82) is multiplied by a single power of ξ. The matrix elements $V_{\ell m}$ are supposed to be small, and ξ is included to remind us of that smallness. No harm is done by inserting ξ, since eventually we set $\xi = 1$. Equation (6.82) now appears as

$$\left[\sum_j \xi^j E_n^{(j)} - E_n^{(0)}\right] \sum_i \xi^i C_{n\ell}^{(i)} = \xi \sum_m V_{\ell m} \sum_k \xi^k C_{nm}^{(k)} \tag{6.86}$$

There is some confusion in the notation. The eigenvalue of H_0 is called $E_n^{(0)}$, but the same term appears as the first term in the series in (6.84). This confusion will soon disappear, since we will soon show that the two quantities are identical.

The terms in (6.86) with like powers of ξ are collected:

$$0 = \sum_p \xi^p D_p \tag{6.87}$$

$$D_0 = \left(E_n^{(0)} - E_\ell^{(0)}\right) C_{n\ell}^{(0)} \tag{6.88}$$

$$D_1 = \left(E_n^{(0)} - E_\ell^{(0)}\right) C_{n\ell}^{(1)} + E_n^{(1)} C_{n\ell}^{(0)} - \sum_m V_{\ell m} C_{nm}^{(0)} \tag{6.89}$$

$$D_2 = \left(E_n^{(0)} - E_\ell^{(0)}\right) C_{n\ell}^{(2)} + E_n^{(1)} C_{n\ell}^{(1)} + E_n^{(2)} C_{n\ell}^{(0)} - \sum_m V_{\ell m} C_{nm}^{(1)} \tag{6.90}$$

Each coefficient of ξ^p is separately set to zero ($D_p = 0$). This procedure generates a set of equations that can be solved to obtain the eigenvalues and eigenfunctions. There is no deep reason or deep mystery to setting each coefficient to zero. It is just a method of defining the terms in the perturbation series. Each order of perturbation theory provides successive approximations to the eigenvalues and eigenfunctions.

The first equation is

$$0 = D_0 = (E_n^{(0)} - E_\ell^{(0)}) C_{n\ell}^{(0)} \tag{6.91}$$

There are two possible circumstances:

1. $n = \ell$. Then the expression is zero since the two energy terms are identical and cancel. This is the point where the eigenvalues of H_0 are set equal to the first term in the expansion in eqn. (6.84).
2. $n \neq \ell$. Equation (6.91) can be satisfied only when

$$C_{\ell n}^{(0)} = 0, \quad \text{if} \quad E_n^{(0)} \neq E_\ell^{(0)} \tag{6.92}$$

States with different n can have the same eigenvalue $E_n^{(0)}$. An example is the hydrogen atom where different orbital states with the same principle quantum number are degenerate. The coefficient $C_{\ell n}^{(0)}$ is zero between states with different energies.

Some care must be taken when treating degenerate states. The smallness parameter in (6.83) diverges in this case. Degenerate states are best handled using the matrix method of the prior section. The first step in perturbation theory should be to treat degenerate states. Use the matrix method to find the new eigenvalues and eigenfunctions caused by the perturbation. Then use these states in treating the other states using perturbation theory. For example, if two states are degenerate, call them ϕ_1 and ϕ_2. The Hamiltonian matrix was given earlier:

$$H = \begin{pmatrix} E_1 + V_{11} & V_{12} \\ V_{12}^* & E_1 + V_{11} \end{pmatrix} \tag{6.93}$$

The eigenvalues and eigenfunctions are

$$E_{\pm} = E_1 + V_{11} \pm |V_{12}| \tag{6.94}$$

$$\phi_{\pm} = \frac{1}{\sqrt{2}}[\phi_1 \pm e^{\mp i\theta}\phi_2] \tag{6.95}$$

$$V_{12} = e^{i\theta}|V_{12}| \tag{6.96}$$

Then use the new functions $\phi_{\pm}(\mathbf{r})$ with eigenvalues $E_{\pm}$ for doing perturbation theory on the other states.

Another set of equations besides (6.87)–(6.90) are needed for perturbation theory. The additional equations come from the normalization of the eigenfunctions. Using the initial equation (6.6) gives

$$1 = \int d^3r|\psi_n(\mathbf{r})|^2 = \sum_m C^*_{nm}C_{n\ell}\int d^3r\phi^*_m\phi_\ell \tag{6.97}$$

$$1 = \sum_m |C_{nm}|^2 \tag{6.98}$$

The coefficients C_{nm} are expanded using (6.85), and again the terms are collected in like powers of ξ:

$$1 = \sum_m |C^{(0)}_{nm} + \xi C^{(1)}_{nm} + \xi^2 C^{(2)}_{nm} + \cdots|^2 \tag{6.99}$$

$$\begin{aligned} &= \sum_m \{\xi^0|C^{(0)}_{nm}|^2 + 2\xi\Re[C^{(0)*}_{nm}C^{(1)}_{nm}] \\ &\quad + \xi^2\Re[2C^{(0)*}_{nm}C^{(2)}_{nm} + |C^{(1)}_{nm}|^2] + O(\xi^3)\} \end{aligned} \tag{6.100}$$

where $\Re$ denotes the real part of the complex function. Again each power of ξ^n is assumed to obey this equation separately:

$$1 = \sum_m |C^{(0)}_{nm}|^2 \tag{6.101}$$

$$0 = \sum_m \Re[C^{(0)*}_{nm}C^{(1)}_{nm}] \tag{6.102}$$

$$0 = \sum_m \Re[2C^{(0)*}_{nm}C^{(2)}_{nm} + |C^{(1)}_{nm}|^2] \tag{6.103}$$

Equation (6.101) has a "1" on the left of the equal sign, which comes from the "1" in (6.100). It goes with the first equation since $\xi^0 = 1$. Earlier we showed that $C^{(0)}_{n\ell}$ is zero unless $n = \ell$. Equation (6.101) shows that

$$C^{(0)}_{n\ell} = \delta_{n\ell} \tag{6.104}$$

This result completes zeroth-order perturbation theory. The first term in the expansion of E_n is the eigenvalue of H_0, which is $E^{(0)}_n$. The first term in the expansion of the eigenfunction is $\psi_n = \phi_n$. So zeroth-order perturbation theory says that the eigenvalues and eigenfunctions of H are those of H_0.

First-order perturbation theory is based on two equations (6.89) and (6.102) derived above:

$$0 = D_1 = (E_n^{(0)} - E_\ell^{(0)})C_{n\ell}^{(1)} + E_n^{(1)}C_{n\ell}^{(0)} - \sum_m V_{\ell m}C_{nm}^{(0)} \tag{6.105}$$

$$0 = \sum_m \Re[C_{nm}^{(0)*}C_{nm}^{(1)}] \tag{6.106}$$

Two quantities need to be determined: $E_n^{(1)}$ and $C_{nm}^{(1)}$. The eigenvalue is found by setting $n = \ell$ and remembering that $C_{nm}^{(0)} = \delta_{nm}$ to find

$$E_n^{(1)} = \sum_m V_{nm}\delta_{nm} = V_{nn} \tag{6.107}$$

The first-order correction to the eigenvalue is the diagonal matrix element V_{nn}. This result is obvious by inspecting the Hamiltonian matrix in (6.14). The diagonal elements are obvious corrections to the eigenvalue.

The first-order corrections to the eigenfunctions are found from the above equations. From eqn. (6.102) and $C_{nm}^{(0)} = \delta_{nm}$ we find that

$$0 = \Re\{C_{nn}^{(1)}\} \tag{6.108}$$

Since the coefficients are real, this asserts that the diagonal contribution $C_{nn}^{(1)} = 0$. The off-diagonal term is found from (6.105), which is solved for $n \neq \ell$:

$$0 = (E_n^{(0)} - E_\ell^{(0)})C_{n\ell}^{(1)} - V_{\ell n} \tag{6.109}$$

$$C_{n\ell}^{(1)} = \frac{V_{\ell n}}{E_n^{(0)} - E_\ell^{(0)}} \tag{6.110}$$

The above expression is valid only when $E_n^{(0)} - E_\ell^{(0)} \neq 0$. For degenerate states, one should not use this formula, but should just diagonalize the Hamiltonian matrix that contains all of the degenerate states.

Equation (6.110) is recognized as the expansion parameter λ defined earlier in (6.83). Rapid convergence of perturbation theory is obtained when this parameter is small. Comparing zeroth- and first-order perturbation theory gives for the eigenfunctions

$$\psi_n^{(0)} = \phi_n \tag{6.111}$$

$$\psi_n^{(1)} = \phi_n + \sum_{m \neq n} \frac{\phi_m V_{mn}}{E_n^{(0)} - E_m^{(0)}} \tag{6.112}$$

Perturbation theory is useful if successive terms become smaller. They need not be smaller at each **r**-point, but at a majority of such points. A more quantitative statement of convergence is obtained by examining the eigenvalue contributions. The ratio of successive contributions $E_n^{(m)}/E_n^{(m-1)}$ should be smaller than λ/m. There is no simple prescription that tells whether perturbation theory is converging. One does the calculation to several orders and examines whether the eigenvalues and eigenfunctions are converging.

Second-order perturbation theory is derived from the two equations (6.90) and (6.103):

$$0 = D_2 = (E_n^{(0)} - E_\ell^{(0)})C_{n\ell}^{(2)} + E_n^{(1)}C_{n\ell}^{(1)} + E_n^{(2)}C_{n\ell}^{(0)} - \sum_m V_{\ell m}C_{nm}^{(1)}$$

$$0 = \sum_m \Re[2C_{nm}^{(0)*}C_{nm}^{(2)} + |C_{nm}^{(1)}|^2]$$

The second-order energy is found from the top equation for $n=\ell$ and remembering the results for $C_{nm}^{(0)}$ and $C_{nm}^{(1)}$:

$$E_n^{(2)} = \sum_m V_{nm} C_{nm}^{(1)} = \sum_{m \neq n} \frac{V_{nm} V_{mn}}{E_n^{(0)} - E_m^{(0)}} \tag{6.113}$$

$$E_n = E_n^{(0)} + V_{nn} + \sum_{m \neq n} \frac{|V_{nm}|^2}{E_n^{(0)} - E_m^{(0)}} + \cdots \tag{6.114}$$

The last equation summarizes the first three orders of perturbation theory. More terms can be derived as needed. Second-order perturbation theory is sufficient for many problems in physics. The derivation of the second-order eigenfunction coefficient $C_{nm}^{(2)}$ is assigned as a homework problem. This completes the formal discussion of perturbation theory.

6.4.2 Harmonic Oscillator in Electric Field

An important problem in quantum mechanics is the Hamiltonian for a charged particle that has two potential terms—a harmonic oscillator and a fixed electric field:

$$H = \frac{p^2}{2m} + \frac{K}{2}x^2 + Fx \tag{6.115}$$

where $F \equiv -eE$ if E is the electric field. The classical problem is a charged particle of mass m connected to a harmonic spring with constant K. The electric field causes a force $F = -eE$. The classical equation of motion is

$$m\ddot{x} = -Kx - F = -K(x + F/K) \tag{6.116}$$

The harmonic oscillation is centered at $x_F = -F/K$

$$x(t) = A\sin(\omega t + \delta) - F/K \tag{6.117}$$

$$\omega^2 = \frac{K}{m} \tag{6.118}$$

The quantum mechanical problem may also be solved exactly and the physics is identical to the classical solution. In (6.115) the x-terms are collected together by completing the square:

$$H = \frac{p^2}{2m} + \frac{K}{2}\left(x + \frac{F}{K}\right)^2 - \frac{F^2}{2K} \tag{6.119}$$

The center of the quantum oscillation is at $x_F = -F/K$. The new variable $x' = x + F/K$ has the commutation relation $[x', p] = i\hbar$. The eigenfunctions and eigenvalues are

$$\psi_n(x) = \phi_n(x + F/K) \tag{6.120}$$

$$E_n = \hbar\omega(n + 1/2) - \frac{F^2}{2K}, \quad \omega^2 = \frac{K}{m} \tag{6.121}$$

where $\phi_n(x)$ are the usual harmonic oscillator eigenfunctions. The exact solution is easy. The classical and quantum solutions describe similar behavior.

Now solve the same problem as an example of perturbation theory. The perturbation is $V = Fx$. The unperturbed Hamiltonian is

$$H_0 = \frac{p^2}{2m} + \frac{K}{2}x^2 \tag{6.122}$$

$$E_n^{(0)} = \hbar\omega\left(n + \frac{1}{2}\right), \quad \phi_n^{(0)} = \phi_n(x) \tag{6.123}$$

The first step in perturbation theory is to evaluate the matrix elements. Recall from section 2.3 that the matrix elements of x are nonzero only between adjacent energy levels:

$$V_{nm} = F\langle n|x|m\rangle = \begin{cases} FX_0\sqrt{n} & m = n-1 \\ FX_0\sqrt{n+1} & m = n+1 \\ 0 & \text{otherwise} \end{cases} \tag{6.124}$$

$$X_0 = \sqrt{\frac{\hbar}{2m\omega}} \tag{6.125}$$

This feature makes perturbation theory easy, since there are only one or two terms in the expansion:

- Zeroth-order perturbation theory has $E_n^{(0)} = \hbar\omega(n + \frac{1}{2})$.
- First-order perturbation theory has $E_n^{(1)} = V_{nn} = 0$.
- Second-order perturbation theory has only two nonzero terms in the summation:

$$\begin{aligned} E_n^{(2)} &= \sum_{m \neq n} \frac{|V_{nm}|^2}{E_n^{(0)} - E_m^{(0)}} = \frac{F^2\hbar}{2m\omega}\left[\frac{n+1}{\hbar\omega(n-n-1)} + \frac{n}{\hbar\omega(n-n+1)}\right] \\ &= -\frac{F^2}{2m\omega^2} = -\frac{F^2}{2K} \end{aligned} \tag{6.126}$$

Collecting these results gives the exact eigenvalue. Second-order perturbation theory has produced the exact result. It can be shown that all higher-order terms give zero contribution to the energy. One should not think this happens often. This simple problem is one of the very few cases that second-order perturbation theory gives the exact eigenvalue.

The exact eigenfunction is *not* produced by the result of low-order perturbation theory. The terms to first order are

$$\psi_n(x) = \phi_n(x) + \frac{FX_0}{\hbar\omega}[\sqrt{n}\phi_{n-1}(x) - \sqrt{n+1}\phi_{n+1}(x)] + O(F^2)$$

The nth order of perturbation theory gives a contribution to the eigenfunction of $O(F^n)$.

Examine the exact eigenfunction in (6.120) for $n = 0$:

$$\psi_0(x) = \phi_0\left(x + \frac{F}{K}\right) = N_0 \exp\left[-\frac{1}{4X_0^2}\left(x + \frac{F}{K}\right)^2\right] \tag{6.127}$$

Expanding this expression in a power series in F will produce a series with all powers of F^n. It has an infinite number of terms. All orders of perturbation theory are needed to give the exact eigenfunction. It is curious that the eigenvalue is given exactly in second order, but not the eigenfunction.

6.4.3 Continuum States

So far the discussion of matrix methods and of perturbation theory has assumed that all of the eigenstates are discrete. For continuum states, the eigenstates are continuous functions of energy. These cases require a slight change in the procedures that is described in this subsection.

Summation over states shown in (6.112) for the eigenfunction and (6.114) for the eigenvalue are taken over all states: both discrete and continuous. As an example, a particle in a short-range attractive potential has both bound and continuum states. When summing over states, the continuum states provide a significant portion of the answer, so their contribution cannot be neglected.

Our treatment of continuum states uses box normalization. Eventually the volume of the box Ω is allowed to go to infinity. The plane wave states then become continuous. This method is discussed in chapter 2 as part of the normalization of eigenfunctions.

As the volume Ω goes to infinity, eigenstates can be classified as either bound or continuum. The continuum states have a normalization prefactor of $1/\sqrt{\Omega}$. Such factors are denoted explicitly in the discussion of perturbation theory:

$$\lim_{\Omega\to\infty} \phi_n(\mathbf{r}) \to \begin{cases} \dfrac{\phi(\mathbf{k},\mathbf{r})}{\sqrt{\Omega}} & \text{continuum} \\ \\ \phi_\lambda(\mathbf{r}) & \text{bound} \end{cases} \tag{6.128}$$

The exact form of $\phi(\mathbf{k}, \mathbf{r})$ depends on the Hamiltonian. Free particles have $\phi(\mathbf{k}, \mathbf{r}) = \exp[i\mathbf{k}\cdot\mathbf{r}]$. If $V(r)$ has a central potential at small r and is constant at large r, then $\phi(\mathbf{k}, \mathbf{r})$ has plane wave behavior only at large r. Continuum eigenfunctions extend over all parts of the box. The bound states $\phi_\lambda(\mathbf{r})$ are localized to the region of the attractive potential, and $\sqrt{\Omega}\phi_\lambda$ diverges at large volume.

How do the matrix elements depend on Ω? Each matrix element contain two eigenstates and either or both may be bound or continuum states. There are three possible circumstances: both are bound states ($V_{nm} \sim \Omega^0$), both are continuum states $V_{nm} \sim 1/\Omega$, or one of each ($V_{nm} \sim 1/\sqrt{\Omega}$):

$$\lim_{\Omega\to\infty} V_{nm} \to \begin{cases} V_{\lambda\lambda'} = \int d^3r \phi_\lambda^*(\mathbf{r}) V(\mathbf{r}) \phi_{\lambda'}(\mathbf{r}) \\ \\ \dfrac{V(\mathbf{k},\mathbf{k}')}{\Omega} = \frac{1}{\Omega}\int d^3r \phi^*(\mathbf{k},\mathbf{r}) V(\mathbf{r}) \phi(\mathbf{k}',\mathbf{r}) \\ \\ \dfrac{V_\lambda(\mathbf{k})}{\sqrt{\Omega}} = \frac{1}{\sqrt{\Omega}}\int d^3r \phi^*(\mathbf{k},\mathbf{r}) V(\mathbf{r}) \phi_\lambda(\mathbf{r}) \end{cases} \tag{6.129}$$

In each case the integral over d^3r is assumed to converge to a value that is independent of the volume. This convergence is automatically satisfied for the two integrals involving bound states. The integral for $V(\mathbf{k}, \mathbf{k}')$ will converge if $V(\mathbf{r})$ is a central potential that vanishes at large r. The one case where the continuum integals are independent of volume is when the potential is also a plane wave, say

$$V(\mathbf{r}) = \mathbf{v}e^{i\mathbf{q}\cdot\mathbf{r}} \tag{6.130}$$

$$\frac{V(\mathbf{k},\mathbf{k}')}{\Omega} = \frac{v}{\Omega}\int d^3r \exp\left[i\mathbf{r}\cdot(\mathbf{k}'+\mathbf{q}-\mathbf{k})\right]$$
$$= v\delta_{\mathbf{k}=\mathbf{k}'+\mathbf{q}} \tag{6.131}$$

The integral gives a Kronecker delta multiplied by the volume of the box, and V_{nm} is independent of volume.

Once the matrix elements are understood, the perturbation theory of continuum states is rather easy. First consider the eigenvalue of bound states. In second-order perturbation theory, the summation over states $m \neq n$ in (6.114) is divided into sums over bound and continuum states:

$$E_\lambda = E_\lambda^{(0)} + V_{\lambda\lambda} + \sum_{\lambda'\neq\lambda}\frac{|V_{\lambda\lambda'}|^2}{E_\lambda^{(0)} - E_{\lambda'}^{(0)}} + \frac{1}{\Omega}\sum_{\mathbf{k}}\frac{|V_\lambda(\mathbf{k})|^2}{E_\lambda^{(0)} - E^{(0)}(\mathbf{k})} \tag{6.132}$$

In the limit of infinite volume, the summation over continuum states is changed to an integral over continuum states (this procedure is discussed in chapter 2):

$$\lim_{\Omega\to\infty}\frac{1}{\Omega}\sum_{\mathbf{k}} = \int\frac{d^3k}{(2\pi)^3} \tag{6.133}$$

Then the second-order eigenvalue for a bound state is

$$E_\lambda = E_\lambda^{(0)} + V_{\lambda\lambda} + \sum_{\lambda'\neq\lambda}\frac{|V_{\lambda\lambda'}|^2}{E_\lambda^{(0)} - E_{\lambda'}^{(0)}} + \int\frac{d^3k}{(2\pi)^3}\frac{|V_\lambda(\mathbf{k})|^2}{E_\lambda^{(0)} - E^{(0)}(\mathbf{k})} \tag{6.134}$$

There are two terms in the second-order contribution: one is a summation over bound states and the other is an integral over continuum states. Both terms usually make important contributions to the eigenvalue.

The bound-state eigenfunction has a similar summation in first-order perturbation theory. The summation over continuum states is also expressed as a continuum integral:

$$\psi_\lambda(\mathbf{r}) = \phi_\lambda(\mathbf{r}) + \sum_{\lambda'\neq\lambda}\frac{V_{\lambda'\lambda}\phi_{\lambda'}(\mathbf{r})}{E_\lambda^{(0)} - E_{\lambda'}^{(0)}} + \int\frac{d^3k}{(2\pi)^3}\frac{V_\lambda(\mathbf{k})\phi(\mathbf{k},\mathbf{r})}{E_\lambda^{(0)} - E^{(0)}(\mathbf{k})} \tag{6.135}$$

For bound states, neither the eigenvalue (6.134) nor the eigenfunction (6.135) depends on the volume of the box Ω.

Consider the same expressions for continuum states. The eigenvalue expression up to second-order of perturbation theory is

$$E(\mathbf{k}) = E^{(0)}(\mathbf{k}) + \frac{1}{\Omega}\left\{V(\mathbf{k},\mathbf{k}) + \sum_\lambda\frac{|V_\lambda(\mathbf{k})|^2}{E^{(0)}(\mathbf{k}) - E_\lambda^{(0)}}\right.$$
$$\left. + \int\frac{d^3k'}{(2\pi)^3}\mathcal{P}\frac{|V(\mathbf{k},\mathbf{k}')|^2}{E^{(0)}(\mathbf{k}) - E^{(0)}(\mathbf{k}')}\right\} \tag{6.136}$$

The symbol $\mathcal{P}$ denotes the principal part of the denominator. Every correction term to the energy is proportional to $1/\Omega$. They all vanish in the limit that $\Omega \to \infty$. For continuum states one has

$$E(\mathbf{k}) = E^{(0)}(\mathbf{k}) \tag{6.137}$$

There are no corrections to the particle energy when $V(\mathbf{r})$ is a local potential.

This result is expected and should not be surprising. In the prior chapters, the eigenfunctions and eigenvalues have been solved for a number of central potentials. For continuum states, the presence of the central potential changes the eigenfunctions by introducing additional curvature in the regions of potentials. However, the central potential does not alter the eigenvalue of the continuum states. For plane waves it is always $E(\mathbf{k}) = \hbar^2 k^2/2m$. The central potential changes the eigenfunction but not this eigenvalue. It is reassuring that the same result is found in perturbation theory.

The eigenfunction of the continuum state is certainly affected by perurbation theory. For continuum states, the two terms in (6.112) give

$$\psi(\mathbf{k}, \mathbf{r}) = \frac{1}{\sqrt{\Omega}} \left\{ \phi(\mathbf{k}, \mathbf{r}) + \sum_{\lambda} \frac{\phi_\lambda(\mathbf{r}) V_\lambda(\mathbf{k})}{E(\mathbf{k}) - E_\lambda^{(0)}} \right.$$
$$\left. + \int \frac{d^3k'}{(2\pi)^3} \mathcal{P} \frac{V(\mathbf{k}, \mathbf{k}')\phi(\mathbf{k}', \mathbf{r})}{E(\mathbf{k}) - E(\mathbf{k}')} \right\} \tag{6.138}$$

The expression in curly brackets is independent of Ω. The perturbed eigenfunction $\psi(\mathbf{k}, \mathbf{r})$ is proportional to $1/\sqrt{\Omega}$ when $\Omega \to \infty$. The continuum eigenfunction is affected by the perturbation V, although the eigenvalue is not.

As an example, let H_0 describe plane waves so there are no bound states. Then $\phi(\mathbf{k}, \mathbf{r})$ $\exp(i\mathbf{k}\cdot\mathbf{r})$. The perturbation is a Yukawa potential:

$$V(r) = g \frac{e^{-k_s r}}{r} \tag{6.139}$$

where g and k_s are constants. The first step, always, is to calculate the matrix elements. For plane wave eigenfunctions, the matrix element is just the Fourier transform of the potential. An overbar notation is used to denote the transform:

$$\bar{V}(\mathbf{k} - \mathbf{k}') = \int d^3 r e^{-i\mathbf{r}\cdot(\mathbf{k} - \mathbf{k}')} g \frac{e^{-k_s r}}{r} \tag{6.140}$$

The Yukawa potential occurs often in physics, and its Fourier transform is often needed. The integral is evaluated by first doing the angular integrals. Define $\nu = \cos\theta$, where θ is the angle between $\mathbf{r}$ and $\mathbf{Q} = \mathbf{k} - \mathbf{k}'$:

$$\bar{V}(Q) = 2\pi g \int_0^\infty r dr e^{-k_s r} \int_{-1}^{1} d\nu e^{-irQ\nu} \tag{6.141}$$

$$\int_{-1}^{1} d\nu e^{-irQ\nu} = \frac{1}{iQr}(e^{iQr} - e^{-iQr}) = \frac{2\sin(Qr)}{Qr} \tag{6.142}$$

$$\bar{V}(Q) = \frac{4\pi g}{Q} \int_0^\infty dr e^{-k_s r} \sin(Qr) = \frac{4\pi g}{Q^2 + k_s^2} \tag{6.143}$$

The remaining integral is the Laplace transform of the sine function, which is found in standard tables:

$$V(r) = \frac{g e^{-k_s r}}{r} \qquad \bar{V}(Q) = \frac{4\pi g}{Q^2 + k_s^2} \tag{6.144}$$

The Coulomb potential is a special case of Yukawa with $k_s = 0$ and $g = Ze^2$. The Fourier transform of the Coulomb potential is

$$V(r) = \frac{Ze^2}{r} \qquad \bar{V}(Q) = \frac{4\pi Z e^2}{Q^2} \tag{6.145}$$

This completes the discussion of the matrix element.

The eigenvalue is unaffected by the potential. The changes of the eigenfunction are found from perturbation theory. There are no bound states, so we start with

$$\psi(\mathbf{k}, \mathbf{r}) = \frac{1}{\sqrt{\Omega}} \left\{ e^{i\mathbf{k}\cdot\mathbf{r}} + \int \frac{d^3k'}{(2\pi)^3} \mathcal{P} \frac{V(\mathbf{k}-\mathbf{k}') e^{i\mathbf{k}'\cdot\mathbf{r}}}{E(\mathbf{k}) - E(\mathbf{k}')} \right\} \tag{6.146}$$

Change the integration variable in the last term to $\mathbf{Q} = \mathbf{k} - \mathbf{k}'$. Then the integrand has a factor of $\exp(i\mathbf{k}\cdot\mathbf{r})$ that is brought outside of the bracket:

$$\psi(\mathbf{k}, \mathbf{r}) = \frac{e^{i\mathbf{k}\cdot\mathbf{r}}}{\sqrt{\Omega}} \left\{ 1 + \int \frac{d^3Q}{(2\pi)^3} \mathcal{P} \frac{V(Q) e^{i\mathbf{Q}\cdot\mathbf{r}}}{E(\mathbf{k}) - E(\mathbf{k}+\mathbf{Q})} \right\} \tag{6.147}$$

The last term, with the integral, is now just a local function of $\mathbf{r}$ that has a nonzero value near the origin. It gives the change in the eigenfunction in first-order perturbation theory. The integral is hard to evaluate for arbitrary values of $\mathbf{k}$ and $\mathbf{r}$. A considerable simplification is achieved by picking the special value of $\mathbf{k} = 0$:

$$\psi(0, \mathbf{r}) = \frac{1}{\sqrt{\Omega}} \left\{ 1 - 4\pi g \int \frac{d^3Q}{(2\pi)^3} \frac{e^{i\mathbf{Q}\cdot\mathbf{r}}}{(Q^2 + k_s^2) E(Q)} \right\} \tag{6.148}$$

$$= \frac{1}{\sqrt{\Omega}} \left\{ 1 - \frac{4mg}{\pi \hbar^2 r} \int_0^\infty \frac{dQ}{Q} \frac{\sin(Qr)}{Q^2 + k_s^2} \right\} \tag{6.149}$$

The angular integral was done according to (6.142). The remaining integral is evaluated by contour integration or by looking up the result in tables:

$$\int_0^\infty \frac{dQ}{Q} \frac{\sin(Qr)}{Q^2 + k_s^2} = \frac{\pi}{2k_s^2} (1 - e^{-k_s r}) \tag{6.150}$$

$$\psi(0, r) = \frac{1}{\sqrt{\Omega}} \left\{ 1 - \frac{2mg}{\hbar^2 k_s^2 r} (1 - e^{-k_s r}) \right\} \tag{6.151}$$

This equation is sketched in figure 6.3 as a function of $k_s r$ with the potential strength $2mg/(\hbar^2 k_s) = 0.5$. The solid line is $\sqrt{\Omega}\psi(0, r)$ and the dashed line is the result of zero-order perturbation theory. The result is interesting since the perturbed eigenfunction varies according to $1/r$ at long distance. One might think that a short-ranged Yukawa potential

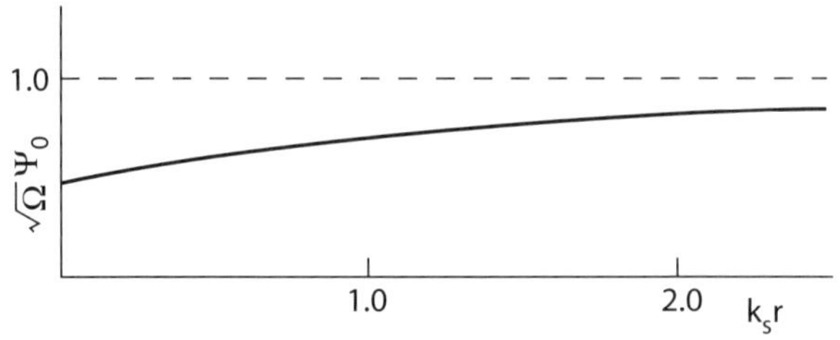

FIGURE 6.3. A plot of $\sqrt{\Omega}\psi(0,r)$ as a function of $k_s r$.

would change the eigenfunction only at short-range. The present example shows that idea is incorrect.

6.4.4 Green's Function

The first-order change in the eigenfunction for continuum states can be evaluated by an alternative procedure that uses classical Green's functions. In the present section we define this Green's function as

$$G(k,\mathbf{r}) = \int \frac{d^3p}{(2\pi)^3} e^{i\mathbf{p}\cdot\mathbf{r}} \mathcal{P}\frac{1}{E(k)-E(p)} \tag{6.152}$$

where $E(k) = \hbar^2k^2/2m$ and $\mathcal{P}$ denotes principal part. The Green's function enters the first-order change in the eigenfunction. It is written as

$$\sqrt{\Omega}\psi^{(1)}(\mathbf{k},\mathbf{r}) = \int \frac{d^3k'}{(2\pi)^3} \mathcal{P}\frac{V(\mathbf{k}-\mathbf{k}')e^{i\mathbf{k}'\cdot\mathbf{r}}}{E(\mathbf{k})-E(\mathbf{k}')} \tag{6.153}$$

$$V(\mathbf{k}-\mathbf{k}') = \int d^3r' e^{i\mathbf{r}'\cdot(\mathbf{k}-\mathbf{k}')} V(\mathbf{r}') \tag{6.154}$$

These two integrals are nested and their order is reversed:

$$\begin{aligned}\sqrt{\Omega}\psi^{(1)}(\mathbf{k},\mathbf{r}) &= \int d^3r' V(\mathbf{r}')e^{i\mathbf{k}\cdot\mathbf{r}'} \int \frac{d^3k'}{(2\pi)^3} e^{i\mathbf{k}'\cdot(\mathbf{r}-\mathbf{r}')} \mathcal{P}\frac{1}{E(k)-E(k')} \\ &= \int d^3r' V(\mathbf{r}')e^{i\mathbf{k}\cdot\mathbf{r}'} G(k,\mathbf{r}-\mathbf{r}')\end{aligned} \tag{6.155}$$

Equation (6.155) is an efficient way to evaluate the first-order change in the eigenfunction. The Green's function is the same for all problems and needs to be evaluated only once.

The angular integral in the Green's function is evaluated as in (6.142):

$$G(k,r) = -\frac{m}{\pi^2\hbar^2 r}\int_0^\infty k'dk' \mathcal{P}\frac{\sin(k'r)}{k^{2\prime}-k^2} \tag{6.156}$$

This integal can be found in tables or evaluated by contour integration. The final result is

$$G(k,r) = -\frac{m}{2\pi\hbar^2 r}\cos(kr) \tag{6.157}$$

The form of the answer very much depends on taking the principal part of the energy denominator. In chapter 10 on scattering theory, we evaluate the same Green's function without the principal part and get a different result.

The first-order change in the eigenfunction is

$$\sqrt{\Omega}\psi^{(1)}(\mathbf{k},\mathbf{r}) = -\frac{m}{2\pi\hbar^2}\int d^3r' V(\mathbf{r}')e^{i\mathbf{k}\cdot\mathbf{r}'}\frac{\cos(k|\mathbf{r}-\mathbf{r}'|)}{|\mathbf{r}-\mathbf{r}'|} \tag{6.158}$$

We evaluate this for the Yukawa potential. Again the integral is hard to evaluate for general values of **k** and **r**. One simple result can be shown: the behavior at large r. In the limit that $r >> r'$ then

$$|\mathbf{r}-\mathbf{r}'| = \sqrt{r^2 + r^{2'} - 2\mathbf{r}\cdot\mathbf{r}'} = r\sqrt{1+(r'/r)^2 - 2\mathbf{r}\cdot\mathbf{r}'/r^2}$$

$$\approx r - \hat{r}\cdot\mathbf{r}' \tag{6.159}$$

$$\frac{1}{|\mathbf{r}-\mathbf{r}'|} = \frac{1}{r} + O\left(\frac{\mathbf{r}'}{r^2}\right) \tag{6.160}$$

$$G(k,\mathbf{r}-\mathbf{r}') = -\frac{m}{2\pi\hbar^2 r}\cos(kr - \tilde{k}\cdot\mathbf{r}'), \quad \tilde{k} = k\hat{r} \tag{6.161}$$

The first-order eigenfunction for large r is

$$\begin{aligned}\sqrt{\Omega}\psi^{(1)}(\mathbf{k},\mathbf{r}) &= -\frac{m}{4\pi\hbar^2 r}\int d^3r' V(\mathbf{r}')e^{i\mathbf{k}\cdot\mathbf{r}'}\left[e^{ikr - i\tilde{k}\cdot\mathbf{r}'} + e^{-ikr + i\tilde{k}\cdot\mathbf{r}'}\right] \\ &= -\frac{m}{4\pi\hbar^2 r}\left[e^{ikr}\tilde{V}(\mathbf{k}-\tilde{k}) + e^{-ikr}\tilde{V}(\mathbf{k}+\tilde{k})\right]\end{aligned} \tag{6.162}$$

The symbol $\tilde{V}$ denotes the Fourier transform of the potential. This result applies for any potential. It shows that the first-order change in the eigenfunction always goes as $O(1/r)$ at large distance.

6.5 The Polarizability

The electrical polarizability of an object has the same definition in classical and quantum mechanics. The definition of polarizability is easily applied to objects that have no permanent dipole moment. Every object will develop an induced dipole moment when subjected to a uniform electric field $\vec{F}$. The polarizability $\alpha_{\mu\nu}$ is the constant ratio between induced moment $\vec{P}$ and applied field:

$$P_\mu = \sum_\nu \alpha_{\mu\nu} F_\nu \tag{6.163}$$

where $\mu = (x, y, z)$ are coordinates. The polarization P_μ may not point in the same direction as the field F_ν, so $\alpha_{\mu\nu}$ is a tensor quantity. The electric field has the units of charge over area. The polarization is charge times length and the polarizability has the units of volume. The units give the correct magnitude, since the polarizability of an atom or molecule is usually proportional to its volume.

6.5.1 Quantum Definition

An equivalent way to define polarizability uses the ground-state energy of the object in the external field. The change in energy from a change in field is

$$\Delta E = -\sum_{\mu} P_\mu(\Delta F_\mu) \tag{6.164}$$

Using (6.163) for P_μ, and doing the integral gives

$$\Delta E = -\int \vec{P}\cdot d\vec{F} = -\frac{1}{2}\sum_{\mu\nu} F_\mu \alpha_{\mu\nu} F_\nu \tag{6.165}$$

For spherically symmetric systems the polarizability is isotropic, $\alpha_{\mu\nu} = \alpha\delta_{\mu\nu}$, and the energy change simplifies:

$$\Delta E = -\frac{\alpha}{2}F^2 - \frac{\gamma}{4}F^4 - O(F^6) \tag{6.166}$$

For isotropic objects, the change in energy is a power series in F^2, and the first term is $\alpha/2$. The next term has a constant γ defined as the *hyperpolarizability*. It arises from nonlinear terms in (6.163):

$$\vec{P} = \vec{F}[\alpha + \gamma F^2 + O(F^4)] \tag{6.167}$$

This discussion is entirely classical, but the results in quantum mechanics are identical. Quantum mechanics provides formulas to calculate the coefficients such as α and γ.

In quantum mechanics, the electric field is introduced through the perturbation $V = -e\vec{F}\cdot\mathbf{r}$. Perturbation theory is used to calculate how the eigenvalues change with field. The nth order of perturbation theory gives energy terms depending on $V^n \propto F^n$. The polarizability is the coefficient of the term F^2 and is given by second-order perturbation theory:

$$E_n(F) = E_n(0) - eF\langle n|z|n\rangle + e^2F^2\sum_{m\neq n}\frac{|\langle n|z|m\rangle|^2}{E_n - E_m} \tag{6.168}$$

Dirac notation is used for the matrix elements. The first-order perturbation result contains the matrix element $\langle n|z|n\rangle$. It vanishes in systems with inversion symmetry $[V(-\mathbf{r}) = V(\mathbf{r})]$. However, many molecules lack inversion symmetry and have a fixed dipole moment—H_2O is an example. In these systems the matrix element $\langle n|z|n\rangle \neq 0$ and there is a first-order term in the energy. This result agrees with classical physics, since an object with a fixed dipole moment $\mathbf{p}$ has an energy $-\mathbf{p}\cdot\vec{F}$ in a static field.

The term of $O(F^2)$ in (6.168) is the quantum mechanical definition of the polarizability of a particle in the state n:

$$\alpha_n = -2e^2\sum_{m\neq n}\frac{|\langle n|z|m\rangle|^2}{E_n - E_m} \tag{6.169}$$

There is no approximation in this definition. It is the exact definition of polarizabillity. Perturbation theory has been used to identify the energy term proportional to F^2, and that expression is the exact polarizability. This formula is valid for any eigenstate n. Usually it is applied to a particle in the ground-state configuration. For the ground state $E_m > E_n$, the energy denominator is negative and the polarizability is always positive. However, the polarizability could be negative for particles in excited states.

For an anisotropic system it is

$$\alpha_{n,\mu\nu} = -2e^2 \sum_{m \neq n} \frac{\langle n|r_\mu|m\rangle\langle m|r_\nu|n\rangle}{E_n - E_m} \tag{6.170}$$

Equation (6.169) for the polarizabiltiy is valid for noninteracting particles. The formula has to be modified in systems with many particles, due to the screening from particle–particle interactions.

Another useful expression is the polarizability as a function of frequency. The formula is derived in chapter 8 and is

$$\alpha_n(\omega) = 2e^2 \sum_{m \neq n} \frac{|\langle n|z|m\rangle|^2 (E_m - E_n)}{(E_m - E_n)^2 - (\hbar\omega)^2} \tag{6.171}$$

Table 6.1 shows some polarizabilities of spherical atoms. Units are cubic angstroms. Both the rare gases in the first column and the alkali ions in the last column are closed shells. The middle column is the neutral alkali atoms, and the outer *ns* electron makes a very large contribution to the polarizability.

6.5.2 *Polarizability of Hydrogen*

The simplest atom is hydrogen. This subsection discusses the polarizability of an electron in the ground state of the hydrogen atom. The exact answer is derived later, but is

$$\alpha = \frac{9}{2} a_0^3 \tag{6.172}$$

Table 6.1 Dipole Polarizabilities in Units of Å^3

Atom[a]	α	Atom[a]	α	Ion[b]	α
He	0.21	Li	17.4	Li^+	0.03
Ne	0.40	Na	16.7	Na^+	0.15
Ar	1.64	K	37.3	K^+	0.81
Kr	2.48	Rb	34.4	Rb^+	1.35
Xe	4.04	Cs	46.2	Cs^+	2.42

[a] R.R. Teachout and R.T. Pack, *Atomic Data*, **3**, 195 (1971).
[b] G.D. Mahan, *J. Chem. Phys.* **76**, 493 (1982).

where a_0 is the Bohr radius. Our discussion begins by considering some approximate methods of finding the polarizability. The approximate methods are easy and can be applied to other atoms and molecules. The exact derivation is given at the end of the subsection.

We need to evaluate the summation in (6.169):

$$\alpha = 2e^2 \sum_{\lambda \neq 1s} \frac{|\langle 1s|z|\lambda\rangle|^2}{E_\lambda - E_{1s}} \tag{6.173}$$

where n is the $1s$ state of hydrogen, and the summation λ is over other eigenstates, including continuum states. Here λ means all quantum numbers: (n, ℓ, m) for bound states and (k, ℓ, m) for continuum states. The ground state is $(n=1, \ell=0, m=0)$.

Examine the matrix element in (6.173). The $1s$ state is isotropic and has no angular dependence. Since $z = r\cos(\theta)$, the only nonzero matrix elements are to those states whose angular dependence is also $\cos(\theta) \sim P_1(\theta)$, $m = 0$. The summation in (6.173) is confined to states λ with $\ell = 1$, $m = 0$. One still must sum over all $n > 1$ and all k.

The first term in the summation over bound states is $2p_z$. A popular approximation is to replace the summation over states by just the first term in the summation $\lambda = 2p_z$. The matrix element was derived in chapter 5:

$$\langle 1s|z|2p_z\rangle = a_0 \frac{2^{15/2}}{3^5} \tag{6.174}$$

The energy denominator is $E_2 - E_1 = (\frac{3}{4})E_{\text{Ry}}$, where $E_n = -E_{R_y}/n^2$. Using $E_{R_y} = e^2/(2a_0)$, the first term in the summation is

$$\alpha \approx 2e^2 \frac{a_0^2 2^{15}/3^{10}}{3e^2/(8a_0)} = a_0^3 \frac{2^{19}}{3^{11}} = 2.96 a_0^3 \tag{6.175}$$

The result is about $3a_0^3$, to be compared with the exact result of $4.5a_0^3$. The first excited state gives two-thirds of the answer. One might think that taking the next states $3p_z$, $4p_z$, etc., would converge to the proper answer. That does not happen since the continuum states make a significant contribution.

The most successful methods of approximating the summation in (6.173) are those that include all states, bound and continuum. One approximate way is to replace the energy denominator $E_\lambda - E_1$ by a constant $\bar{E} - E_1$. It is the energy denominator that makes the summation difficult. Removing the denominator from the summation leaves only the factors in the numerator, which are easy to sum.

The completeness relation is given by a summation over all states. The summation over states in the polarizability excludes the term $\lambda = 1s$. It is easy to add this term back to the summation:

$$\alpha \approx \frac{2e^2}{\bar{E} - E_1} \sum_{m \neq 1} \langle 1s|z|m\rangle\langle m|z|1s\rangle \tag{6.176}$$

$$\approx \frac{2e^2}{\bar{E} - E_1} \{\langle 1s|z(\sum_m |m\rangle\langle m|)z|1s\rangle - |\langle 1s|z|1s\rangle|^2\} \tag{6.177}$$

$$\approx \frac{2e^2}{\bar{E} - E_1} \langle 1s|z^2|1s\rangle \tag{6.178}$$

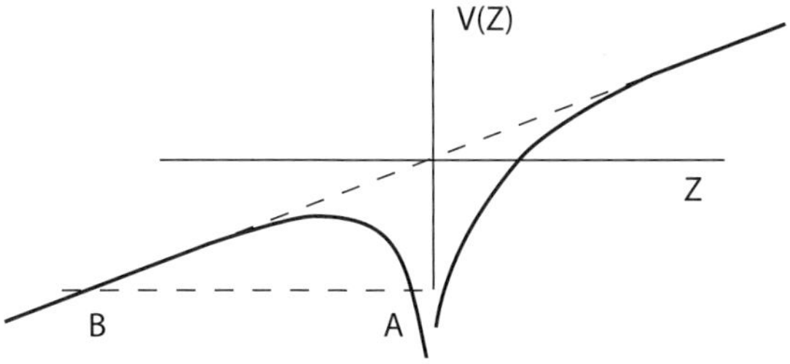

FIGURE 6.4. Potential energy $V(z)$ for an electron in the potential of the nucleus and an electric field.

The matrix element $\langle 1s|z|1s\rangle = 0$, due to parity arguments, so the final answer has only the expectation of z^2:

$$\langle 1s|z^2|1s\rangle = \langle r^2\rangle\langle\cos^2(\theta)\rangle = a_0^2 \tag{6.179}$$

where $\langle r^2\rangle = 3a_0^2$ and $\langle\cos^2(\theta)\rangle = \frac{1}{3}$. The last step is to choose the average energy $\bar{E}$. Since the summation runs over bound ($E_\lambda < 0$) and continuum ($E(k) > 0$) states, choose $\bar{E} = 0$. Then the approximate polarizability is

$$\alpha \approx \frac{2e^2 a_0^2}{e^2/(2a_0)} = 4a_0^3 \tag{6.180}$$

The result of $4a_0^3$ is getting closer to the exact answer of $4.5a_0^3$. Usually the approximation of an average energy denominator is superior to taking just the first term in the summation over states.

The remaining part of this section provides an exact derivation of the polarizability of atomic hydrogen. Hydrogen is the only atom for which the polarizability can be derived exactly. Very accurate numerical values are known for many atoms and ions. The exact result for hydrogen has been known since 1926. The original solution used the feature that Schrödinger's equation for the atom in an electric field is separable in parabolic coordinates. This fancy coordinate system can be used to obtain the dependence of the eigenvalues on the electric field. Similar coordinate systems can be used for related problems such as an electron bound to two protons separated by a distance d—the problem of H_2^+.

An important physical point is that there are no rigorous bound states for an atom in an electric field. Figure 6.4 shows a graph of the potential energy along the z-direction ($x = 0$, $y = 0$) for the potential energy

$$V(z) = -\frac{e^2}{|z|} - eFz \tag{6.181}$$

when $e < 0$, $F > 0$. The proton is at the origin, where the potential is very attractive. If the bound state of the atom has an energy A, the electron can tunnel through the barrier to the point B and escape the nucleus. The rate of tunneling can be calculated using WKBJ. For small fields, the tunneling rate is infinitesimally small, but for larger fields it becomes an important process. A strong dc electric field can strip electrons from atoms. Our interest is small electric fields, where we wish to find the change in energy levels due to the electric field F.

An exact result for the polarizability is obtained using the formula for the first-order change in the eigenfunction:

$$\psi_{1s}^{(1)}(\mathbf{r}) = \sum_{m \neq 1} \frac{|m\rangle\langle m|(-eFz)|1s\rangle}{E_1 - E_m} \tag{6.182}$$

Earlier we showed how the summation over states $|m\rangle$ can be evaluated by approximate methods. Now we give an exact method of doing the summation.

Operate on both sides of eqn. (6.182) by $(H_0 - E_1)$, where H_0 is the Hamiltonian of the hydrogen atom without the applied field F. Since $H_0|\mathrm{m}\rangle = E_m|m\rangle$, we find

$$(H_0 - E_1)\psi_{1s}^{(1)}(\mathbf{r}) = \sum_{m \neq 1} \frac{(E_m - E_1)|m\rangle\langle m|(-eFz)|1s\rangle}{E_1 - E_m} \tag{6.183}$$

$$= -\sum_{m \neq 1} |m\rangle\langle m|(-eFz)|1s\rangle \tag{6.184}$$

$$(H_0 - E_1)\psi_{1s}^{(1)}(\mathbf{r}) = eFz\phi_{1s}(\mathbf{r}) \tag{6.185}$$

Equation (6.185) is an inhomogeneous differential equation for the unknown function $\psi^{(1)}(\mathbf{r})$. It can be solved. It is going to be proportional to eF because of the source term on the right. Define

$$\psi_{1s}^{(1)}(\mathbf{r}) = eF\phi_{1s}^{(1)}(\mathbf{r}) \tag{6.186}$$

$$\left[-\frac{\hbar^2}{2m}\nabla^2 - \frac{e^2}{r} + \frac{e^2}{2a_0}\right]\phi_{1s}^{(1)} = z\phi_{1s}(\mathbf{r}) \tag{6.187}$$

The induced polarization in the atom is

$$P_z = \int d^3r \psi_{1s}(\mathbf{r})^2 (ez) \tag{6.188}$$

$$= e\int d^3rz[\phi_{1s} + eF\phi_{1s}^{(1)} + O(F^2)]^2 \tag{6.189}$$

$$= 2e^2F\int d^3r\phi_{1s}(\mathbf{r})z\phi^{(1)}(\mathbf{r}) + O(F^2) \tag{6.190}$$

$$\alpha = \frac{P_z}{F} = 2e^2\int d^3r\phi_{1s}(\mathbf{r})z\phi^{(1)}(\mathbf{r}) \tag{6.191}$$

The derivation first solves eqn. (6.187) for $\phi_{1s}^{(1)}$ and then uses this function in the integral (6.191) for the polarizability.

The right-hand side of (6.187) is proportional to $z = r\cos(\theta)$. This makes $\phi_{1s}^{(1)}$ also be proportional to $\cos(\theta)$, which is $P_1(\theta)$. It has angular momentum $\ell = 1$. This feature should not be surprising. In the earlier, approximate, method of doing ths summation, we concluded that only states with $\ell = 1$, $m = 0$ are needed. We are finding the same angular momentum states in the exact solution. The next step is to go to atomic units, where the dimensionless variable is $\rho = r/a_0$. Then define an unknown radial function $G(\rho)$ according to

$$\phi_{1s}^{(1)}(r) = \frac{\cos(\theta)}{E_{R_y}\sqrt{\pi a_0}} G(\rho) \tag{6.192}$$

and the differential equation (6.187) is

$$\left[\frac{d^2}{d\rho^2} + \frac{2}{\rho}\frac{d}{d\rho} + \frac{\mathcal{L}}{\rho^2} + \frac{2}{\rho} - 1\right]\phi^{(1)} = -\frac{\rho\cos(\theta)}{E_{R_y}\sqrt{\pi a_0}} e^{-\rho} \tag{6.193}$$

$$\mathcal{L}\cos(\theta) = -\ell(\ell+1)\cos(\theta) = -2\cos(\theta)$$

$$\left[\frac{d^2}{d\rho^2} + \frac{2}{\rho}\frac{d}{d\rho} - \frac{2}{\rho^2} + \frac{2}{\rho} - 1\right] G(\rho) = -\rho e^{-\rho} \tag{6.194}$$

where $\mathcal{L}$ is the angular derivatives in ∇^2. Since $G(\rho)$ is determined by the right-hand side, one can guess that it must be a polynomial in ρ times $\exp(-\rho)$:

$$G(\rho) = e^{-\rho} \sum_{n=0}^{\infty} a_n \rho^n \tag{6.195}$$

This guess does not limit the form of $G(\rho)$ as long as the coefficients a_n are found exactly. Insert the above form for $G(\rho)$ into (6.194) and evaluate the first and second derivatives. Then collect together all terms with the same power of ρ^n:

$$\frac{dG}{d\rho} = e^{-\rho} \sum_n [-a_n \rho^n + n a_n \rho^{n-1}] \tag{6.196}$$

$$= e^{-\rho} \sum_n \rho^n [(n+1)a_{n+1} - a_n] \tag{6.197}$$

The term with ρ^{n-1} was relabeled $n' = n-1$ and became $(n'+1)a_{n'+1}\rho^{n'}$ and then we dropped the primes. The second derivative is

$$\frac{d^2G}{d\rho^2} = e^{\rho} \sum_n [-\rho^n + n\rho^{n-1}][(n+1)a_{n+1} - a_n]$$

$$= e^{-\rho} \sum_n \rho^n [(n+2)(n+1)a_{n+2} - 2(n+1)a_{n+1} + a_n] \tag{6.198}$$

and eqn. (6.194) is

$$e^{-\rho} \sum_n \rho^n (n+1)[(n+4)a_{n+2} - 2a_{n+1}] = -\rho e^{-\rho} \tag{6.199}$$

The above result is rather simple, since many terms canceled. If this equation is valid for all values of ρ, then each term ρ^n must be satisfied separately. Cancel the factor of $\exp(-\rho)$ from both sides, and then evaluate the left-hand side for increasing values of $n = -2, -1, 0, 1, \ldots$:

$$n = -2 \qquad 2a_0 = 0 \tag{6.200}$$

$$n = -1 \qquad 0 \cdot [3a_1 - 2a_0] = 0 \tag{6.201}$$

$$n = 0 \qquad 4a_2 - 2a_1 = 0 \tag{6.202}$$

$$n = 1 \qquad 2(5a_3 - 2a_2) = -1 \tag{6.203}$$

$$n = 2 \qquad 3(6a_4 - 2a_3) = 0 \tag{6.204}$$

The solution to these equations is

$$a_0 = 0, \quad a_1 = \frac{1}{2}, \quad a_2 = \frac{1}{4} \tag{6.205}$$

$$a_\ell = 0, \quad \text{for} \quad \ell \geq 3 \tag{6.206}$$

$$G(\rho) = \frac{\rho}{2}\left(1 + \frac{\rho}{2}\right)e^{-\rho} \tag{6.207}$$

Now that $G(\rho)$ is known, use (6.192) to find

$$\phi_{1s}^{(1)}(\rho, \theta) = \frac{\cos(\theta)}{2E_{R_y}\sqrt{\pi a_0}}\rho\left(1 + \frac{\rho}{2}\right)e^{-\rho} \tag{6.208}$$

An exact analytical result has been derived for the first-order change in the eigenfunction due to the external electric field. Insert this result into the integral (6.191) for the polarizability. The radial integral is put into dimensionless form:

$$\alpha = \frac{2e^2 a_0^4}{2(e^2/2a_0)\sqrt{\pi^2 a_0^4}} \int_0^\infty d\rho \rho^4 \left(1 + \frac{\rho}{2}\right)e^{-2\rho} \int d\Omega \cos^2(\theta)$$

The angular integral is the average of $\cos^2(\theta)$, which is $4\pi/3$. The radial integral is

$$\int_0^\infty d\rho \rho^4 \left(1 + \frac{\rho}{2}\right)e^{-2\rho} = \frac{4!}{2^5}\left(1 + \frac{5}{4}\right) = \frac{27}{16} \tag{6.209}$$

$$\alpha = \frac{9}{2}a_0^3 \tag{6.210}$$

The various factors are collected to give the final result for α.

The trick to obtaining the exact result is to avoid evaluating the direct summation over states in (6.169). Instead, the first-order change in the eigenfunction is defined by an inhomogeneous differential equation (6.185). The exact solution to this equation gives an exact result for the polarizability. Similar methods are applied to other atoms, but the equations must be solved numerically.

6.6 van der Waals Potential

The van der Waals potential energy between two objects i and j has the form

$$V_{ij}(R) = -\frac{C_{6,ij}}{R^6} \tag{6.211}$$

The constant $C_{6,ij}$ is different for each pair of objects. It is an important force between neutral molecules and atoms, and contributes to the binding energy in ionic, rare-gas, and molecular solids. The force is generally rather weak. Almost any other interaction is larger. If the atoms or molecules are charged or if they have a fixed dipole or quadrupole moment, then there are Coulomb forces, dipole–dipole forces, etc. The van der Waals potential also exists between ions and charged molecules, but is usually ignored since it is much weaker than the forces from Coulomb's law. For neutral objects, that have no long-range Coulomb interaction, the van der Waals interaction provides the longest-range force.

A quantum mechanical expression is derived for $C_{6,ij}$, where (ij) are two neutral, spherically symmetric objects. Prototype systems are atoms with only closed shells, or *s*-shells. The centers of the two atoms are separated by the distance R, which is assumed to be large enough that bound electron orbitals do not overlap. Electron densities outside of atoms decline according to $\exp(-\alpha_i r)$, and the overlap energy between two atoms goes as $\exp[-R(\alpha_i+\alpha_j)]$. This latter term is assumed to be small compared to (6.211). The van der Waals force exists beyond the range of chemical bonding.

The two objects, or atoms, are denoted A and B. The total Hamiltonian of the system is denoted as

$$H = H_A + H_B + H_{AB} \tag{6.212}$$

where H_A is the total Hamiltonian of object A if it were completely isolated in space. H_B is the Hamiltonian of object B in the same circumstance. The term H_{AB} is the interaction between A and B when they are nearby, but sufficiently separated that overlap of orbitals is neglected. When the objects are atoms or molecules in space, the only long-range forces are Coulombic. Since it is assumed they are neutral, then A has Z_A electrons and a nuclear charge of Z_A. B has Z_B electrons and nuclear charge Z_B. Let $\mathbf{r}_{jA}$ and $\mathbf{r}_{iB}$ denote the electron positions for the two atoms, with respect to their individual nuclei. The interaction $H_{AB} = V$ is the Coulomb potential between the two nuclei, between the nucleus of one and the electrons of the other, and between the two sets of electrons:

$$\begin{aligned} H_{AB} = e^2 \Bigg\{ & \frac{Z_A Z_B}{R} - Z_B \sum_j \frac{1}{|\mathbf{R}+\mathbf{r}_{jA}|} - Z_A \sum_i \frac{1}{|\mathbf{R}-\mathbf{r}_{iB}|} \\ & + \sum_{ij} \frac{1}{|\mathbf{R}+\mathbf{r}_{jA}-\mathbf{r}_{iB}|} \Bigg\} \end{aligned} \tag{6.213}$$

The interactions within a single atom are part of H_A and H_B. The average value of H_{AB} is zero, since it is the Coulomb interaction between two neutral objects. The instantaneous value of H_{AB} need not be zero, and these fluctuations cause the van der Waals interaction.

The ratios $\mathbf{r}_{jA}/R$ and $\mathbf{r}_{iB}/R$ are less than one and are both expanded in a Taylor series about their respective origins. We carry the expansion to second-order and denote the dipole–dipole interaction by $\phi_{\mu\nu}(\mathbf{R})$:

$$\phi_{\mu\nu}(\mathbf{R}) = \frac{\delta_{\mu\nu}}{R^3} - 3\frac{R_\mu R_\nu}{R^5} \tag{6.214}$$

$$\begin{aligned} H_{AB} = e^2 \Bigg\{ & \frac{Z_A Z_B}{R} - Z_B \sum_j^{Z_A} \left[\frac{1}{R} - \frac{\mathbf{r}_{jA}\cdot\mathbf{R}}{R^3} - \frac{1}{2}\mathbf{r}_{jA}\cdot\phi\cdot\mathbf{r}_{jA} \right] \\ & - Z_A \sum_i^{Z_B} \left[\frac{1}{R} + \frac{\mathbf{r}_{iB}\cdot\mathbf{R}}{R^3} - \frac{1}{2}\mathbf{r}_{iB}\cdot\phi\cdot\mathbf{r}_{iB} \right] \\ & + \sum_{ij} \left[\frac{1}{R} - \frac{(\mathbf{r}_{jA}-\mathbf{r}_{iB})\cdot\mathbf{R}}{R^3} - \frac{1}{2}(\mathbf{r}_{jA}-\mathbf{r}_{iB})\cdot\phi\cdot(\mathbf{r}_{jA}-\mathbf{r}_{iB}) \right] \Bigg\} \end{aligned} \tag{6.215}$$

Both series are convergent as long as R is large compared with the size of the atom or object. The terms in the series are the multipole expansion. All of the terms in the above expansion that are shown cancel, except for one:

$$H_{AB} = e^2\left(\sum_j \mathbf{r}_{jA}\right) \cdot \phi(\mathbf{R}) \cdot \left(\sum_i \mathbf{r}_{iB}\right) + O(R^{-4}) \tag{6.216}$$

The first term on the right becomes the interaction term in the perturbation series to derive the van der Waals formula. It is possible to retain additional terms in this series, and they lead to additional interaction terms that go as R^{-n}, $n = 8$, 10, etc. The above interaction is the first nonzero term in the Coulomb expansion and gives the van der Waals interaction, which is the leading term in the long-range potential energy.

Perturbation theory uses

$$H_0 = H_A + H_B \tag{6.217}$$

$$V = e^2\left(\sum_j \mathbf{r}_{jA}\right) \cdot \phi(\mathbf{R}) \cdot \left(\sum_i \mathbf{r}_{iB}\right) \tag{6.218}$$

First, find the eigenfunctions and eigenvalues of H_0, which are the separate problems of H_A and H_B. Generally, finding these functions for atoms of large atomic number is very complicated. Here we are making a formal derivation, so assume that such functions exist and form a complete set:

$$H_A|m, A\rangle = E_{mA}|m, A\rangle \tag{6.219}$$

$$H_B|n, B\rangle = E_{nB}|n, B\rangle \tag{6.220}$$

$$H_0|m, A\rangle|n, B\rangle = (E_{mA} + E_{nB})|m, A\rangle|n, B\rangle \tag{6.221}$$

The ground state of each atom is denoted as $|g, A\rangle$ and $|g, B\rangle$, $|g\rangle \equiv |g, A\rangle|g, B\rangle$, and the ground-state energy is $E_g^{(0)} = E_{gA} + E_{gB}$.

The use of a product eigenfunction, such as $|m, A\rangle|n, B\rangle$, is seldom valid in quantum mechanics. Usually the many-particle eigenfunction must be antisymmetric under exchange of any two electrons. This requirement is discussed in chapter 9. The product eigenfunction is a valid representation whenever the orbitals are separated in space and do not overlap. These conditions are assumed in the present example.

It is further assumed that the atom or molecule is spherically symmetric. It has no permanent dipole moment. In that case the expectation value of the dipole operator is zero:

$$\langle g, A|\left(\sum_j \mathbf{r}_{jA}\right)|g, A\rangle = 0 \tag{6.222}$$

$$\langle g, B|\left(\sum_i \mathbf{r}_{iB}\right)|g, B\rangle = 0 \tag{6.223}$$

In the perturbation series for the ground-state energy,

$$E_g^{(0)} = E_{gA} + E_{gB} \tag{6.224}$$

$$E_g^{(1)} = \langle g, B|\langle g, A|V|g, A\rangle|g, B\rangle$$
$$= e^2 \langle g, A|\sum_j \mathbf{r}_{jA}|g, A\rangle \cdot \phi \cdot \langle g, B|\sum_i \mathbf{r}_{iB}|g, B\rangle = 0 \tag{6.225}$$

The first-order energy is zero. The van der Waals potential energy is derived from second-order perturbation theory:

$$E_g^{(2)} = -\sum_{n,m} \frac{|\langle m, A|\langle n, B|V|g\rangle|^2}{E_{mA} + E_{nB} - E_g^{(0)}} = -\frac{C_{6,AB}}{R^6} \tag{6.226}$$

Each matrix element contains the dipole–dipole interaction $\phi \sim R^{-3}$. The net factor of R^{-6} can be extracted from the summation. The remaining terms determine the coefficient:

$$C_{6,AB} = e^4 \sum_{m,n} \frac{|M_{mn}|^2}{E_{mA} + E_{nB} - E_g^{(0)}}$$
$$M_{mn} = \langle m, A|\sum_j r_{jA,\mu}|g, A\rangle(\delta_{\mu\nu} - 3R_\mu R_\nu/R^2)\langle n, B|\sum_i r_{iB,\nu}|g, B\rangle \tag{6.227}$$

This expression is called the *London equation* [1]. Although it was derived using perturbation theory, it is the exact coefficient of R^{-6} and the exact expression for C_6. Higher-order terms in perturbation theory give interaction potentials $\sim R^{-n}$ with powers of n larger than six. Van der Waals is not the exact potential between atoms or molecules. It is the leading term in the perturbation series and the largest term at large R.

The units of C_6 are joules (length)^6. The values for atoms are usually expressed in units of $e^2 a_0^5$, where e is the unit of charge and a_0 is the Bohr radius. For example, $C_6 = 6.47 e^2 a_0^5$ for the van der Waals potential between two hydrogen atoms. The coefficient 6.47 is obtained from numerical solution.

A simple formula is obtained from (6.227) if the two atoms are spherically symmetric. The dipole–dipole interaction in the matrix element M_{mn} has the spatial variables

$$\mathbf{r}_i \cdot \mathbf{r}_j - 3z_i z_j = x_i x_j + y_i y_j - 2z_i z_j \tag{6.228}$$

When the matrix element is squared, the cross-terms vanish. The matrix element of $x_i \equiv x_{iB}$ is nonzero only if $\langle m, B|$ is a state with x-symmetry. The matrix element of $y_i \equiv y_{iB}$ is nonzero only if $\langle m, B|$ has y-symmetry. The cross-terms require that $\langle m, B|$ have both x- and y-symmetry (but not xy). The double requirement cannot be satisfied. So the square of the matrix element is

$$|\langle mx, A|x_i|g, A\rangle|^2 |\langle nx, B|x_j|g, B\rangle|^2 + |\langle my, A|y_i|g, A\rangle|^2 |\langle ny, B|y_j|g, B\rangle|^2$$
$$+ 4|\langle mz, A|z_i|g, A\rangle|^2 |\langle nz, B|z_j|g, B\rangle|^2 \tag{6.229}$$

If the atom or molecule is spherically symmetric, all of the matrix elements for each atom are alike. In this case (6.227) can be replaced by

$$C_{6,AB} = 6e^4 \sum_{m,n} \frac{|\langle mz, A|z_i|g, A\rangle|^2 |\langle nz, B|z_j|g, B\rangle|^2}{E_{mA} + E_{nB} - E_g^{(0)}} \tag{6.230}$$

This formula is obviously easier to work with.

Equation (6.230) is rather complicated to evaluate. A good approximation exists if the electrons in one atom are much more bound than those in the other [3]. For example, if $E_{mA}-E_{gA}<<E_{mB}-E_{gB}$, then neglect the energy difference $E_{mA}-E_{gA}$ and find

$$C_{6,AB}=e^2 M_A M_B \tag{6.231}$$

$$M_A=3\langle g,A|z_i\sum_m(|m,A\rangle\langle m,A|)Z_i|g,A\rangle \tag{6.232}$$

$$=3\langle g,A|z_i^2|g,A\rangle=\langle g,A|r^2|g,A\rangle \tag{6.233}$$

$$M_B=2e^2\sum_n\frac{|\langle nz,B|z_j|g,B\rangle|^2}{E_{nB}-E_{gB}}=\alpha_B \tag{6.234}$$

$$C_{6,AB}=e^2\alpha_B\langle g,A|r^2|g,A\rangle \tag{6.235}$$

where α_B is the polarizability of atom B. This approximate formula can be quite useful.

The best approximation was introduced by London. It is simple and gives good results. The method starts with introducing the concept of imaginary frequency. It cannot be achieved experimentally, but is useful theoretically. Start from the formula (6.171) for the polarizability as a function of frequency, and replace $\hbar\omega$ by iu, where u is the imaginary frequency times Planck's constant:

$$\alpha_g(iu)=2e^2\sum_{m\neq n}\frac{|\langle g|z|m\rangle|^2(E_m-E_g)}{(E_m-E_g)^2+u^2} \tag{6.236}$$

where g denotes ground state. In the ground state, E_m-E_g is always positive, so $\alpha_g(iu)$ is always a positive function. Next consider the integral

$$\int_0^\infty\frac{du}{(u^2+A^2)(u^2+B^2)}=\frac{1}{A^2-B^2}\int_0^\infty du\left[\frac{1}{u^2+B^2}-\frac{1}{u^2+A^2}\right]$$
$$=\frac{1}{A^2-B^2}\frac{\pi}{2}\left(\frac{1}{B}-\frac{1}{A}\right)=\frac{\pi/2}{AB(A+B)}$$

This integral allows us to multiply two functions like (6.236) for atoms A and B and then do the integral over u. The energy denominators and matrix elements come out just like

$$\int_0^\infty du\alpha_A(iu)\alpha_B(iu)=2\pi e^4\sum_{m,n}\frac{|\langle mz,A|z_i|g,A\rangle|^2|\langle nz,B|z_j|g,B\rangle|^2}{E_{mA}+E_{nB}-E_g^{(0)}}$$

The right-hand side of this expression is identical to eqn. (6.230) for the van der Waals coefficient, except the prefactor is wrong. Multiplying the above integral by $3/\pi$ gives

$$C_{6,AB}=\frac{3}{\pi}\int_0^\infty du\alpha_A(iu)\alpha_B(iu) \tag{6.237}$$

The van der Waals coefficient between two atoms or molecules equals the convolution of the polarizabilities when integrated over complex energy.

London proposed a simple approximation for the polarizability:

$$\alpha_i(iu)=\frac{\alpha_i(0)}{1+u^2/E_i^2} \tag{6.238}$$

The numerator contains the static polarizability $\alpha_i(0) \equiv \alpha_i$. The u-dependence of the polarizability can be approximated by a simple characteristic energy E_i that is different for each atom, molecule, or ion. Using this approximation in the integral in (6.237) gives the London formula:

$$C_{6,AB} = \frac{3}{2}\alpha_A\alpha_B\frac{E_A E_B}{E_A + E_B} \quad (6.239)$$

$$C_{6,AA} = \frac{3}{4}E_A\alpha_A^2 \quad (6.240)$$

The characteristic energy E_i can be found from the last equation. Both C_6 and α can be measured, so the values for E_i are found experimentally:

$$E_A = \frac{4}{3}\frac{C_{6,AA}}{\alpha_A^2} \quad (6.241)$$

We have tested the London formula extensively and found that it is accurate to within 1% for all cases, even between very dissimilar atoms. Table 6.2 shows experimental values for the rare gas atoms.

The physics of the the van der Waals potential can be understood in a simple way. First consider an atom in an electric field. In the section on polarizability it was shown that the eigenfunctions of the atom distort since the electrons gain energy by being able to adjust to the attraction of the field. Next consider a neutral atom outside of a perfectly conducting mirror. The atom exerts no force on itself as long as it remains perfectly spherical. However, a small distortion in the charge distribution sets up a local electric field. Through the perfect mirror, the field can act back on the atom. The atom finds it is energetically favorable to distort slightly, which lowers the energy by the self-energy of interaction through the mirror. The result is an attractive potential between the atom and the mirror. A similar attraction exists between all atoms and metal surfaces and is also a form of van der Waals interaction.

The third example is between an atom and a nearby metal sphere. If the atom has a slight distortion in its charge density, the resulting fields induce multipoles on the metal sphere. These multipoles on the sphere create their own fields that act back on the atom,

Table 6.2 Dipole Polarizabilies in Units of a_0^3, van der Waals Coefficient $C_{6,ii}$ in Units of $e^2a_0^5$, and London Energy E_i in Rydberg Units $e^2/2a_0$.

Atom	α_i	$C_{6,ii}$	E_i
He	1.38	1.47	2.05
Ne	2.70	6.4	2.34
Ar	11.07	65	1.41
Kr	16.74	131	1.25

inducing a self-interaction. The longest-range potential between the atom and the perfect sphere is just the van der Waals potential, with C_6 given:

$$C_6 = e^2 \alpha_S \langle g, A|r^2|g, A\rangle \tag{6.242}$$

where α_S is the polarizability of the sphere. Compare with eqn. (6.235). The van der Waals interaction between two atoms is caused by the same type of self-energies. A charge fluctuation on one atom causes an electric field at the other atom. This field induces a dipole moment at the other atom that acts back on the first atom. These mutual polarizations on the two atoms can lower the energy of the system, which is the van der Waals potential energy.

6.7 Spin–Orbit Interaction

Relativistic quantum mechanics is discussed in chapter 11. Most relativistic phenomena are important only when particles have velocities approaching the speed of light. However, there are several phenomena that are relativistic in origin, yet are quite observable for particles going ordinary velocities. Particles with spin will have a magnetic moment. For electrons, the moment is expressed in terms of the Bohr magneton:

$$\mu_0 = \frac{|e|\hbar}{2mc} = 0.927 \cdot 10^{-27} \text{ergs /gauss} = 0.927 \cdot 10^{-30} \frac{J}{T} \tag{6.243}$$

where m is the electron mass and c is the speed of light. The actual magnetic moment $\vec{\mu}$ of the electron is aligned with particle spin $\vec{s} = (\hbar/2)\vec{\sigma}$:

$$\vec{\mu} = \frac{g\mu_0}{\hbar}\vec{s} = \frac{g\mu_0}{2}\vec{\sigma} \tag{6.244}$$

$$V = \vec{\mu} \cdot \vec{B}, \quad g = 2.00232 \tag{6.245}$$

The g-factor would be two except for field theoretical corrections. They are often ignored, except where great accuracy is required, and then $\vec{\mu} \approx \mu_0\vec{\sigma}$. The interaction between the magnetic field $\vec{B}$ and the electron spin in (6.245) is called the *Pauli term*. In our notation, angular momenta with vector arrows above have the dimensions of Planck's constant, where the same quantities without vector symbols are dimensionless. For example, $\vec{s} \cdot \vec{s} = \hbar^2 s(s+1)$.

The spin–orbit interaction is another relativistic term. It provides a correction to the energy of particles that have both a spin and orbital motion. The form of the interaction is

$$V_{\text{so}} = \frac{1}{2m^2c^2}\vec{s} \cdot \left(\vec{\nabla} V \times \mathbf{p}\right) \tag{6.246}$$

A thorough derivation of (6.245) and (6.246) is provided in chapter 11. A very unrigorous derivation of V_{so} is given here, to make the result plausible.

A particle with charge e moving with velocity $\vec{v}$ in a magnetic field $\vec{B}$ experiences a Lorentz force of $e\vec{v} \times \vec{B}/c$. The force is an effective electric field generated by the motion in a magnetic field. In a symmetric way, motion in an electric field $\vec{E}$ generates an equivalent magnetic field of $\vec{v}\times\vec{E}/c$. A possible physical model is to visualize an electron in orbit around an atom. Transform coordinates to the rest frame of the electron so the atom revolves around it. The swirling charge generates magnetic fields that interact with the magnetic moment of the stationary electron. Using $g = 2$ and eqn. (6.245) then

$$V'_{so} = \frac{1}{c}\vec{\mu}\cdot(\vec{v}\times\vec{E}) = \frac{|e|}{mc^2}\vec{s}\cdot(\vec{v}\times\vec{E}) \tag{6.247}$$

The electric field is related to the gradient of the potential energy $e\vec{E} = -\vec{\nabla}V$, and the velocity is replaced by the momentum $\vec{v} = \vec{p}/m$. The factors of $\mathbf{p}$ and $\vec{\nabla}V$ are interchanged so $\vec{p} = -i\hbar\vec{\nabla}$ does not operate on V:

$$V'_{so} = -\frac{1}{m^2c^2}\vec{s}\cdot(\vec{\nabla}V\times\vec{p}) \tag{6.248}$$

The result is similar to (6.246). They differ by $-\frac{1}{2}$. The relativistic derivation provides this factor.

6.7.1 Spin–Orbit in Atoms

The spin–orbit interaction is important in the description of electron states in atoms. There the potential energy $V(\mathbf{r})$ is usually well approximated as a central potential, even when the electronic shells are partially filled. For spherically symmetric potentials, the gradient of $V(\mathbf{r})$ has only a radial derivative,

$$\vec{\nabla}V(r) = \hat{r}\frac{dV}{dr} \tag{6.249}$$

$$\vec{\nabla}V\times\vec{p} = \frac{1}{r}\frac{dV}{dr}(\mathbf{r}\times\mathbf{p}) \tag{6.250}$$

$$= \frac{1}{r}\frac{dV}{dr}\vec{\ell}, \quad \vec{\ell} = \mathbf{r}\times\mathbf{p} \tag{6.251}$$

The angular momentum is $\vec{\ell} = \mathbf{r}\times\mathbf{p}$. The spin–orbit interaction can be written as

$$V_{so} = \zeta(r)\frac{\vec{s}\cdot\vec{\ell}}{\hbar^2} \tag{6.252}$$

$$\zeta(r) = \frac{\hbar^2}{2m^2c^2}\frac{1}{r}\frac{dV}{dr} \tag{6.253}$$

For spherically symmetric potentials, the spin–orbit interaction can be written as a coupling between the spin and orbital angular momentum of the electron. The interaction is nonzero only for states with nonzero orbital angular momentum: there is no spin–orbit interaction for atomic s-states.

For an orbital state $\phi_i(\mathbf{r}) = |i\rangle$, first-order perturbation theory gives

$$\langle i|V_{so}|i\rangle = \frac{\zeta_i}{\hbar^2}\langle\vec{s}_i\cdot\vec{\ell}_i\rangle \tag{6.254}$$

$$\zeta_i = \int d^3r|\phi_i(\mathbf{r})|^2\zeta(r) \tag{6.255}$$

ζ_i has the units of energy. For atoms with large atomic number, then ζ_i is as large as an electron-volt, even for electrons in the outer shell of the atom. Generally, ζ_i is a positive number. Although $V(r) < 0$, it becomes less negative at large distance, so its radial derivative is positive.

Equation (6.254) also contains the factor of $\langle \vec{s}_i \cdot \vec{\ell}_i \rangle$. This factor is easy to evaluate if there is only one electron in the atomic shell. This simple case is assumed here. For example, it could be the $3p$ electron in sodium, or the $4p$ electron in potassium.

Section 6.1 discussed the coupling of two spins. The same technique is used here for the coupling of spin and angular momentum. Since $s = \hbar/2$ for electron spins, the problem is the coupling of two angular momentum: one of ℓ and one of $\frac{1}{2}$. The summation of these two contributions is the total angular momentum $\vec{j}_i$:

$$\vec{j}_i = \vec{\ell}_i + \vec{s}_i \tag{6.256}$$

$$\vec{j}_i \cdot \vec{j}_i = (\vec{\ell}_i + \vec{s}_i) \cdot (\vec{\ell}_i + \vec{s}_i) \tag{6.257}$$

$$\hbar^2 j_i(j_i+1) = \hbar^2 \ell_i(\ell_i+1) + \hbar^2 s_i(s_i+1) + 2\langle \vec{s}_i \cdot \vec{\ell}_i \rangle \tag{6.258}$$

$$\langle \vec{s}_i \cdot \vec{\ell}_i \rangle = \frac{\hbar^2}{2}[j_i(j_i+1) - \ell_i(\ell_i+1) - s_i(s_i+1)] \tag{6.259}$$

For electrons in atoms, $s_i = \frac{1}{2}$ and $j_i = \ell_i \pm \frac{1}{2}$. Equation (6.254) gives

$$\langle i|V_{so}|i\rangle = \frac{\zeta_i}{2}\left[j_i(j_i+1) - \ell_i(\ell_i+1) - \frac{3}{4}\right] \tag{6.260}$$

The two possible states are

$$j_i = \ell_i + \frac{1}{2} \qquad \langle i|V_{so}|i\rangle = \frac{1}{2}\zeta_i \ell_i \tag{6.261}$$

$$j_i = \ell_i - \frac{1}{2} \qquad \langle i|V_{so}|i\rangle = -\frac{1}{2}\zeta_i(\ell_i+1) \tag{6.262}$$

As an example, p-states have $\ell_i = 1$ and $j_i = \frac{3}{2}, \frac{1}{2}$. The eigenvalues of the spin–orbit interaction for these states are

$$\langle i|V_{so}|i\rangle = \begin{cases} \zeta_i/2 & j = \frac{3}{2} \\ -\zeta_i & j = \frac{1}{2} \end{cases} \tag{6.263}$$

Figure 6.5 shows how the energy state E_{np} is split into two levels by the spin–orbit interaction. The original p-states are sixfold degenerate, with a threefold degeneracy from the orbital motion ($m = -1, 0, 1$) and a twofold degeneracy from the spin. The combined system also has six states: four for $j = \frac{3}{2}$ and two for $j = \frac{1}{2}$.

The *center of gravity* is defined as the summation of all of the energy levels, multiplied by their degeneracy. It is the sum of all eigenvalues of all states. For this example, multiply the $j = \frac{3}{2}$ energy ($\zeta_i/2$) by four and add it to the $j = \frac{1}{2}$ energy ($-\zeta_i$) by two to get

$$\mathrm{CG} = \zeta_i\left[4\left(\frac{1}{2}\right) + 2(-1)\right] = 0 \tag{6.264}$$

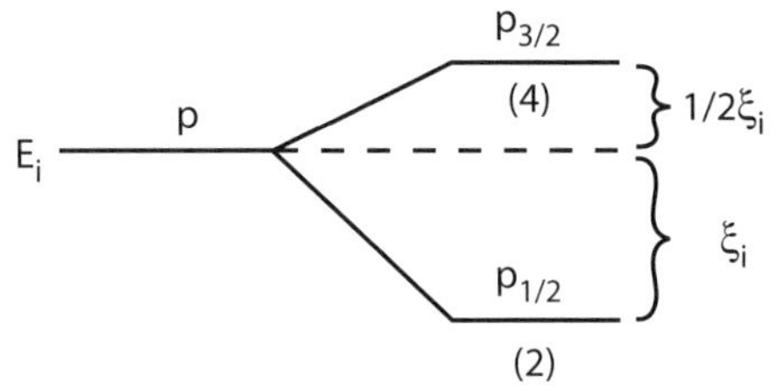

FIGURE 6.5. Spin–orbit splitting of an atomic p-state.

The result is zero. The same zero is obtained for any value of ℓ_i. In (6.261) the state with $j_i = \ell_i + \frac{1}{2}$ has a degeneracy of $(2j_i + 1) = 2(\ell_i + 1)$, and in (6.262) the state with $j_i = \ell_i - \frac{1}{2}$ has a degeneracy of $(2j_i + 1) = 2\ell_i$. The center of gravity is

$$CG = \xi[2(\ell_i + 1)\ell_i + 2\ell_i(-\ell_i - 1)] = 0 \tag{6.265}$$

The center of gravity, as calculated in first-order perturbation theory, is not changed by the spin–orbit interaction. Higher orders of perturbation theory can usually be neglected since ζ_i is small.

The eigenfunctions of the atomic states, that are split by the spin–orbit interaction, are easily found from table 4.1 of Clebsch-Gordon coefficients. They are just the states with $j = \ell \pm \frac{1}{2}$. For p-states, let (α, β) denote spin-up and spin-down, and $|1, m\rangle$ denote orbital angular momentum.

1. The eigenstates for $j = \frac{3}{2}$ are

$$\begin{aligned}
\left|\frac{3}{2}, \frac{3}{2}\right\rangle &= \left|1, 1\right\rangle \alpha \\
\left|\frac{3}{2}, \frac{1}{2}\right\rangle &= \frac{1}{\sqrt{3}}\left|1, 1\right\rangle \beta + \sqrt{\frac{2}{3}}\left|1, 0\right\rangle \alpha \\
\left|\frac{3}{2}, -\frac{1}{2}\right\rangle &= \frac{1}{\sqrt{3}}\left|1, -1\right\rangle \alpha + \sqrt{\frac{2}{3}}\left|1, 0\right\rangle \beta \\
\left|\frac{3}{2}, -\frac{3}{2}\right\rangle &= \left|1, -1\right\rangle \beta
\end{aligned} \tag{6.266}$$

2. The eigenstates for $j = 1/2$ are

$$\begin{aligned}
\left|\frac{1}{2}, \frac{1}{2}\right\rangle &= \sqrt{\frac{2}{3}}\left|1, 1\right\rangle \beta - \sqrt{\frac{1}{3}}\left|1, 0\right\rangle \alpha \\
\left|\frac{1}{2}, -\frac{1}{2}\right\rangle &= \sqrt{\frac{2}{3}}\left|1, -1\right\rangle \alpha - \sqrt{\frac{1}{3}}\left|1, 0\right\rangle \beta
\end{aligned} \tag{6.267}$$

These states are usually denoted as $p_{3/2}, p_{1/2}$. Chapter 5 described states p_z, p_+, p_-. Later we used p_x, p_y, p_z. Now we have the set $p_{3/2}, p_{1/2}$. These different representations are all linear combinations of each other. Each combination is useful for a specific problem. Table 6.3 shows values of the spin–orbit parameter ζ_i for outer p-shell of the rare gas atoms. Since the energy difference in (6.263) is $3\zeta_i/2$, the values in table 6.3 are $(\frac{2}{3})$ of the energy difference between the ionization energies for the $p_{3/2}$ and $p_{1/2}$ states.

Table 6.3 Spin–Orbit Splittings between Lowest $p_{3/2}$ and $p_{1/2}$ States in Singly Ionized Rare Gas Atoms and Neutral Alkali Atoms.

Ion	ζ_i	Atom	ζ_i
NeII	782	Na	17.2
ArII	1432	K	57.7
KrII	5371	Rb	237.6
XeII	10537	Cs	554.1

Note. Units are cm^{-1}. Data from C.E. Moore *Atomic Energy Levels* (NBS).

Another way to determine the influence of the spin–orbit interaction is to diagonalize the matrix. The matrix method produces the same eigenvalues and eigenvectors as the above method of vector addition. The technique is similar to the solution of the coupled spin problem following (6.32). The matrix method is required when the spin–orbit interaction is solved along with another interaction such as an electric field or a magnetic field.

The $\vec{\ell}\cdot\vec{s}$ term is written in terms of raising and lowering operators:

$$\vec{\ell}\cdot\vec{s}=\ell_z s_z+\frac{1}{2}\left[\ell^{(+)}s^{(-)}+\ell^{(-)}s^{(+)}\right] \tag{6.268}$$

We solve the spin–orbit interaction for the six p-states using the $q_j(m)$ factors from chapter 4. Some sample results are

$$\vec{\ell}\cdot\vec{s}|1,1\rangle\beta=\hbar^2\left[-\frac{1}{2}|1,1\rangle\beta+\frac{1}{\sqrt{2}}|1,0\rangle\alpha\right] \tag{6.269}$$

$$\vec{\ell}\cdot\vec{s}|1,0\rangle\alpha=\frac{\hbar^2}{\sqrt{2}}|1,1\rangle\beta \tag{6.270}$$

$$\vec{\ell}\cdot\vec{s}|1,0\rangle\beta=\frac{\hbar^2}{\sqrt{2}}|1,-1\rangle\alpha \tag{6.271}$$

All the matrix elements can be found for the 6×6 pairs of states. The states are arranged in decreasing value of their total value of magnetic quantum number $M=m_s+m_\ell$:

$$\begin{vmatrix} \frac{\zeta}{2} & 0 & 0 & 0 & 0 & 0 \\ 0 & \frac{-\zeta}{2} & \frac{\zeta}{\sqrt{2}} & 0 & 0 & 0 \\ 0 & \frac{\zeta}{\sqrt{2}} & 0 & 0 & 0 & 0 \\ 0 & 0 & 0 & 0 & \frac{\zeta}{\sqrt{2}} & 0 \\ 0 & 0 & 0 & \frac{\zeta}{\sqrt{2}} & -\frac{\zeta}{2} & 0 \\ 0 & 0 & 0 & 0 & 0 & \frac{\zeta}{2} \end{vmatrix} \begin{bmatrix} |1,1\rangle\alpha \\ |1,1\rangle\beta \\ |1,0\rangle\alpha \\ |1,0\rangle\beta \\ |1,-1\rangle\alpha \\ |1,-1\rangle\beta \end{bmatrix} \tag{6.272}$$

The matrix has block form. There are two blocks of 2×2 in the center, and two blocks of 1×1 at the ends. The latter have eigenvalue $\lambda = \zeta/2$. The 2×2 blocks need to be diagonalized. They are identical, so just one needs to be done:

$$0 = \det\begin{vmatrix} -\frac{\zeta}{2}-\lambda & \frac{\zeta}{\sqrt{2}} \\ \frac{\zeta}{\sqrt{2}} & -\lambda \end{vmatrix} \tag{6.273}$$

$$0 = \lambda^2 + \frac{\zeta}{2}\lambda - \frac{\zeta^2}{2} \tag{6.274}$$

$$\lambda = \frac{\zeta}{2}, -\zeta \tag{6.275}$$

In total, there are four states with eigenvalue $\lambda = \zeta/2$, and two states with $\lambda = -\zeta$. The same eigenvalues were found in (6.263). The degeneracy is also correct for the $j = \frac{3}{2}$ and $j = \frac{1}{2}$ states. The matrix method provides the correct result.

6.7.2 Alkali Valence Electron in Electric Field

A static electric field acting on an atom causes a variety of phenomena. If the field is strong enough, the atom can be ionized: an electron is pulled directly off the atom. Very large fields are required for this stripping process. An estimate from dimensional analysis is that $eFd = E_B$, where $d = \langle r \rangle$ is the radius of the orbit, and E_B is the binding energy. The formula predicts a required field of 10^{10} volts per meter. Experimental fields are seldom that large, and field ionization is a rare occurance.

Usually, the influence of the external static electric field is to change the energy levels of the atom. Several discussions of the phenomena have been given in this chapter: The Stark effect for hydrogen is treated in section 6.2, and the polarizability is discussed in section 6.5. The present section presents a more complicated and realistic example of the role of electric fields. There is no linear term in electric field at small fields for most atoms since none of the energy levels are degenerate. The spin–orbit interaction also changes the response to the field.

The competition between the electric field and the spin–orbit interaction is pedagogically interesting. It is not obvious which is the best basis set for calculating the matrix elements. A good choice of basis set usually reduces the effort in the calculation. The matrix has fewer off-diagonal elements and is easier to diagonalize. The spin–orbit interaction is diagonal in a basis using the eigenstates of $|j, m_j\rangle$, such as $p_{3/2}$ and $p_{1/2}$. The electric field is easy to diagonalize in a basis set using $|\ell, m_\ell\rangle\chi$, where χ is α or β for the spin states. The states $|\ell, m_\ell\rangle$ are eigenstates of the parity operator and render most matrix elements zero. Whichever basis is selected, one has to find all the matrix elements and diagonalize the matrix.

Here we discuss the application of an electric field for the outer electron in an alkali atom. There is only one electron, since we assume the closed shell is harmless. We include in our basis set eight states: the ground state (*ns*) with spin-up and spin-down, and the first excited (*np*) state with all six spin and orbital possibilities.

We select the states $|j, m_j\rangle$ as the basis set for the matrix. This choice makes the spin–orbit interaction diagonal. The states have mixed parity, so that quantum number cannot be used set to zero some of the matrix elements. But the magnetic quantum number $m_j = m_s + m_\ell$ is a valid quantum number. It can be used to set to zero many matrix elements. Since the interaction term $V_F = -eFz$ has no spin dependence,

$$\langle j, m_j | V_F | j', m_{j'} \rangle = D(j, j') \delta_{m_j m_{j'}} \tag{6.276}$$

Let H_0 be the Hamiltonian of the atom without spin–orbit or electric field. The (ns) and (np) states $|s\rangle$, $|1, m_\ell\rangle$ have eigenvalue E_s or E_p for the eight states in our basis. Introduce the following matrix elements:

$$\begin{aligned} \Delta &= eF\langle s|z|1,0\rangle \\ 0 &= eF\langle s|z|1,\pm 1\rangle \\ 0 &= eF\langle 1,m|z|1,m'\rangle \end{aligned} \tag{6.277}$$

The second one vanishes due to the conservation of magnetic quantum number. The third one vanishes due to parity.

When evaluating the matrix elements of V_F using the basis set $|j, m_j\rangle$, we need to pick out the factors in (6.277). Also, the spins have to be in alignment. The $j = \frac{1}{2}$ states from p-orbitals are denoted as $|\frac{1}{2}, \pm\frac{1}{2}\rangle$, while those from s-orbitals are $|s, \alpha\rangle$, $|s, \beta\rangle$. Some typical matrix elements are

$$\left\langle s, \alpha | eFz | \frac{3}{2}, \frac{3}{2} \right\rangle = eF\langle\alpha|\alpha\rangle\langle s|z|1,1\rangle = 0 \tag{6.278}$$

$$\begin{aligned} \left\langle s, \alpha | eFz | \frac{3}{2}, \frac{1}{2} \right\rangle &= eF\left[\sqrt{\frac{1}{3}}\langle\alpha|\beta\rangle\langle s|z|1,1\rangle + \sqrt{\frac{2}{3}}\langle\alpha|\alpha\rangle\langle s|z|1,0\rangle\right] = \sqrt{\frac{2}{3}}\Delta \\ \left\langle s, \alpha | eFz | \frac{3}{2}, -\frac{1}{2} \right\rangle &= eF\left[\sqrt{\frac{2}{3}}\langle\alpha|\beta\rangle\langle s|z|1,0\rangle + \sqrt{\frac{1}{3}}\langle\alpha|\alpha\rangle\langle s|z|1,-1\rangle\right] = 0 \\ \left\langle s, \alpha | eFz | \frac{3}{2}, -\frac{3}{2} \right\rangle &= 0 \end{aligned} \tag{6.279}$$

Orthogonal spin combinations $\langle\alpha|\beta\rangle = 0$. Writing out the 8×8 matrix gives, using the notation $\Delta_1 = \sqrt{\frac{1}{3}}\Delta$, $\Delta_2 = \sqrt{\frac{2}{3}}\Delta$:

$$\left|\begin{matrix} E_p + \frac{\zeta}{2} & 0 & 0 & 0 & 0 & 0 & 0 & 0 \\ 0 & E_p + \frac{\zeta}{2} & 0 & 0 & 0 & 0 & \Delta_2 & 0 \\ 0 & 0 & E_p + \frac{\zeta}{2} & 0 & 0 & 0 & 0 & \Delta_2 \\ 0 & 0 & 0 & E_p + \frac{\zeta}{2} & 0 & 0 & 0 & 0 \\ 0 & 0 & 0 & 0 & E_p - \zeta & 0 & -\Delta_1 & 0 \\ 0 & 0 & 0 & 0 & 0 & E_p - \zeta & 0 & \Delta_1 \\ 0 & \Delta_2 & 0 & 0 & -\Delta_1 & 0 & E_s & 0 \\ 0 & 0 & \Delta_2 & 0 & 0 & \Delta_1 & 0 & E_s \end{matrix}\right| \left[\begin{matrix} |\frac{3}{2}, \frac{3}{2}\rangle \\ |\frac{3}{2}, \frac{1}{2}\rangle \\ |\frac{3}{2}, -\frac{1}{2}\rangle \\ |\frac{3}{2}, -\frac{3}{2}\rangle \\ |\frac{1}{2}, \frac{1}{2}\rangle \\ |\frac{1}{2}, -\frac{1}{2}\rangle \\ |s, \alpha\rangle \\ |s, \beta\rangle \end{matrix}\right]$$

Nonzero matrix elements exist only between eigenstates with the same value of magnetic quantum number m_j. Most of the matrix elments are zero since they exist only between the two s-states and the six p-states.

Now that we have derived the matrix, the next step is to change the order of the eigenfunctions, to reduce the matrix to block-diagonal form. Rather than write it all out, we note that $|\frac{3}{2}, \pm\frac{3}{2}\rangle$ has no nonzero matrix elements with any other state. So these two states have the eigenvalue $\lambda = E_p + \zeta/2$. The three states with $m_j = \frac{1}{2}$ mutual interact. The 3×3 block is

$$\begin{vmatrix} E_p + \frac{\zeta}{2} & 0 & \Delta_2 \\ 0 & E_p - \zeta & -\Delta_1 \\ \Delta_2 & -\Delta_1 & E_s \end{vmatrix} \begin{bmatrix} \left|\frac{3}{2}, \frac{1}{2}\right\rangle \\ \left|\frac{1}{2}, \frac{1}{2}\right\rangle \\ |s, \alpha\rangle \end{bmatrix} \tag{6.280}$$

Find the eigenvalues of this block using

$$0 = \det\begin{vmatrix} E_p + \frac{\zeta}{2} - \lambda & 0 & \Delta_2 \\ 0 & E_p - \zeta - \lambda & -\Delta_1 \\ \Delta_2 & -\Delta_1 & E_s - \lambda \end{vmatrix} \tag{6.281}$$

After some algebra, the cubic eigenvalue equation is

$$\begin{aligned} 0 = \lambda^3 - \lambda^2\left(2E_p + E_s - \frac{\zeta}{2}\right) + \lambda\left[E_s\left(2E_p + \frac{\zeta}{2}\right) + E_p\left(E_p - \frac{\zeta}{2}\right) - \Delta^2 - \frac{\zeta^2}{2}\right] \\ - E_s(E_p - \zeta)\left(E_p + \frac{\zeta}{2}\right) + \Delta^2\left(E_p - \frac{\zeta}{2}\right) \end{aligned} \tag{6.282}$$

The other 3×3 block, for $m_j = -\frac{1}{2}$, has exactly the same matrix and the same eigenvalue equation. Each of these states is doubly degenerate.

Any cubic equation has an analytical solution. The present one is too messy to present here. The important physics is obtained by first setting to zero the spin–orbit interaction ($\zeta = 0$). Then the cubic equation factors into a simple root ($\lambda = E_p$) times a quadratic equation:

$$0 = (\lambda - E_p)[\lambda^2 - \lambda(E_s + E_p) + E_s E_p - \Delta^2] \tag{6.283}$$

$$\lambda = E_p \tag{6.284}$$

$$\lambda = \frac{1}{2}[E_s + E_p \pm \sqrt{(E_s - E_p)^2 + 4\Delta^2}] \tag{6.285}$$

The results are plotted in figure 6.6a. The parameter Δ is proportional to the electric field. The sixfold degenerate p-states are split into two that depend on electric field and four that are unaffected.

The s-states decrease in energy. At low electric field the change in energy goes as $O(F^2)$. The coefficient of the F^2 term defines the polarizability:

$$\lambda = \frac{1}{2}\left[E_s + E_p - (E_p - E_s)[1 + \frac{2\Delta^2}{(E_p - E_s)^2} + \cdots]\right] \tag{6.286}$$

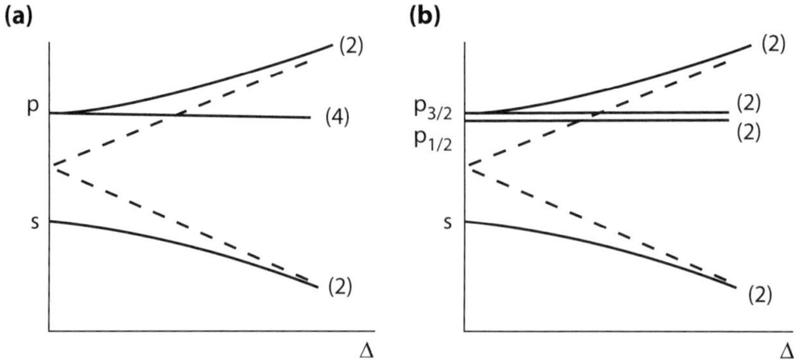

FIGURE 6.6. (a) How eigenvalues depend on electric field $\Delta \propto F$ when neglecting the spin–orbit interaction. (b) The dependence when including the spin–orbit interaction.

$$\lambda \approx E_s - \frac{\Delta^2}{E_p - E_s} + O(F^4) \tag{6.287}$$

$$\alpha = 2e^2 \frac{|\langle 1, 0|z|s\rangle|^2}{E_p - E_s} \tag{6.288}$$

The polarizability contains the square of the a single matrix element between the *s*- and *p*-state, divided by their energy difference. That is exactly the correct result for α when the basis set contains only one pair of *s*- and *p*-states.

At large values of field, when $\Delta > E_p - E_s$, the energy change asymptotically becomes linear with the field. The dashed line is the Stark effect for a set of degnerate levels ($E_s = E_p$). The energy levels change quadratically with electric field at small fields and linearly at high fields.

Including the spin–orbit interaction causes little change in these results. At zero electric field ($\Delta = 0$), the spin–orbit interaction splits the sixfold degenerate *p*-state into the four $j = \frac{3}{2}$ states and the two $j = \frac{1}{2}$ states. Figure 6.6b shows the behavior at nonzero electric fields, including the spin–orbit interaction, obtained by solving eqn. (6.282). The fourfold $p_{3/2}$ state is split by the electric field. Two states increase with field: they come from the components with $m_j = \pm\frac{1}{2}$. Two are unaffected: they come from the $m_j = \pm\frac{3}{2}$ components. The two states in $p_{1/2}$ are are slightly affected: the spin–orbit splitting gradually vanishes at very large field. The *s*-state has the same variation with field as in figure 6.6a.

This example of an alkali atom in an electric field is typical of many atomic physics experiments that measure the response of eigenvalues to external perturbations such as electric or/and magnetic fields.

6.8 Bound Particles in Magnetic Fields

The behavior of free particles in magnetic fields is discussed in the last chapter. There the quantum of magnetic energy was found to be $\hbar\omega_c = |e|\hbar B/mc$. This energy is very small for magnetic fields ($B \sim 1$T) typical of most laboratory experiments. If m is the electron mass, the energy $\hbar\omega_c$ is less than one millielectron volt. This energy is very small compared to the energies that bind electrons to atoms. So the magnetic field has only a small influence on the energy states, which can be accurately evaluated in perturbation theory.

For the discussion of electron states in atoms, a static magnetic field may be treated as a constant over the space of the atom. The magnetic field enters the Hamiltonian as a vector potential, exactly as in chapter 5. We also include the Pauli term:

$$H = \frac{1}{2m}\sum_j \left[\mathbf{p}_j - \frac{e}{c}A(\mathbf{r}_j)\right]^2 + \left(\sum_j \vec{\mu}_j\right)\cdot\mathbf{B} + V \tag{6.289}$$

$$V = -Ze^2\sum_j \frac{1}{r_j} + \frac{e^2}{2}\sum_{ij}\frac{1}{r_{ij}}$$

The potential V is the electron–nucleus and electron–electron interaction. In atoms, we divide these terms as

$$H = H_0 + V_B \tag{6.290}$$

$$H_0 = \frac{1}{2m}\sum_j p_j^2 + V \tag{6.291}$$

$$V_B = \sum_j\left[\frac{|e|}{mc}A(\mathbf{r}_j)\cdot\mathbf{p}_j + \frac{e^2}{2mc^2}A(\mathbf{r}_j)^2\right] + \left(\sum_j \vec{\mu}_j\right)\cdot\mathbf{B} \tag{6.292}$$

The Hamiltonian H_0 is that of the atom and generates all of the atomic eigenfunctions and eigenvalues. The potential energy V_B has all the terms generated by the applied magnetic field. This section discusses how V_B changes the eigenvalues of electron states in atoms. The first term in (6.292) is called "p-dot-A" and the second term is called "A-squared."

The vector potential A_μ commutes with p_ν when $\mu = \nu$. Although $A_\mu(\mathbf{r})$ is a function of position, it does not contain r_μ. So it does not matter in what order we write $\mathbf{A}\cdot\mathbf{p}$ or $(\mathbf{p}\cdot\mathbf{A})$ in (6.292). We use the symmetric gauge for the vector potential:

$$\mathbf{A} = \frac{1}{2}\mathbf{B}\times\mathbf{r}, \quad \mathbf{B} = \vec{\nabla}\times\mathbf{A} \tag{6.293}$$

The vector identity $\mathbf{A}\cdot(\mathbf{B}\times\mathbf{C}) = (\mathbf{A}\times\mathbf{B})\cdot\mathbf{C}$ is used to rearrange the terms in the p-dot-A interaction. Also remember that the orbital angular momentum is $\ell = \mathbf{r}\times\mathbf{p}$, and the magnetic moment is $\vec{\mu}_j = g\mu_0\vec{s}_j/\hbar$:

$$\mathbf{A}\cdot\mathbf{p} = \frac{1}{2}(\mathbf{B}\times\mathbf{r})\cdot\mathbf{p} = \frac{1}{2}\mathbf{B}\cdot(\mathbf{r}\times\mathbf{p}) = \frac{1}{2}\mathbf{B}\cdot\vec{\ell} \tag{6.294}$$

$$V_B = \frac{|e|}{2mc}\mathbf{B}\cdot\sum_j(\vec{\ell}_j + g\vec{s}_j) + \frac{e^2}{8mc^2}\sum_j(\mathbf{B}\times\mathbf{r}_j)^2 \tag{6.295}$$

A sign change occurred in the first term since the electron charge is negative $e = -|e|$. The magnetic moment μ_0 is also positive. The summation over all orbital angular momentum is $\vec{L}$, and over all spins is $\vec{S}$:

$$V_B = \frac{\mu_0}{\hbar}\mathbf{B}\cdot(\vec{L} + g\vec{S}) + \frac{e^2}{8mc^2}\sum_j(\mathbf{B}\times\mathbf{r}_j)^2 \tag{6.296}$$

Three different terms depend on the magnetic field. They have standard names:

$$\begin{aligned} &\text{Pauli term} && \frac{\mu_0}{\hbar} g\vec{S}\cdot\mathbf{B} \\ &\text{Landau term} && \frac{\mu_0}{\hbar}\vec{L}\cdot\mathbf{B} \\ &\text{Diamagnetic term} && \frac{e^2B^2}{8mc^2}\sum_j (\mathbf{r}_{\perp j})^2 \end{aligned} \tag{6.297}$$

The notation $\mathbf{r}_\perp$ denotes the component perpendicular to the magnetic field. The Landau term is also called the *paramagnetic term*. These three potential terms contain the influence of the magnetic field on the electron states in the atom. They are the starting point for perturbation theory.

6.8.1 Magnetic Susceptibility

Let χ denote the magnetic susceptibility of the atom. It plays the role for the magnetic field that is analogous to the electronic polarizability α for electric fields. The change in energy ΔE from the field and the magnetic moment $\vec{M}$ are both determined by χ:

$$\Delta E = -\frac{1}{2}\chi B^2 \tag{6.298}$$

$$\vec{M} = \chi\mathbf{B} \tag{6.299}$$

Both formulas assume the atom has no intrinsic magnetic moment in the absence of the magnetic field **B**. Of course, there are many atoms that do have magnetic moments: those with partially filled *d*- and *f*-shells. The present theory does not apply to them. Materials are called *paramagnetic* if $\chi > 0$, and *diamagnetic* if $\chi < 0$. The names for the interaction terms (6.297) derive from these definitions.

In perturbation theory, the diamagnetic term in (6.297) gives an energy shift that can be found from first-order perturbation theory:

$$E_g^{(1)} = \frac{e^2B^2}{8mc^2}\sum_j \langle g|\mathbf{r}_{\perp j}^2|g\rangle = -\frac{1}{2}\chi B^2 \tag{6.300}$$

$$\chi_d = -\frac{e^2}{4mc^2}\sum_j \langle g|\mathbf{r}_{\perp j}^2|g\rangle \tag{6.301}$$

The diamagnetic potential in (6.297) gives a diamagnetic susceptibility. Atoms with all electron shells closed have $\vec{L}=0$, $\vec{S}=0$. Then the Landau and Pauli interactions are zero and there is only the diamagnetic term. Atoms with all closed shells, such as the rare gases helium, neon, and argon, all have diamagnetic susceptibilities. $\chi < 0$ means $\vec{M}$ points in the opposite direction of $\vec{B}$. The atom sets up eddy currents that oppose the external magnetic field.

Similarly, the paramagnetic (Landau) term is so-named because it gives a susceptibility that is positive. Usually the contribution from first-order perturbation theory is zero. If the magnetic field is in the $\hat{z}$ direction, the interaction depends on $L_z B$ and the eigenvalues of L_z are $\hbar m_\ell B$. But m_ℓ is equally likely to have plus or minus values, and its average is zero. The first-order term is zero, and the paramagnetic susceptibility comes from second-order perturbation theory. Define the direction of magnetic field as $\hat{z}$, $\chi = \chi_d + \chi_p$, and then

$$E_g^{(2)} = \frac{\mu_0^2 B^2}{\hbar^2} \sum_m \frac{|\langle m|(L_z + gS_z)|g\rangle|^2}{E_g - E_m} \tag{6.302}$$

$$\chi_p = 2\frac{\mu_0^2}{\hbar^2} \sum_m \frac{|\langle m|(L_z + gS_z)|g\rangle|^2}{E_m - E_g} \tag{6.303}$$

Since $E_m > E_g$, the energy denominator is positive in the formula for χ_p, as is the entire expression. For electron states that are partially filled, the susceptibility has contributions from both diamagnetic and paramagnetic terms. Paramagnetic terms are usually larger than diamagnetic ones, so usually $\chi > 0$.

6.8.2 Alkali Atom in Magnetic Field

A simple example of magnetic phenomena is provided by the alkali atoms. The neutral atoms have a single electron outside of a closed shell. For the low-energy perturbations from magnetic fields, one can ignore the closed shell and consider only the response of the single outer electron.

We examine how *p*-states are perturbed by the magnetic field, along with the spin–orbit interaction. Each is solved easily alone, but together they are complicated. Again, the technique is to construct a matrix and find its eigenvalues. Both the magnetic energy and the spin–orbit interaction are small, and the mixing of states from different atomic shells can usually be neglected. The two energies are of similar size in moderate magnetic fields.

For moderate magnetic fields the diamagnetic interaction is quite small and may be neglected compared to the other interactions. In this section we neglect terms of $O(B^2)$. The three interactions under consideration are spin–orbit, Pauli, and Landau. Set $g = 2$ in the Pauli term:

$$H_{\text{int}} = \frac{\zeta_i}{\hbar^2} \vec{\ell} \cdot \vec{s} + \frac{\mu_0}{\hbar} (\vec{\ell} + 2\vec{s}) \cdot \mathbf{B} \tag{6.304}$$

The total magnetic moment is defined as $\vec{\mu}_T = \mu_0(\vec{\ell} + 2\vec{s})/\hbar$. It includes both spin and orbital motion. There are two energy scales in this problem: the spin–orbit energy ζ_i is fixed by the choice of atom, while the magnetic energy $\Delta = \mu_0 B$ can be controlled experimentally.

Only the factor of $g = 2$ prevents exact diagonalizing of the perturbation in eqn. (6.304). If this factor were absent ($g = 1$), then the magnetic interaction contains $\vec{j} = \vec{\ell} + \vec{s}$ and $\vec{\mu} = \mu_0 \vec{j}$. The interaction energy is simply $m_j \Delta$. The basis set of total angular momentum $|j, m_j\rangle$ diagonalizes both terms in (6.304), and the entire problem is solved easily. The factor of $g = 2$ prevents this simple solution. When $g \neq 1$ the vector of total magnetic moment is not parallel to the vector of angular momentum. Then the solution is complicated.

The spin–orbit interaction is diagonalized in the basis set of angular momentum $|j, m_j\rangle$. It vanishes for an *s*-shell. For a *p*-shell, it has eigenvalues of $\zeta_i/2$ and $-\zeta_i$ for $j=\frac{3}{2}$ and $j=\frac{1}{2}$. The magnetic interaction is not diagonal in this basis set. Its natural basis set are the orbital–spin products $|\ell, m_\ell\rangle\alpha$, $|\ell, m_\ell\rangle\beta$. Then the eigenvalue is $\Delta(m_\ell + 2m_s)$. Again there is a problem of two perturbations, where each is naturally solved in a different basis set. To solve the combined problem, pick one of the basis sets and calculate the matrix elements. Here we chose the latter basis set. Then the spin–orbit interaction generates off-diagonal matrix elements, which are the same as in (6.272):

$$\begin{vmatrix} 2\Delta + \frac{\zeta}{2} & 0 & 0 & 0 & 0 & 0 \\ 0 & \frac{-\zeta}{2} & \frac{\zeta}{\sqrt{2}} & 0 & 0 & 0 \\ 0 & \frac{\zeta}{\sqrt{2}} & \Delta & 0 & 0 & 0 \\ 0 & 0 & 0 & -\Delta & \frac{\zeta}{\sqrt{2}} & 0 \\ 0 & 0 & 0 & \frac{\zeta}{\sqrt{2}} & -\frac{\zeta}{2} & 0 \\ 0 & 0 & 0 & 0 & 0 & -2\Delta + \frac{\zeta}{2} \end{vmatrix} \begin{bmatrix} |1,1\rangle\alpha \\ |1,1\rangle\beta \\ |1,0\rangle\alpha \\ |1,0\rangle\beta \\ |1,-1\rangle\alpha \\ |1,-1\rangle\beta \end{bmatrix} \tag{6.305}$$

The matrix has block-diagonal form. There are two 1×1 blocks for the states $|1, 1\rangle\alpha$, $|1, -1\rangle\beta$ with eigenvalues:

$$\lambda = 2\Delta + \frac{\zeta_i}{2}, \quad \lambda = -2\Delta + \frac{\zeta_i}{2} \tag{6.306}$$

The other four rows and columns divide into two 2×2 blocks. For the first one, the states $|1, 1\rangle\beta$, $|1,0\rangle\alpha$ both have $m = +\frac{1}{2}$. Its eigenvalues are found from

$$0 = \det\begin{vmatrix} -\lambda - \zeta/2 & \zeta/\sqrt{2} \\ \zeta/\sqrt{2} & -\lambda + \Delta \end{vmatrix} \tag{6.307}$$

The eigenvalues are

$$\lambda = \frac{1}{2}\left[\Delta - \frac{\zeta}{2} \pm \sqrt{(\Delta + \zeta/2)^2 + 2\zeta^2}\right] \tag{6.308}$$

The other 2×2 block in (6.305) has the same structure, but the sign of Δ is changed. Its eigenvalues are

$$\lambda = -\frac{1}{2}\left[\Delta + \frac{\zeta}{2} \pm \sqrt{(\Delta - \zeta/2)^2 + 2\zeta^2}\right] \tag{6.309}$$

The six eigenvalues in (6.306, 6.308, 6.309) provide the energy levels of the *p*-state as a function of the applied magnetic field. Figure 6.7 shows the energy splitting in a magnetic field for the 6*p* state of ^{199}Hg, according to Kaul and Latshaw [4]. At zero field the splitting

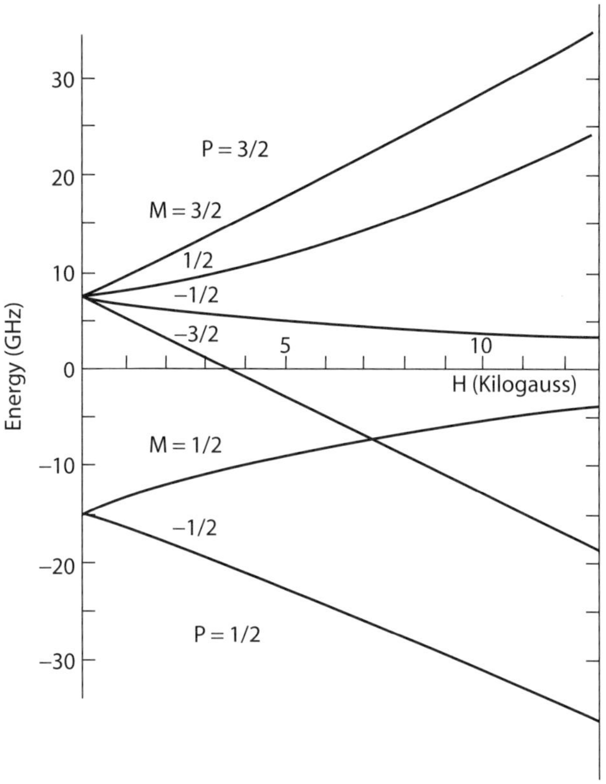

FIGURE 6.7. *p*-state energy splittings of Hg in a magnetic field. From Kaul and Latshaw [4].

is due to the spin–orbit interaction. At high values of magnetic field the splitting is due to the magnetic energy.

6.8.3 Zeeman Effect

The weak field response is called the Zeeman effect. At low values of B the magnetic energy is very small compared to the spin–orbit interaction $\zeta_i >> \Delta$. The spin–orbit interaction is the "unperturbed" Hamiltonian H_0, and the magnetic interaction is the perturbation. The basis set that diagonalizes the spin–orbit interaction is $|j, m_j\rangle$.

In (6.304) write $\vec{\ell} + 2\vec{s} = \vec{j} + \vec{s}$. The expectation value of this interaction is

$$\langle j, m_j | H_{int} | j, m_j\rangle = m_j\Delta + \frac{\zeta}{\hbar^2}\langle j, m_j|\vec{\ell}\cdot\vec{s}\,|j, m_j\rangle + \frac{\mu_0}{\hbar}\langle\, j, m_j|\vec{s}\cdot\mathbf{B}|j, m_j\rangle$$

$$\langle j, m_j|\vec{\ell}\cdot\vec{s}\,|j, m_j\rangle = \frac{\hbar^2}{2}[\,j(j+1) - \ell(\ell+1) - s(s+1)] \tag{6.310}$$

The latter identity was derived earlier in this chapter. We need to evaluate the factor of $\langle\, j, m_j|\vec{s}\cdot\mathbf{B}|\, j, m_j\rangle$. One's first impulse is to write it as $m_s\hbar B$. That is improper since m_s is not a valid quantum number in the $|j, m_j\rangle$ basis set. The only valid magnetic quantum number is m_j. The vectors $\vec{s}$ and $\vec{j}$ point in different directions. Evaluate the average value of $\vec{s}$ by its projection on $\vec{j}$:

$$\vec{s} \rightarrow \frac{\vec{j}(\vec{j} \cdot \vec{s})}{\vec{j} \cdot \vec{j}} \tag{6.311}$$

$$\frac{\mu_0}{\hbar}\langle j, m_j|\vec{s} \cdot \mathbf{B}| j, m_j\rangle = m_j \Delta \frac{\langle \vec{s} \cdot \vec{j}\rangle}{\langle \vec{j} \cdot \vec{j}\rangle}$$

We evaluate $\langle \vec{s} \cdot \vec{j}\rangle$ by writing $\vec{\ell} = \vec{j} - \vec{s}$ so

$$\vec{\ell} \cdot \vec{\ell} = (\vec{j} - \vec{s})^2 = \hbar^s[j(j+1) + s(s+1)] - 2\vec{s} \cdot \vec{j} \tag{6.313}$$

$$\langle \vec{s} \cdot \vec{j}\rangle = \frac{\hbar^2}{2}[j(j+1) + s(s+1) - \ell(\ell+1)] \tag{6.314}$$

All of the magnetic terms are now proportional to $m_j\Delta$. The constant of proportionality is called the *Landé g-factor*:

$$g = \frac{3}{2} + \frac{1}{2j(j+1)}[s(s+1) - \ell(\ell+1)] \tag{6.315}$$

$$\langle j, m_j|H_{int}| j, m_j\rangle = g m_j \Delta + \frac{\zeta}{2}[j(j+1) - \ell(\ell+1) - s(s+1)]$$

The first term is the magnetic energy, and the second term is the spin–orbit energy. The Landé g-factor has the same symbol g as the moment enhancement of the electron spin, but they are different quantities. Some values are given in table 6.4, assuming that $s = \frac{1}{2}$. The Landé g-factor gives the splitting of energy levels in a small magnetic field. This behavior is evident from the low-field region of figure 6.7.

The exact diagonalization of the matrix is given in eqns.(6.306), (6.308), and (6.309). Expanding these solutions for $\zeta >> \Delta$ gives

Table 6.4 Landé g-Factors for Atomic Orbitals

State	ℓ	j	g
s	0	$\frac{1}{2}$	2
p	1	$\frac{1}{2}$	$\frac{2}{3}$
p	1	$\frac{3}{2}$	$\frac{4}{3}$
d	2	$\frac{3}{2}$	$\frac{4}{5}$
d	2	$\frac{5}{2}$	$\frac{6}{5}$

$$E_{3/2} = \frac{\zeta}{2} + \begin{cases} 2\Delta & m_j = \frac{3}{2} \\ \frac{2\Delta}{3} & m_j = \frac{1}{2} \\ -\frac{2\Delta}{3} & m_j = -\frac{1}{2} \\ -2\Delta & m_j = -\frac{3}{2} \end{cases} \tag{6.316}$$

$$E_{1/2} = -\zeta + \begin{cases} \frac{\Delta}{3} & m_j = \frac{1}{2} \\ -\frac{\Delta}{3} & m_j = -\frac{1}{2} \end{cases} \tag{6.317}$$

The coefficients of the magnetic term agree perfectly with those predicted using Landé g-factors.

6.8.4 *Paschen-Back Effect*

The limit of very large magnetic field produces the Paschen-Back effect. By "large" magnetic field is meant the limit that $\Delta >> \zeta$. Here one treats the magnetic interaction as the dominant term and uses the basis set $|\ell, m_\ell\rangle\alpha/\beta$ that diagonalizes this interaction. The spin–orbit interaction is evaluated in this basis set using perturbation theory. It is exactly the mirror reverse of the procedure for the Zeeman effect.

Given an atomic orbital (ℓ) and the electron spin (s), the two quantum numbers are m_ℓ, m_s. The interaction Hamiltonian (6.304) is evaluated as

$$\langle \ell, m_\ell, m_s | H_{int} | \ell, m_\ell, m_s \rangle = \Delta(m_\ell + 2m_s) + \frac{\zeta}{\hbar^2}\langle \vec{\ell} \cdot \vec{s}\,\rangle \tag{6.318}$$

The spin–orbit interaction is found using first-order perturbation theory

$$\vec{\ell} \cdot \vec{s} = \ell_z s_z + \frac{1}{2}\left[\ell^{(+)} s^{(-)} + \ell^{(-)} s^{(+)}\right] \tag{6.319}$$

$$\langle \ell, m_\ell, m_s | \vec{\ell} \cdot \vec{s} | \ell, m_\ell, m_s \rangle = \hbar^2 m_\ell m_s \tag{6.320}$$

$$E^{(1)}(m_\ell, m_s) = \Delta(m_\ell + 2m_s) + \zeta m_\ell m_s \tag{6.321}$$

The only contribution from the spin–orbit interaction is the diagonal term from $\ell_z s_z$. Table 6.5 shows the Paschen-Back energies for an atomic p-shell. They would give the high-field limit in figure 6.7. The magnetic field in this figure is not large enough to attain this limit of $\Delta >> \zeta$. Finally, if one expands the eigenvalues in eqns. (6.306)–(6.308) for $\Delta >> \zeta$, they give the results of table 6.5.

References

1. E. Eisenschitz and F. London, *Z. Phys.* **60**, 491 (1930)
2. J.A. Barker and P.J. Leonard, *Phys. Lett.* **13**, 127 (1964)
3. G.D. Mahan, *J. Chem. Phys.* **48**, 950 (1968)
4. P.D. Kaul and W.S. Latshaw, *J. Opt. Soc. Am.* **62**, 615 (1972)

Table 6.5 Paschen-Back Energy Levels for an Atomic *p*-Shell

m_ℓ	m_s	$E^{(1)}$
-1	$-1/2$	$2\Delta+\zeta/2$
0	$-1/2$	Δ
-1	$1/2$	$-\zeta/2$
1	$-1/2$	$-\zeta/2$
0	$1/2$	$-\Delta$
1	$1/2$	$-2\Delta+\zeta/2$

Homework

1. Find the eigenvalues of a 3×3 matrix for which all matrix elements are identical: $E_j^{(0)}=A$, $V_{\ell m}=V$.

2. Evaluate (6.23) for the case that $\vec{s}_1$ has spin-1 and $\vec{s}_2$ has spin-$\frac{1}{2}$. Do the problem two ways: (i) using the trick method, and (ii) using the matrix method.

3. Find the eigenvalues of the Hamiltonian of three interacting spin-$\frac{1}{2}$ particles:

$$H=A[\vec{s}_1\cdot\vec{s}_2+\vec{s}_1\cdot\vec{s}_3+\vec{s}_3\cdot\vec{s}_2] \tag{6.322}$$

4. A benzene molecule (C_6H_6) has a ring of six carbon atoms. The $2p_z$ orbitals of carbon are perpendicular to the plane. They form a periodic system of 6 identical states for an electron. Use the first-neighbor tight binding model.

   a. What are the six different eigenvalues and eigenvectors of this system?
   b. Six electrons occupy these states in the neutral molecule. Including spin degeneracy, which states are occupied, assuming $V>0$?

5. Assume a hypothetical molecule composed of six carbon atoms in a line. Use the same matrix elements as in the prior problem. What are the energies of the six orbital states for the $2p_z$ system. What is the total energy of six electrons in these states. Is it higher or lower than for the ring?

6. A tetrahedron is a polygon with 4 vertices equidistant from each other (e.g., CH_4). An electron on a vertex has a site energy E_0 and a matrix element V for hopping to any other vertex.

a. Write down the Hamiltonian matrix for this problem.
b. Give its eigenvalues and eigenvectors.

7. Work out the first-order Stark effect for the $n=3$ shell of atomic hydrogen.

8. In the perturbation theory solution, derive the second-order correction $C_{\ell m}^{(2)}$ to the wave function.

9. In the perturbation theory solution, derive the third-order energy $E_n^{(3)}$. Verify that it gives zero for the perturbation $V = Fx$ on the harmonic oscillator.

10. The exact eigenvalue spectrum in one dimension for the potential

$$V(x) = \frac{M}{2}[\omega_0^2 + \Omega^2]x^2 \tag{6.323}$$

is

$$E_n = \hbar\sqrt{\omega_0^2 + \Omega^2}\left(n + \frac{1}{2}\right) \tag{6.324}$$

Divide the Hamiltonian acccording to

$$H = H_0 + V \tag{6.325}$$

$$H_0 = \frac{p^2}{2M} + \frac{M}{2}\omega_0^2 x^2, \quad V = \frac{M}{2}\Omega^2 x^2 \tag{6.326}$$

Treat V as a perturbation to H_0. Find the contribution to the energy in first- and second-order perturbation theory. Successive terms should correspond to expanding the exact result:

$$E_n = \hbar\sqrt{\omega_0^2 + \Omega^2}\left(n + \frac{1}{2}\right) = \hbar\omega_0\left(n + \frac{1}{2}\right)\left[1 + \frac{\Omega^2}{2\omega_0^2} + \cdots\right]$$

11. For the helium atom, write

$$H_0 = -\frac{\hbar^2}{2m}\left[\nabla_1^2 + \nabla_2^2\right] - 2e^2\left[\frac{1}{r_1} + \frac{1}{r_2}\right] \tag{6.327}$$

$$V = \frac{e^2}{|\vec{r}_1 - \vec{r}_2|} \tag{6.328}$$

Estimate the ground-state energy from first-order perturbation theory.

12. Consider a hydrogen atom that has an added delta-function potential $V(r) = \lambda\delta(r)/r^2$. If λ is small, how does this perturb the energy levels of the atom? Are states with different angular momenta affected differently?

13. Consider the change in the continuum wave function using first-order perturbation theory for the case that $V(r)$ is a repulsive square well ($V(r) = V_0$ for $r < a$). Find $\tilde{V}(\vec{Q})$. Evaluate (6.147) when $\vec{k} = 0$.

14. A small charge $Q(Q << e)$ is placed a long distance $R(R >> a_0)$ from a hydrogen atom. What is the leading term, in powers of $1/R$, of the energy change in the system? Compare it to the classical result.

15. Calculate the exact polarizability α of an electron bound to an alpha-particle.

16. How much does the $3p_z$ state contribute to the polarizability of hydrogen?

17. Estimate C_6 between two hydrogen atoms by the two approximate methods: (i) Replacing the summation over states by just the lowest nonzero matrix element, and (ii) summing over all states using the constant energy denominator approximation. The numerical answer is $6.47e^2a_0^5$.

18. Use the data from table 6.2 and London's approximation to estimate the van der Waals interaction between an argon and a krypton atom.

19. Estimate the van der Waals coefficient between a neutral potassium atom and the potassium ion (K^+).

20. What is the diamagnetic susceptibility of a helium atom? Use the variational eigenfunction to evaluate any integrals over coordinates. Give the answer in terms of fundamental constants.

21. Consider a spinless ($s=0$) particle in the $n=2$ state of a hydrogen atom. Assume that a magnetic and electric field both perturb the eigenstates:

$$V = \mu_0 \vec{\ell} \cdot \vec{H}_0 + |e| \vec{F} \cdot \vec{r} \tag{6.329}$$

Find the energy levels for the two cases that (i) $H_0 \parallel F$ and (ii) $H_0 \perp F$.

22. Consider an atom with a closed shell plus one d-electron ($\ell=2$). The d-electron state is perturbed by a magnetic field and the spin–orbit interaction. Discuss only how the two states $m = (\frac{5}{2}, \frac{3}{2})$ behave as a function of magnetic fields in the two limits of Zeeman and Paschen-Back.

23. A lithium atom has three electrons: two are tightly bound in (1s) orbitals, while the third one is confined to the 2s or 2p states. The outer electron is perturbed by (i) the spin–orbit interaction, (ii) an electric field, and (iii) a magnetic field:

$$V = \frac{\zeta}{\hbar^2} \vec{\ell} \cdot \vec{s} + |e| \vec{F} \cdot \vec{r} + \mu_0 (\vec{\ell} + 2\vec{s}) \cdot \vec{H}_0 \tag{6.330}$$

Derive the Hamiltonian matrix for this set of perturbations. Which matrix elements are nonzero? Assume $\vec{F} \parallel \vec{H}_0$.

7 Time-Dependent Perturbations

7.1 Time-Dependent Hamiltonians?

Many perturbations depend on time. This dependence is expressed as a perturbation $V(\mathbf{r}, t)$ that has explicit time dependence. This chapter is devoted to solving Hamiltonians that can be written in the form

$$H = H_0 + V(\mathbf{r}, t) \tag{7.1}$$

$$i\hbar \frac{\partial}{\partial t}\psi(\mathbf{r}, t) = H\psi(\mathbf{r}, t) \tag{7.2}$$

Of course, this approach is actually nonsense. From a rigorous viewpoint, Hamiltonians cannot depend on time.

As an example, consider the problem of a very fast alpha-particle that whizzes by an atom. The proper Hamiltoniian for this problem would contain the complete Hamiltonian for the atom, the kinetic energy of the alpha-particle, and the Coulomb interaction between the alpha-particle, the nucleus, and the electrons in the atoms. We omit the nuclear forces that bind the nucleons into the alpha-particle. This complete Hamiltonian is hard to solve. However, suppose that your interest is in how the atom is affected by the passage of the alpha-particle. Are some electrons in the atom excited out of the ground-state configuration? For a very fast alpha-particle, it is a good approximation to replace it in the Hamiltonian of the atom by just a time-dependent potential:

$$\phi(\mathbf{r}, t) = \frac{2e}{|\mathbf{r}-\mathbf{R}(t)|} \tag{7.3}$$

where $\mathbf{r}$ are the positions of electrons, and $\mathbf{R}(t)$ is the classical trajectory of the alpha-particle. One makes, at the outset, the approximation that the alpha-particle follows a classical path, and then solves the atomic problem of how the electrons react to this time-dependent potential. Obviously, this approach does not provide information about the quantum corrections to the motion of the alpha-particle. But it may provide a satisfactory way of solving the atomic problem of the motion of the electrons. Using a Hamiltonian that has a time-dependent potential is always an approximation, but often one that is useful and satisfactory.

Another example of a popular, time-dependent Hamiltonian is the interaction of charged particles with an electromagnetic field. It is common to treat the electromagnetic field classically. The electric field is written as

$$\mathbf{E}(\mathbf{r}, t) = \vec{E}_0 e^{i\mathbf{q}\cdot\mathbf{r}-i\omega t} \tag{7.4}$$

The oscillating electric field can be introduced into the Hamiltonian as a vector potential, which produces a Hamiltonian of the form (7.2). Again, this procedure is an approximation. The electric field is generated by photons, and the correct Hamiltonian contains terms that describe the photons, the charged particles, and their mutual interaction. This complete Hamiltonian is difficult to solve. The complete Hamiltonian is required for a discussion of how lasers operate. Treating the electric field classically is a good approximation for linear optics: how the charged particles absorb radiation. The full Hamiltonian must be used for nonlinear optics, such as light scattering. The full discussion of the electromagnetic fields is given in chapter 8.

Equation (7.2) is hard to solve whenever the potential $V(\mathbf{r}, t)$ is arbitrary and never ending. In many physics problems, the form of $V(\mathbf{r}, t)$ is simple and the Hamiltonian is easier to solve. One case is when the potential only exists for a time interval. An example is when the alpha-particle whizzes by the atom: the potential is "on" only while the alpha-particle is near the atom. For a potential of finite temporal duration, two separate limits are

1. The time is short.
2. The time is long.

The characteristic time t_c for an atom is given by the uncertainty principle that relates it to the energy interval between atomic states:

$$t_c \sim \frac{\hbar}{E_n - E_m} \tag{7.5}$$

"Short" time is $t << t_c$, and "long" time is $t > t_c$. Another typical situation, which is easy to solve, is the case that the perturbation is weak but steady: it lasts a long time. An example is the constant illumination by an electric field. Some of these situations have standard names:

- *Adiabatic limit* is when the perturbation is very long (slow) compared to t_c. An example is when the alpha-particle goes by the atom very slowly. The important feature of the adiabatic limit is that the system evolves very slowly in time. The atom can slowly polarize, and depolarize, as the alpha goes by, but the electrons never change their atomic states.

- *Sudden approximation* is when the perturbation is very short compared to t_c. An example is when the alpha-particle goes rapidly by the atom. Then it is very likely that some electrons in the atom are excited to other orbital states.

- *Golden Rule* is when the perturbation is weak but steady. The perturbation can cause the atom or particle to change its state. The rate of change is such a useful formula that Fermi called it the "Golden Rule."

These three cases are each discussed in detail.

7.2 Sudden Approximation

The sudden approximation applies when the potential changes rapidly compared to the characteristic time of the system. A simple example that is easy to solve mathematically is the case where the perturbation is a step function in time:

$$\Theta(t) = \begin{cases} 1 & t > 0 \\ 0 & t < 0 \end{cases} \tag{7.6}$$

Of course, it is never possible to switch on (or off) a potential infinitely fast. Here the step function has the meaning implied in the first sentence of this paragraph: the potential is switched during a time interval much shorter than the response time of the system.

An important example is the Hamiltonian of the form

$$H = H_i + \Theta(t)V \tag{7.7}$$

This Hamiltonian is easy to solve when it is viewed as two problems. The first is when $t < 0$. Then the Hamiltonian is H_i, where the subscript i denotes the initial state. This Hamiltonian has a set of eigenstates $\phi_n^{(i)}$ and eigenvalues $E_n^{(i)}$ that obey

$$H_i\phi_n^{(i)} = E_n^{(i)}\phi_n^{(i)} \tag{7.8}$$

For $t < 0$ each eigenstates has an amplitude a_n. The method of choosing or determining these amplitudes is discussed briefly in section 1.2. For now it is only necessary to know they exist. Their values are assumed known, perhaps because the state has been prepared by an experimentalist. Then the most general form of the wave function for $t < 0$ is given in chapter 1:

$$\psi_i(t) = \sum_n a_n\phi_n^{(i)} \exp[-itE_n^{(i)}/\hbar] \tag{7.9}$$

This wave function is an exact solution to Schrödinger's equation (7.2) for $t < 0$. The a_n are independent of time.

For $t > 0$ the new Hamiltonian is

$$H_f = H_i + V \tag{7.10}$$

It also has an exact solution in terms of eigenfunctions $\phi_n^{(f)}$ and eigenvalues $E_n^{(f)}$. These are usually different than the initial set. In analogy with (7.9), the final-state wave function must have the form

$$\psi_f(t) = \sum_m b_m\phi_m^{(f)} \exp[-itE_m^{(f)}/\hbar] \tag{7.11}$$

This equation is also a solution to (7.2) with constant coefficients b_m.

At time $t = 0$ the Hamiltonian switches from H_i to H_f. The wave function $\psi_i(t)$ applies when $t < 0$, and $\psi_f(t)$ applies for $t > 0$. At $t = 0$ they must be the same wave function. Set them equal at this one point in time:

$$\sum_n a_n \phi_n^{(i)} = \sum_m b_m \phi_m^{(f)} \tag{7.12}$$

The eigenfunctions $\phi_m^{(f)}$ are orthogonal. An expression for b_m is obtained by multiplying the above equation by $\int d\tau \phi_\ell^{(f)*}$ and doing the integral for each term, where $d\tau$ signifies the relevant phase space. All terms vanish on the right except the one with $m = \ell$. An expression is obtained for b_ℓ:

$$b_\ell = \sum_n a_n \langle \bar{\ell} | n \rangle \tag{7.13}$$

$$\langle \bar{\ell} | n \rangle = \int d^3 r \phi_\ell^{(f)*} \phi_n^{(i)} \tag{7.14}$$

The overbar denotes final state. All of the quantities on the right-hand side of (7.13) are known. This equation determines the coefficients b_ℓ of the final state in terms of those of the initial state a_n and the overlap matrix elements $\langle \bar{\ell} | n \rangle$. Since $\phi_n^{(i)}$ and $\phi_\ell^{(f)}$ are eigenstates of different Hamiltonians, they are not orthogonal, and many overlap matrix elements $\langle \bar{\ell} | n \rangle$ are nonzero.

The sudden approximation is an exact solution to the Hamiltonian (7.7). The derivation applies to any kind of system with any number of particles. The Hamiltonian could be for a one-particle system or for a many-particle system. For many particles, the overlap matrix element $\langle \bar{\ell} | n \rangle$ has multidimensional integrals.

7.2.1 *Shake-up and Shake-off*

As an example of the sudden approximation, consider the evolution of atomic states during nuclear beta-decay. When a nucleus undergoes beta-decay, both an electron and a neutrino are emitted. The nucleus changes its charge state from Z to $Z+1$ as a result of losing one electron. In effect, a neutron has been converted to a proton. If the nucleus is part of an atom, the electrons in the atomic orbitals have their nuclear potential changed from $-Ze^2/r$ to $-(Z+1)e^2/r$. The initial Hamiltonian H_i describes an atom with nuclear charge Z, while the final-state Hamiltonian H_f describes the atom with a nuclear charge of $Z+1$.

Usually the electrons in the atomic orbitals are in the ground state: the states with the lowest energy, consistent with the Pauli principle. The most probable final configuration after the beta-decay is that the electrons are still in the ground state of the "new" atom. They are more tightly bound because of the increase in nuclear charge. The ground-state to ground-state transition is indicated schematically in figure 7.1a. There is a small probability that one or more of the electrons end up in excited states of the final atom. This process is called *shake-up*, and is shown in figure 7.1b. Another possibility is that an electron, initially bound, is excited to a final continuum state and leaves the atom. This process is called *shake-off*, and is shown in Figure 7.1c. Overall energy is conserved, since the kinetic

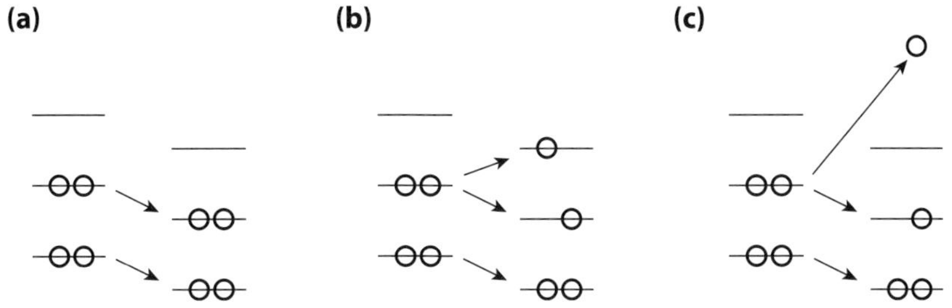

FIGURE 7.1. (a) Ground-state to ground-state transition during nuclear beta-decay. (b) A shake-up process. (c) A shake-off process.

energies of the departing particles, electron and neutrino, make small adjustments to account for the change in energy of the bound electrons.

The sudden approximation can usually be used to calculate the rate of shake-up and shake-off. In a typical beta-decay the electron has a kinetic energy of 5 MeV. Since this is 10 rest-mass energies, it departs the nucleus with a velocity close to the speed of light $v \sim 10^8$m/s. The length scale of the atom is $a_0 \sim 0.1$ nm. The time scale for the change in nuclear potential is $t \sim a_0/v \sim 10^{-18}$ seconds. The outer electrons are usually involved in shake-up, and their energy spacing is typically $\Delta E \sim 1$ eV $\sim 10^{-19}$ joule. The characteristic time for the bound electrons is $t_c \sim \hbar/\Delta E \sim 10^{-15}$ seconds. The potential switches in a time (10^{-18} s) much faster than the response time (10^{-15}s) of the atom. Then it is appropriate to use the sudden approximation.

For most atoms, the wave functions for the many-electron eigenstates are quite complicated and usually computer generated. They must have the proper symmetry, as discussed in chapter 9. The overlap matrix element $\langle \bar{\ell}|n\rangle$ between initial and final wave functions in (7.13) is evaluated on the computer.

A simple example is for an atom with only one electron. Then one can use hydrogen eigenfunctions. The change in potential is spherically symmetric and has no angular dependence. So the angular quantum numbers (ℓ, m) cannot change during the shake-up: s-states go to s-states and p-states go to p-states. Only radial integrals need to be evaluated.

The $1s$ radial eigenfunction for nuclear charge Z is

$$|1s\rangle = 2\left(\frac{Z}{a_0}\right)^{3/2} \exp\left(-\frac{Zr}{a_0}\right) \tag{7.15}$$

The probability that the atomic electron remains in the $1s$ state in going from $Z \rightarrow Z+1$ is

$$\begin{aligned}\langle \bar{1s}|1s\rangle &= 2^2\left(\frac{Z}{a_0}\right)^{3/2}\left(\frac{Z+1}{a_0}\right)^{3/2}\int_0^\infty r^2 dr \exp\left[-\frac{(2Z+1)r}{a_0}\right] \\ &= \frac{[4Z(Z+1)]^{3/2}}{(2Z+1)^3} = \left[1-\frac{1}{(2Z+1)^2}\right]^{3/2}\end{aligned} \tag{7.16}$$

$$|\langle \bar{1s}|1s\rangle|^2 = \left[1-\frac{1}{(2Z+1)^2}\right]^3 \tag{7.17}$$

For $Z = 1$ it is $\left(\frac{8}{9}\right)^3$, and for $Z = 2$ it is $\left(\frac{24}{25}\right)^3$. So the ground-state to ground-state transition becomes very probable at larger values of Z.

7.2.2 Spin Precession

Another example of the sudden approximation is provided by the precession of a spin in a magnetic field. Assume that the magnetic field is initially in the z-direction for $t < 0$, and the Hamiltonian contains only the Pauli term:

$$H_i = -\frac{g\mu_0 B}{\hbar}\hat{z}\cdot\vec{s} \tag{7.18}$$

The g-factor is approximately two, and the electron spin operator is given in terms of the Pauli spin matrices:

$$\vec{s} = \frac{\hbar}{2}\vec{\sigma} \tag{7.19}$$

$$H_i = -\mu_0 B\sigma_z \tag{7.20}$$

The exact eigenstates have spin-up or spin-down in relation to the z-axis. They are called $a(t)$ and $b(t)$, respectively. They are represented by a two-component spinor:

$$\psi_i(t) = \begin{pmatrix} a(t) \\ b(t) \end{pmatrix} \tag{7.21}$$

As a simple example, assume that initially ($t < 0$) the electron has spin-up ($a = 1, b = 0$), so the eigenstates of the Hamiltonian are

$$\psi_i(t) = \begin{pmatrix} 1 \\ 0 \end{pmatrix} e^{-i\omega_i t} \tag{7.22}$$

$$\varepsilon_i = \hbar\omega_i = -\mu_0 B \tag{7.23}$$

This completes the description of the initial state.

At $t = 0$ it is assumed that the magnetic field is suddenly rotated by 90°, say to the x-direction. For $t > 0$ the final-state Hamiltonian is taken to have the form of (7.20), except σ_z is replaced by σ_x because of the rotation of the magnetic field:

$$H_f = -\mu_0 B\sigma_x \tag{7.24}$$

The problem is to find the quantum mechanical solution for the motion of the spin for $t > 0$.

The classical solution is quite simple. As shown in figure 7.2, a classical spin perpendicular to a magnetic field will precess in a circle in the plane perpendicular to the field direction. The same behavior is found in quantum mechanics.

The final-state Hamiltonian (7.24) may be solved by assuming that the x-axis is the new direction for defining spin-up and spin-down. However, this basis is inconvenient for

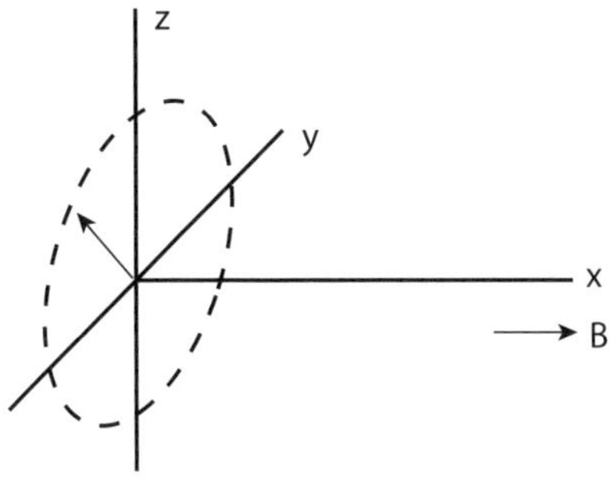

FIGURE 7.2. Classical spin precession.

expressing the initial conditions. The easiest way to solve the problem is to retain the z-axis as the basis for spin-up and spin-down. In this basis, the time-dependent Schrödinger's equation becomes

$$H_f \psi_f = i\hbar \frac{\partial}{\partial t} \psi_f \tag{7.25}$$

$$-\mu_0 B \begin{pmatrix} 0 & 1 \\ 1 & 0 \end{pmatrix} \begin{pmatrix} a_f(t) \\ b_f(t) \end{pmatrix} = i\hbar \frac{\partial}{\partial t} \begin{pmatrix} a_f(t) \\ b_f(t) \end{pmatrix} \tag{7.26}$$

$$\hbar\omega \equiv \mu_0 B \tag{7.27}$$

The precession frequency ω is defined in the last line. The matrix equation provides a pair of coupled equations for the two components:

$$i \frac{\partial}{\partial t} a_f = -\omega b_f \tag{7.28}$$

$$i \frac{\partial}{\partial t} b_f = -\omega a_f \tag{7.29}$$

This pair of equations is easily solved. Take a second time derivative of $a_f(t)$:

$$i^2 \frac{\partial^2}{\partial t^2} a_f = -i\omega \frac{\partial}{\partial t} b_f = (-\omega)^2 a_f \tag{7.30}$$

$$\frac{\partial^2}{\partial t^2} a_f + \omega^2 a_f = 0 \tag{7.31}$$

The second-order differential equation has solutions in terms of sines and cosines:

$$a_f = C_1 \cos(\omega t) + C_2 \sin(\omega t) \tag{7.32}$$

$$b_f = -\frac{i}{\omega} \frac{\partial}{\partial t} a_f = iC_1 \sin(\omega t) - iC_2 \cos(\omega t) \tag{7.33}$$

The initial conditions are applied at $t = 0$. At that moment, $\psi_f(t = 0) = \psi_i(t = 0)$. The initial conditions are $a_f(0) = 1$, $b_f(0) = 0$. Then the constant coefficients above are $C_1 = 1$, $C_2 = 0$. The exact wave function for the spin precession is

$$\psi_f(t) = \begin{pmatrix} \cos(\omega t) \\ i\sin(\omega t) \end{pmatrix} \tag{7.34}$$

The wave function is not an exact eigenstate of H_f. Instead, it is a linear combination of two eigenfunctions, with spin parallel or antiparallel to the x-axis. They have eigenvalues $\pm\hbar\omega$. There are two terms in eqn. (7.11).

The wave function is properly normalized. At any time ($\theta = \omega t$) one has that

$$\langle\psi_f^\dagger|\psi_f\rangle = (\cos\theta,\ -i\sin\theta)\begin{pmatrix} \cos\theta \\ i\sin\theta \end{pmatrix} = \cos^2\theta + \sin^2\theta = 1 \tag{7.35}$$

The solution is complete.

The classical picture has the spin precessing in a circle in the yz plane. The quantum mechanical picture is the same, which is shown by evaluating the three components of spin angular momentum:

$$\langle s_j\rangle = \frac{\hbar}{2}\langle\psi_f^\dagger|\sigma_j|\psi_j\rangle \tag{7.36}$$

The three components give

$$\langle s_x\rangle = \frac{\hbar}{2}(\cos\theta,\ -i\sin\theta)\begin{pmatrix} 0 & 1 \\ 1 & 0 \end{pmatrix}\begin{pmatrix} \cos\theta \\ i\sin\theta \end{pmatrix} = 0 \tag{7.37}$$

$$\langle s_y\rangle = \frac{\hbar}{2}(\cos\theta,\ -i\sin\theta)\begin{pmatrix} 0 & -i \\ i & 0 \end{pmatrix}\begin{pmatrix} \cos\theta \\ i\sin\theta \end{pmatrix} \tag{7.38}$$

$$= \hbar\cos\theta\sin\theta = \frac{\hbar}{2}\sin(2\theta) \tag{7.39}$$

$$\langle s_z\rangle = \frac{\hbar}{2}(\cos\theta,\ -i\sin\theta)\begin{pmatrix} 1 & 0 \\ 0 & -1 \end{pmatrix}\begin{pmatrix} \cos\theta \\ i\sin\theta \end{pmatrix} \tag{7.40}$$

$$= \frac{\hbar}{2}[\cos^2\theta - \sin^2\theta] = \frac{\hbar}{2}\cos(2\theta) \tag{7.41}$$

The expectation value of s_x is zero. The results for $\langle s_y\rangle = (\hbar/2)\ \sin(2\omega t)$, $\langle s_z\rangle = (\hbar/2)$ $\cos(2\omega t)$, show the spin does precess in a circle in the yz plane. The quantum solution agrees with the classical one.

Why is the frequency 2ω? Since the energy of spin-up is $\hbar\omega$ and spin-down is $-\hbar\omega$, the change in energy during the spin flip is $2\hbar\omega$. The characteristic time for spin precession is $t_c \approx 2\pi/\omega$. The sudden approximation is valid as long as the field is rotated by 90° during a time short compared to t_c.

7.3 Adiabatic Approximation

The adiabatic approximation is where the perturbation is turned on very slowly compared to the response time of the system. In this case the most likely outcome is that the system

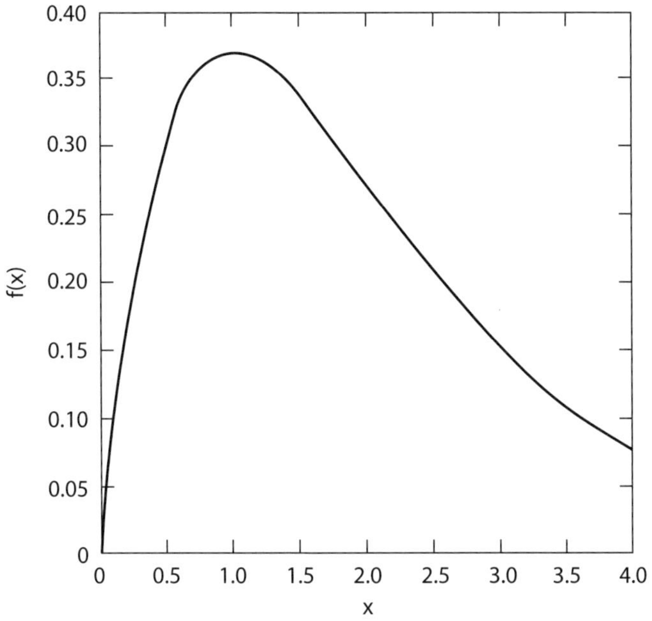

FIGURE 7.3. Time dependence of electric field.

stays in its ground state. The ground state evolves slowly as the pertubation is initiated. The system stays in the ground state as this state evolves in time.

Consider the example of an atom in an electric field. The initial ground state has no applied electric field. Apply an electric field $V(\mathbf{r}, t) = -e\vec{E}(t) \cdot \vec{r}$, which is increased very slowly. The new ground state of the electrons is that of the atom in the electric field. In the adiabatic approximation, all electrons remain in this ground state.

Consider the hydrogen atom as a simple example. The electric field has the time dependence

$$\vec{E}(t) = E_0 \hat{z} f(t/\tau_0) \tag{7.42}$$

$$f(t/\tau_0) = \frac{t}{\tau_0} \exp\left[-\frac{t}{\tau_0}\right] \Theta(t) \tag{7.43}$$

The theta-function $\Theta(t)$ ensures that the perturbation starts at $t = 0$. Its maximum value is $f(1) = 1/e$ regardless of τ_0. The constant τ_0 regulates whether the electric field is turned on rapidly or slowly. This function is graphed in figure 7.3. The constant τ_0 determines the time scale over which the field is turned on.

The time-dependent Schrödinger's equation is

$$i\hbar \frac{\partial}{\partial t} \psi(\mathbf{r}, t) = H\psi(\mathbf{r}, t) \tag{7.44}$$

$$H = H_0 + V(\mathbf{r}, t) \tag{7.45}$$

$$\psi(\mathbf{r}, t) = \sum_n a_n(t) \phi_n(\mathbf{r}) \exp[-itE_n/\hbar] \tag{7.46}$$

If H_0 is the Hamiltonian of the hydrogen atom, then its eigenfunctions $\phi_n(\mathbf{r})$ and eigenvalues $E_n = -E_{Ry}/n^2$ are well known. Initially, for $t < 0$, before the electric field is turned on, the atom is in its ground state:

$$a_{1s}(t) = 1, \quad t < 0 \tag{7.47}$$

$$a_n(t) = 0, \quad t < 0, \quad n \neq 1s \tag{7.48}$$

Multiply eqn. (7.44) by $-i/\hbar$. Then multiply by $\phi_{2pz}(\mathbf{r})^*$ and integrate over all space. Because the eigenfunctions are orthogonal, this step eliminates all terms in $\psi(\mathbf{r}, t)$ except the $2p_z(n = 2, \ell = 1, m = 0)$ hydrogenic state:

$$\left[\frac{\partial}{\partial t} a_{2pz}(t) - i\frac{E_{2p}}{\hbar} a_{2pz}\right] e^{-i\omega_2 t} = -i\frac{E_{2p}}{\hbar} a_{2pz} e^{-i\omega_2 t} - i\frac{e}{\hbar} E_0 f(t/\tau_0) \sum_n a_n \langle 2p_z|z|n\rangle e^{-i\omega_n t} \tag{7.49}$$

$$\langle 2p_z|z|n\rangle = \int d^3 r \phi_{2pz}(\mathbf{r})^* z \phi_n(\mathbf{r}) \tag{7.50}$$

where $\omega_n = E_n/\hbar$. On the right-hand side of the above equation, the largest term in the series is $a_n = a_{1s}$. In the adiabatic approximation, this amplitude changes little. So keep only this term. Define $\omega = \omega_2 - \omega_1$

$$\frac{\partial}{\partial t} a_{2pz}(t) = -i\frac{e}{\hbar} E_0 f(t/\tau_0) a_{1s} \langle 2p_z|z|1s\rangle e^{i\omega t} \tag{7.51}$$

$$a_{2pz}(\infty) = -i\frac{e}{\hbar} E_0 \langle 2p_z|z|1s\rangle \int_0^\infty dt f(t/\tau_0) e^{i\omega t} \tag{7.52}$$

$$= -i\frac{e}{\hbar} E_0 \langle 2p_z|z|1s\rangle \frac{\tau_0}{(1 - i\omega\tau_0)^2} \tag{7.53}$$

$$P_{2pz} = |a_{2pz}|^2 = \frac{[eE_0\langle z\rangle]^2 \tau_0^2}{\hbar^2 (1 + \omega^2\tau_0^2)^2} \tag{7.54}$$

The adiabatic limit is when $\omega\tau_0 \gg 1$, so the perturbation is turned on slowly with respect to the response time of the system. In this limit, the probability of being excited to the $2p_z$ state is

$$P_{2pz} = \frac{[eE_0\langle z\rangle]^2}{(\hbar\omega)^2} \frac{1}{\omega^2\tau_0^2} \ll 1 \tag{7.55}$$

The $2p_z$ state is the one most likely to be excited in a shake-up process. Its probability is very small. When the electric field is turned on over a long time scale, then the system stays in its ground state. This is the adiabatic limit.

7.4 Transition Rates: The Golden Rule

The most general time dependence is to have a potential function $V(\mathbf{r}, t)$, where the time variation is on the same scale as the response time of the particles that experience the potential. A common experimental arrangement is to have the potential $V(\mathbf{r}, t)$ start at some time t_i. The potential may be switched on either suddenly or gradually. In either case, assume $V(\mathbf{r}, t) = 0$ for $t < t_i$. Since time is relative, set $t_i = 0$. The Hamiltonian is

$$H = H_i + \Theta(t) V(\mathbf{r}, t) \tag{7.56}$$

The Hamiltonian H_i is assumed to be independent of time: all time dependence is put into $V(\mathbf{r}, t)$. The Hamiltonian H_i has eigenstates $\phi_n(\mathbf{r})$ and eigenvalues E_n. For $t < 0$ the exact solution to Schrödinger's equation is given by eqn. (7.11):

$$H_i \psi_i(\mathbf{r}, t) = i\hbar \frac{\partial}{\partial t} \psi_i(\mathbf{r}, t) \tag{7.57}$$

$$\psi_i(\mathbf{r}, t) = \sum_n a_n \phi_n(\mathbf{r}) e^{-i\omega_n t} \tag{7.58}$$

$$\hbar\omega_n \equiv E_n \tag{7.59}$$

This solution is standard. How does it change for $t > 0$ when the potential is turned on?

The eigenfunctions ϕ_n are a complete set and can be used to define any function of $\mathbf{r}$. The exact wave function $\psi(\mathbf{r}, t)$ for $t > 0$ is expanded in this same set of eigenfunctions:

$$[H_i + V(\mathbf{r}, t)]\psi(\mathbf{r}, t) = i\hbar \frac{\partial}{\partial t} \psi(\mathbf{r}, t) \tag{7.60}$$

$$\psi(\mathbf{r}, t) = \sum_n b_n(t) \phi_n(\mathbf{r}) e^{-i\omega_n t} \tag{7.61}$$

The coefficients $b_n(t)$ depend on time, due to the time dependence of the potential function.

The present procedure is different than in the sudden approximation. There the basis for expansion for $t > 0$ was the eigenfunctions $\phi_n^{(f)}$ of H_f. In the present case, with $V(\mathbf{r}, t)$ some arbitrary function of t, it is assumed that there is no natural H_f in the problem. If there is no H_f, there is no $\phi_n^{(f)}$ to serve as the basis set for the wave function. In some cases it might happen at very large times that the potential function becomes $V(\mathbf{r}, \infty)$, and then $H_f = H_i + V(\mathbf{r}, \infty)$. Then it is useful to solve H_f for its basis set. Here we are assuming that this is not the case.

One may use any basis set for expanding $\psi(\mathbf{r}, t)$. Using ϕ_n usually reduces the amount of work and increases the insight into the behavior of the system. The wave function (7.61) is put into Schrödinger's equation: (7.60)

$$i\hbar \sum_n [\dot{b}_n - i\omega_n b_n] \phi_n e^{-i\omega_n t} = \sum_\ell b_\ell [E_\ell + V(\mathbf{r}, t)] \phi_\ell e^{-i\omega_\ell t} \tag{7.62}$$

Each side contains the factor of $E_n b_n$, which cancels:

$$i\hbar \sum_n \dot{b}_n \phi_n e^{-i\omega_n t} = \sum_\ell b_\ell V(\mathbf{r}, t) \phi_\ell e^{-i\omega_\ell t} \tag{7.63}$$

A formal solution to the last equation is obtained by multiplying each term by ϕ_m^* and then integrating over all $\mathbf{r}$. Since the eigenfunctions are orthogonal, the terms on the left are zero unless $n = m$:

$$i\hbar\dot{b}_m = \sum_\ell V_{m\ell}(t) b_\ell e^{-i\omega_{\ell m} t} \tag{7.64}$$

$$V_{m\ell}(t) = \langle m|V(\mathbf{r}, t))|\ell\rangle \tag{7.65}$$

$$\hbar\omega_{\ell m} = E_\ell - E_m \tag{7.66}$$

A factor of $\exp(-i\omega_m t)$ was transferred across the equal sign. Equation (7.64) is the basis of all transition rate theory.

At $t = 0$ the system can usually be described by having a small set of initial amplitudes b_n different from zero in (7.58). At $t = 0$, when the potential function $V(\mathbf{r}, t)$ is started, then $b_n(t = 0) = b_n$. This result is true because the same basis set ϕ_n is used for $t < 0$ and $t > 0$. Equation (7.64) describes how each amplitude $b_m(t)$ evolves in time. One has a set of coupled differential equations between all of the amplitudes in the basis set. There are an infinite number of amplitudes b_m. If all are coupled together by the matrix elements $V_{m\ell}$, then formidable efforts are required to solve the simple equation (7.64).

There are several situations where approximations can be employed to simplify the calculation. One case is where the potential is only "on" for a finite duration in time, after which it returns to zero. Furthermore, it is assumed that the potential is weak, in that most of the amplitudes b_m change little. Then the initial set of nonzero $b_n(0)$ can be assumed to retain their initial value, while the other set of b_m, $m \neq n$ remain small. In this case the values of $b_m(t)$ are

$$\dot{b}_m(t) = -\frac{i}{\hbar}\sum_n b_n(0) V_{mn}(t) e^{-i\omega_{nm} t} \tag{7.67}$$

This equation is derived from (7.64) by assuming the only values of ℓ with nonnegligible values of $b_\ell(t)$ are the set b_n.

Equation (7.67) is solved by doing the time integral, with the initial condition $b_m(0) = 0$:

$$b_m(t) = -\frac{i}{\hbar}\sum_n b_n(0) \int_0^t dt' V_{mn}(t') e^{-i\omega_{nm} t'} \tag{7.68}$$

The probability $P_m = |b_m|^2$ of finding the particle in the state m is

$$P_m = \frac{1}{\hbar^2}\left|\sum_n b_n(0) \int_0^t dt' V_{mn}(t') e^{-i\omega_{nm} t'}\right|^2 \tag{7.69}$$

This formula is valid only when the probability P_m is very small. If P_m is found to be close to one, then the initial assumptions are wrong and one has to solve the more accurate eqn. (7.64).

Another standard situation is when the perturbation, after it is started, is small and oscillatory. The discussion is simplified by assuming that the potential oscillates with a single frequency. The integral in (7.69) contains the total frequency $\Omega = \omega + \omega_n - \omega_m$:

$$V_{mn}(t') = V_{mn} e^{-i\omega t'} \tag{7.70}$$

$$P_m(t) = \frac{1}{\hbar^2}\left|\sum_n V_{mn} b_n(0) \frac{i}{\Omega}(e^{-i\Omega t} - 1)\right|^2 \tag{7.71}$$

For most values of Ω, the expression on the right oscillates with time, and the terms average to zero. In this case there is no steady occupation of the state m. The exception is provided whenever $\Omega = 0$, since there is a term of $O(t^2)$. Since $\Omega = \omega + \omega_n - \omega_m$, then $\Omega = 0$ means $\omega_m = \omega + \omega_n$. Usually, no sets or only one set of frequencies (ω_n, ω_m) satisfy this identity. In the above series, say it is obeyed by only one value of $n = \ell$. Then the above expression is

$$b_m(t) = t^2 \left| \frac{V_{m\ell} b_\ell(0)}{\hbar} \right|^2 + 2t V_{m\ell} b_\ell(0) \sum_{n \neq \ell} b_n(0) V_{n\ell} \frac{\sin(t\Omega)}{\Omega}$$
$$+ \frac{1}{\hbar^2} \left| \sum_{n \neq \ell} b_n(0) \frac{i}{\Omega} (e^{-i\Omega t} - 1) \right|^2 \tag{7.72}$$

This formula predicts that $P_m(t)$ is proportional to t^2 whenever $\omega_m = \omega + \omega_\ell$.

The t^2 behavior is unphysical and its prediction must be discarded. Our intuition is that $P_m(t)$ should increase linearly with t, not quadratically. The relation $E_m = E_\ell + \hbar\omega$ is interpreted as implying that the particle has made a transition from ℓ to m. The energy difference $E_m - E_\ell$ is provided by the quantum of energy $\hbar\omega$ from the oscillating potential. The probability per unit time is the quantity that is constant in time. After the potential $V(\mathbf{r}, t)$ is started, the system should quickly lose memory of the start-up transients. The transition between the states ℓ and m will be equally likely to occur in any time interval. The longer one waits, the more likely the transition. In the limit of long time, one postulates that $P_m(t)$ should be proportional to t and not t^2. A dependence on t^2 requires the probability per unit time increases with time, which is unphysical.

Define $w_{m\ell}$ as the transition rate per unit time, in the limit that $t \to \infty$, between states ℓ and m. The strict definition is

$$w_{m\ell} = \lim_{t \to \infty} \frac{d}{dt} \left| \frac{V_{m\ell}}{\hbar\Omega} (e^{-i\Omega t} - 1) \right|^2 \tag{7.73}$$

$$= \left| \frac{V_{m\ell}}{\hbar} \right|^2 \frac{d}{dt} \left[\frac{\sin(\Omega t/2)}{\Omega/2} \right]^2 \tag{7.74}$$

$$\lim_{t \to \infty} P_m(t) = t \sum_n w_{mn} |b_n(0)|^2 \tag{7.75}$$

The expression for $w_{m\ell}$ is the time derivative of a single term in (7.71). The cross-terms are omitted since they oscillate and average to zero. The relative frequency Ω is zero only for one term in the series, so only this one term is important. If there are two or more states with the same values of $(\omega_m - \omega_\ell)$, their matrix elements should be added before taking the magnitude squared.

The time derivative in (7.74) gives

$$\frac{d}{dt} \left[\frac{\sin(\Omega t/2)}{\Omega/2} \right]^2 = \frac{2}{(\Omega/2)} \sin(\Omega t/2) \cos(\Omega t/2) = \frac{2}{\Omega} \sin(\Omega t) \tag{7.76}$$

The important functional dependence is $\sin(\Omega t)/\Omega$. This expression is plotted in figure 7.4 as a function of Ω. It is evaluated in the limit that $t \to \infty$. At $\Omega = 0$ it equals t, and

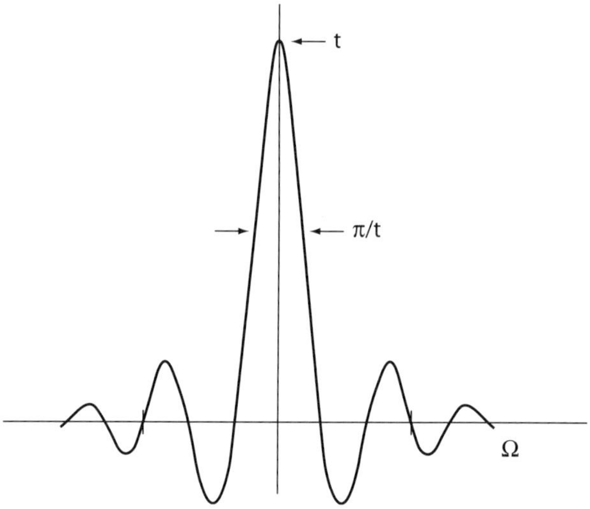

FIGURE 7.4. A plot of sin $(\Omega t)/\Omega$ for large t and small Ω.

its half-width is given by π/t. As $t \to \infty$ the peak becomes higher and narrower. It is a delta-function [$\sim\delta(\Omega)$] in the limit that $t \to \infty$. The area under the peak is independent of t and equals π. In the limit of large times, the probability per time contains a delta-function:

$$\lim_{t\to\infty}\left[\frac{2}{\Omega}\sin(\Omega t)\right] = 2\pi\delta(\Omega) = 2\pi\hbar\delta(E_m - E_\ell - \hbar\omega) \tag{7.77}$$

$$w_{mn} = \frac{2\pi}{\hbar}|V_{m\ell}|^2\delta(E_m - E_\ell - \hbar\omega) \tag{7.78}$$

This extremely important expression is called the *Golden Rule*. Technically, it is the Born approximation to the Golden Rule. The delta-function has the feature that energy conservation is automatically built into the transition rate. Only transitions that conserve energy occur. The frequency dependence of the potential function $V(\mathbf{r}, t)$ provides the quantum of energy $\hbar\omega$ that permits transitions between states of different energy. There are also cases where $\omega = 0$ and the Golden Rule gives the scattering between two states of the same energy.

Equation (7.78) is an approximate expression for the transition rate. The exact expression has the same form, but with a different matrix element:

$$w_{mn} = \frac{2\pi}{\hbar}|T_{m\ell}|^2\delta(E_m - E_\ell - \hbar\omega) \tag{7.79}$$

The T-matrix $T_{m\ell}$ is defined in chapter 10.

7.5 Atomic Excitation by a Charged Particle

Charged particles are slowed down by passage through matter, such as a gas, solid, or liquid. The particle can collide with a nucleus and recoil at large angle. Since the mass of nuclei are large compared to most particles, the recoil is nearly elastic. Such collisions are unlikely if the particle is neutral, since nuclei are small. If the particle is charged, it can scatter from the nucleus, but again the scattering is nearly elastic.

A charged particle can lose energy by exciting the electronic states of the atom, which is the important mechanism of energy loss. The passage of a fast particle causes a potential $V(\mathbf{r}, t)$ at the atom that varies rapidly with time. This potential can cause some of the electrons in the atom to be excited: to change their state to one of higher energy. The energy to make this electronic excitation comes from the charged particle, so it is slowed down by the energy transfer. A fast, charged particle going through matter leaves a wake of excited and ionized atoms, which causes it to slow down.

The calculation is done for one atom assuming that the charged particle goes by with a straight trajectory of constant velocity. This model is appropriate if the particle is very energetic. Then, by the uncertainty principle, its trajectory can be defined closely. Also, the energy loss is a small fraction of the energy of the particle, which does not cause a significant deflection of its path. In a typical experiment, the particle might be a proton or an alpha-particle with a kinetic energy of kilovolts, while the electronic excitations are of the order of electron-volts.

The geometry is shown in figure 7.5. The particle has charge Q, mass m, and velocity $\vec{v}$. The atom A is at rest, and D is the closest point between the atom and the trajectory of the particle. D is called the *impact parameter*. The angular momentum is mvD. An x–y coordinate system is shown in the figure. The particle has a trajectory given by $\vec{R}(t) = (vt, -D)$, where $t = 0$ is the closest point. The atom is at the origin, with nuclear charge Z and N electrons: the ion charge is $Z-N$. The Coulomb potential between the particle and the atom is

$$V(\mathbf{R}) = eQ\left[\frac{Z}{R} - \sum_i^N \frac{1}{|\mathbf{R}-\mathbf{r}_i|}\right] \tag{7.80}$$

The recoil of the atom is neglected. For excitations between the electronic states n and m, the matrix element is

$$V_{nm}(t) = \langle n|V(\mathbf{R}(t))|m\rangle = -eQ\int d^3r \frac{\phi_n^*(\mathbf{r})\phi_m(\mathbf{r})}{|\mathbf{R}(t)-\mathbf{r}|} \tag{7.81}$$

Constant terms such as Z/R have a zero matrix element if $n \neq m$. The discussion is simplified by assuming that the distance D is large compared with the dimensions of the atom. The Coulomb interaction is expanded in a multipole expansion. Only the first non-zero term is retained, which is the dipole interaction:

$$\frac{1}{|\mathbf{R}-\mathbf{r}|} = \frac{1}{R} + \frac{\mathbf{r}\cdot\mathbf{R}}{R^3} + O\left(\frac{1}{R^3}\right) \tag{7.82}$$

$$V_{nm}(t) \approx -eQ\frac{x_{nm}vt - y_{nm}D}{[(vt)^2 + D^2]^{3/2}} \tag{7.83}$$

$$x_{nm} = \langle n|x|m\rangle, \; y_{nm} = \langle n|y|m\rangle \tag{7.84}$$

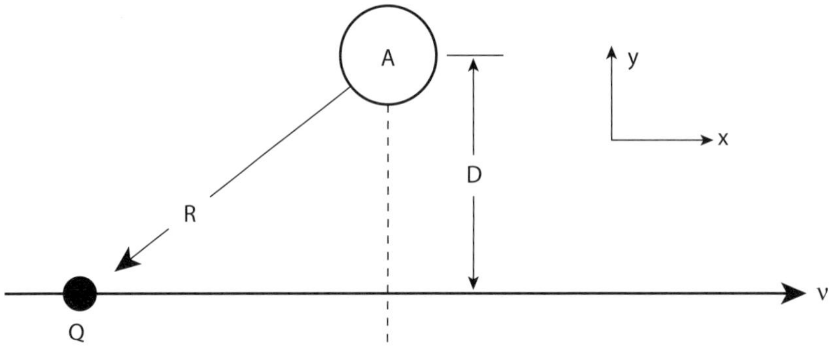

FIGURE 7.5. The circle is the atom A, while the straight line is the trajectory of the charged particle.

In the dipole approximation, the matrix element $V_{nm}(t)$ is given in terms of x_{nm}, y_{nm}. The probability of an electronic excitation between the states m and n at time t is given by eqn. (7.69):

$$P_{nm}(t) = \left(\frac{eQ}{\hbar}\right)^2 \left| \int_{-\infty}^{t} dt' e^{-i\omega_{nm}t'} \frac{x_{nm}vt' - y_{nm}D}{[(vt')^2 + D^2]^{3/2}} \right|^2 \tag{7.85}$$

The integral is evaluated by defining two dimensionless quantities: the integration variable $\tau = t'v/D$, and the parameter $S = D\omega_{nm}/v$. Then the exponent has $\omega_{nm}t' = (t'v/D)(D\omega_{nm}/v) = \tau S$. The probability of excitation is rewritten in these new variables:

$$P_{nm}(t) = \left(\frac{eQ}{\hbar v D}\right)^2 \left| \int_{-\infty}^{tv/D} d\tau e^{-iS\tau} \frac{y_{nm} - x_{nm}\tau}{[\tau^2 + 1]^{3/2}} \right|^2 \tag{7.86}$$

The fine-structure constant is dimensionless and small:

$$\frac{e^2}{\hbar c} = \frac{1}{137.04} \tag{7.87}$$

The prefactor in (7.86) has a pair of such constants: $e^2/\hbar v$ and $Q^2/\hbar v$. Since $Q \sim e$, these factors are small as long as $v > c/10$. The other dimensionless factors are x_{nm}/D and y_{nm}/D. The are also small if D is larger than the size of the atom. The probability $P_{nm}(t)$ is small as long as v is large and the integral is well behaved.

Now evaluate the τ-integral in (7.86). First examine the case where $S \gg 1$. The ratio D/v is the typical time associated with the passage of the charged particle by the atom. The quantity ω_{nm}^{-1} is the time associated with the response of the atom. The inequality $S = D\omega_{nm}/v \gg 1$ means that the time of passage is long compared with the response time of the atom. This case is defined in section, 7.1 as the adiabatic limit. When S is very large, the factor of $\exp(-iS\tau)$ oscillates rapidly with τ and the integral averages to a small number. The denominator is hard to integrate in a simple way. However, the exponent $\frac{3}{2}$ is midway between 1 and 2, and both of these cases can be integrated:

$$\int_{-\infty}^{\infty} \frac{d\tau}{1+\tau^2} e^{-iS\tau} = \pi e^{-S} \tag{7.88}$$

$$\int_{-\infty}^{\infty} \frac{d\tau}{(1+\tau^2)^2} e^{-iS\tau} = \frac{\pi}{2} e^{-S}(1+S) \tag{7.89}$$

These integrals suggest that in the limit of $S \gg 1$ the excitation probability P_{nm} is approximately

$$P_{nm}(\infty) \approx \pi \left(\frac{eQ}{\hbar v D}\right)^2 (y_{nm}^2 + x_{nm}^2) e^{-S} \tag{7.90}$$

The factor of $\exp(-S)$ is negligibly small when $S \gg 1$. This conclusion is correct in the adiabatic limit. When the charged particle goes by the atom slowly, it is unlikely that any of the electrons alter their electronic state. Instead, all states evolve slowly in time. The atom remains in the ground state. The ground state is altered by the temporary presence of the charged particle, but so slowly that excitations are exceedingly unlikely. As an absurd example, consider moving the charged particle at the velocity of one millimeter per year. This speed is unlikely to excite anything.

The other limit is when S is very small. Then the factor of $\exp(-i\tau S)$ is neglected in the integrand of (7.86). Here the charged particle goes by the atom rapidly compared with the response time of the atom. The sudden approximation could be used if the potential were turned on rapidly and stayed on. However, here it is rapidly turned on and then turned off. In the limit $t \rightarrow \infty$ the term involving x_{nm} averages to zero because of the linear factor of τ. The final probability of a transition for a fast passage by the particle is

$$P_{nm}(\infty) = \left(\frac{eQy_{nm}}{\hbar v D}\right)^2 \left| \int_{-\infty}^{\infty} \frac{d\tau}{[1+\tau^2]^{3/2}} \right|^2 \tag{7.91}$$

The integral equals two. The final result for the probability of excitation of a single atom from state n to m is

$$P_{nm}(\infty) = 4\left(\frac{eQy_{nm}}{\hbar v D}\right)^2 \tag{7.92}$$

This probability is for a particle of charge Q and velocity v that passes a single atom at a distance D from the nucleus. The atomic properties are contained in the dipole matrix element. The probability is a dimensionless number less than unity. After the charged particle has gone, the probability $P_{nm}(\infty)$ gives the likelyhood that the atom is excited.

A typical experiment involves a charged particle going through matter. The particle interacts with a number of atoms during its passage. The rate of energy loss is found by computing the average energy loss for each atom, multiplied by the rate that atoms are encountered by the particle.

As an example, the rate of energy loss is found for a gas. The gas has a density n_A of atoms per unit volume. They are assumed to be randomly located in space. Since the charged particle is fast, one can neglect the motion of the atoms. The product $n_A v$ has the dimensions of particle flux: number of atoms/(m^2 sec). In the rest frame of the charged

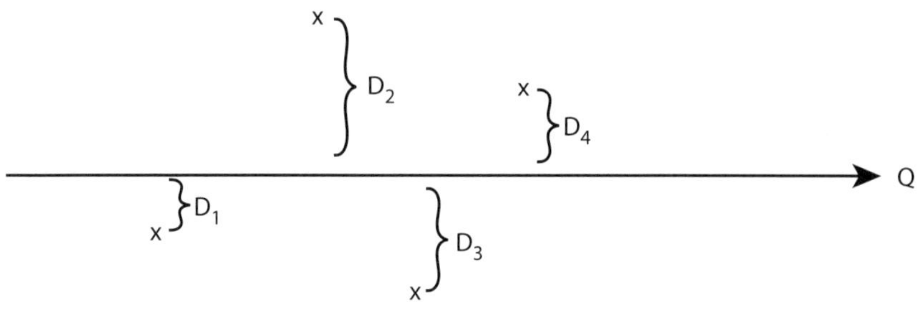

FIGURE 7.6. Particle Q encounters atoms at a variety of impact distances D_j.

particle, the flux is the number of atoms going by it per second per unit area. When the atoms are fixed and the particle is moving, it is still the rate at which the charged particle encounters the atoms.

Figure 7.6 shows the particle Q going through a random array of atoms, which are denoted by the symbol $\times$. Each atom has a different impact parameter D_j. The unit of area dA at a distance D is $dA = 2\pi D dD$. The product of $n_A v dA$ is the number of atoms per second encountered at a distance D. Multiply this by P_{nm} in (7.92), which is the excitation probability as a function of D. Integrating over all D gives the number of transitions, n to m, in the gas per second, which is the transition probability:

$$w_{nm} = 2\pi n_A v \int D dD P_{nm}(D) \tag{7.93}$$

The term $P_{nm}(D) \propto O(1/D^2)$. The other factors can be brought out of the integral.

What are the limits of the D-integral? The lower limit is taken to be the dimension of the atom, which is called a. The present theory is inadequate if the particle bashes directly into the atom. It applies only for external passage, so a is a reasonable cutoff. The upper limit on D is given by the condition that $S < 1$, which is $D < v/\omega_{nm}$. For larger values of D, the particle passes the atom so far away that the adiabatic limit applies and there is negligible chance of excitation. The integral in (7.93) is now evaluated:

$$w_{nm} = \frac{8\pi e^2 Q^2 \gamma_{nm}^2}{\hbar^2 v} n_A \int_a^{v/\omega_{nm}} \frac{dD}{D} \tag{7.94}$$

$$= \frac{8\pi e^2 Q^2 \gamma_{nm}^2}{\hbar^2 v} n_A \ln\left(\frac{v}{a\omega_{nm}}\right) \tag{7.95}$$

Equation (7.95) is Bethe's formula for the number of excitations per second.

Multiplying this formula by $\hbar\omega_{nm}$ gives the rate of energy loss (joules/second) by the charged particle. Divide by v to get the energy loss per distance (joule/meter):

$$\frac{dE}{dx} = \frac{\hbar\omega_{nm} w_{nm}}{v} \tag{7.96}$$

There are now two factors of v in the denominator that can be converted to the kinetic energy of the particle $v^2 = 2E/M$. The experimental data are usually presented by dividing by n_A, which is the stopping power per atom (joules meter^2):

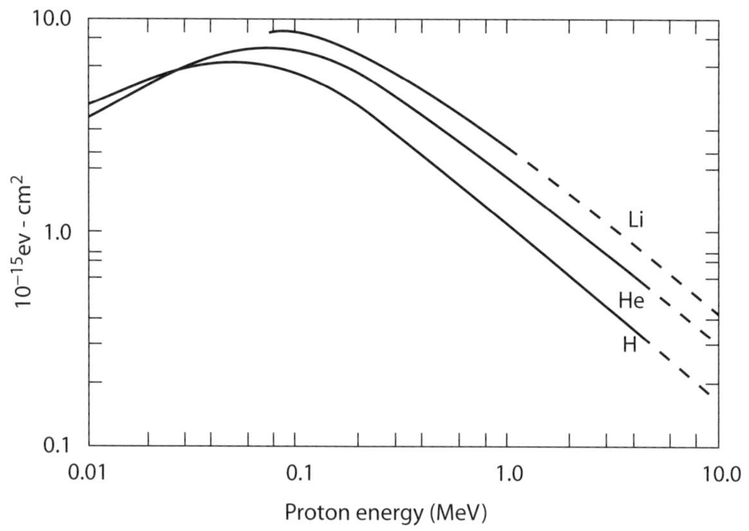

FIGURE 7.7. Stopping power of protons, from W. Whaling in *Handbuch der Physik* **34**, 201 (1958), edited by S. Flugge.

$$\frac{1}{n_A}\frac{dE}{dx} = \frac{4\pi e^2 Q^2 M y_{nm}^2 \omega_{nm}}{\hbar E} \ln\left(\frac{\sqrt{2E/M}}{a\omega_{nm}}\right) \tag{7.97}$$

Figure 7.7 shows the experimental data for the stopping power (eV-cm^2) of protons in H_2, He, and Li. The steady falloff for $E > 0.1$ MeV is due to the $1/E$ dependence. The data in this region of energy are well described by the above formula.

7.6 Born Approximation to Scattering

When a particle approaches a scattering center, such as an atom or nuclei, there is a mutual force between the particle and the center. The force may cause the particle to change its direction of motion: it is deflected from its original path. This process is called "scattering." The term *elastic scattering* is used whenever the particle does not alter its total energy, but only alters the direction of motion. Inelastic scattering occurs when the kinetic energy of the particle is changed by giving or gaining energy from the scattering center. An example of inelastic scattering is presented in the previous section. Now the rate of elastic scattering is calculated.

The physical system under consideration has a scattering center fixed at the origin. It exerts a potential $V(\mathbf{r})$ on the particle being scattered. The scattering center is assumed not to recoil, or else the scattering is not elastic in the laboratory frame. For a particle scattering from a free atom or nuclei, one should go into center-of-mass coordinates. For central forces between the particle and the atom [$V(\mathbf{r}) = V(r)$], the center-of-mass momentum does not change. Then the scattering is elastic in relative coordinates, but not in the laboratory frame.

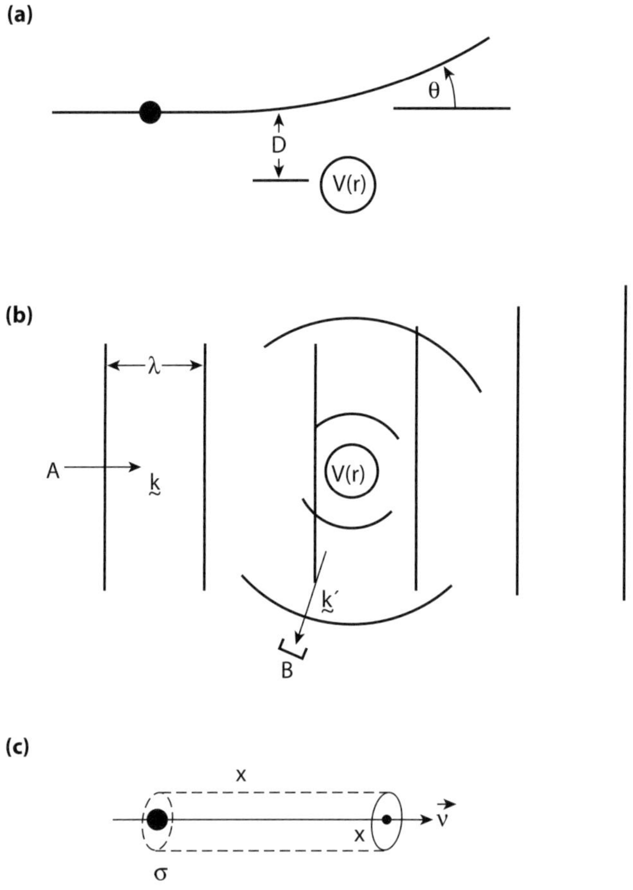

FIGURE 7.8. (a) A particle with impact distance D scatters through an angle θ. (b) The wave description of scattering. (c) The product $v\sigma$ gives the volume swept out per second.

The particle approaches the scattering center at an impact distance D. Figure 7.8a shows that the potential $V(\mathbf{r})$ causes a deflection of the particle through an angle θ. A classical description of the scattering derives a formula $\theta(D)$ for the angle as a function of D. The angle also depends on the initial particle velocity v and the potential $V(r)$.

The quantum mechanical formulation is different. The impact parameter is ill defined by the uncertainly principle, since distances cannot be determined precisely. The classical idea of impact distance is replaced in quantum mechanics by the angular momentum. The classical momentum is $\tilde{\ell} = mvD$. In quantum mechanics, the classical $\tilde{\ell}$ is replaced by $\hbar\ell$, where ℓ is a dimensionless integer. Angular momentum is quantized.

7.6.1 Cross Section

In classical mechanics, the particle trajectory and the final direction after scattering are known as accurately as one can determine the initial values of m, v, D. In quantum mechanics, events are not predicted precisely, but are described in terms of probabilities. The rate of scattering is determined by a quantity called the *cross section*, which is denoted as σ and has the units of area. The particle has a velocity v and is going through a density n_A of scattering centers. The quantity $v\sigma$ has the units of volume per second and represents the volume swept out by the particle each second. This is shown in figure7.8c. The cross section gives the area being swept out for that type of scattering center. The figure shows atoms at the points $\times$, where scattering occurs. Multiplying $v\sigma$ by n_A, to get $v\sigma n_A$, gives the average number of scattering events per second. This quantity is also given by Fermi's Golden Rule:

$$w = v\sigma n_A \tag{7.98}$$

Quantum mechanics is used to compute the cross section from the Golden Rule. It can also be measured and the theory compared with experiment.

According to wave–particle duality, particle motion can be viewed as waves. This viewpoint provides additional insight into the scattering process, as shown in figure 7.8b. The particle enters the scattering region at point A, as a plane wave with wave vector $\mathbf{k}$. The parallel vertical lines represent wave crests separated by the wavelength. The scattering center causes part of the wave to be transformed into a spherical outgoing wave, although most of the wave proceeds undisturbed. A detector at B measures only the spherical component of the scattered wave. It is assumed that the potential is weak and only a small fraction of the wave amplitude is scattered. The same amount of scattering occurs in any unit of time. The important physical quantity must be the scattering rate per unit time, which is the Golden Rule of (7.74).

The Golden Rule is used to calculate the rate of scattering $w(\mathbf{k}, \mathbf{k}')$ from an initial state $\mathbf{k}$ to a final state $\mathbf{k}'$. The result is summed over all final states $\mathbf{k}'$ and muliplied by the number of scattering centers N_A to find the rate of scattering out of state $\mathbf{k}$:

$$\tilde{w}(\mathbf{k}) = N_A \sum_{\mathbf{k}'} w(\mathbf{k}, \mathbf{k}') = v_k \sigma(k) n_A \tag{7.99}$$

Since $n_A = N_A/\Omega$ for volume Ω, we obtain a formula for the cross section:

$$\sigma(k) = \frac{\Omega}{v_k} \sum_{\mathbf{k}'} w(\mathbf{k}, \mathbf{k}') = \frac{2\pi\Omega}{\hbar v_k} \sum_{\mathbf{k}'} |T(\mathbf{k}, \mathbf{k}')|^2 \delta[E(k) - E(k')] \tag{7.100}$$

Equation (7.100) is the basic result, although the right-hand side benefits from some manipulations. We have used the exact Golden Rule using the T-matrix in eqn. (7.79). In the present example, the Golden Rule from (7.78) is used, where the matrix element is the Fourier transform of the potential. Since the plane wave eigenfunctions are $\psi(\mathbf{k}, \mathbf{r}) = \exp(i\mathbf{k} \cdot \mathbf{r})/\sqrt{\Omega}$, the approximate matrix element is

$$T(\mathbf{k}, \mathbf{k}') \approx \frac{1}{\Omega} \tilde{V}(\mathbf{k} - \mathbf{k}') \tag{7.101}$$

$$\tilde{V}(\mathbf{k} - \mathbf{k}') = \int d^3 r e^{i\mathbf{r}\cdot(\mathbf{k}-\mathbf{k}')} V(\mathbf{r}) \tag{7.102}$$

This approximation is called the *first Born approximation* or just the *Born approximation.* In this approximation the cross section is

$$\sigma(k) = \frac{2\pi}{\Omega \hbar v_k} \sum_{\mathbf{k}'} |\tilde{V}(\mathbf{k}-\mathbf{k}')|^2 \delta[E(k)-E(k')] \tag{7.103}$$

The next step is to let the volume Ω of the box go to infinity. The summation over $\mathbf{k}'$ become a continuous integral, as discussed in chapter 2:

$$\lim_{\Omega \to \infty} \frac{1}{\Omega} \sum_{\mathbf{k}'} \to \int \frac{d^3k'}{(2\pi)^3} \tag{7.104}$$

$$\sigma(k) = \frac{1}{(2\pi)^2 \hbar v} \int d^3k' |\tilde{V}(\mathbf{k}-\mathbf{k}')|^2 \delta[E(k)-E(k')] \tag{7.105}$$

Equation (7.105) is the final formula for the cross section of elastic scattering in the Born approximation. It depends on neither the volume Ω nor the atom density n_A. It depends on the form of the potential $V(\mathbf{r})$ and the particle wave vector k, where $v_k = \hbar k/m$. The integral on the right is well defined for any potential function $V(\mathbf{r})$ based on physics.

The quantity $\sigma(k)$ is the *total cross section* for a particle in the state $\mathbf{k}$, for scattering to all possible final states k'. The d^3k' integral is expressed as $d^3k' = (k')^2 dk' d\Omega'$, where $d\Omega' = \sin(\theta) d\theta d\phi$ is the unit of solid angle. Here θ is the angle between the two vectors $\hat{k} \cdot \hat{k}' = \cos(\theta)$. The delta-function of energy $\delta[E(k) - E(k')]$ serves to eliminate the integral over dk'. It forces $|\mathbf{k}| = |\mathbf{k}'|$:

$$\int dk'(k')^2 \delta[E(k)-E(k')] = \frac{mk}{\hbar^2} = \frac{m^2 v_k}{\hbar^3} \tag{7.106}$$

$$\sigma(k) = \left(\frac{m}{2\pi\hbar^2}\right)^2 \int d\Omega' |\tilde{V}(\mathbf{k}-\mathbf{k}')|^2 \tag{7.107}$$

$$|\mathbf{k}-\mathbf{k}'| = 2k|\sin(\theta/2)| \tag{7.108}$$

Equation (7.107) for the cross section is very simple. The cross section is the square of the Fourier transform of the potential integrated over all solid angles and multiplied by a constant prefactor. The dimensions of $\tilde{V}$ are joule meter^3 so the dimensions of

$$\frac{m\tilde{V}}{\hbar^2} \sim \text{meter} \tag{7.109}$$

The cross section does have the dimensions of area. Given a simple potential, there is little work in finding the Fourier transform, so the formula for the cross section is rather easy to use. Equation (7.107) is only the Born approximation to scattering. The exact formula is given in chapter 10.

The quantity $\sigma(k)$ is the total cross section for all elastic scattering events. Quite often the experiment measures the scattering as a function of angle, or energy, or both. These measurements are of the *differential cross section,* for which there are several options. In the experiment shown in figure 7.8b, the detector at B measures only the scattered particles at a small solid angle $\Delta\Omega$. If the detector has an area A at a distance L from the target, then $\Delta\Omega = A/L^2$. The differential cross section as a function of solid angle is obtained from (7.107) by not integrating over solid angle:

$$\frac{d\sigma}{d\Omega} = \left| \frac{m}{2\pi\hbar^2} \tilde{V}[2k \sin(\theta/2)] \right|^2 \tag{7.110}$$

$$\sigma = \int d\Omega \frac{d\sigma}{d\Omega} \tag{7.111}$$

The differential cross section is found by measuring the scattering as a function of angle. In the Born approximation, this measurement provides direct information about the Fourier transform of the scattering potential. In the experimental arrangement shown in figure 7.8b, the average number of particles per second measured at B would be

$$vn_i N_A \left(\frac{d\sigma}{d\Omega} \right) \Delta\Omega \tag{7.112}$$

where vn_i is the flux of incoming particles.

Another possibility is that the experiment measures the energy of the scattered particles. The differential cross section as a function of energy is found from (7.105) by *not* integrating over energy. Write $d^3k' = mk'dE'd\Omega'/\hbar^2$ and move dE' to the left, to obtain

$$\frac{d\sigma}{dE} = \left(\frac{m}{2\pi\hbar^2} \right)^2 \delta(E-E') \int d\Omega' |\tilde{V}(\mathbf{k}-\mathbf{k}')|^2 \tag{7.113}$$

$$\sigma = \int dE' \frac{d\sigma}{dE} \tag{7.114}$$

For elastic scattering, the energy measurement finds all particles in the final state at the same energy as the initial-state kinetic energy. The result is a delta-function, which would be broadened by the instrumental resolution.

Another possibility is that the experiment measures simultaneously the energy and angular dependence of the scattering. The double differential cross section is

$$\frac{d^2\sigma}{dEd\Omega} = \left[\frac{m}{2\pi\hbar^2} \tilde{V} \left[2k \sin\left(\frac{\theta}{2} \right) \right] \right]^2 \delta(E-E') \tag{7.115}$$

$$\sigma = \int dE'\Omega' \frac{d^2\sigma}{dEd\Omega} \tag{7.116}$$

This completes the general definition of cross section.

7.6.2 Rutherford Scattering

As an example, consider the differential cross section for a proton scattering from a nucleus of charge Z. The Fourier transform of the Coulomb potential is

$$V(r) = \frac{Ze^2}{r}, \quad \tilde{V}(q) = \frac{4\pi Ze^2}{q^2} \tag{7.117}$$

$$\frac{d\sigma}{d\Omega}=\left(\frac{m}{2\pi\hbar^2}\right)^2\left(\frac{4\pi Ze^2}{4k^2\sin^2(\theta/2)}\right)^2 \tag{7.118}$$

$$=\frac{Za_0^2}{[2(ka_0)^2\sin^2(\theta/2)]^2} \tag{7.119}$$

where the Bohr radius is $a_0=\hbar^2/me^2$. This formula is recognized as the Rutherford formula for the scattering from a Coulomb potential. He derived it for the scattering of alpha-particles from nuclei rather than protons, which changes the prefactor by four. Otherwise, it is the same expression. At small angle, the differential cross section goes like $O(1/\theta^4)$. The total cross section in (7.111) diverges due to the behavior at small angle. This divergence is due to the long range of the Coulomb potential. The Coulomb potential is actually screened by the electrons in the atom. When they are included, the cross section no longer diverges.

7.6.3 Electron Scattering from Hydrogen

The screening by the electrons in the atom are well illustrated by considering the scattering of an electron from a hydrogen atom. The electron scatters from the nucleus, as well as from the bound electron. We will calculate the scattering in the Born approximation using the above formulas.

There is an important process we are ignoring in this calculation. The two electrons are identical and indistinguishable. In processes called exchange scattering they trade places during the scattering event. The incoming electron becomes bound and the initially bound electron gets knocked out and appears as the scattered electron. These exchange scattering events are discussed in chapter 10. An alternative is to have the incoming particle be a proton, but then it has exchange processes with the nucleus of the hydrogen atom.

The incoming electron has a potential $V(\mathbf{r})$ given by

$$V(\mathbf{r},\mathbf{r}_e)=e^2\left[-\frac{1}{r}+\frac{1}{|\mathbf{r}-\mathbf{r}_e|}\right] \tag{7.120}$$

The nucleus is the origin. The incoming electron is at $\mathbf{r}$ and the bound electron is at $\mathbf{r}_e$. The first term in V is the interaction with the proton and the second term is interaction with the bound electron. The latter is usually averaged over its allowed positions:

$$V(\mathbf{r})=e^2\left[-\frac{1}{r}+\int d^3r_e\frac{\phi_{1s}(r_e)^2}{|\mathbf{r}-\mathbf{r}_e|}\right] \tag{7.121}$$

The scattering cross section depends on the Fourier transform of this potential:

$$\tilde{V}(q)=-\frac{4\pi e^2}{q^2}[1-F(q)] \tag{7.122}$$

$$F(q)=\int d^3r_e\phi_{1s}(r_e)^2e^{i\mathbf{q}\cdot\mathbf{r}_e} \tag{7.123}$$

$$=\frac{1}{\pi a_0^3}\int d^3r_e e^{-2r/a_0}e^{i\mathbf{q}\cdot\mathbf{r}_e} \tag{7.124}$$

where $F(q)$ is the Fourier transform of the bound electron charge distribution and is called the "form factor." Evaluate the integral in spherical coordinates: $d^3r_e = r_e^2 dr_e \sin(\theta) d\theta d\phi$, where $\hat{q} \cdot \hat{r}_e = \cos(\theta)$:

$$F(q) = \frac{4}{a_0^3 q} \int_0^\infty r dr e^{-2r/a_0} \sin(qr) = \frac{1}{[1 + (qa_0/2)^2]^2} \tag{7.125}$$

Note that $F(q=0) = 1$, so $\tilde{V}(q=0)$ does not diverge in (7.122). The matrix element goes to a constant in this limit. The differential cross section does not diverge at small angle and the total cross section is integrated to a finite number. These results are altered by the inclusion of exchange scattering.

7.7 Particle Decay

A radioactive nucleus can decay by emitting alpha-, beta-, or gamma-particles. Often the nucleus will decay into several fragments. A measurement of the energy distribution of the fragments can reveal information about the nature of the decay process.

Consider a radioactive nucleus A that decays into two fragments B and C. Solve the kinematics in the rest frame of A, so its initial momentum is $\mathbf{p}_A = 0$. Let E_{j0} be the energy at rest of particle j. For a simple mass it would be $m_j c^2$, while for a composite particle there is additional binding energy. The calculation is kept simple by assuming nonrelativisitic kinematics. The conservation of momentum and the conservation of energy are

$$0 = \mathbf{p}_B + \mathbf{p}_C \tag{7.126}$$

$$E_{A0} = E_{B0} + E_{C0} + \frac{1}{2}\left[\frac{p_B^2}{M_B} + \frac{p_C^2}{M_C}\right] \tag{7.127}$$

The momentum $\mathbf{p}_B$ is equal in magnitude and opposite in direction to $\mathbf{p}_C$. Use this fact in the energy equation, and the kinetic energy depends on the reduced mass:

$$\frac{1}{\mu} = \frac{1}{M_B} + \frac{1}{M_C} \tag{7.128}$$

$$\frac{p_C^2}{2\mu} = \frac{p_B^2}{2\mu} = E_{A0} - E_{B0} - E_{C0} \tag{7.129}$$

$$KE_B = \frac{p_B^2}{2M_B} = \frac{\mu}{M_B}(E_{A0} - E_{B0} - E_{C0}) \tag{7.130}$$

$$KE_C = \frac{p_C^2}{2M_C} = \frac{\mu}{M_C}(E_{A0} - E_{B0} - E_{C0}) \tag{7.131}$$

Each final particle has a unique value of kinetic energy that is determined by the conservation of momentum and energy. The excess rest energy $(E_{A0} - E_{B0} - E_{C0})$ is converted into kinetic energy, and the conservation of momentum dictates the fraction of that kinetic energy assigned to each fragment. The conclusion is the same when using rel-

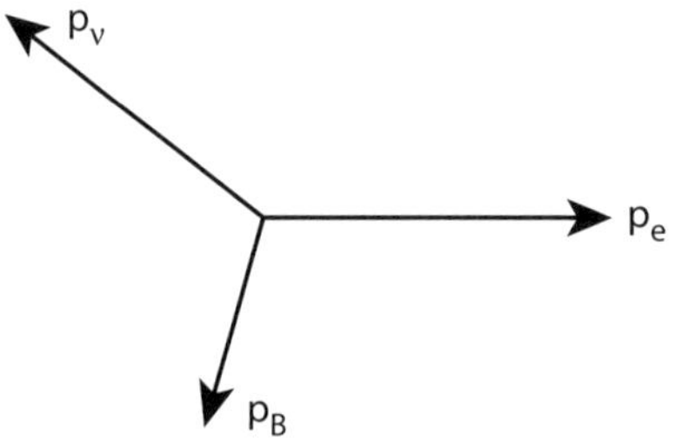

FIGURE 7.9. Conservation of momentum in beta-decay.

ativisitic kinematics. If the experiment was conducted so that all of the A particles were at rest, then all of the B particles would be found to have the same kinetic energy. The A nuclei would be all at rest if they were part of atoms in a solid or liquid.

The experiments on beta-decay showed that the electrons are emitted with a continuous distribution of kinetic energies. This was puzzling, since only two decay fragments were observed, electron and nucleus, which implied that all of the electrons should have the same kinetic energy.

Pauli conjectured that a third particle must be emitted as part of the decay process. Besides the electron and the nucleus, he suggested a chargeless, massless particle. He was right, and the particle is called the *neutrino*. Fermi did the first calculation of the energy distribution of the fragments. Since the electron is the easiest particle to observe in the laboratory, its energy distribution is calculated below. Fermi assumed the neutrino has zero mass, and we will describe his derivation using that same assumption. Now it is known that the neutrino has a very small mass.

An initial nucleus A decays into a final nucleus B plus an electron and a neutrino. Solve the problem in the rest frame of $A(\mathbf{p}_A = 0)$. Figure 7.9 shows the final three particles leaving the origin. The electron must be treated relativistically, but the much heavier nucleus can be treated nonrelativistically:

$$E_e = \sqrt{m_e^2 c^4 + p_e^2 c^2} \tag{7.132}$$

$$E_B = E_{B0} + \frac{p_B^2}{2M_B} \tag{7.133}$$

$$E_\nu = p_\nu c \tag{7.134}$$

Assuming the neutrino is massless, it has momentum p_ν and energy $p_\nu c$. The conservation of energy and momentum gives

$$E_{A0} = E_{B0} + \frac{p_B^2}{2M_B} + \sqrt{m_e^2 c^4 + p_e^2 c^2} + p_\nu c \tag{7.135}$$

$$0 = \vec{p}_B + \vec{p}_e + \vec{p}_\nu \tag{7.136}$$

Now there are four equations and six unknowns (since the three vectors are in the same plane), so there is not a unique solution for the momentum and kinetic energy of each fragment. Each fragment is emitted with a distribution of kinetic energies.

These energy distributions are calculated using the Golden Rule. The total decay rate w in (7.99) is given by the product of the total number of initial A particles N_A times the summation over all final states f:

$$w = N_A \sum_f w_{if} \tag{7.137}$$

The summation over final variables means, in this case, the integrals over the three momenta $\mathbf{p}_B$, $\mathbf{p}_e$, $\mathbf{p}_\nu$. There are two ways of writing the summation over final states. The first way has integrals over all three sets of final momenta, plus the delta-function for conservation of momentum:

$$w = N_A \int \frac{d^3p_B}{(2\pi)^3} \int \frac{d^3p_e}{(2\pi)^3} \int \frac{d^3p_\nu}{(2\pi)^3} (2\pi)^3 \delta^3(\mathbf{p}_B + \mathbf{p}_e + \mathbf{p}_\nu) W(\mathbf{p}_B, \mathbf{p}_e, \mathbf{p}_\nu)$$

The second way of writing this expression is to use the delta-function to eliminate one set of momentum integrals, say d^3p_B:

$$w = N_A \int \frac{d^3p_e}{(2\pi)^3} \int \frac{d^3p_\nu}{(2\pi)^3} W(-\mathbf{p}_e - \mathbf{p}_\nu, \mathbf{p}_e, \mathbf{p}_\nu) \tag{7.138}$$

The second method (7.138) is preferred since it involves less writing. The first method has the advantage that all three momentum variables are treated equally, while (7.138) seems to discriminate against $\mathbf{p}_B$. The complete expression, including energy conservation, is

$$w = \frac{2\pi N_A}{\hbar} \int \frac{d^3p_e}{(2\pi)^3} \int \frac{d^3p_\nu}{(2\pi)^3} |M|^2 \times \delta[E_{A0} - E_B(-\mathbf{p}_e - \mathbf{p}_\nu) - E_e(\mathbf{p}_e) - E_\nu(\mathbf{p}_\nu)] \tag{7.139}$$

This formula is the basis of the discussion of beta-decay. It includes the matrix element M, which plays a key role in the theory of beta-decay.

Our treatment follows Fermi and takes $|M|^2$ to be a constant. However, considerable effort has been expended over the years to show that it is not a constant. It contains, for example, angular variables so the probability of beta-decay is weighted according to the angles between the three momenta. These matrix elements require a relativistic description and are discussed in chapter 11.

The kinetic energy of the B-particle becomes negligible as its mass M_B becomes large. It is convenient to neglect this term. Setting $p_B^2/2M_B = 0$ reduces the algebraic complexity tenfold. So the expression in (7.139) is simplified to

$$w = \frac{N_A |M|^2}{\hbar(2\pi)^5} \int d^3p_e \int d^3p_\nu \delta\left[E_{A0} - E_{B0} - \sqrt{m_e^2c^4 + p_e^2c^2} - p_\nu c\right]$$

Each momentum integral is written as $d^3p = p^2 dp d\Omega$. Each angular integral $\int d\Omega = 4\pi$, since the integrand does not depend on angles. The dp_e and dp_ν momentum variables are changed to energy variables:

$$p_e^2 dp_e = \frac{E_e}{c^3} \sqrt{E_e^2 - m_e^2c^4}\, dE_e \tag{7.140}$$

$$p_\nu^2 dp_\nu = \frac{E_\nu^2}{c^3} dE_\nu \tag{7.141}$$

$$\Delta E = E_{A0} - E_{B0} \tag{7.142}$$

$$w = \frac{N_A|M|^2}{2\pi^3\hbar c^6} \int E_e\sqrt{E_e^2 - m_e^2c^4}dE_e \int E_\nu^2 dE_\nu \delta(\Delta E - E_e - E_\nu)$$

$$w = \frac{N_A|M|^2}{2\pi^3\hbar c^6} \int_{m_ec^2}^{\Delta E} E_e\sqrt{E_e^2 - m_e^2c^4}dE_e(\Delta E - E_e)^2 \tag{7.143}$$

The integral over dE_ν eliminates the delta-function and sets $E_\nu = \Delta E - E_e$. There remains the integral over dE_e. Evaluating this integral gives the decay rate (number of events per second) for all electron final energies. The distribution of final electron energies is found by not doing this integral over dE_e. Technically, one is taking a functional derivative

$$\frac{dw}{dE_e} = \frac{N_A|M|^2}{2\pi^3\hbar c^6} E_e\sqrt{E_e^2 - m_e^2c^4}(\Delta E - E_e)^2 \tag{7.144}$$

where the range of E_e is $m_ec^2 \leq E_e \leq \Delta E$. The quantity dw/dE_e is the number of decay events per second per unit of electron energy. The minimum value of E_e is its rest energy m_ec^2. When an electron has this energy, the neutrino is carrying away the excess kinetic energy. The maximum value of E_e is ΔE. When an electron has this energy, the neutrino has none, since we assume zero mass. Between these limits the electron and neutrino each leave with some kinetic energy.

Equation (7.144) is plotted in figure 7.10. The shape of this curve has two interesting characteristics. At the lower end, it rises from zero with a shape dominated by the square root $\sqrt{E_e^2 - m_e^2c^4}$. This square-root dependence is typical of particles with a nonzero rest mass. At the upper end of the curve, the shape is dominated by the quadratic form $(\Delta E - E_e)^2$. This quadratic shape occurs because we assumed the neutrino has no rest mass. Since the neutrino has a small mass, this end will also have a square-root shape. The shape of the energy spectrum provides information about the masses of the particles involved in the decay process. These results are slightly altered by the energy dependence of the matrix element $|M|^2$.

Sometimes the experimentalists express their results in terms of the momentum of the electron, rather than the energy. The number of electrons emitted per second per unit of momentum is

$$\frac{dw}{dp_e} = \frac{N_A|M|^2}{2\pi^3\hbar c^3} p_e^2(\Delta E - E_e)^2 \tag{7.145}$$

A common way to express the experimental result is to divide both sides of this equation by p_e^2 and then take the square root:

$$\left[\frac{1}{p_e^2}\frac{dw}{dp_e}\right]^{1/2} = \left[\frac{N_A|M|^2}{2\pi^3\hbar c^3}\right]^{1/2} (\Delta E - E_e) \tag{7.146}$$

The right-hand side becomes a linear function of $\Delta E - E_e$ as long as the neutrino has no mass. The linearity of the experimental plot is a test of this hypothesis. Such a curve is

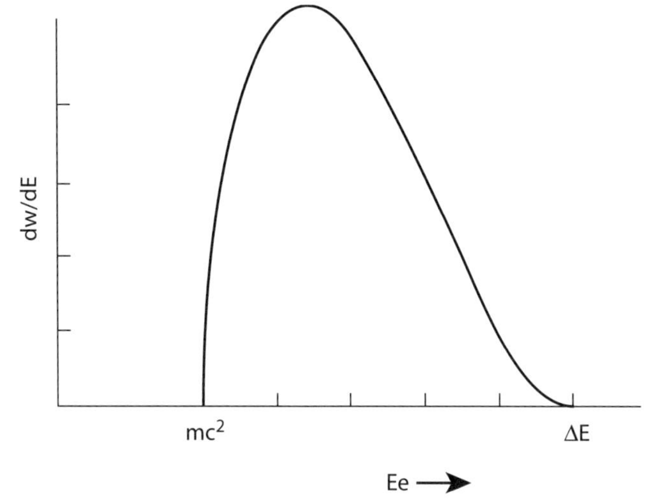

FIGURE 7.10. Electron energy distribution in beta-decay.

called a *Fermi plot*. The small mass of the neutrino will not show up on this large energy scale.

Figure 7.11 shows a Fermi plot of the beta-decay of tritium (^{3}H) into ^{3}He. The vertical axis is the factor

$$\left[\frac{1}{p_e^2 F}\frac{dw}{dp_e}\right]^{1/2} \tag{7.147}$$

where F is a simple function, part of $|M|^2$, that corrects for the Coulomb potential between the electron and the nucleus. The plot is indeed linear as a function of energy as it approaches zero at the high-energy end. The two kinetic energy scales show $E = E_e - m_e c^2$ in keV, as well as $W = E_e/(m_e c^2)$.

Equation (7.144) and figure 7.10 are the result of a *phase-space* calculation. The phase space available to a decay product is $d^3p_e d^3p_\nu$. The assumption of a constant value for $|M|^2$ is equivalent to the assumption that every part of phase space has equal probability. Of course, the available space is limited by the conservation of energy. Often a phase-space calculation gives a good answer, since typically matrix elements do not vary much with energy.

Homework

1. In the sudden approximation, the probability that a system changes from initial state i to final state n is defined P_{in}. Prove that

$$\sum_n P_{in} = 1 \tag{7.148}$$

2. Calculate analytically and numerically for $1 \leq Z \leq 5$ the probability of an electron staying in the 1s state of a one-electron ion during beta-decay.

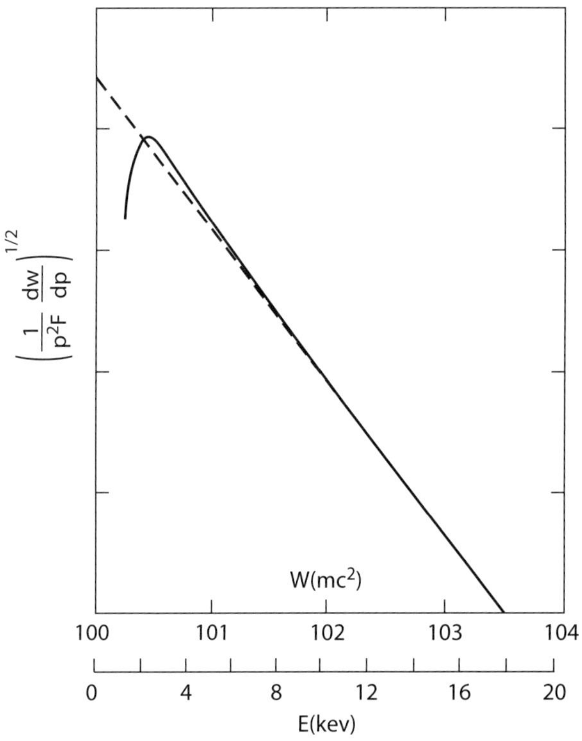

FIGURE 7.11. Fermi plot of the beta-decay of tritium. From Langer and Moffat, *Phys. Rev.* **88**, 689–694 (1952).

3. For the beta-decay $(Z=1) \rightarrow (Z=2)$, what is the probability that an electron initially in the 1s state ends up in the 2s state or the 3s state? What is the total probability for the three events:
   (i) $1s \rightarrow 1s$, (ii) $1s \rightarrow 2s$, (iii) $1s \rightarrow 3s$ when $Z=1$?

4. A particle in one dimension is confined to a box $0<x<L$. At $t=0$, in the sudden approximation, the length of the box is doubled to $0<x<2L$. If n is the quantum number of the initial state, and m is the quantum number of the final state, derive the probability P_{mn}. Give numerical values for the case that $n=1$ and $m=1,2,3$.

5. A particle with mass m is bound to a delta-function potential $[V(x)=-\lambda\delta(x)]$. The strength is suddenly doubled $(\lambda \rightarrow 2\lambda)$.

   a. What is the probability that the particle remains bound?
   b. What is the probability that the particle becomes unbound?

6. A mass m is connected to two springs in a linear array, where the other ends of the two springs are connected to rigid walls. For $t<0$ the system is in the ground state of the one-dimensional harmonic oscillator. At $t=0$ one spring is cut. What is the probability that the particle remains in ground state?

7. A harmonic oscillator in one dimension has an electric field suddenly switched on:

$$H = \frac{p^2}{2m} + \frac{Kx^2}{2} + eFx\Theta(t) \tag{7.149}$$

a. What is the probability that the particle remains in the ground state after the field is switched on?
b. What is the probability that the particle has a shake-up to the $n = 2$ final state?

8. In one dimension a harmonic oscillator of frequency ω is initially in its ground state ($n = 0$). During the time interval $0 < t < t_0$ a potential $V(x) = Fx$ is suddenly switched on (at $t = 0$), and then off (at $t = t_0$). Describe the time evolution of the system for (a) $t < 0$, (b) $0 < t < t_0$, and (c) $t > t_0$.

9. A neutron is traveling up the y-axis from minus to plus infinity at a fixed velocity v. Assume it passes between two fixed charges, an electron and a positron, located at the points $\pm a_0$ (Bohr radius) on the x-axis. If the magnetic moment of the neutron initially points in the ($+z$) direction, what is the final probabillity it points in the ($-z$) direction? The neutron has spin-1/2 with a nuclear moment μ_N. The interaction is just the Pauli term. Find the time-dependent magnetic field from the Lorentz force.

10. Another neutron has a velocity $v = 10^6$m/s and a magnetic moment of $\mu_N = 0.966 \times 10^{-26}$ J/T. It goes between the pole faces of a magnet ($B = 1.0$ tesla) a distance of $L = 1.0$ cm. If the neutron spin is initially perpendicular to both $\vec{v}$ and $\vec{B}$ (and $\vec{v} \perp \vec{B}$), what is the probability that the spin flips after going through the pole faces of the magnet?

11. What is the differential cross section for the scattering of a charged particle from a helium atom? Evaluate in the Born approximation. Use the variational wave function for the electrons in helium.

12. For the Yukawa potential (k_s is a constant),

$$V(r) = A\frac{e^{-k_s r}}{r} \tag{7.150}$$

a. Derive the differential cross section $d\sigma/d\Omega$ in the Born approximation.
b. Obtain the total cross section by integrating (a) over all solid angle.

13. Consider a hypothetical nuclear reaction $A \rightarrow B + e^+ + e^-$ (electron plus positron). Calculate the distribution of final energies of the electron using a phase space calculation, assuming constant matrix elements. Use relativistic kinematics for the electron and positron, and neglect the kinetic energy of the nucleon.

14. Positronium decays by emitting two photons, whose energies sum to $E_0 = 2mc^2$. Derive the distribution of energies of the emitted photons based on phase space, assuming a constant matrix element and an initial center-of-mass momentum of positronium equal to $\vec{p}$.

8 Electromagnetic Radiation

Electromagnetic radiation is composed of elementary particles called photons. They are massless, chargeless, and have spin -1. They have a momentum $\mathbf{p} = \hbar k$, energy cp, and a frequency $\omega = cp/\hbar = ck$. The light has two choices of polarization, which are given the label $\lambda(\lambda = 1,2)$. The frequency is the same one that is associated with an oscillating electric field. The correspondence between the classical electric field and the quantum picture of photons is that the intensity of light is proportional to the number of photons $n_{\mathbf{k}\lambda}$. For an electric field of classical amplitude E_0, wave vector $\mathbf{k}$, and frequency $\omega = ck$, the energy flux is given by the Poynting vector to be

$$S = \frac{c}{4\pi} E_0^2 = \frac{c}{\Omega} n_{k\lambda} \hbar \omega_k \tag{8.1}$$

where Ω is the volume of the box. In the photon picture, the particles have an energy flux given by the product of their velocity c, the energy $\hbar\omega$, and the density $n_{\mathbf{k}\lambda}/\Omega$. The refractive index n_r is assumed to be unity. The speed of light c should be replaced by c/n_r when n_r differs from one. From the above equation, we deduce a quantum mechanical definition of electric field:

$$E_0^2 = 4\pi \frac{n_{k\lambda}}{\Omega} \hbar \omega_k \ (\text{ergs/cm}^3) \tag{8.2}$$

The main difference between the classical and quantum picture is that in classical physics the electric field E_0 can have any value. In quantum physics the number of quanta $n_{\mathbf{k}\lambda}$ in volume Ω must be an integer, so that the light intensity is not continuous but comes in multiples of $4\pi\hbar\omega/\Omega$. This difference is unobservable on a macroscopic basis where the number of quanta $n_{\mathbf{k}\lambda}$ is very large. It can be quite noticable for very small fields, where $n_{\mathbf{k}\lambda}$ is a small integer.

The photon picture applies to electromagnetic radiation of all freqencies, from the microwave to hard x-rays. The only difference is the wavelength. The Hamiltonian for photons interacting with charge particles is derived in the next section.

8.1 Quantization of the Field

Electromagnetic theory is based on Maxwell's equations, which are

$$\nabla \cdot \mathbf{B} = 0 \tag{8.3}$$

$$\nabla \cdot \mathbf{E} = 4\pi\rho \tag{8.4}$$

$$\nabla \times \mathbf{B} = \frac{1}{c}\frac{\partial}{\partial t}\mathbf{E} + \frac{4\pi}{c}\mathbf{j} \tag{8.5}$$

$$\nabla \times \mathbf{E} = -\frac{1}{c}\frac{\partial}{\partial t}\mathbf{B} \tag{8.6}$$

Here ρ and $\mathbf{j}$ are the charge density and the current density. These formulas should be well known to the reader. The problem at hand is to convert the above classical equations into a quantum mechanical description. In particular, it is necessary to derive the Hamiltonian that is equivalent to Maxwell's equations.

The correct answer is

$$H = \sum_j \frac{[\mathbf{p}_j - e_j\mathbf{A}(\mathbf{r}_j)]^2}{2m_j} + \frac{1}{2}\sum_{i \neq j} \frac{e_i e_j}{r_{ij}} + \sum_{\mathbf{k}\lambda} \hbar\omega_{\mathbf{k}\lambda}\left(a^{\dagger}_{\mathbf{k}\lambda}a_{\mathbf{k}\lambda} + \frac{1}{2}\right) \tag{8.7}$$

A set of charged particles of mass m_j, momentum $\mathbf{p}_j$, charge e_j, and position $\mathbf{r}_j$ are interacting with each other through the Coulomb term $e_i e_j / r_{ij}$, and with the photons through the vector potential $\mathbf{A}(\mathbf{r}_j)$. The last term, involving raising and lowering operators, is explained below. The term $\sum \hbar\omega_{\mathbf{k}} a^{\dagger}_{\mathbf{k}\lambda} a_{\mathbf{k}\lambda}$ represents its unperturbed photon Hamiltonian in the absence of charges. This Hamiltonian treats the charged particles nonrelativistically. The photons are always treated relativistically. The relativistic Hamiltonian for electrons is given in chapter 11. The present section is devoted to deriving the above equation.

The radiation field is represented by the vector potential:

$$\frac{1}{c}A_\mu(\mathbf{r}, t) = \sum_{\mathbf{k}\lambda}\left(\frac{2\pi}{\Omega\omega_{\mathbf{k}}}\right)^{1/2} \xi_\mu(\mathbf{k}, \lambda)\left(a_{\mathbf{k}\lambda}e^{i(\mathbf{k}\cdot\mathbf{r} - \omega_{\mathbf{k}}t)} + \text{h.c.}\right) \tag{8.8}$$

$$= \frac{1}{\sqrt{\Omega}}\sum_{\mathbf{k}} e^{i\mathbf{k}\cdot\mathbf{r}} A_\mu(\mathbf{k}, t) \tag{8.9}$$

The unit polarization vector is ξ_μ. One feature of this Hamiltonian is the term

$$\frac{1}{2}\sum_{i \neq j}\frac{e_i e_j}{|\mathbf{r}_i - \mathbf{r}_j|} \tag{8.10}$$

which is the Coulomb interaction between charges. The Coulomb interaction is instantaneous in time, since the potential has no retardation or speed of light built into it. The lack of retardation is not an approximation but is rigorously correct in the Coulomb gauge. Of course, there is retardation in the total interaction, which arises through the vector potential fields.

8.1.1 Gauges

Although the topic of gauges is treated correctly in a number of texts, it still seems to be poorly understood by students. It seems appropriate to start at the beginning and reproduce some standard material.

There is an important theorem that any vector function of position can be written as the sum of two terms. One is the gradient of a potential and the other is the curl of a vector:

$$\mathbf{S}(\mathbf{r}) = \nabla g + \nabla \times \mathbf{m}(\mathbf{r}) = \mathbf{S}_\ell + \mathbf{S}_t \tag{8.11}$$

$$\mathbf{S}_\ell = \nabla g \tag{8.12}$$

$$\mathbf{S}_t = \nabla \times \mathbf{m} \tag{8.13}$$

The term $\mathbf{S}_\ell$ is called the *longitudinal* part of $\mathbf{S}$, and $\mathbf{S}_t$ is called the *transverse* part. If $\mathbf{B}(\mathbf{r})$ is assumed to have this form, then Eq. (8.3) becomes

$$\nabla \cdot (\nabla g + \nabla \times \mathbf{A}) = \nabla^2 g = 0 \tag{8.14}$$

Usually $g = 0$, so that the vector potential is defined as

$$\mathbf{B} = \nabla \times \mathbf{A} \tag{8.15}$$

However, this expression does not uniquely define $\mathbf{A}(\mathbf{r})$. The definition $\mathbf{B} = \nabla \times A$ is put into (8.6):

$$\nabla \times \left[\mathbf{E} + \frac{1}{c}\frac{\partial}{\partial t}\mathbf{A} \right] = 0 \tag{8.16}$$

Now the factor in brackets is also the sum of a longitudinal and transverse part:

$$\nabla \times [\nabla \psi + \nabla \times \mathbf{M}] = 0 = \nabla \times (\nabla \times \mathbf{M}) \tag{8.17}$$

The equation is satisfied if $\mathbf{M} = 0$, so the electric field is

$$\mathbf{E} = -\frac{1}{c}\frac{\partial \mathbf{A}}{\partial t} - \nabla \psi \tag{8.18}$$

where ψ is the scalar potential. When these two forms for $\mathbf{B}(\mathbf{r})$ and $\mathbf{E}(\mathbf{r})$ are put into (8.4) and (8.5), the equations for the scalar and vector potentials are

$$\nabla^2 \psi + \frac{1}{c}\frac{\partial}{\partial t}\nabla \cdot \mathbf{A} = -4\pi\rho \tag{8.19}$$

$$\nabla \times (\nabla \times \mathbf{A}) + \frac{1}{c^2}\frac{\partial^2}{\partial t^2}\mathbf{A} + \nabla \frac{\partial}{c\partial t}\psi = \frac{4\pi}{c}\mathbf{j} \tag{8.20}$$

At first it appears that there are four equations for the four unknowns (A_x, A_y, A_z, ψ): Eq. (8.19) and the three vector components of (8.20). If this assertion were true, the four unknown functions would be determined uniquely in terms of the sources ($\mathbf{j}$,ρ). However, these four equations are not linearly independent; only three of them are independent. The linear dependence is shown by operating on (8.19) by $(1/c)(\partial/\partial t)$ and on (8.20) by ∇. Then subtract the two equations and find

$$\nabla\cdot[\nabla\times(\nabla\times\mathbf{A})]=\frac{4\pi}{c}\left(\frac{\partial\rho}{\partial t}+\nabla\cdot\mathbf{j}\right) \tag{8.21}$$

The left-hand side is zero because it is the gradient of a curl. The right-hand side vanishes since it is the equation of continuity. The four equations are not independent. There are only three equations for the four unknowns.

Therefore, the four unknown functions (A_x, A_y, A_z, ψ) are not uniquely determined. It is necessary to stipulate one additional condition or constraint on their values. It is called the *gauge condition*. The imposed condition is that the Coulomb field $\psi(\mathbf{r},t)$ shall act instantaneously, which is accomplished by insisting that

$$\nabla\cdot\mathbf{A}=0 \tag{8.22}$$

Equation (8.22) defines the *Coulomb gauge*, sometimes called the *transverse gauge*. The latter name arises because (8.22) implies that $\nabla\cdot\mathbf{A}_\ell=0$, so that $\mathbf{A}$ is purely transverse. One should realize that any arbitrary constraint may be imposed as long as one can satisfy (8.19) and (8.20). As long as these two equations are satisfied, one always obtains the same value for $\mathbf{E}(\mathbf{r})$ and $\mathbf{B}(\mathbf{r})$. The arbitrary choice of gauge does not alter the final value of observable quantities.

Define an arbitrary scalar function $\chi(\mathbf{r},t)$. A *gauge transformation* is to alter the vector and scalar potentials as

$$\mathbf{A}\to\mathbf{A}+\nabla\chi \tag{8.23}$$

$$\psi\to\psi-\frac{1}{c}\frac{\partial\chi}{\partial t} \tag{8.24}$$

$$\mathbf{B}\to\nabla\times[\mathbf{A}+\nabla\chi]=\nabla\times\mathbf{A} \tag{8.25}$$

$$\mathbf{E}\to-\frac{1}{c}\frac{\partial}{\partial t}[\mathbf{A}+\nabla\chi]-\nabla\left[\psi-\frac{1}{c}\frac{\partial\chi}{\partial t}\right] \tag{8.26}$$

$$=-\frac{1}{c}\frac{\partial}{\partial t}\mathbf{A}-\nabla\psi \tag{8.27}$$

The electric and magnetic fields are unaffected by the gauge transformation. However, the Hamiltonian operator contains the vector and scalar potentials and is affected by the gauge transformation. Next consider the change

$$\psi(\mathbf{r},t)\to\psi(\mathrm{r},t)\exp\left[i\frac{e}{\hbar c}\chi\right] \tag{8.28}$$

The kinetic energy term in the Hamiltonian contains the vector potential

$$\frac{1}{2}mv^2=\frac{p^2}{2m}\to\frac{1}{2m}\left(p-\frac{e}{c}A\right)^2 \tag{8.29}$$

Consider how the momentum operates on the revised wave function:

$$\mathbf{p}\left(\psi(\mathbf{r},t)\exp\left[i\frac{e}{\hbar c}\chi\right]\right)=\exp\left[i\frac{e}{\hbar c}\chi\right]\left(\mathbf{p}+\frac{e}{c}\nabla_\chi\right)\psi(\mathbf{r},t) \tag{8.30}$$

If the gauge transformation in eqn. (8.23) is accompanied by the change in eigenfunction (8.28) then the Hamiltonian does not change!

Each Hamiltonian we work with is a particular choice of gauge. Any observable quantity is unaffected by the choice of gauge.

In the Coulomb gauge eqn. (8.19) simplifies to

$$\nabla^2\psi = -4\pi\rho \tag{8.31}$$

which is easily solved to give

$$\psi(\mathbf{r}, t) = \int d^3r' \frac{\rho(\mathbf{r}', t)}{|\mathbf{r}-\mathbf{r}'|} \tag{8.32}$$

The potential $\psi(\mathbf{r},t)$ is instantaneous and is not retarded. This result is not an approximation but is an exact consequence for our choice of gauge. Later it is shown that a different choice of gauge leads to a retarded scalar potential.

Now evaluate the other equation (8.20). The following identity is useful:

$$\nabla\times(\nabla\times\mathbf{A}) = -\nabla^2\mathbf{A} + \nabla(\nabla\cdot\mathbf{A}) \tag{8.33}$$

The second term vanishes in the Coulomb gauge, which gives

$$\nabla^2\mathbf{A} - \frac{1}{c^2}\frac{\partial^2}{\partial t^2}\mathbf{A} = -\frac{4\pi}{c}\mathbf{j} + \nabla\frac{\partial\psi}{c\partial t} \tag{8.34}$$

It is useful to operate a bit on the second term on the right-hand side. Using (8.32), this term is

$$\nabla\frac{\partial\psi}{c\partial t} = \frac{1}{c}\nabla\int d^3r' \frac{(\partial/\partial t)\rho(\mathbf{r}', t)}{|\mathbf{r}-\mathbf{r}'|} \tag{8.35}$$

$$\nabla\frac{\partial\psi}{c\partial t} = -\frac{1}{c}\nabla\int d^3r' \frac{1}{|\mathbf{r}-\mathbf{r}'|}\nabla'.\mathbf{j}(\mathbf{r}', t) \tag{8.36}$$

where the last identity uses the equation of continuity. Integrate by parts in the last term,

$$\nabla\frac{\partial\psi}{c\partial t} = \frac{1}{c}\nabla\int d^3r'\mathbf{j}(\mathbf{r}', t).\nabla'\frac{1}{|\mathbf{r}-\mathbf{r}'|} \tag{8.37}$$

and then pull the gradient out by letting it operate on **r** instead of **r**′; the latter step requires a sign change. Also operate on the current term itself in (8.34) by using the identity

$$4\pi\mathbf{j}(\mathbf{r}, t) = -\nabla^2\int d^3r' \frac{\mathbf{J}(\mathbf{r}', t)}{|\mathbf{r}-\mathbf{r}'|} \tag{8.38}$$

By combining these results

$$\nabla^2\mathbf{A} - \frac{1}{c^2}\frac{\partial^2}{\partial t^2}\mathbf{A} = (\nabla^2 - \nabla\nabla)\frac{1}{c}\int d^3r' \frac{\mathbf{j}(\mathbf{r}', t)}{|\mathbf{r}-\mathbf{r}'|} \tag{8.39}$$

Finally, using the identity (8.33),

$$\nabla^2\mathbf{A} - \frac{1}{c^2}\frac{\partial^2}{\partial t^2}\mathbf{A} = -\frac{1}{c}\nabla\times\left[\nabla\times\int d^3r' \frac{\mathbf{j}(\mathbf{r}', t)}{|\mathbf{r}-\mathbf{r}'|}\right] \tag{8.40}$$

The point of this exercise is that the right-hand side of (8.40) is now a transverse vector, since it is the curl of something. The current is written as a longitudinal plus a transverse part,

$$\mathbf{j} = \mathbf{j}_l + \mathbf{j}_t \tag{8.41}$$

Then the vector potential obeys the equation

$$\nabla^2 \mathbf{A} - \frac{1}{c^2}\frac{\partial^2}{\partial t^2}\mathbf{A} = -\frac{4\pi}{c}\mathbf{j}_t(\mathbf{r}, t) \tag{8.42}$$

where $\mathbf{j}_t(\mathbf{r}, t)$ is defined by the right-hand side of (8.40). The final equation for **A** is very reasonable. Since the vector potential **A** is purely transverse, it should respond only to the transverse part of the current. If it were to respond to the longitudinal part of the current, it would develop a longitudinal part. The longitudinal component of **A** does not occur in the Coulomb gauge.

As a simple example, consider a current of the form

$$\mathbf{j}(\mathbf{r}) = \mathbf{J}_0 e^{i\mathbf{k}\cdot\mathbf{r}} \tag{8.43}$$

whose transverse part is

$$\mathbf{j}_t = -\frac{1}{k^2}\mathbf{k}\times(\mathbf{k}\times\mathbf{J}_0)e^{i\mathbf{k}\cdot\mathbf{r}} \tag{8.44}$$

and $\mathbf{j}_t$ is the component of **j** that is perpendicular to **k**; see figure 8.1. In free space, the transverse and longitudinal parts are just those components that are perpendicular and parallel to the wave vector **k** of the photon. However, solids are not homogeneous but periodic. Along major symmetry directions in the crystal, it is often true that transverse components are perpendicular to the wave vector. However, it is generally not true for arbitrary wave vectors in the solid, even in cubic crystals. The words *transverse* and *longitudinal* do not necessarily mean perpendicular and parallel to **k**.

Two charges interact by the sum of the two interactions: Coulomb plus photon. The net interaction may not have a component that is instantaneous. In fact, for a frequency-dependent charge density at distances large compared to c/ω, one finds that the photon part of the interaction produces a term $-e^2/r$, which exactly cancels the instantaneous Coulomb interaction. The remaining parts of the photon contribution are the net retarded interaction. Solid-state and atomic physics are usually concerned with interactions over short distances. Then retardation is unimportant for most problems. In the study of the homogeneous electron gas, for example, the photon part is small and may be neglected. In real solids, the photon part causes some crystal field effects, which is an unexciting many-body effect. The main effect of retardation is the polariton effects at long wavelength [5]. In general, the Coulomb gauge is chosen because the instantaneous Coulomb interaction is usually a large term that forms a central part of the analysis, while the photon parts are usually secondary. Like most generalizations, this one has its exceptions.

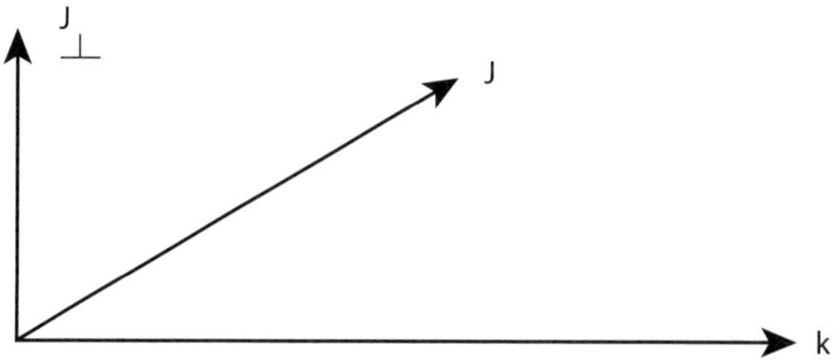

FIGURE 8.1. Perpendicular component of the current.

The Coulomb gauge $\nabla \cdot A = 0$ is not the only gauge condition popular in physics. Many physicists use the *Lorentz gauge*:

$$\nabla \cdot \mathbf{A} + \frac{\partial \psi}{c \partial t} = 0 \tag{8.45}$$

which causes (8.19) and (8.20) to have the forms

$$\nabla^2 \psi - \frac{1}{c^2} \frac{\partial^2}{\partial t^2} \psi = -4\pi\rho \tag{8.46}$$

$$\nabla^2 \mathbf{A} - \frac{1}{c^2} \frac{\partial^2}{\partial t^2} \mathbf{A} = -\frac{4\pi}{c} \mathbf{j} \tag{8.47}$$

Now both the vector and scalar potentials obey the retarded wave equation. They combine to produce a four-vector that is invariant under a Lorentz transformation. The Lorentz invariance is useful in many branches of physics, and the Lorentz gauge is used frequently. Both the scalar and vector parts are retarded.

Another gauge that is often used, but that does not have a formal name, is the condition that $\psi = 0$. The scalar potential is set equal to zero. In this case it is found that the longitudinal vector potential is not zero but now plays an important role. In fact, the longitudinal part of the vector potential leads to an interaction between charges, which is just the instantaneous Coulomb interaction. When $\psi = 0$ the longitudinal part of the vector potential plays a role that is identical to that of the scalar potential in the Coulomb gauge.

The Hamiltonian (8.7) is written in a gauge that has the scalar potential acting instantaneously, so that the Coulomb interaction is unretarded. This form of the Hamiltonian is consistent with either gauge $\nabla \cdot \mathbf{A} = 0$ or $\psi = 0$. One gets a different Hamiltonian for other choices of gauge.

8.1.2 Lagrangian

So far it has been proved that the Coulomb gauge makes the vector potential transverse and that the scalar potential acts instantaneously. Next it is shown that the Hamiltonian has the form indicated in eqn.(8.7), which is done by starting from the following Lagrangian (Schiff, 1968):

$$L=\frac{1}{2}\sum_i mv_i^2+\int\frac{d^3r}{8\pi}[\mathbf{E}(\mathbf{r})^2-\mathbf{B}(\mathbf{r})^2]-\sum_i e_i[\psi(\mathbf{r}_i)-\frac{1}{c}\mathbf{v}_i\cdot\mathbf{A}(\mathbf{r}_i)]$$

$$\sum_i e_i\psi(\mathbf{r}_i)=\int d^3r\rho(\mathbf{r})\psi(\mathbf{r}),\ \rho(\mathbf{r})=\sum_i e_i\delta(\mathbf{r}-\mathbf{r}_i) \tag{8.48}$$

$$\sum_i \frac{e_i}{c}\mathbf{v}_i\cdot\mathbf{A}(\mathbf{r}_i)=\frac{1}{c}\int d^3r\mathbf{j}(\mathbf{r})\cdot\mathbf{A}(\mathbf{r}),\ \mathbf{j}(r)=\sum_i e_i\ v_i\delta(\mathbf{r}-\mathbf{r}_i) \tag{8.49}$$

This Lagrangian is chosen to produce Maxwell's equations as well as the classical equations of motion for a particle at $\mathbf{r}_i$, with charge e_i and velocity $\mathbf{v}_i$, in a magnetic and electric field. There are $4+3N$ variables η, which are ψ, $\mathbf{A}$, and $\mathbf{r}_i$ for $i=1, N$. In terms of these variables the electric and magnetic fields are: $\mathbf{E}=-\mathring{\mathbf{A}}/c-\nabla\psi$, $\mathbf{B}=\nabla\times\mathbf{A}$. Each variable η is used to generate an equation from

$$\frac{d}{dt}\frac{\delta L}{\delta\dot{\eta}}+\sum_{\alpha=1}^{3}\frac{d}{dr_\alpha}\frac{\partial L}{\partial(\partial\eta/\partial r_\alpha)}-\frac{\delta L}{\delta\eta}=0 \tag{8.50}$$

where the conjugate momentum is

$$P\eta=\frac{\partial L}{\partial\dot{\eta}} \tag{8.51}$$

When the scalar potential ψ is chosen as the variable η,

$$\frac{\delta L}{\delta\psi}=-\rho(\mathbf{r})=-\sum_i e_i\delta(\mathbf{r}-\mathbf{r}_i) \tag{8.52}$$

$$\frac{\partial L}{\partial(\partial\psi/\partial x)}=-\frac{E_x}{4\pi} \tag{8.53}$$

$$P_\psi=\frac{\delta L}{\delta\dot{\psi}}=0 \tag{8.54}$$

This choice produces the Maxwell equation (8.5) from (8.50):

$$\nabla\cdot\mathbf{E}=4\pi\rho \tag{8.55}$$

When one component of the scalar potential such as A_x is chosen as the variable in the Lagrangian,

$$c\frac{\delta L}{\delta A_x}=j_x=\sum_i e_i v_{ix}\delta(\mathbf{r}-\mathbf{r}_i) \tag{8.56}$$

$$\frac{\delta L}{\delta\dot{A}_x}=-\frac{E_x}{4\pi c} \tag{8.57}$$

$$\frac{\delta L}{\delta(\partial A_x/\partial y)}=-\frac{B_z}{4\pi} \tag{8.58}$$

When this equation is used in Lagrange's equation (8.50), the x-component of the Maxwell equation (8.6) is

$$\nabla \times B = \frac{1}{c}\frac{\partial \mathbf{E}}{\partial t} + \frac{4\pi}{c}\mathbf{j} \tag{8.59}$$

Choosing either A_y or A_z as the active variable will generate the y- and z-components of this equation. The Lagrangian does generate the two Maxwell equations, which depend on particle properties.

Another important feature is that the momentum conjugate to the scalar potential is zero:

$$P_\psi = 0 \tag{8.60}$$

The momentum variable conjugate to the vector potential is

$$\mathbf{P}_A = -\frac{\mathbf{E}}{4\pi c} \tag{8.61}$$

which is just proportional to the electric field. This relationship is important because the quantization of the fields will require that the vector potential no longer commute with the electric field, since they are conjugate variables.

The other equations generated by this Lagrangian are the equations for particle motion in electric and magnetic fields. Here the variable in the Lagrangian is the the particle coordinate r_α:

$$\frac{\delta L}{\delta r_{i\alpha}} = -e_i\left[\nabla_\alpha \psi(\mathbf{r}_i) - \frac{v_{i\delta}}{c}\nabla_\alpha A_\delta(\mathbf{r}_i)\right] \tag{8.62}$$

$$\frac{\delta L}{\delta v_{i\alpha}} = p_{i\alpha} = m v_{i\alpha} + \frac{e_i}{c}A_\alpha(\mathbf{r}_i) \tag{8.63}$$

In the last term of (8.62), repeated indices over δ imply summation. The equation deduced from Lagrange's equation is

$$0 = \frac{d}{dt}\left[m\mathbf{v}_i + \frac{e_i}{c}\mathbf{A}(\mathbf{r}_i)\right] + e_i\nabla\left[\psi - \frac{1}{c}\mathbf{v}\cdot\mathbf{A}(\mathbf{r})\right]$$

The total time derivative on the position dependence of the vector potential is interpreted as a hydrodynamic derivative:

$$\frac{d}{dt}\mathbf{A} = \frac{\partial}{\partial t}\mathbf{A} + \mathbf{v}_i\cdot\nabla\mathbf{A}$$

Then the equation may be rearranged into

$$m\frac{d}{dt}\mathbf{v}_i = e_i\{\mathbf{E}(\mathbf{r}_i) + \frac{1}{c}[\nabla(\mathbf{v}\cdot\mathbf{A}) - \mathbf{v}\cdot\nabla\mathbf{A}]\} = e_i\left[\mathbf{E}(\mathbf{r}_i) + \frac{1}{c}\mathbf{v}_i\times\mathbf{B}\right]$$

The last term is just the Lorentz force on a particle in a magnetic field since

$$\mathbf{v}\times\mathbf{B} = \mathbf{v}\times(\nabla\times\mathbf{A}) = \nabla(\mathbf{v}\cdot\mathbf{A}) - (\mathbf{v}\cdot\nabla)\mathbf{A} \tag{8.64}$$

The equation for $m\dot{\mathbf{v}}$ is Newton's law for a spinless particle in an electric and magnetic field. The Lagrangian is a suitable starting point for the quantization of the interacting system of particles and fields.

8.1.3 Hamiltonian

The Hamiltonian is derived from

$$H=\sum_i \dot{\eta}_i P_\eta - L \tag{8.65}$$

where the first summation is over all of the variables and their conjugate moments. In the present problem this summation includes the vector potential and the particle momentum; the scalar potential has no momentum. The Hamiltonian has the form

$$\begin{aligned} H=&\frac{1}{m}\sum_i \mathbf{p}_i\cdot\left[\mathbf{p}_i-\frac{e_i}{c}\mathbf{A}(\mathbf{r}_i)\right]+\int d^3r\left[\frac{1}{4\pi c}\dot{\mathbf{A}}\cdot\left(\frac{1}{c}\dot{\mathbf{A}}+\nabla\psi\right)-\frac{E^2-B^2}{8\pi}\right]\\ &+\sum_i e_i\psi(\mathbf{r}_i)-\frac{1}{2m}\sum_i\left[\mathbf{p}_i-\frac{e_i}{c}\mathbf{A}\right]^2-\sum_i\frac{e_i}{mc}\mathbf{A}\cdot\left[\mathbf{p}_i-\frac{e_i}{c}\mathbf{A}\right] \end{aligned} \tag{8.66}$$

It may be collected into the form

$$\begin{aligned} H=&\frac{1}{2m}\sum_i\left[\mathbf{p}_i-\frac{e_i}{c}\mathbf{A}(\mathbf{r}_i)\right]^2+\sum_i e_i\psi(\mathbf{r}_i)\\ &+\int\frac{d^3r}{8\pi}\left[\left(\frac{\dot{\mathbf{A}}}{c}\right)^2+B^2-(\nabla\psi)^2\right] \end{aligned} \tag{8.67}$$

Terms that are the cross product between the scalar and vector potential parts of the electric field always vanish after an integration by parts,

$$\int d^3r(\nabla\psi\cdot\mathbf{A})=-\int d^3r\psi(\nabla\cdot\mathbf{A})=0 \tag{8.68}$$

since the Coulomb gauge is used, wherein $\nabla\cdot\mathbf{A}=0$. The terms involving the square of the scalar potential may also be reduced to an instantaneous interaction between charges using (8.32):

$$\begin{aligned} \int\frac{d^3r}{4\pi}\nabla\psi\cdot\nabla\psi&=-\int\frac{d^3r}{4\pi}\psi\nabla^2\psi=\int d^3r\psi(\mathbf{r})\rho(\mathbf{r})=\sum_i e_i\psi(\mathbf{r}_i)\\ &=\sum_{ij}\frac{e_ie_j}{r_{ij}} \end{aligned} \tag{8.69}$$

With these simplifications, the Hamiltonian is written as

$$H=\frac{1}{2m}\sum_i\left[\mathbf{p}_i-\frac{e_i}{c}\mathbf{A}(\mathbf{r}_i)\right]^2+\int\frac{d^3r}{8\pi}\left[\left(\frac{\dot{\mathbf{A}}}{c}\right)^2+B^2\right]+\frac{1}{2}\sum_{ij}\frac{e_ie_j}{r_{ij}} \tag{8.70}$$

The factor of $\frac{1}{2}$ is in front of the electron–electron interaction since the summation over all (ij, $i\neq j$) counts each pair twice.

The electric field is the conjugate momentum density of the vector potential. These two field variables must obey the following equal time commutation relation:

$$\left[A_\alpha(\mathbf{r},t),\ -\frac{1}{4\pi c}E_\beta(\mathbf{r}',t)\right]=i\hbar\delta_{\alpha\beta}\delta(\mathbf{r}-\mathbf{r}') \tag{8.71}$$

This commutation relation is satisfied by defining the vector potential in terms of creation and destruction operators in the form

$$A_\alpha(\mathbf{r},\, t) = \sum_{k\lambda} \left(\frac{2\pi\hbar c^2}{\Omega\omega_{\mathbf{k}}}\right)^{1/2} \xi_\alpha(\mathbf{k},\, \lambda) e^{ik\cdot r} (a_{k\lambda} e^{-i\omega_{\mathbf{k}} t} + a^\dagger_{-k\lambda} e^{i\omega_{\mathbf{k}} t}) \tag{8.72}$$

There are two transverse modes, one for each transverse direction, and λ is the summation over these two modes. The unit polarization vector ξ_α gives the direction for each mode. The operators obey the commutation relations

$$[a_{\mathrm{k}\lambda},\, a^\dagger_{\mathbf{k}'\lambda'}] = \delta_{\mathbf{kk}'}\delta_{\lambda\lambda'} \tag{8.73}$$

$$[a_{\mathrm{k}\lambda},\, a_{\mathbf{k}'\lambda'}] = 0 \tag{8.74}$$

The time derivative of the vector potential is

$$\dot{A}_\alpha(\mathbf{r},\, t) = -i \sum_{k\lambda} \left(\frac{2\pi\hbar c^2\omega_{\mathbf{k}}}{\Omega}\right)^{1/2} \xi_\alpha(\mathbf{k},\, \lambda) e^{i\mathbf{k}\cdot\mathbf{r}} (a_{k\lambda} e^{-i\omega_{\mathbf{k}} t} - a^\dagger_{-k\lambda} e^{i\omega_{\mathbf{k}} t})$$

The electric field is

$$E_\alpha(\mathbf{r},\, t) = -\frac{1}{c}\dot{A}_\alpha - \nabla_\alpha \psi \tag{8.75}$$

The scalar potential is not expressed in terms of operators, since its conjugate momentum is zero. It does not influence the commutator in (8.71). The commutation relation (8.71) is the commutator of the vector potential and its time derivative:

$$[A_\alpha(\mathbf{r},\, t),\, \dot{A}_\beta(\mathbf{r}',\, t)] = 4\pi c^2 \hbar i \delta_{\alpha\beta}\delta(\mathbf{r}-\mathbf{r}') \tag{8.76}$$

The vector potential in (8.72) has been chosen to give almost this result:

$$\begin{aligned}
[A_\alpha(\mathbf{r},\, t),\, \dot{A}_\beta(\mathbf{r}',\, t)] &= i \sum_{k\lambda,\mathbf{k}'\lambda'} \left(\frac{2\pi\hbar c^2\omega_{\mathbf{k}}}{\Omega}\right)^{1/2} \left(\frac{2\pi\hbar c^2}{\Omega\omega_{\mathbf{k}}'}\right)^{1/2} \xi_\alpha(\mathbf{k},\, \lambda)\xi_\beta(\mathbf{k}',\, \lambda') \\
&\quad \times e^{i(\mathbf{k}\cdot\mathbf{r}+\mathbf{k}'\cdot\mathbf{r}')} 2\delta_{\mathbf{k},\, -\mathbf{k}'}\delta_{\lambda\lambda'} \\
&= 4\pi i c^2 \hbar \Phi_{\alpha\beta}(\mathbf{r}-\mathbf{r}')
\end{aligned} \tag{8.77}$$

$$\Phi_{\alpha\beta}(\mathbf{R}) = \frac{1}{\Omega}\sum_{\mathbf{k}} \mathbf{e}^{i\mathbf{k}\cdot\mathbf{R}} \left[\delta_{\alpha\beta} - \frac{\mathbf{k}_\alpha \mathbf{k}_\beta}{\mathbf{k}^2}\right] \tag{8.78}$$

$$= \delta_{\alpha\beta}\delta^3(\mathbf{R}) + \nabla_\alpha\nabla_\beta \int \frac{d^3k}{(2\pi)^3 k^2} e^{i\mathbf{k}\cdot\mathbf{R}} \tag{8.79}$$

$$\Phi_{\alpha\beta}(\mathbf{R}) = \delta_{\alpha\beta}\delta^3(\mathbf{R}) + \nabla_\alpha\nabla_\beta \frac{1}{4\pi\mathbf{R}} \tag{8.80}$$

The last term on the right is needed to satisfy the feature that the vector potential is transverse.

The various factors of 2π and $\omega_{\mathbf{k}}$ that enter (8.72) are selected so that the commutation relation (8.71) is satisfied for the usual commutation relations for the operators $a_{\mathrm{k}\lambda}$

and $a^{\dagger}_{\mathbf{k}\lambda}$. The expressions for the energy density of electric and magnetic fields have the form

$$\int d^3r \frac{1}{8\pi c^2}\dot{\mathbf{A}}^2(\mathbf{r},t) = -\sum_{k\lambda}\frac{\hbar\omega_{\mathbf{k}}}{4}\left(a_{\mathbf{k}\lambda}a_{-\mathbf{k}\lambda}e^{-2it\omega_{\mathbf{k}}} + a^{\dagger}_{\mathbf{k}\lambda}a^{\dagger}_{-\mathbf{k}\lambda}e^{2it\omega_{\mathbf{k}}}\right.$$
$$\left. - a_{\mathbf{k}\lambda}a^{\dagger}_{\mathbf{k}\lambda} - a^{\dagger}_{\mathbf{k}\lambda}a_{\mathbf{k}\lambda}\right) \tag{8.81}$$

$$\int d^3r \frac{1}{8\pi}(\nabla\times\mathbf{A})^2 = \sum_{\mathbf{k}\lambda}\frac{\hbar c^2}{4\omega_{\mathbf{k}}}(\mathbf{k}\times\xi)^2\left(a_{\mathbf{k}\lambda}a_{-\mathbf{k}\lambda}e^{-2it\omega_{\mathbf{k}}} + a^{\dagger}_{-k\lambda}a^{\dagger}_{-\mathbf{k}\lambda}e^{2it\omega_{\mathbf{k}}}\right.$$
$$\left. + a_{\mathbf{k}\lambda}a^{\dagger}_{\mathbf{k}\lambda} + a^{\dagger}_{\mathbf{k}\lambda}a_{\mathbf{k}\lambda}\right) \tag{8.82}$$

In the second term, the photon energy is $\omega_{\mathbf{k}} = ck$. This term may be added to the first, and the aa and $a^{\dagger}a^{\dagger}$ terms both cancel. Combining the results from (8.70) gives

$$H = \sum_i \frac{1}{2m}\left[\mathbf{p}_i - \frac{e_i}{c}\mathbf{A}(\mathbf{r}_i)\right]^2 + \frac{1}{2}\sum_{i\neq j}\frac{e_ie_j}{|\mathbf{r}_i - \mathbf{r}_j|} + \sum_{\mathbf{k}\lambda}\hbar\omega_{\mathbf{k}}\left[a^{\dagger}_{\mathbf{k}\lambda}a_{\mathbf{k}\lambda} + \frac{1}{2}\right]$$

This form of the Hamiltonian is just the result that was asserted in the beginning in (8.7). The vector potential in (8.72) is expressed in terms of the creation and destruction operators of the photon field. The photon states behave as bosons, as independent harmonic oscillators. Each photon state of wave vector $\mathbf{k}$ and polarization λ has eigenstates the form

$$|n_{\mathbf{k}\lambda}\rangle = \frac{(a^{\dagger}_{\mathbf{k}\lambda})^{n_{\mathbf{k}\lambda}}}{(n_{\mathbf{k}\lambda}!)^{1/2}}|0\rangle \tag{8.83}$$

where $n_{\mathbf{k}\lambda}$ is an integer that is the number of photons in that state. The state $|0\rangle$ is the photon vacuum. The total energy in the free photon part of the Hamiltonian is

$$H_0|n_{k_1\lambda_1}n_{k_2\lambda_2}\ldots n_{k_n\lambda_n}\rangle = \sum_{\mathbf{k}\lambda}\hbar\omega_{\mathbf{k}}\left(n_{\mathbf{k}\lambda} + \frac{1}{2}\right)|n_{\mathbf{k}_1\lambda_1}n_{k_2\lambda_2}\ldots n_{\mathbf{k}_n\lambda_n}\rangle \tag{8.84}$$

The Hamiltonian (8.7) has been derived for spinless, nonrelativistic particles. Certainly the most important relativistic term is the spin–orbit interaction. The effects of spin also enter through the direct interaction of the magnetic moment with an external magnetic field.

That photons obey the statistics of a harmonic oscillator is important in understanding how lasers operate. When the system emits a photon of wave vector $\mathbf{k}$, the operator algebra is $a^{\dagger}_{\mathbf{k}}|n_{\mathbf{k}}\rangle = \sqrt{n_{\mathbf{k}+1}}|n_{\mathbf{k}} + 1\rangle$. The matrix element contains the square of this factor

$$M^2 \sim n_{\mathbf{k}} + 1 \tag{8.85}$$

The factor of $n_{\mathbf{k}}$ means that the emitted photon is more likely to go into the state $\mathbf{k}$ if that state is occupied by other photons. An emitted photon can be created in any state with the same energy. In practice, this means that the magnitude of the wave vector is fixed, but not the direction. The photon can be emitted with a wave vector that points in any direction in space. If one has an optical cavity, so that some photons are trapped in one

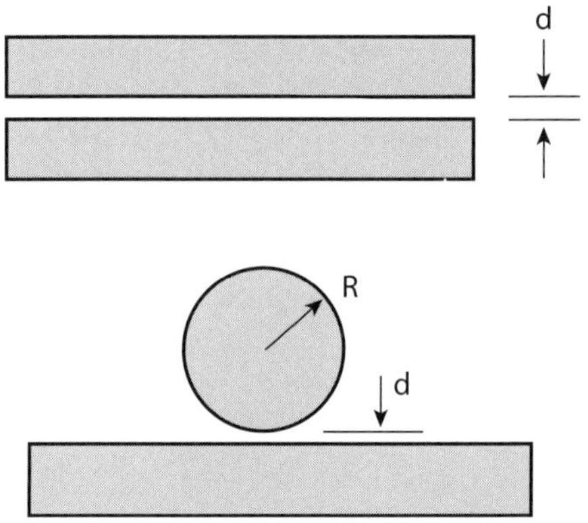

FIGURE 8.2. (*Top*) Two parallel conducting plates separated by a distance d. (*Bottom*) A sphere of radius R separated from a plate by distance d.

state, then the photons in that state stay in the system. Over time, the number in this single state builds up, which is the lasing state. This process would not happen without the above matrix element.

8.1.4 Casimir Force

Henrik Casimir made a remarkable prediction in 1948. He predicted a new force between macroscopic objects such as metal plates and metal spheres. His prediction has been confirmed and is now called the Casimir force. Figure 8.2 shows two plates separated by a small distance d. The Casimir force between these two plates depends on the area A, the separation d, and two fundamental constants:

$$F(d) = -\frac{\pi^4}{240}\frac{\hbar c}{d^4}A \tag{8.86}$$

The force depends on the fourth power of the separation between plates. The strength of the force depends on $\hbar c$. It is a type of van der Waals interaction, but does not depend on the polarizability of the material nor on other parameters such as electron charge or mass. It is a true quantum effect.

The derivation of this formula starts with a term we have been ignoring: the zero-point energy of the photon gas:

$$E(ZPE) = \frac{1}{2}\sum_{\mathbf{k},\lambda} \hbar c|\mathbf{k}| \tag{8.87}$$

The right-hand side of this expression diverges, since there is no maximum value of wave vector. It is actually infinity. The photon states are changed in the region between the plates. Parallel plates are a type of wave guide, and one has to determine the electromagnetic modes with appropriate boundary conditions at the surface of the plates. Both transverse electric (TE) and transverse magnetic (TM) modes must be evaluated. Then

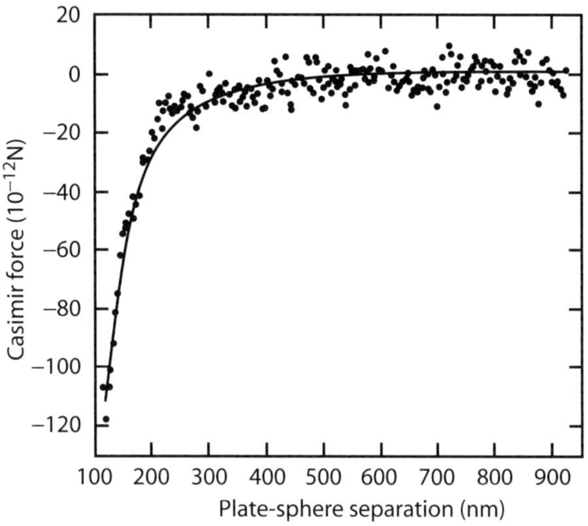

FIGURE 8.3. Casimir force as measured by Mohideen and Roy, *Phys. Rev. Lett.* **81**, 4551 (1998).

Casimir calculated the zero-point energy of this confined system and subtracted the result from the above formula. He subtracted two formulas, which are each infinity. The difference was a finite energy that was a function of d. He then differentiated this energy to get the force.

Surface science has advanced recently to where it is now possible to measure this force. Figure 8.3 shows the results of Mohideen and Roy. Their measurements were between a flat plate and a sphere of radius R. The distance d is between the plate and the closest point of the sphere. When $R >> d$ the formula for the force is changed to

$$F = -\frac{\pi^3}{360} R \frac{\hbar c}{d^3} A \tag{8.88}$$

Figure 8.3 shows their measurements agree well with this formula.

Casimir made his prediction based on several assumptions:

- Perfect conductors
- Zero temperature
- Perfect vacuum

Actual experiments are done on

- Real metals, with resistance
- Nonzero temperatures
- Laboratory vacuum

Whether these laboratory conditions change Casimir's formula by a small percentage is much debated. Nevertheless, it is a remarkable macroscopic quantum effect.

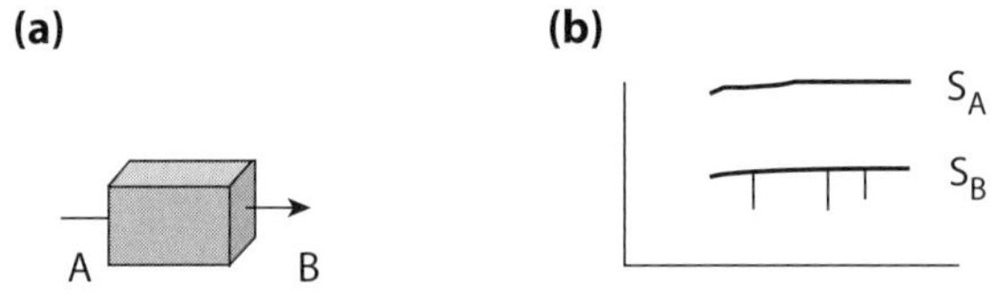

FIGURE 8.4. (a) Experimental geometry. (b) Frequency spectrum $S_A(\mathbf{k})$ of photons entering gas and $S_B(\mathbf{k})$ when leaving gas. The latter has spikes due to absorption.

8.2 Optical Absorption by a Gas

Electromagnetic radiation is absorbed by a gas of atoms. Different mechanisms cause the absorption, and they tend to affect different frequency ranges. This section considers one process: the absorption of radiation by excitation of electronic states. The photon is destroyed in the absorption step. Its energy $\hbar\omega$ is used to excite the electron from energy state E_i to a final electronic state with energy $E_f = E_i + \hbar\omega$. Not all photons are absorbed. Only those are absorbed whose frequency is exactly the right amount to make up the energy difference $E_f - E_i$ between the electronic levels.

Figure 8.4 shows a schematic view of the experiment. The box holds the gas. Light enters at A and exits at B. The energy fluxes $S_{A,B}$ are shown at both locations. The spectrum at B is similar to A, except that $S_B(\mathbf{k})$ is reduced at occasional points of frequency. These spikes represent the optical absorption. Spikes occur because both E_i and E_f have discrete values, since both are energies of bound states. If E_f is a continuum state, then the absorption is a continuous function of frequency.

The term "optical absorption" does not imply that the radiation has to be in the range of visible frequency. The same process and the same theory apply to any frequency, from microwave to x-ray.

There are other processes whereby gases can absorb electromagnetic radiation:

1. The collisions of the atoms cause a weak absorption.
2. A gas of molecules absorbs low-frequency photons and converts them to excitations that are rotational or vibrational.
3. There are nonlinear processes such as Raman scattering, where the photons are inelastically scattered.
4. Two-photon absorption uses two photons to have an electron change its energy levels.

The present calculation is kept simple by making a few drastic assumptions. The first is that the atoms do not move, so that Doppler broadening is ignored. The second is that collisions and other multiatom events are ignored. The absorption is calculated for a single atom, and the results for the box are multiplied by the number N_A of atoms in the box. The "box" can be viewed as the laboratory room where the experiment is performed or as the box of gas atoms in the path of the photon beam.

The calculation is rather simple, since it uses two formulas that have been derived previously. The first is the Golden Rule from chapter 7. It gives the number of events w per second. In this case each event is the absorption of one photon. The result is multiplied by the N_A since the total number of absorption events is proportional to the number of atoms doing the absorbing:

$$w = \frac{2\pi}{\hbar} N_A \sum_{if} |V_{if}|^2 \delta(E_i + \hbar\omega - E_f) \tag{8.89}$$

The summation is taken over initial (i) and final (f) state. The initial states are occupied: they must initially have an electron in them. The final states cannot be occupied: they must have room to add an electron.

The matrix element $V_{if} = \langle f|V|i\rangle$ is the important quantity in the Golden Rule. The perturbation V must describe the absorption of photons, which requires that it contain both electron and photon operators. It is found from the Hamiltonian (8.7) for the system of photons and charged particles. This Hamiltonian is rewritten as

$$H = H_{0A} + H_{0P} + V \tag{8.90}$$

$$H_{0A} = \sum_i \left[\frac{p_i^2}{2m} - \frac{Ze^2}{r_i}\right] + \frac{e^2}{2}\sum_{i \neq j} \frac{1}{|\mathbf{r}_i - \mathbf{r}_j|} \tag{8.91}$$

$$H_{0P} = \sum_{\kappa\lambda} \hbar\omega_{\kappa\lambda}\left[a^\dagger_{\kappa\lambda} a_{\kappa\lambda} + \frac{1}{2}\right] \tag{8.92}$$

$$V = -\frac{e}{mc}\sum_i \mathbf{p}_i \cdot \mathbf{A}_{(\mathbf{r}_i)} + \frac{e^2}{2mc^2}\sum_i A(\mathbf{r}_i)^2 \tag{8.93}$$

The Hamiltonian of the atom alone is H_{0A}. It contains the kinetic energy of the electrons, the interaction between the electrons and the nucleus, and the interaction between the electrons. The factor of $\frac{1}{2}$ in the latter term occurs because the summation over all ($i \neq j$) counts each pair twice. These terms are the standard ones for a nonrelativistic theory of electron states in atoms. The Hamiltonian H_{0P} are for the unperturbed photon states. The perturbation V contains the interaction between the electrons and the photon field. The first term in V is called "p-dot-A," while the second term is called "A-squared."

The A-squared term in V contains the factor of A^2. The two factors of A require that two photons are involved. This interaction term is important in light scattering, such as the Compton effect. The $\mathbf{p} \cdot \mathbf{A}$ term contains only one power of the vector potential. Only one photon is involved. This interaction is the only one that contributes to optical absorption or emission of a single photon. For example, in linear optics one assumes that the rate of absorption, by the Golden Rule, must be proportional to the energy flux (Poynting vector) $S_{\kappa\lambda} \propto w$. Since A is proportional to the electric field, then $S_{\kappa\lambda} \propto E_0^2 \propto A^2 \propto |V|^2$. Linear optics requires that $V \propto A$, which is achieved only using the $\mathbf{p} \cdot \mathbf{A}$ term. The A-squared term is discarded in the discussion of linear optics.

Equation (8.89) is evaluated using the $\mathbf{p} \cdot \mathbf{A}$ term for V:

$$V = -\frac{e}{mc}\sum_i \mathbf{p}_i \cdot \mathbf{A}(\mathbf{r}_i) = -\frac{1}{c}\sum_i \mathbf{j}(\mathbf{r}_i) \cdot \mathbf{A}(\mathbf{r}_i) \tag{8.94}$$

$$\mathbf{A}_\mu(\mathbf{r}) = \sum_{\kappa\lambda} A_k \xi_\mu(\kappa\lambda) e^{i\mathbf{k}\cdot\mathbf{r}} (a_{\kappa\lambda} + a^\dagger_{-\kappa\lambda}) \tag{8.95}$$

$$A_k = \sqrt{\frac{2\pi\hbar c^2}{\omega_k \Omega}} \tag{8.96}$$

The combination of $e\mathbf{p}_i/m$ is really the electric current $\mathbf{j}$ of particle i, so the $\mathbf{p}\cdot\mathbf{A}$ interaction is really the scalar product of the current and the vector potential divided by the speed of light c. The vector potential is taken at $t=0$. The time dependence of this operator has already been used in the Golden Rule to provide the energy term $\hbar\omega_k$ in the delta-function for energy conservation.

The matrix element for the perturbation V is evaluated between initial and final states. The initial state has an electron in orbital ϕ_i and spin s_i, and $n_{\kappa\lambda}$ photons in state $\kappa\lambda$. After the absorption process, the electron is in a final orbital state ϕ_f, spin $s_f = s_i$, and there are $n_{\kappa\lambda}-1$ photons in the state $\kappa\lambda$. This matrix element is written as

$$\langle f|V|i\rangle = -\frac{e}{mc}\langle n_{\kappa\lambda}-1|a_{\kappa\lambda}|n_{\kappa\lambda}\rangle A_k \hat{\xi}(\boldsymbol{\kappa}\lambda)\cdot\mathbf{p}_{fi} \tag{8.97}$$

$$\mathbf{p}_{fi} = -i\hbar\int d^3r\,\phi_f^*(\mathbf{r})\nabla\left(e^{i\mathbf{k}\cdot\mathbf{r}}\phi_i(\mathbf{r})\right)\langle s_f|s_i\rangle \tag{8.98}$$

The last factor of $\langle s_f|s_i\rangle$ is the matrix element between the initial and final spin states of the electron. The $\mathbf{p}\cdot\mathbf{A}$ interaction does not change the spin of the electron, so the initial and final electron states must have the same spin configuration. This factor is usually omitted from the notation, with the understanding that the spin is unchanged during the optical transition. The photon matrix element gives

$$\langle n_{\kappa\lambda}-1|a_{\kappa\lambda}|n_{\kappa\lambda}\rangle = \sqrt{n_{\kappa\lambda}} \tag{8.99}$$

In the electron orbital part of the matrix element, the momentum operator $\mathbf{p} = -i\hbar\nabla$ could operate on the factor of $e^{i\mathbf{k}\cdot\mathbf{r}}$, which produces the factor of $\hat{\xi}\cdot\mathbf{k} = 0$. This factor vanishes since we assumed that $\nabla\cdot\mathbf{A} = 0$. The only nonzero contribution comes from the momentum operating on the initial state orbital:

$$\mathbf{p}_{fi} = -i\hbar\int d^3r\,e^{i\mathbf{k}\cdot\mathbf{r}}\phi_f^*(\mathbf{r})\nabla\phi_i(\mathbf{r}) \tag{8.100}$$

In most applications, the factor of $e^{i\mathbf{k}\cdot\mathbf{r}}$ can be replaced by unity. For example, visible light has $k \approx 10^5\text{cm}^{-1}$, while a typical atomic distance is $r \approx 10^{-8}$ cm, so that $kr \approx 10^{-3}$. However, there are two types of situations where this factor must be retained.

1. For the absorption of x-rays, k is larger and $kr \sim 1$.
2. If ϕ_i and ϕ_f are states of the same parity, the matrix element vanishes when $\mathbf{k}=0$, but does not vanish for nonzero $\mathbf{k}$:

$$\mathbf{p}_{fi} \approx -i\hbar\int d^3r[1+i\mathbf{k}\cdot\mathbf{r}]\phi_f^*(\mathbf{r})\nabla\phi_i(\mathbf{r}) \tag{8.101}$$

$$\approx \hbar\mathbf{k}\cdot\int d^3r\,\mathbf{r}\phi_f^*(\mathbf{r})\nabla\phi_i(\mathbf{r}) \tag{8.102}$$

The integral $\int\phi_f^*\nabla\phi_i$ vanishes by parity arguments. The second integral $\int\mathbf{r}\phi_f^*\nabla\phi_i$ does not vanish. In this case, a small amount of absorption can take place by using the factor of $i\mathbf{k}\cdot\mathbf{r}$ to make the integral d^3r nonzero.

This second term contains the term that gives optical transitions by (i) electric quadrupole, and (ii) magnetic dipole.

The matrix element contains the factor $\sum_\mu \xi_\mu p_\mu \sum_\nu k_\nu r_\nu$, where $\hat{\xi}$ is the optical polarization. Write

$$p_\mu r_\nu = \frac{1}{2}[p_\mu r_\nu + p_\nu r_\mu] + \frac{1}{2}[p_\mu r_\nu - p_\nu r_\mu] \tag{8.103}$$

The first term on the right is the quadrupole interaction. The second is the magnetic dipole; the antisymmetric commutator is a component of $\vec{r} \times \vec{p} = \vec{L}$, the angular momentum. The magnetic dipole matrix element contains the factor

$$\frac{i}{2}(\vec{k} \times \hat{\xi}) \cdot \vec{L} = \frac{i}{2} k\hat{\xi}' \cdot \vec{L} \tag{8.104}$$

where $\hat{\xi}'$ is the polarization vector perpendicular to $\vec{k}$ and $\hat{\xi}$.

The rate of optical absorption is large if ϕ_i and ϕ_f are states of different parity or if parity is an invalid quantum number. In this case the orbital matrix element is nonzero at $\mathbf{k} = 0$. That case is assumed here. Then the important matrix element for optical absorption is, for photons of wave vector $\mathbf{k}$,

$$V_{fi} = -\frac{e}{mc}\sqrt{n_{\kappa\lambda}} A_k \hat{\xi} \cdot \mathbf{p}_{fi}, \quad \frac{1}{c} A_k = \sqrt{\frac{2\pi\hbar}{\Omega\omega_k}} \tag{8.105}$$

The formula for the Golden Rule contains a summation over final states. In the present example, this summation is over the final states of the electron and of the photon of wave vector $\mathbf{k}$ and polarization λ. The summation is taken over these variables. In the Golden Rule, one calculates the transition to each states and sums over states at the end. That is, one evaluates $\sum_{\kappa\lambda} | V(\kappa\lambda) |^2$ and not $|\sum_{\kappa\lambda} V(\kappa\lambda)|^2$. The reason is that each state has a different energy, and the time averaging for the derivation of the Golden Rule will cause interference between states of different energy. For different final states of the same energy, one puts the summation outside the $|V|^2$ expression. This feature means that states with the same wave vector $\mathbf{k}$ but different polarization (λ) should have the matrix element squared before their contributions are added.

The expression for the transition rate is now

$$w = \left(\frac{2\pi e}{m}\right)^2 \frac{N_A}{\Omega} \sum_{if\kappa\lambda} \frac{n_{\kappa\lambda}}{\omega_k} (\hat{\xi} \cdot \mathbf{p}_{fi})^2 \delta(E_i + \hbar\omega_{\kappa\lambda} - E_f) \tag{8.106}$$

An important parameter is the flux $F_{\kappa\lambda}$ of photons in the gas, which is proportional to the Poynting vector. It has the dimensional units of number of photons/(cm^{2}s). The transition rate $w_{\kappa\lambda}$ for photons in state $\boldsymbol{\kappa}\lambda$ is given by

$$F_{\kappa\lambda} = \frac{S(\kappa\lambda)}{\hbar\omega_{\kappa\lambda}} = \frac{cn_{\kappa\lambda}}{\Omega} \tag{8.107}$$

$$w = \sum_{\kappa\lambda} w_{\kappa\lambda} \tag{8.108}$$

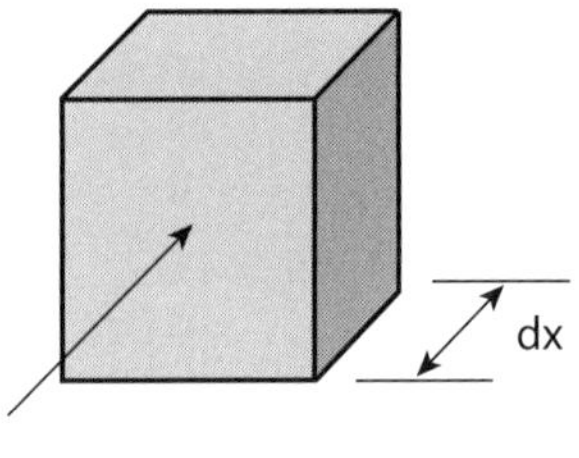

FIGURE 8.5. Geometry for Beer's law.

$$w_{\kappa\lambda} = \left(\frac{2\pi e}{m}\right)^2 \frac{N_A}{c\omega_k} F_{\kappa\lambda} \sum_{if} (\hat{\xi} \cdot \mathbf{p}_{fi})^2 \delta(E_i + \hbar\omega_{\kappa\lambda} - E_f) \tag{8.109}$$

Let the box of atoms be a thin slab, of area $\mathcal{A}$ and thickness dx. The photons are going in the x-directions, as shown in figure 8.5 The product of $F_{\kappa\lambda}$ and $\mathcal{A}$ is the number of photons per second going through the slab with wave vector $\mathbf{k}$ and polarization λ:

$$F_{\kappa\lambda}\mathcal{A} = \text{number of photons/second} \tag{8.110}$$

The absorption of photons reduces the value of $F_{\kappa\lambda}$. Define $dF_{\kappa\lambda} < 0$ as the change in the number of photons in the slab due to absorption. The product $dF_{\kappa\lambda}\mathcal{A}$ is the number of photons absorbed in the slab per second, which is just the quantity given by the Golden Rule:

$$dF_{\kappa\lambda}\mathcal{A} = -w_{\kappa\lambda} \tag{8.111}$$

$$\frac{dF_{\kappa\lambda}}{dx}(dx\mathcal{A}) = -w_{\kappa\lambda} \tag{8.112}$$

In the second line there is a multiplication and division by the slab thickness dx. The part in parentheses is just the volume $\Omega = \mathcal{A}dx$. Define $n_A = N_A/\Omega$ as the number of gas atoms per unit volume. Then we can rearrange the above equation to read

$$\frac{dF_{\kappa\lambda}}{dx} = -\alpha(\kappa\lambda)F_{\kappa\lambda} \tag{8.113}$$

$$\alpha(\mathbf{k}\lambda) = \left(\frac{2\pi e}{m}\right)^2 \frac{n_A}{c\omega_k} \sum_{if} (\hat{\xi} \cdot \mathbf{p}_{fi})^2 \delta(E_i + \hbar\omega_{\kappa\lambda} - E_f) \tag{8.114}$$

These two equations are quite important. Equation (8.113) is called *Beer's law of absorption*. It states that the rate of change of the photon flux dF/dx is proportional to this flux and another constant α. This differential equation has an easy solution:

$$F_{\kappa\lambda}(x) = F_0 e^{-\alpha x} \tag{8.115}$$

The flux decays exponentially with increasing path x into the box of gas atoms. This expression agrees with classical optics for the decrease in beam intensity in an absorbing medium.

The quantity $\alpha(\omega)$ is the *absorption coefficient*. It has the units of inverse length. It is a summation of delta-functions of energy when i and f are both bound states. If f is a continuum state, then the summation over f is converted to an integral over the electron wave vector $\mathbf{q}$, and the absorption becomes a continuous function of frequency. In a continuum state, then, $E_f = \hbar^2 q^2/2m$ and $E_i < 0$. The matrix element will have a factor of $1/\sqrt{\Omega}$, which comes from the normalization of the final-state continuum wave function:

$$\phi_f(\mathbf{r}) = \frac{\eta_{\mathbf{q}}(r)}{\sqrt{\Omega}} \tag{8.116}$$

$$V_{fi} = \frac{v_{i\mathbf{q}}}{\sqrt{\Omega}}, \quad v_{i\mathbf{q}} = -\frac{e}{mc}\sqrt{n_{\kappa\lambda}} A_k \hat{\xi} \cdot \tilde{\mathbf{p}}_{i\mathbf{q}} \tag{8.117}$$

$$\tilde{\mathbf{p}}_{i\mathbf{q}} = -i\hbar \int d^3 r \eta_{\mathbf{q}}^*(\mathbf{r}) \nabla \phi(\mathbf{r}) \tag{8.118}$$

The photon wave vector is $\mathbf{k}$ and the electron wave vector is $\mathbf{q}$. They do not have to be equal in this case, since wave vector is not conserved when the transition happens at a local region in space, such as an atom. The factor of $1/\sqrt{\Omega}$ is squared when we square the matrix element. Then it is used to convert the summation over $\mathbf{q}$ to a continuous integral:

$$\frac{1}{\Omega}\sum_{\mathbf{q}} = \int \frac{d^3 q}{(2\pi)^3} \tag{8.119}$$

$$\alpha(\boldsymbol{\kappa}\lambda) = \left(\frac{2\pi e}{m}\right)^2 \frac{n_A}{c\omega_k} \sum_i \int \frac{d^3 q}{(2\pi)^3} (\hat{\xi} \cdot \tilde{\mathbf{p}}_{i\mathbf{q}})^2 \delta\left(E_i + \hbar\omega_{\kappa\lambda} - \frac{\hbar^2 q^2}{2m}\right)$$

The quantity α/n_A is the absorption coefficient per density of atoms. It depends on the atom through the energies and wave functions of the initial and final states. This quantity $(\sigma(\omega) = \alpha/n_A)$ has the dimensional units of area. In fact it is the imaginary part of the cross section. The real part of the cross section describes the scattering of photons by atoms, which is discussed in a later section. The imaginary part of the cross section is the absorption.

The absorption coefficient is related to important quantities in classical optics. A classical wave propagates in the x-direction with an amplitude given by $\exp(ikx)$. The wave vector k is a complex function of frequency with real and imaginary parts: $k = k_r + ik_I$. The amplitude decays with distance as $\exp(ik_r x - k_I x)$. The intensity decays with distance as $\exp(-2k_I x)$. The absorption coefficient is simply $\alpha = 2k_I$. The solutions to Maxwell's equation show that complex wave vector is related to the complex dielectric function $\varepsilon(\omega)$:

$$k^2 = (k_r + ik_I)^2 = \left(\frac{\omega}{c}\right)^2 \varepsilon(\omega) \tag{8.120}$$

$$k = \frac{\omega}{c}\sqrt{\varepsilon} = \frac{\omega}{c}(n_r + in_I) \tag{8.121}$$

where the complex refractive index is $n = n_r + in_I$. The absorption coefficient is related to the dielectric function. The reader should be cautioned that we have been assuming that the real part of the refractive index n_r is unity. When $n_r \neq 1$ the formulas have additional factors of n_r.

The classical equations of optics are still valid. Quantum mechanics provides explicit formulas that can be used to calculate the dielectric function and refractive index.

As an example, consider the absorption of a photon by the excitation of an electron from the 1s to the 2p state of atomic hydrogen. The photon energy must be $\hbar\omega = E_{\rm Ry}\left(1-\frac{1}{4}\right) = 10.2$ eV. The matrix element is independent of the direction of the polarization of light. One can call any direction $\hat{z}$, so make it the direction of the light polarization $\hat{\xi} = z$. The matrix element is then

$$\hat{\xi}\cdot\mathbf{p}_{fi} = \langle 2p_z|p_z|1s\rangle = \frac{2^{9/2}}{3^4}\frac{\hbar}{ia_0} \tag{8.122}$$

which was evaluated in an earlier chapter. The absorption coefficient for this one transition is

$$\alpha_{1s,\,2p}(\omega) = n_A\alpha_0\delta(\hbar\omega - 10.2eV) \tag{8.123}$$

$$\alpha_0 = 4\pi^2\frac{e^2\hbar}{m^2c(3E_{\rm Ry}/4)}\left(\frac{2^9\hbar^2}{3^8a_0^2}\right) \tag{8.124}$$

$$= \frac{2^{15}}{3^9}\pi^2\alpha_f E_{\rm Ry}a_0^2, \quad \alpha_f = \frac{e^2}{\hbar c} \tag{8.125}$$

The formula introduces the dimensionless *fine-structure constant* $\alpha_f = e^2/\hbar c$, which has an approximate numerical value of $\alpha_f \approx 1/137$. The other constants are the Rydberg energy and the Bohr radius.

The measurement of optical emission is a complimentary experiment to absorption. The atom must first be excited so that some of the electrons are in excited states. This step is usually accomplished by optical absorption or by bombarding the atoms with a beam of energetic electrons. After an electron is in an excited state, it can return to the ground state by the emission of a photon. The emitted photons are observed to have a single frequency $\hbar\omega = E_i - E_f$. In this case the intial state E_i lies higher in energy than the final state E_f.

An important quantity is the *radiative lifetime* τ of an electron in the excited state. How long, on the average, does it take for the electron in i to emit the photon and jump to the final state f? The inverse time is again given by the Golden Rule:

$$\frac{1}{\tau_i} = w = \frac{2\pi}{\hbar}\sum_f |V_{if}|^2\delta(E_i - \hbar\omega - E_f) \tag{8.126}$$

This formula looks just like (8.89). The factor of N_A is missing, since we want the transition rate for each atom. The matrix elements for the two cases are similar, with one very important difference. Since the photon is being emitted, the photon part of the matrix element looks like

$$\langle n_{\kappa\lambda}+1|a^\dagger_{\kappa\lambda}|n_{\kappa\lambda}\rangle = \sqrt{n_{\kappa\lambda}+1} \tag{8.127}$$

$$V_{fi} = -\frac{e}{mc}\sqrt{n_{\kappa\lambda}+1}A_k\hat{\xi}\cdot\mathbf{p}_{fi} \tag{8.128}$$

$$\frac{1}{\tau_i} = \left(\frac{2\pi e}{m}\right)^2\frac{1}{\Omega}\sum_{f\kappa\lambda}\frac{n_{\kappa\lambda}+1}{\omega_k}(\hat{\xi}\cdot\mathbf{p}_{fi})^2\delta(E_i - \hbar\omega_{\kappa\lambda} - E_f)$$

The emission has the factor of $(n_{\kappa\lambda}+1)$ compared to the absorption with the factor of $n_{\kappa\lambda}$. All of the other factors are identical. This is called the *principle of detailed balance*. The process of emission and absorption must have the same matrix elements (except for the photon occupation numbers) to maintain thermal equilibrium. If one matrix element were larger, say emission, the gas of atoms would emit more light than it absorbs, and it would spontaneously cool down, which violates the second law of thermodynamics.

The right-hand side is nonzero even if initially $n_{\kappa\lambda}=0$. The summation on the right is over all final states f and all photon states $\boldsymbol{\kappa}\lambda$.

Another feature of the emission formula is that a photon is more likely to emit into a state $\boldsymbol{\kappa}\lambda$ if that state already has photons. For a fixed frequency, the photon can be emitted in any direction and with either polarization. It preferentially creates the photon in the state $\boldsymbol{\kappa}\lambda$ if that state already has a large value of $n_{\kappa\lambda}$. It is this process that makes lasers coherent, i.e., the eventual buildup of photons in one quantum state.

Most emission experiments are done with an apparatus that is designed to minimize stray light. That is, the experimentalists wants $\langle n_{\kappa\lambda}\rangle \approx 0$. Only the photons emitted from the gas are observed.

In the summation over final states, there are sums over available electron states and also over final photon states. The available electron states are usually few in number, since an excited electron usually has a limited choice of available states of lower energy. The available states must differ in parity to have a nonzero electric dipole matrix element. The summation over photon states $\kappa\lambda$ is important, since any photon state is allowed that is consistent with energy conservation. The expression for the Golden Rule for emission is now

$$\frac{1}{\Omega}\sum_{\mathbf{k}} = \int \frac{d^3k}{(2\pi)^3} \tag{8.129}$$

$$\frac{1}{\tau_i} = \left(\frac{2\pi e}{m}\right)^2 \sum_{f\lambda} \int \frac{d^3k}{(2\pi)^3} \frac{n_{\kappa\lambda}+1}{\omega_k} (\hat{\xi}\cdot\mathbf{p}_{fi})^2 \delta(E_i - \hbar\omega_{\kappa\lambda} - E_f)$$

The summation over photon wave vector has been changed to a continuous integral. The integral over the magnitude of k is eliminated by the delta-function for energy conservation:

$$\int_0^\infty \frac{k^2 dk}{\omega_k} \delta(\hbar ck + E_f - E_i) = \frac{E_i - E_f}{\hbar^2 c^3} \tag{8.130}$$

The next step is to perform the angular integrals over the directions of the photon wave vector. The matrix element $\mathbf{p}_{fi}$ is assumed to point in some direction, and that is defined as the z-axis. The vector $\mathbf{k}$ makes an angle θ with the $\hat{z}$ axis. The two polarization vectors can be chosen in the following manner, consistent with $\hat{\xi}_1 \cdot \hat{\xi}_2 = 0$:

1. $\hat{\xi}_1$ is perpendicular to both $\mathbf{k}$ and $\hat{z}$, so $\hat{\xi}_1 \cdot \hat{z} = 0$.
2. $\hat{\xi}_2$ is perpendicular to $\mathbf{k}$ but is in the plane defined by $\mathbf{k}$ and $\hat{z}$. The angle between $\hat{\xi}_2$ and $\hat{z}$ is $\pi/2 - \theta$, so that $\hat{\xi}_2 \cdot \hat{z} = \cos(\pi/2 - \theta) = \sin(\theta)$.

Therefore, we find that

$$\sum_\lambda (\hat{\xi}\cdot\mathbf{p}_{fi})^2 = \sin^2(\theta) p_{if}^2 \tag{8.131}$$

The angular average of $\sin^2(\theta)$ over 4π of solid angle is $\frac{2}{3}$. Collecting all of these terms gives

$$\frac{1}{\tau_i} = \frac{4}{3}\frac{e^2}{m^2\hbar^2c^3}\sum_f p_{if}^2(E_i - E_f) \tag{8.132}$$

Note that the collection of constants can be grouped as

$$\frac{1}{\tau_i} = \frac{4}{3\hbar}\frac{\alpha_f}{mc^2}\sum_j \frac{p_{if}^2}{m}(E_i - E_f) \tag{8.133}$$

Since p_{if} has the dimensions of momentum, then p_{if}^2/m has the units of energy. All of the factors are either the fine structure constant or an energy. The prefactor of inverse Planck's constant shows that the units of this expression are indeed inverse seconds.

As an example, consider an electron in the $2p_z$ state of atomic hydrogen. It can only decay to the $1s$ state. The matrix element is known. The radiative lifetime is $\tau \sim 10^{-9}$ seconds.

According to the uncertainty principle, the energy width of the $2p_z$ state is

$$\Delta E = \frac{\hbar}{\tau} = 10^{-25}\text{joule} \approx 10^{-6}\ \text{eV} \tag{8.134}$$

This energy uncertainty contributes a linewidth. The delta-function of energy is replaced by

$$2\pi\delta(\hbar\omega_k + E_{1s} - E_{2p}) \rightarrow \frac{2\Delta E}{(\hbar\omega_k + E_{1s} - E_{2p})^2 + (\Delta E)^2} \tag{8.135}$$

Neither the emission nor the absorption spectra have delta-function linewidths. The shape is always a narrow Lorentzian with a width given by the intrinsic radiative lifetime of the state higher in energy. The energy width ΔE is also increased by collisions among the atoms and Auger events.

There are two common formulas for the optical matrix elements. They look different but are formally equivalent. The first one is $\mathbf{p}_{if}$, which has been used so far. The second one is derived by expressing the momentum operator as the commutator of $\mathbf{r}$ with the Hamiltonian, since H always has a term $p^2/2m$:

$$p_\mu = i\frac{m}{\hbar}[H, r_\mu] \tag{8.136}$$

Take the matrix element of the above expression between initial and final electronic states. Assume these states are eigenstates of the Hamiltonian:

$$H\phi_i = E_i\phi_i, \quad H\phi_f = E_f\phi_f \tag{8.137}$$

$$\langle f|p_\mu|i\rangle = i\frac{m}{\hbar}\langle f|Hr_\mu - r_\mu H|i\rangle \tag{8.138}$$

$$= i\frac{m}{\hbar}(E_f - E_i)\langle f|r_\mu|i\rangle = i\frac{m}{\hbar}(E_f - E_i)r_{\mu,fi} \tag{8.139}$$

In the brackets of eqn. (8.138), the first factor of H operates to the left and produces E_f. The second factor of H operates to the right and produces E_i. The matrix element of momentum is proportional to the matrix element of position. This expression is exact in a one-electron system such as the hydrogen atom or the harmonic oscillator. In atoms with more than one electron, the electron–electron interactions make it impossible to calculate an exact wave function for the many-electron system. Even calculations by large, fast, computers produce wave functions that are only approximate. Then one finds, with approximate wave functions, that the momentum matrix element and the length matrix element do not quite obey the above relationship. Usually, one takes the average.

Replace the matrix element p_{if} in (8.133) with the one for position. This formula can be rewritten as

$$\frac{1}{\tau_i} = \frac{4}{3}\alpha_f \sum_f \omega_{if} \left(\frac{\omega_{if} r_{ij}}{c}\right)^2 \tag{8.140}$$

where $\hbar\omega_{if} = E_f - E_i$. The factor of $(\omega\, r/c)$ is dimensionless and usually small. This way of writing the radiation lifetime is more popular than eqn. (8.133). It is exactly the same as the classical formula for the same quantity.

To derive the classical formula, assume there is a dipole of length $\vec{d}$ that sits at the origin and oscillates with a frequency ω. The far field radiation pattern (i.e., Jackson) is given by the electric field:

$$\vec{E} = \left(\frac{\omega}{c}\right)^2 \hat{\xi}(\hat{\xi} \cdot \vec{d}) \frac{e^{ikR}}{R} \left[1 + O\left(\frac{1}{R}\right)\right] \tag{8.141}$$

The energy flux is given by the Poynting vector, which is written as eqn. (8.1):

$$S = \frac{c}{4\pi R^2} \left(\frac{\omega}{c}\right)^4 (\hat{\xi} \cdot \vec{d})^2 \tag{8.142}$$

The units are watts/m^2. The result is averaged over the surface of a sphere at the distance R. Multiply by R^2 and average $(\hat{\xi} \cdot \vec{d})^2 = \sin^2(\theta) d^2$ over solid angle, which gives $(2/3) \times 4\pi$:

$$\langle S \rangle = R^2 \int d\phi \; \sin(\theta) d\theta S = \frac{2c}{3} \left(\frac{\omega}{c}\right)^4 d^2 \tag{8.143}$$

The units are ergs/s. To compare with (8.140), divide by the energy of each photon $\hbar\omega$, which gives the number of photons escaping from the dipole each second:

$$\frac{\langle S \rangle}{\hbar\omega} = \frac{2}{3\hbar} \left(\frac{\omega}{c}\right)^3 d^2 \tag{8.144}$$

The last step is to set $d = \sqrt{2} e r_{if}$, and then the above classical expression is identical to the quantum formula (8.140).

The classical and quantum formulas give the same expression for the rate at which photons leave the atom. In classical physics, the classical dipole emits these quanta in a continuous stream of energy at this average rate. In quantum physics, each excited atom can emit only one photon and then it is finished radiating until again excited to the high-energy orbital state. The one photon can be emitted at any time, and τ is the average time it takes for the emission.

Equation (8.140) applies for electric dipole emission. If the dipole emission is forbidden by parity, then the next term is magnetic dipole emission. In this case the lifetime is

$$\frac{\hbar}{\tau_i} = \frac{\alpha_f}{3} \sum_f \frac{(\hbar\omega_{if})^3}{(mc^2)^2} |M_{ij}|^2 \tag{8.145}$$

$$M_{ij} = \frac{1}{\hbar} \langle j | L_z + 2S_z | i \rangle. \tag{8.146}$$

The magnetic dipole matrix element was discussed above and expressed in terms of $\hat{\xi} \cdot \vec{L}$. The Pauli interaction $(\vec{\mu} \cdot \vec{B})$ provides the spin term in the matrix element.

8.2.1 *Entangled Photons*

Some electronic decay processes occur by the emission of two photons. Generally this happens whenever the decay by one photon is forbidden by parity or other reasons. A well-known example is the decay of positronium. The positron is the antiparticle of the electron. The positron has a charge of $+|e|$ and the same mass m as the electron. The electron and positron can bind together in a hydrogen-like bound state. In relative coordinates, their Hamiltonian is

$$H = -\frac{\hbar^2}{2\mu} \nabla^2 - \frac{e^2}{r} \tag{8.147}$$

where $\mu = m/2$ is the reduced mass. The binding energy in the ground state is $E_{\text{Ry}}/2$, where the factor of two comes from the reduced mass. Since the electron and positron are antiparticles, they can annihilate each other, and the Rydberg state has a short lifetime. Energy is conserved by emitting two photons, whose total energy must be $E = 2mc^2 - E_{\text{Ry}}/2 = \hbar c(k_1 + k_2)$. If the positronium is initially at rest and momentum is conserved, the two photons are emitted with wave vectors $\pm\mathbf{k}$. Each photon has an energy of $\hbar ck = mc^2 - E_{\text{Ry}}/4$.

Parity arguments dictate that the two photons must have polarizations $\hat{\xi}_j$ that are perpendicular: $\hat{\xi}_1 \cdot \hat{\xi}_2 = 0$. However, each photon can be polarized in either direction. Recall that a photon going in the direction $\mathbf{k}$ has two possible polarizations that are perpendicular to $\mathbf{k}$, which are called $\hat{\xi}_1(\mathbf{k})$, $\hat{\xi}_2(\mathbf{k})$, where we use the convention that $\hat{\xi}_1 \times \hat{\xi}_2 = \hat{k}$. Then $\hat{\xi}_1(-\mathbf{k}) = \hat{\xi}_1(\mathbf{k})$, $\hat{\xi}_2(-\mathbf{k}) = -\hat{\xi}_2(\mathbf{k})$.

If photon $\mathbf{k}$ has polarization $\hat{\xi}_1$, the other in $-\mathbf{k}$ has polarization $\hat{\xi}_2(-\mathbf{k})$. If $\mathbf{k}$ has polarization $\hat{\xi}_2(\mathbf{k})$, then the one in $-\mathbf{k}$ has polarization $\hat{\xi}_1(-\mathbf{k})$. Two photons with this property are said to be *entangled*.

Consider an experiment where three sites are in a line. In the center of the line at the site S sits the source of the entangled photons. It could be a source of positrons from beta-decay that is used to make positronium. At the right end of the line at the point A, Alice is measuring the polarizations of the x-ray photons that come from S. If the direction $\hat{k} \equiv \hat{z}$, she finds that half of the photons have x-polarization and half have y-polarization. She measures first. To the left of S at site B, Bob is measuring the polarizations of the photons that come his direction $-\mathbf{k}$. He also finds that half of his photons have x-polarization and half have y-polarization. Later, he compares results with Alice and find that whenever she had x he had y, and vice versa. This experiment does not violate the theory of relativity, since Alice and Bob do not exchange information during the experiment.

This experiment has been performed by several research groups [1, 8, 9]. In some cases the distances were 10 km between source and measurement. In all cases the results agree with quantum mechanics and the theory of entangled photons. Such entangled photons are used in quantum encryptation and possibly in quantum computers.

The experiment could also be done not with photons, but with two electrons whose spins are correlated because they started in an initial singlet state $|S=0, M=0\rangle$. However, it is hard to have an electron beam go 10 km without interference. It is easier in optics, using glass fibers to carry the signal.

Albert Einstein won his Nobel Prize for his contribution to quantum mechanics. He showed that photons have a quantum of energy equal to $E=\hbar ck$, which explained the photoelectric effect. Yet Einstein never accepted the Copenhagen (e.g., Bohr) interpretation in terms of particles whose probability obeyed the wave equation. In 1935 he published an article with two coauthors (Einstein, Podolsky, and Rosen [3]), which argued that quantum mechanics must act locally and that states 20 km apart could not be correlated. In their view, entangled states did not make physical sense. If Bob is measuring polarization while 20 km from Alice, then his results should not depend on hers. As we said, the experiments were done and the predictions of quantum mechanics were upheld.

Many physicists have pondered over the idea that quantum mechanics could be governed by *hidden variables* that are not directly observable. John Bell proposed a set of experiments that could test this hypothesis. The experiments with Alice and Bob mentioned above were done to test Bell's hypothesis. The experiments showed no evidence for hidden variables. A description of Bell's ideas are given by Greenstein and Zajonic [4].

8.3 Oscillator Strength

The oscillator strength is a dimensionless number that describes the relative intensity of an optical transition. The important quantity for the determination of the intensity of an absorption or emission line is the matrix element of momentum $\mathbf{p}_{if}$ between initial and final state. This matrix element enters the absorption coefficient as the factor $(\hat{\xi}\cdot\mathbf{p}_{\text{if}})^2$. To obtain a dimensionless quantity, this factor is divided by $mE_{fi}=m(E_f-E_i)$. It is conventional to add a factor of two, so the definition of oscillator strength is

$$f_{if,\,\mu\nu}=\frac{2}{mE_{fi}}p^*_{if,\,\mu}p_{if,\,\nu} \tag{8.148}$$

$$\sigma(\omega)=\frac{\alpha(\omega)}{n_A}=2\pi^2\alpha_f\frac{\hbar^2}{m}\sum_{if}(\hat{\xi}\cdot\underline{f}_{if}\cdot\hat{\xi})\delta(\hbar\omega-E_{fi}) \tag{8.149}$$

The second equation is the absorption cross section. The underline on $(\underline{f})$ denotes a tensor quantity.

The quantity f_{if} contains all of the information that pertains to the strength of the absorption or emission intensity. The other dimensional factors are the same for every atom and for every transition.

The absorption coefficient between pairs of bound states are just a series of sharp absortion lines. They have a Lorentzian shape with an energy width given by the radiative lifetime. The oscillator strength gives the relative intensity of each absorption line. Intense absorption lines have values of f about unity, while weak lines have much smaller values. Forbidden transitions (e.g., forbidden by parity, or spin) have a zero oscillator strength. In tables of optical transitions, it is sufficient to list two numbers: E_{fi}, f_{if}.

Another expression for the oscillator strength is found by using the length formula for the optical matrix element. Instead of (8.148),

$$f_{if,\mu\nu} = \frac{2m}{\hbar^2} E_{fi} r^*_{if,\mu} r_{if,\nu} \tag{8.150}$$

This formula is easy to remember, since everything of importance goes into the numerator. Equation (8.148) is called the *velocity formula*, while eqn. (8.150) is called the *length formula*. They would give the same numerical result if the eigenfunctions were known precisely. They give different results for computationally generated wave functions for actual many-electron atoms. Figure 8.6 shows an early theory and experiment for the photoionization of neon. Many-electron wave functions are discussed in chapter 9. There it is shown that exact wave functions are impossible, even with large, fast computers. Recent calculations have better wave functions than Henry and Lipsky, and the difference between velocity and length calculations is smaller.

In the photoionization experiment, shown in figure 8.6, a photon is absorbed by an electron and the electron is changed from a bound to a continuum state. The continuum state has a wave vector of $\mathbf{q}$, which is not identical to the wave vector of the photon. Momentum is not conserved, since we fixed the atom, which means it can change arbitrary amounts of momentum. For continuum states, the absorption cross section is written as

$$\sigma(\omega) = \left(\frac{2\pi e}{m}\right)^2 \frac{1}{c\omega_k} \sum_i \int \frac{d^3q}{(2\pi)^3} (\hat{\xi} \cdot \tilde{\mathbf{p}}_{iq})^2 \delta\left(E_i + \hbar\omega_{\kappa\lambda} - \frac{\hbar^2 q^2}{2m}\right)$$

The total amount of oscillator strength of an atom is limited. If the system has Z electrons, then the total amount of oscillator strength in all transitions of all electrons equals Z. This result is called *the f-sum rule*, or alternately *the Thomas-Kuhn sum rule*. For an atom or molecule with Z electrons,

$$\sum_{if} f_{if,\,\mu\nu} = Z\delta_{\mu\nu} \tag{8.151}$$

The summation over all initial and final states includes all electrons. The final state can be in either a bound or a continuum state.

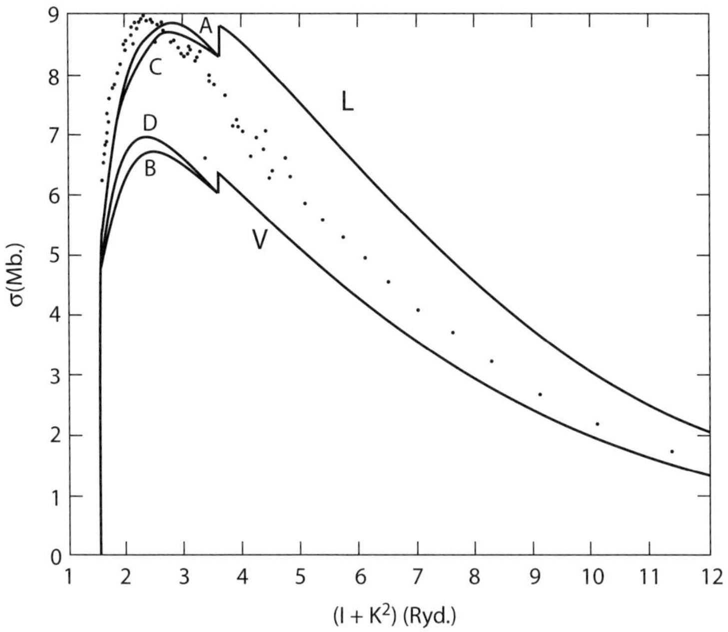

FIGURE 8.6. Dotted line is photoionization cross section of atomic neon. Two solid lines are length (*L*) and velocity (*V*) formulas for this cross section. From R.J. W. Henry and O. Lipsky, *Phys. Rev. 153*, 51 (1967).

The f-sum rule is proven by starting with the velocity formula and then replacing one momentum matrix element with the length formula. Make it two terms, make it symmetric in $(\mu\nu)$, and substitute a different one in each term:

$$f_{if,\mu\nu} = \frac{1}{mE_{fi}}[p^*_{if,\,\nu}p_{if,\,\mu} + p^*_{if,\,\mu}p_{if,\,\nu}] \tag{8.152}$$

$$= \frac{1}{mE_{fi}}[-p^*_{if,\,\nu}r_{if,\,\mu} + r^*_{if,\,\mu}p_{if,\,\nu}]\frac{imE_{fi}}{\hbar} \tag{8.153}$$

The sign change in the second term is from the complex conjugate of $i^* = -i$. Note that the factors of mE_{fi} cancel. Use the Dirac notation to write

$$\vec{r}_{if} = \langle i|\vec{r}|f\rangle, \quad \vec{r}^*_{if} = \langle f|\vec{r}|i\rangle \tag{8.154}$$

$$\sum_f f_{ij,\,\mu\nu} = \frac{i}{\hbar}\sum_f [\langle i|p_\nu|f\rangle\langle f|r_\mu|i\rangle - \langle i|r_\mu|f\rangle\langle f|p_\nu|i\rangle] \tag{8.155}$$

$$= \frac{i}{\hbar}\langle i|[p_\nu r_\mu - r_\mu p_\nu]|i\rangle \tag{8.156}$$

The factor in brackets is the commutator $[p, r] = -i\hbar$ for the same components. So we have proved that

$$\sum_f f_{ij,\,\mu\nu} = \delta_{\mu\nu}\langle i|i\rangle = \delta_{\mu\nu} \tag{8.157}$$

The summation over all i then gives the number of electrons Z.

The rigorous proof uses the fact that the above symbols $\vec{p}$ and $\vec{r}$, in a many-electron system, actually stand for $\sum_i \vec{p}_i$, $\sum_j \vec{r}_j$. This feature arises from the $\mathbf{p}\cdot\mathbf{A}$ interaction, which has the form

$$-\frac{e}{mc}\sum_j \mathbf{p}_j\cdot\mathbf{A}(\mathbf{r}_j) \approx -\frac{e}{mc}\mathbf{A}(0)\cdot\sum_j \mathbf{p}_j \tag{8.158}$$

$$\mathbf{p}_{gf} = \langle g|\sum_j \mathbf{p}_j|f\rangle \tag{8.159}$$

In this notation, $|g\rangle$, $|f\rangle$ are many-electron wave functions: g denotes the ground state. In the steps to prove the f-sum rule, we have the commutator

$$\sum_{i=1}^{Z}\sum_{j=1}^{Z}\langle g|[p_{i\mu}, r_{j\nu}]|g\rangle = -i\hbar\sum_{ij}^{Z}\delta_{\mu\nu}\delta_{ij} = -i\hbar Z\delta_{\mu\nu} \tag{8.160}$$

The factor of Z comes from the summation over all of the electrons. These steps complete the proof.

An example is the oscillator strength between the $1s$ and $2p_z$ state of atomic hydrogen. The matrix element was worked out earlier:

$$f_{1s,\,2p} = \frac{2}{3E_{\text{Ry}}/4}\frac{2^9}{3^8}\frac{\hbar^2}{ma_0^2} = \frac{2^{13}}{3^9} = 0.416 \tag{8.161}$$

Next, calculate the oscillator strength in the Stark effect. Recall that an electric field splits the four states $[2p(3)+2s]$ into three energies: two at $E_2 \pm 3ea_0F$ and two remain at E_2. What are the oscillator strength for transitions to these different states? Neglect spin–orbit and denote the three orbital components of the p-state as $|2p_x\rangle$, $|2p_y\rangle$, $|2p_z\rangle$. It is easy to show that

$$\langle 1s|x|2p_x\rangle = \langle 1s|y|2p_y\rangle = \langle 1s|z|2p_z\rangle \tag{8.162}$$

Put the electric field in the z-direction. Then the four states in the Stark effect are $|2p_x\rangle$, $|2p_y\rangle$, $|+\rangle$, $|-\rangle$, where

$$|+\rangle = \frac{1}{\sqrt{2}}[|2s\rangle + |2p_z\rangle], \quad |-\rangle = \frac{1}{\sqrt{2}}[|2s\rangle + |2p_z\rangle] \tag{8.163}$$

1. If light is polarized $\hat{\xi} \parallel \hat{x}$ the electron goes to the $|2p_x\rangle$ state with an oscillator strength of $f_{1s,\ 2p}$ at a frequency $\hbar\,\omega = (E_2 - E_1)$.
2. If light is polarized $\hat{\xi} \parallel \hat{y}$ the electron goes to the $|2p_y\rangle$ state with an oscillator strength of $f_{1s,\ 2p}$ at a frequency $\hbar\omega = (E_2 - E_1)$.
3. If light is polarized $\hat{\xi} \parallel \hat{z}$ the electron goes to

   - The $|+\rangle$ state with an oscillator strength of $f_{1s,\ 2p}/2$ at a frequency $\hbar\omega = E_2 - E_1 + 3ea_0F$.
   - The $|-\rangle$ state with an oscillator strength of $f_{1s,\ 2p}/2$ at a frequency $\hbar\omega = E_2 - E_1 - 3ea_0F$.

   The first two cases have a single absorption line. The latter case has a double absorption line. The reason the oscillator strength is one-half comes from the matrix element:

$$\langle +|r|1s\rangle = \frac{1}{\sqrt{2}}[\langle 2s|p_z|1s\rangle + \langle 2p_z|p_z|1s\rangle] = \frac{1}{\sqrt{2}}\langle 2p_z|p_z|1s\rangle$$

The factor of one-half comes from squaring this matrix element. The other term $\langle 2s|p_z|1s\rangle = 0$ by parity arguments. For light polarized along the direction of the electric field, the absorption line splits into two components, with each one getting half of the available oscillator strength.

8.4 Polarizability

The application of an electric field to an object will usually cause it to develop an induced dipole moment. If the object does not have a permanent moment, then the induced moment is a function of the field. The constant of proportionality for the linear term is defined as the polarizability:

$$P_\mu = \sum_\nu \alpha_{\mu\nu} E_{nu} \tag{8.164}$$

The polarizability $\alpha_{\mu\nu}$ is a tensor quantity. For atoms with closed atomic shells, it is a diagonal tensor $\alpha_{\mu\nu} = \alpha\delta_{\mu\nu}$. This case is assumed for the present discussion. The field has the units of charge/(length)^2, while the polarization has the units of charge × (length), so the polarizability has the units of volume. The quantum mechanical expression for the polarizability was given in chapter 6. It can be written in terms of the oscillator strength:

$$\alpha = \frac{e^2}{m}\sum_{if} \frac{f_{if}}{\omega_{if}^2} \tag{8.165}$$

$$f_{if} = \frac{2m\omega_{if} r_{if}^2}{\hbar} \tag{8.166}$$

This formula applies at zero frequency.

The present section will derive the polarizability as a function of frequency. Again we use the definition that the dipole moment is the expectation value of the product of the charge and positions of the electrons:

$$\vec{P} = e\langle g|\sum_j \mathbf{r}_j|g\rangle \tag{8.167}$$

where again $|g\rangle$ is the ground state. The above expectation value is zero if the atom or molecule lacks a fixed dipole moment. A nonzero polarization and polarizability is induced by an applied electric field. The frequency dependence of the polarizability is found from an ac electric field. It is introduced into the Hamiltonian through the vector potential:

$$E_\mu = -\frac{1}{c}\frac{\partial A_\mu}{\partial t} = \frac{i}{c}\sum_{\mathbf{\kappa}\lambda} A_k e^{i\mathbf{k}\cdot\mathbf{r}} \omega_k \xi_\mu(\mathbf{\kappa}\lambda)[a_{\mathbf{\kappa}\lambda} e^{-i\omega_k t} - a_{\mathbf{\kappa}\lambda}^\dagger e^{i\omega_k t}]$$

$$V = -\frac{e}{mc}\mathbf{p}\cdot\mathbf{A} \tag{8.168}$$

The atom is assumed to be at the origin, and the electric field varies little over the scale of the atom, so we can evaluate the vector potential at $\mathbf{r} = 0$.

The response of the atom utilizes the time-dependent perturbation theory developed in chapter 7. The ground state of the system is $|g\rangle$. Turning on the interaction V at $t=0$ causes some electrons to be put into excited states. Define H_0 as the Hamiltonian without V, and its eigenstates are ψ_n, where $H_0\psi_n = \varepsilon_n\psi_n$. For $t > 0$, after starting the interaction, the eigenfunctions are

$$\Psi_g = |\tilde{g}\rangle = \sum_n b_n(t)\psi_n(\mathbf{r})e^{-it\varepsilon_n/\hbar} \tag{8.169}$$

$$b_n(t) = -\frac{i}{\hbar}\sum_{\ell\neq n} b_\ell(0)\int_0^t dt' V_{n\ell}(t')e^{it'\omega_{n\ell}} \tag{8.170}$$

The equation for the coefficient $b_n(t)$ is derived in chapter 7. The initial condition is that at $t=0$ only the ground state is occupied: $b_n(0) = \delta_{n=g}$. Therefore, one can do the time integral:

$$V(t) = -\frac{e}{mc}\sum_{\mathbf{\kappa}\lambda}\hat{\xi}\cdot\mathbf{p}_{ng}A_k(a_{\mathbf{\kappa}\lambda}e^{-i\omega_k t} + a^\dagger_{\mathbf{\kappa}\lambda}e^{i\omega_k t}) \tag{8.171}$$

$$b_n(t) = -\frac{i}{\hbar}\int_0^t dt' V_{ng}(t')e^{it'\omega_{ng}} \tag{8.172}$$

$$= -\frac{e}{mc\hbar}\sum_{\mathbf{\kappa}\lambda}\hat{\xi}\cdot\mathbf{p}_{ng}A_k\left[\frac{a_{\mathbf{\kappa}\lambda}(e^{i(\omega_{ng}-\omega_k)t}-1)}{\omega_{ng}-\omega_k} + \frac{a^\dagger_{\mathbf{\kappa}\lambda}(e^{i(\omega_{ng}+\omega_k)t}-1)}{\omega_{ng}+\omega_k}\right]$$

$$|\tilde{g}\rangle = \psi_g e^{-i\omega_g t} + \sum_{n\neq g} b_n(t)\psi_n(\mathbf{r})e^{-i\omega_n t}$$

The occupation $b_g(0)$ is unchanged from unity if the perturbation is weak. For strong electric fields, the changes in this quantity need to be considered.

The new ground state $|\tilde{g}\rangle$ is used in evaluating the dipole moment (8.167). We only want the value of P linear in the electric field, which comes from the second term in $|g\rangle$. Only the terms are retained that oscillate in time:

$$P_\mu = e\langle\tilde{g}|r_\mu|\tilde{g}\rangle$$

$$= e[\langle g|e^{i\omega_g t} + \sum_n b^*_n e^{it\omega_n}\langle n|]r_\mu[|g\rangle e^{-i\omega_g t} + \sum_m b_m e^{-it\omega_m}|m\rangle] \tag{8.173}$$

$$= e\sum_n [b_n\langle g|r_\mu|n\rangle e^{-it\omega_{ng}} + \text{h.c.}] \tag{8.174}$$

The abbreviation "h.c." means "Hermitian conjugate." The first term in eqn. (8.173) is $\langle g|r_\mu|g\rangle = 0$, since there is no moment in the ground state, by assumption. Similarly, since b_n is first order in the vector potential, we can neglect the terms $b^*_n b_m$ that are second order in the electric field.

The combination $b_n\, r_{\mu,gn}$ contains the factors $p_{\nu,ng}\, r_{\mu,gn} = \hbar i f_{\mu\nu,ng}/2$. So we can write the above expression as

$$b_n r_{gn} e^{-it\omega_{ng}} + \text{h.c.} = -i\frac{e}{2mc} f_{\mu\nu,ng} \sum_{\mathbf{\kappa}\lambda} \xi_\nu A_k \left[a_{\kappa\lambda} e^{-it\omega_k} \left(\frac{1}{\omega_{ng} - \omega_k} - \frac{1}{\omega_{ng} + \omega_k} \right) \right.$$

$$\left. + a^\dagger_{\mathbf{\kappa}\lambda} e^{it\omega_k} \left(\frac{1}{\omega_{ng} + \omega_k} - \frac{1}{\omega_{ng} - \omega_k} \right) \right] \tag{8.175}$$

$$= -i\frac{e}{mc} f_{\mu\nu,ng} \sum_{\mathbf{\kappa}\lambda} \xi_\nu A_k \frac{\omega_k}{\omega_{ng}^2 - \omega_k^2} \left[a_{\mathbf{\kappa}\lambda} e^{-it\omega_k} - a^\dagger_{\mathbf{\kappa}\lambda} e^{it\omega_k} \right]$$

$$= \frac{e}{m} \sum_{\kappa\lambda} \left[\frac{\underline{f}_{ng} \cdot \vec{E}_{\mathbf{\kappa}\lambda}}{\omega_{ng}^2 - \omega^2} \right] \tag{8.176}$$

We have used the previous expression for the electric field. The frequency dependent polarizability is

$$\alpha_{\mu\nu}(\omega) = \frac{e^2}{m} \sum_n \frac{f_{ng,\,\mu\nu}}{\omega_{ng}^2 - \omega^2} \tag{8.177}$$

The dc result is recovered by setting $\omega = 0$.

A simple classical system has a charge q with mass m, which is connected to a harmonic spring of constant K. The spring and mass have a frequency of oscillation given by $\omega_0^2 = K/m$. The classical equations of motion of the charged mass, when subject to an oscillating electric field, are

$$m\ddot{X} + KX = qE_0 \cos(\omega t) \tag{8.178}$$

The most general solution to this differential equation is

$$X(t) = \frac{qE_0 \cos(\omega t)}{m(\omega_0^2 - \omega^2)} + A\sin(\omega_0 t) + B\cos(\omega_0 t) \tag{8.179}$$

There are two homogeneous solutions with constants A and B, which have nothing to do with the electric field. These terms are neglected. The polarizability is defined as the ratio of the dipole moment $qX(t)$ divided by the electric field $E_0 \cos(\omega t)$:

$$\alpha(\omega) = \frac{q^2}{m(\omega_0^2 - \omega^2)} \tag{8.180}$$

This expression is the classical polarizability of a charged mass connected to a spring.

The classical formula for the polarizability has a remarkable similarity to the quantum mechanical expression. For an atom or molecule, the resonance frequencies $\omega_0 = \omega_{ng}$ are the excitation energies divided by Planck's constant. For each resonance, the effective charge is $q = e\sqrt{f_{ng}}$. The quantum mechanical system responds to the ac field in the same fashion as does a collection of springs and masses. It should be emphasized that there has been no "harmonic approximation" in the treatment of the quantum mechanical system. The quantum mechanical derivation of the polarizability is quite rigorous. The origin of the phrase "oscillator strength" is that it is the strength of the equivalent classical oscillator. The classical analogy is valid only for transitions to bound states. In the quantum mechanical definition, the summation over n includes bound and continuum states. The latter is a continuous integral.

The dielectric function $\varepsilon(\omega)$ is the ratio between D and E, $D=\varepsilon E$. The student should be familiar with this expression from a course in E&M. There ε is treated as a dielectric constant. If D and E are oscillating with a frequency ω, then the dielectric function is also a function of frequency. If the electromagnetic wave has the form $Ee^{i\mathbf{k}\cdot\mathbf{r}}$ then the dielectric function depends on ε $(\mathbf{k}, \omega)$. For most optical phenomena the wave vector $\mathbf{k}$ is small and is ignored. However, it is important at x-ray frequencies and also for some phenomena at small wave vectors. Here we ignore the dependence on $\mathbf{k}$. We also assume the system is isotropic, so that the dielectric tensor is a c-number.

For a gas of atoms, with a density of n_A, the dielectric function is given by the polarizability of the atoms:

$$\varepsilon(\omega)=1+\frac{4\pi n_A\alpha(\omega)}{1-\frac{4\pi}{3}n_A\alpha(\omega)} \tag{8.181}$$

Here we assume the polarizability tensor is isotropic. The denominator of the dielectric function contains the Lorentz-Lorenz local field correction, which is derived in courses in E&M. It applies to a gas, and a liquid, as well as to a solid. In dilute gases, the term $4\pi\, n_A\alpha$ is small, and the denominator is approximated as one.

The dielectric function is a complex quantity. Its real and imaginary parts are denoted as ε_1 and ε_2:

$$\varepsilon(\omega)=\varepsilon_1(\omega)+i\varepsilon_2(\omega) \tag{8.182}$$

The imaginary part comes from the imaginary part of the polarizability. Heretofore we have treated the polarizability as a real quantity. It is actually a complex quantity, that is written as

$$\alpha(\omega)=\frac{e^2}{m}\sum_n\frac{f_n}{\omega_n^2-(\omega+i\eta)^2} \tag{8.183}$$

The term η is real, so we have given the frequency an imaginary component. One contribution to η is the radiative lifetime, while another is collisions between the atoms in the gas.

The imaginary part of the polarizability was added to eqn. (8.183) in a particular way. If you treat ω as a complex variable, the poles of this expression are at the two points $\omega=-i\eta\pm\omega_n$. Both poles are in the lower half-plane (LHP) of frequency. This placement is required by causality. Define $\tilde{\alpha}(t)$ as the Fourier transform of the polarizability. In a time formalism, the polarization $P(t)$ is

$$P(t)=\int_0^t dt'\tilde{\alpha}(t-t')E(t') \tag{8.184}$$

The electric field at t' causes a polarization at t. From causality, $t>t'$: the signal cannot arrive before the source. That means that $\tilde{\alpha}(t-t')=0$ if $t<t'$. Now consider the Fourier transform:

$$\tilde{\alpha}(t)=\int_{-\infty}^{\infty}\frac{d\omega}{2\pi}e^{-i\omega t}\alpha(\omega) \tag{8.185}$$

The integral is evaluated by contour integration, where ω is a complex variable. Write $\omega=\omega_r+i\omega_i$. The frequency exponent is $\exp[-it(\omega_r+i\omega_i)]=\exp(t\omega_i)\exp(-it\omega_r)$. The first factor $\exp(\omega_i t)$ controls the contour. Convergence requires that $t\omega_i < 0$. If $t < 0$, then $\omega_i > 0$. One closes the contour in the upper half-plane (UHP). Since there are no poles in the UHP, the integral is zero by Cauchy's theorem. For $t > 0$ then $\omega_i < 0$. Now close the contour in the LHP, where both poles are located. The result is

$$\tilde{\alpha}(t) = \Theta(t)\frac{e^2}{m}e^{-\eta t}\sum_n \frac{f_n}{\omega_n}\sin(\omega_n t) \tag{8.186}$$

where $\Theta(t)$ is the step function. The above proof is quite general: any physical, causal function can not have poles in the UHP of frequency space.

The width η is usually small. Treat it as infinitesimal, and then use the theorem

$$\frac{1}{x+i\eta} = P\frac{1}{x} - i\pi\delta(x) \tag{8.187}$$

where here P denotes principal parts. The complex part of the dielectric function is derived from the complex part of the polarizability:

$$\alpha(\omega) = \frac{e^2}{m}\sum_n \frac{f_n}{2\omega_n}\left[\frac{1}{\omega_n - \omega - i\eta} + \frac{1}{\omega_n + \omega + i\eta}\right] \tag{8.188}$$

$$= \frac{e^2}{m}\sum_n f_n\left[P\frac{1}{\omega_n^2 - \omega^2} + \frac{i\pi}{2\omega_n}[\delta(\omega - \omega_n) - \delta(\omega + \omega_n)]\right]$$

$$\alpha = \alpha_1 + i\alpha_2 \tag{8.189}$$

$$\alpha_2(\omega) = \frac{\pi e^2}{2m}\sum_n \frac{f_n}{\omega_n}\delta(\omega - \omega_n) \tag{8.190}$$

$$\varepsilon_2(\omega) = 4\pi n_A \alpha_2 = \frac{2\pi^2 e^2}{m} n_A \sum \frac{f_n}{\omega_n}\delta(\omega - \omega_n) \tag{8.191}$$

where we have neglected local field corrections in finding the imaginary part of the dielectric function. The frequency is assumed to be positive so that the term $\delta(\omega + \omega_n)$ can be neglected.

The expression for $\varepsilon_2(\omega)$ should be familiar. Aside from the factor of (ω/c) it is the expression for the absorption coefficient $\alpha(\omega)$. Earlier we showed that $k^2 = (\omega/c)^2\varepsilon(\omega)$. Writing out the real and imaginary parts gives

$$k_r^2 - k_I^2 + 2ik_r k_I = \left(\frac{\omega}{c}\right)^2(\varepsilon_1 + i\varepsilon_2) \tag{8.192}$$

$$2k_r k_I = \left(\frac{\omega}{c}\right)^2 \varepsilon_2 \tag{8.193}$$

Recall that $k_r = n_r\omega/c$, $2k_I = \alpha$, so that

$$\alpha = \frac{\omega}{n_r c}\varepsilon_2 \tag{8.194}$$

When the refractive index $n_r = 1$, then the absorption coefficient is related to ε_2. The absorption coefficient is proportional to the imaginary part of the polarizability.

In this chapter the symbol α has been used for three different quantities: (i) the polarizability, (ii) the absorption coefficient, and (iii) the fine-structure constant. I apologize for this confusion. However, there are only a few Greek and Latin letters, and each one is used for many different quantities. The symbol α is particularly overused (alpha-particles, etc.). The symbol α is traditionally used for all of these quantities, and the student should recognize which one is being used by the context of the equation.

8.5 Rayleigh and Raman Scattering

Rayleigh scattering is the elastic scattering of light from an object. The photon approaches and leaves the object with the same frequency, and only the direction of the photon has changed. Raman scattering is defined as the inelastic scattering of light by an object. The photon approaches the object with one frequency and departs with another. The two scattering processes have a common origin and a common theory. They are best treated together. The distinction between the two processes is not sharp. Inelastic scattering is often called Rayleigh scattering if the frequency shifts are small. For example, if light with a single frequency ω_i from a laser is directed toward the sample, the spectrum of scattered light is found to have a "quasielastic" peak. This peak is quite high and has a width of several wave numbers $\Delta\omega \sim 10\text{cm}^{-1}$. The quasielastic peak is called Rayleigh scattering, even for the part that has a small change of frequency.

There are two approaches to the theory of light scattering, and both derive the same formula. The first is classical. Figure 8.7 shows the geometry of the experiment. An incident electromagnetic wave has a direction $\mathbf{k}_i$ and a polarization $\hat{\xi}_i$. At the center is the object doing the scattering, such as an atom, molecule, liquid, or solid. The object responds to the incident EM field by developing an induced dipole moment $\mathbf{p}$ that is proportional to the incident electric field and the polarizability of the object:

$$\mathbf{E}_i = \hat{\xi}_i E_i e^{-i\omega t} \tag{8.195}$$

$$\mathbf{p} = \underline{\alpha}\cdot\mathbf{E}_i = \underline{\alpha}\cdot\hat{\xi}_i E_i e^{-i\omega t} \tag{8.196}$$

The object becomes an oscillating dipole that will radiate EM radiation in nearly all directions. In this fashion, the object causes the wave to scatter. The far-field radiation pattern from the oscillating dipole is

$$\mathbf{E}_f = \left(\frac{\omega}{c}\right)^2 \hat{\xi}_f(\hat{\xi}_f\cdot\mathbf{p})\frac{e^{ikR}}{R}\left[1 + O\left(\frac{1}{R}\right)\right] \tag{8.197}$$

The experiment measures the number of photons that enter a detector. Figure 8.7 shows a detector of area $dA = R^2\, d\Omega$, where $d\Omega = d\phi \sin(\theta) d\theta$ is the unit of solid angle. The energy flux per second through dA is the product of dA and the Poynting vector:

$$dAS_f = R^2 d\Omega \frac{c}{4\pi}\Re(\vec{E}_f^* \times \vec{B}_f) = \frac{cd\Omega}{4\pi}\left(\frac{\omega}{c}\right)^4 (\hat{\xi}_f\cdot\underline{\alpha}\cdot\hat{\xi}_i)^2 |E_i|^2 \tag{8.198}$$

The differential cross section is defined as this quantity divided by the rate of incident energy $S_i = c|E_i|^2/(4\pi)$:

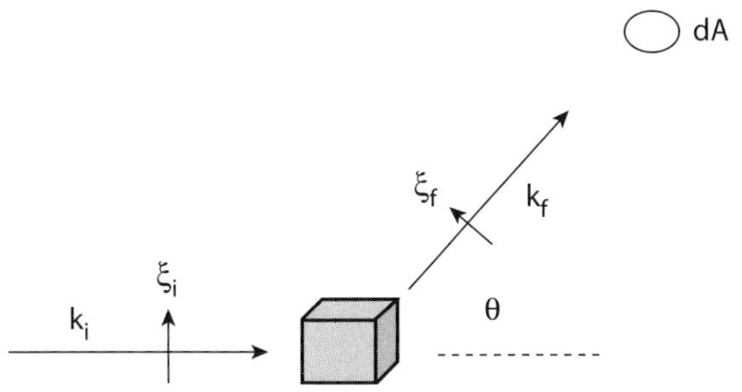

FIGURE 8.7. Experimental arrangment in light scattering.

$$d\sigma = dA\frac{S_f}{S_i} \quad (8.199)$$

$$\frac{d\sigma}{d\Omega} = \left(\frac{\omega}{c}\right)^4 (\hat{\xi}_f \cdot \underline{\alpha} \cdot \hat{\xi}_i)^2 \quad (8.200)$$

The final formula is very simple. The rate of scattering depends on the fourth power of the wave vector ($k = \omega/c$) and the square of the polarizability of the target object. This formula is valid as long as the object is smaller than the wavelength of the light. It is widely used for the scattering of light by atoms. The polarizability depends on frequency.

In most experiments, the scattering target is a collection of atoms in the form of molecules, liquids, or solids. The scattering from a collection of objects is a subtle business, which is treated at length in chapter 10. If there are many identical objects within a distance defined by the wavelength of light, then the scattered radiation from the different atoms must be added coherently. According to the Huygens principle, the scattered waves add coherently into a new wave front. The scattering just slows down the wave, and is the origin of the refractive index. The scattering objects can be treated as independent scattering centers only if they are farther apart, on the average, then the wavelength of light.

The above cross section for Rayleigh scattering can also be derived from quantum mechanics. The same result is obtained in terms of the quantum mechanical definition of the polarizability. The derivation starts with the Hamiltonian for the interaction of electromagnetic radiation with a charged particles. The two interaction terms are

$$V = -\frac{e}{mc}\sum_i \mathbf{p}_i \cdot \mathbf{A}(\mathbf{r}_i) + \frac{e^2}{2mc^2}\sum_i A^2(\mathbf{r}_i) \quad (8.201)$$

The particles that comprise the object have a charge e and a mass m. The rate of scattering is calculated using the Golden Rule, with the interaction V. The cross section is related to the matrix element:

$$\sigma = \frac{2\pi}{\hbar c}\int \frac{d^3k'}{(2\pi)^3} |V_{kk'}|^2 \delta(E_k - E_{k'}) \quad (8.202)$$

The particles are photons with a velocity c and energy $E_k = \hbar ck$. Do the integral by writing $d^3k' = k'^2 dk' d\Omega'$. The k'-integral eliminates the delta-function for energy conservation. Do not do the angular integral over $d\Omega'$, but divide by this quantity:

$$\int k'^2 dk' \delta(\hbar ck - \hbar ck') = \frac{k^2}{\hbar c} = \frac{\omega^2}{\hbar c^3} \tag{8.203}$$

$$\frac{d\sigma}{d\Omega} = \left(\frac{\omega}{2\pi\hbar c^2}\right)^2 |V_{kk'}|^2 \tag{8.204}$$

The only remaining task is to evaluate the matrix element $V_{kk'}$. It is not easy.

The matrix element $V_{kk'}$ is between the initial and final states of the system. The system is composed of charged particles (electrons, protons, etc.) and photons. The initial state has $n_{\mathbf{\kappa}\lambda}$ photons in state $\mathbf{\kappa}\lambda$, $n_{\mathbf{\kappa}'\lambda'}$ in state $\mathbf{\kappa}'\lambda'$ and the charged particles in the ground state $|g\rangle$. The final state has $n_{\mathbf{\kappa}\lambda} - 1$ photons in state $\mathbf{\kappa}\lambda$, $n_{\mathbf{\kappa}'\lambda'} + 1$ in state $\mathbf{\kappa}'\lambda'$, and the charged particles in the ground state $|g\rangle$. The charged particles are assumed not to change their state, since Rayleigh scattering does not change photon energy. Raman scattering occurs when the charged particles do change their state and change their energy. Then the outgoing photon has a different frequency than does the incoming photon. In Rayleigh scattering the two photons have the same frequency: $\omega_{k'} = \omega_k$. The two states are

$$|i\rangle = |g\rangle |n_{\mathbf{\kappa}\lambda}, n_{\mathbf{k}'\lambda'}\rangle \tag{8.205}$$

$$|f\rangle = |g\rangle |n_{\mathbf{\kappa}\lambda} - 1, n_{\mathbf{k}'\lambda'} + 1\rangle \tag{8.206}$$

The number of initial-state photons $n_{\mathbf{\kappa}\lambda}$ is large since it is proportional to the intensity of the laser. The number of final-state photons $n_{\mathbf{\kappa}'\lambda'}$ is usually not large. The scattered photons have many possible directions for scattering.

There are two contributions to the matrix element. The first one comes from the A^2 term in V. The matrix element is

$$V^{(1)} = \langle f | \frac{e^2}{2mc^2} A^2 | i \rangle \tag{8.207}$$

$$= \frac{e^2}{2mc^2} \sum_{\mathbf{q}, \mathbf{q}', \eta\eta'} A_{\mathbf{q}} A_{\mathbf{q}'} \langle g | \sum_i e^{i\mathbf{r}\cdot(\mathbf{q}-\mathbf{q}')} | g \rangle \hat{\xi} \cdot \hat{\xi}' \tag{8.208}$$

$$\times \langle n_{\mathbf{\kappa}\lambda} - 1, n_{\mathbf{k}'\lambda'} + 1 | (a_{\mathbf{q}\eta} + a^{\dagger}_{-\mathbf{q}\eta})(a_{\mathbf{q}'\eta'} + a^{\dagger}_{-\mathbf{q}'\eta'}) | n_{\kappa\lambda}, n_{\mathbf{k}'\lambda'} \rangle$$

In an atom or small molecule, all of the electrons are in a small region of space—small compared to the wavelength of light. Then the exponent $i\mathbf{r}_i \cdot (\mathbf{q} - \mathbf{q}')$ is small and the summation over i just gives the number of electrons Z. In the last matrix element, there are two ways to assign the wave vectors $(\mathbf{q}\eta, \mathbf{q}'\eta')$: either $(\mathbf{q} = \mathbf{k}, \eta = \lambda; \mathbf{q}' = -\mathbf{k}', \eta' = \lambda')$ or $(\mathbf{q} = -\mathbf{k}', \eta = \lambda'; \mathbf{q}' = \mathbf{k}, \eta' = \lambda)$. These two make the identical contributions. So the matrix element is

$$V^{(1)} = \frac{Ze^2}{mc^2} A_k A_{k'} (\hat{\xi}_k \cdot \hat{\xi}_k') \sqrt{n_{\mathbf{\kappa}\lambda}(n_{\mathbf{k}'\lambda'} + 1)} \tag{8.209}$$

$$= \frac{2\pi Z e^2 \hbar}{m\omega_k \Omega} \sqrt{n_{\mathbf{\kappa}\lambda}(n_{\mathbf{k}'\lambda'} + 1)} \tag{8.210}$$

Several factors need to be eliminated since they have already been used in the derivation of the cross section. The first is the factor of $\sqrt{n_{\mathbf{\kappa}\lambda}/\Omega}$. After squaring the matrix element, this factor was used in the definition of the incoming photon flux $F_{\mathbf{\kappa}\lambda}$. Similarly, the remaining factor of $1/\sqrt{\Omega}$, when squared $(1/\Omega)$, is used in converting the summation over final wave vector to a continuous integral d^3k'. Also, we assume the experiment was designed to have $n_{\mathbf{k}'\lambda'} \approx 0$. This quantum correction would never appear in the classical formula. So the effective matrix element is now

$$V^{(1)} = \frac{2\pi Ze^2\hbar}{m\omega_k}(\hat{\xi}_k \cdot \hat{\xi}_{k'}) \tag{8.211}$$

$$\frac{d\sigma}{d\Omega} = \left(\frac{Ze^2}{mc^2}\right)^2 (\hat{\xi}_k \cdot \hat{\xi}_{k'})^2 \tag{8.212}$$

The second line gives the scattering cross section using this matrix element. It does not resemble the classical formula. Another contribution is needed to the matrix element. The prefactor contains $r_0 \equiv e^2/(mc^2) = 2.818 \times 10^{-15}$m, which is called the *radius of the electron*.

There is another process that contributes to the Rayleigh scattering. The $\mathbf{p} \cdot \mathbf{A}$ in-interaction is evaluated in second-order perturbation theory. The formula is

$$V^{(2)} = -\left(\frac{e}{mc}\right)^2 \sum_I \frac{\langle f|\mathbf{p} \cdot \mathbf{A}|I\rangle\langle I|\mathbf{p} \cdot \mathbf{A}|i\rangle}{E_I - E_i} \tag{8.213}$$

The above term can be derived from perturbation theory. It is the wave function coefficient $C^{(2)}_{\ell, m}$ of chapter 6.

It is peculiar that we evaluate the A^2 term in first-order perturbation theory, but the $\mathbf{p} \cdot \mathbf{A}$ term in second order. Yet, as we shall see below, they make a similar contribution to the scattering process. Note that this term also has two powers of the vector potential and a prefactor with $e^2/(mc^2)$.

There are a number of possible choices for the intermediate state $|I\rangle$. First consider the photon system. Two processes must occur: the photon in $(\mathbf{\kappa}\lambda)$ is absorbed, and the photon in $(\mathbf{\kappa}'\lambda')$ is emitted. These steps can occur in either order. There is no requirement that energy be conserved in the intermediate state. The electron has a set of eigenstates $|\ell\rangle$, where $\ell = g$ is the ground state. The electron must go from g to ℓ in the first matrix element, and from ℓ to g in the second matrix element. There are two sequence of matrix elements: the first for the case $\mathbf{\kappa}\lambda$ is absorbed in the first step, and the second when it is absorbed in the second step:

$$V^{(2)} = -\left(\frac{e}{mc}\right)^2 A_k A_{k'} \sqrt{n_{\kappa\lambda}(n_{\mathbf{k}'\lambda'}+1)}$$
$$\times \sum_\ell \left[\frac{(\hat{\xi}_f \cdot \mathbf{p}_{g\ell})(\hat{\xi}_i \cdot \mathbf{p}_{\ell g})}{E_\ell - E_g - \hbar\omega} + \frac{(\hat{\xi}_i \cdot \mathbf{p}_{g\ell})(\hat{\xi}_f \cdot \mathbf{p}_{\ell g})}{E_\ell - E_g + \hbar\omega}\right] \tag{8.214}$$

The denominators have the factor of $\pm\hbar\omega$ for the cases that the photon is absorbed $(-\hbar\omega)$ or emitted $(+\hbar\omega)$ in the intermediate state. In Rayleigh scattering, the initial and final states have the same energy. But the intermediate states can have different energies.

Figure 8.8 shows an energy level diagram for an atom. The electron is initially in the $p_{1/2}$ state of an atom, which is split by the spin–orbit interaction. The first arrow shows the electron absorbing the photon and going to an intermediate *s*-state. The second set of arrows show the electron emitting a photon and going back to the $p_{1/2}$ state (Rayleigh) or

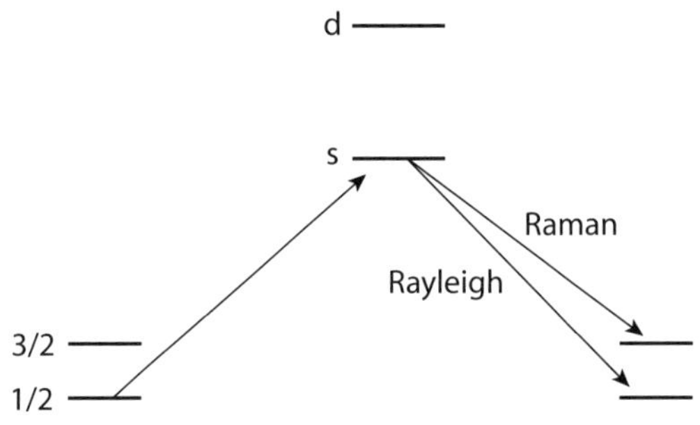

FIGURE 8.8. Sequence of steps in Rayleigh and Raman scattering.

the $p_{3/2}$ state(Raman). The intermediate state could also be a *d*-orbital. The electron does not have the correct energy to complete the transition to the intermediate states, so its occupation is a virtual process.

The factor of $\sqrt{n_{\mathbf{\kappa}\lambda}(n_{\mathbf{k}'\lambda'}+1/\Omega}$ is eliminated using the same reasoning we did for the prior formula. The numerators that contain the matrix elements are the Hermitian conjugates of each other. For bound states we can assume they are real, in which case they are identical. Then we can remove them from the bracket and convert them to an oscillator strength:

$$V^{(2)} = -\frac{\pi e^2\hbar^2}{m\omega}\sum_\ell (\hat{\xi}_f \cdot \underline{f}_{g\ell} \cdot \hat{\xi}_i)\omega_{\ell g}\left[\frac{1}{E_\ell - E_g - \hbar\omega} + \frac{1}{E_\ell - E_g + \hbar\omega}\right]$$

$$= -\frac{2\pi e^2\hbar}{m\omega}\sum_\ell (\hat{\xi}_f \cdot \underline{f}_{g\ell} \cdot \hat{\xi}_i)\frac{\omega_{\ell g}^2}{\omega_{\ell g}^2 - \omega^2} \tag{8.215}$$

Next write the frequency factor as

$$\frac{\omega_{\ell g}^2}{\omega_{\ell g}^2 - \omega^2} = 1 + \frac{\omega^2}{\omega_{\ell g}^2 - \omega^2} \tag{8.216}$$

The term "1" is evaluated using the *f*-sum rule ($\Sigma f = Z$), which gives

$$V^{(2)} = -\frac{2\pi e^2\hbar}{m\omega}\left[Z(\hat{\xi}_f \cdot \hat{\xi}_i) + \sum_\ell (\hat{\xi}_f \cdot \underline{f}_{g\ell} \cdot \hat{\xi}_i)\frac{\omega^2}{\omega_{\ell g}^2 - \omega^2}\right] \tag{8.217}$$

The first term in brackets is equal and opposite to the matrix element $V^{(1)}$ from the A^2 term. These two terms cancel. The remaining term above has a summation over states ℓ that is exactly the polarizability. Thus, we have shown

$$V^{(1)} + V^{(2)} = -2\pi\hbar\omega[\hat{\xi}_f \cdot \underline{\alpha}(\omega) \cdot \hat{\xi}_i] \tag{8.218}$$

The total matrix element gives exactly the classical cross section.

There are several lessons to be learned from this derivation. One is that the classical and quantum theories often agree, as long as we know the quantum definition of quantities such as the polarizability. The second lesson is that in doing perturbation theory, we must keep all terms of the same power of the vector potential. This may involve different orders of perturbation theory for the $\mathbf{p} \cdot \mathbf{A}$ and A^2 terms.

The theory of Raman scattering is rather similar. Electronic Raman scattering is when the light scatters and changes its frequency by exciting an electron to an excited state. The incoming frequency is ω and the outgoing frequency is ω'. The change in energy of the electron must be $\Delta E = \hbar(\omega - \omega')$ since energy is conserved in the final state. The final electron state $\langle f|$ is different than the ground state $|g\rangle$. Energy conservation is $\hbar\omega + E_g = \hbar\omega' + E_f$. The cross section is derived using the same steps for Rayleigh scattering. Now the summation over final photon states gives a factor of ω'^2:

$$\frac{d\sigma}{d\Omega} = \left(\frac{\omega'}{2\pi\hbar c^2}\right)^2 |V_{kk'}|^2 \tag{8.219}$$

The matrix elements are found as before. The A^2 term makes a negligible contribution because of the orthogonality of the initial and final electronic states: $\langle f|A(r)^2|g\rangle \approx A(0)^2 \langle f|g\rangle = 0$. The term $V^{(2)}$, from evaluating the $\mathbf{p}{\cdot}\mathbf{A}$ in second order, gives a result similar to that found in Rayleigh scattering:

$$V^{(2)} = -\frac{2\pi e^2}{m^2\sqrt{\omega\omega'}}\sum_{\ell}\left[\frac{(\hat{\xi}_f\cdot\mathbf{p}_{f\ell})(\hat{\xi}_i\cdot\mathbf{p}_{\ell g})}{\omega_{\ell g}-\omega} + \frac{(\hat{\xi}_i\cdot\mathbf{p}_{f\ell})(\hat{\xi}_f\cdot\mathbf{p}_{\ell g})}{\omega_{\ell g}+\omega'}\right] \tag{8.220}$$

The first denominator has $-\omega$ since the incident photon is absorbed in the first step. The second denominator has $+\omega'$ since this term has the k' emission as the first step. The above matrix element is hard to evaluate. Rayleigh scattering depends only on the polarizability, which can be measured. However, the above matrix element really depends on all of the terms in the summation over ℓ. A special case is where the denominator $\omega_{\ell g} - \omega$ becomes small. Then this term in the matrix element is very large. This process is called *resonance Raman scattering.*

Another important Raman process is the light scattering by the creation or absorption of an optical phonon in a crystal or the vibration of a molecule. In this case the initial and final electronic states are the same, but the system has gained or lost a vibrational quantum. Then $\omega' = \omega \pm \omega_0$. This Raman process is third order in perturbation theory:

$$V = \sum_{I,I'} \frac{\langle f|H'|I'\rangle\langle I'|H''|I\rangle\langle I|H'''|g\rangle}{(E_{I'}-E_g)(E_I-E_g)} \tag{8.221}$$

Two of the interactions $H^{(n)}$ are the $\mathbf{p}{\cdot}\mathbf{A}$ and the third is the electron–vibrational interaction. There are many arrangements of these three steps: photon aborption, photon emission, and vibrational excitation or de-excitation. This matrix element is quite complicated and is rarely evaluated in practice.

8.6 Compton Scattering

Compton scattering is the scattering of an electron by photons. It differs from Raman scattering in that the final state of the electron is in the continuum state:

1. The initial state of the electron could be bound to an atom or molecule. The photon ejects the electron from the bound state. Note that this is a different process than photoionization.

The latter is a one-photon process. Compton is a scattering process. One measures the outgoing photon.

2. The initial state of the electron could be in the continuum state.

Note that a free electron cannot absorb a photon. Since there is no other source of momentum, the electron would acquire the momentum **p** and energy of the photon. For an electron initially at rest, energy conservation is

$$mc^2 + cp = \sqrt{(mc^2)^2 + c^2p^2} \tag{8.222}$$

This equation has no solution unless $p = 0$. One cannot conserve energy and momentum. That is why scattering is required. A photon interacting with a free, charged particle must scatter rather than be absorbed.

We consider the second case here. Assume the electron is initially in the lowest state for a free particle: $\mathbf{k} = 0$. The wave function $\phi_{\mathbf{k}}(\mathbf{r})$ for $\mathbf{k} = 0$ is $\phi_0 = 1/\sqrt{\Omega}$. The $\mathbf{p} \cdot \mathbf{A}$ operator on the state $\phi_{\mathbf{k}}$ gives

$$\mathbf{p} \cdot \mathbf{A} e^{i\mathbf{k}\cdot\mathbf{r}} = \hbar \mathbf{k} \cdot \mathbf{A} e^{i\mathbf{k}\cdot\mathbf{r}} \tag{8.223}$$

which vanishes if $\mathbf{k} = 0$. So the $\mathbf{p} \cdot \mathbf{A}$ interaction makes no contribution. In this case, the matrix element involves only the A^2 term. During the scattering, the electron acquires the momentum change of the photons ($\mathbf{p} = \mathbf{k} - \mathbf{k}'$):

$$V^{(1)} = \frac{e^2}{2mc^2} \langle f | A^2 | i \rangle \tag{8.224}$$

$$= \frac{2\pi e^2 \hbar}{\Omega m \sqrt{\omega \omega'}} (\hat{\xi}_f \cdot \hat{\xi}_i) \int \frac{d^3 r}{\Omega} e^{-i\mathbf{p}\cdot\mathbf{r}} e^{i\mathbf{r}\cdot(\mathbf{k}-\mathbf{k}')}$$

$$= \frac{2\pi e^2 \hbar}{\Omega m \sqrt{\omega \omega'}} (\hat{\xi}_f \cdot \hat{\xi}_i) \delta_{\mathbf{k}-\mathbf{p}-\mathbf{k}'} \tag{8.225}$$

The first factor of the volume Ω comes from the A^2 term. The second factor of volume comes from the normalization of the electron wave functions.

Next write the expressions for energy conservation, assuming momentum conservation. The first is the nonrelativistic theory and the second is the relativistic theory:

$$\hbar ck = \hbar ck' + \frac{\hbar^2}{2m} (\mathbf{k} - \mathbf{k}')^2 \tag{8.226}$$

$$\hbar ck + mc^2 = \hbar ck' + \sqrt{(mc^2)^2 + \hbar^2 c^2 (\mathbf{k} - \mathbf{k}')^2} \tag{8.227}$$

The relativistic equation is actually easier to solve. There are two unknowns in the final state: k' and $\cos(\theta) = \hat{k} \cdot \hat{k}'$. We solve for k'. First rearrange (8.227) and then square it:

$$(mc^2)^2 + \hbar^2 c^2 (\mathbf{k} - \mathbf{k}')^2 = [\hbar c(k - k') + mc^2]^2 \tag{8.228}$$

All of the terms quadratic in k' cancel, which makes the solution easy:

$$k' = \frac{k}{1 + \frac{k\lambda_C}{2\pi}[1 - \cos(\theta)]}, \quad \lambda_C = \frac{h}{mc} \tag{8.229}$$

The collection of fundamental constants $\lambda_C = 2.426 \times 10^{-12}$m is called the *Compton wavelength.*

In the Golden Rule, square the matrix element and sum over all final states. This sum is over electron momentum **p** and photon wave vector and polarization $\mathbf{k}'\lambda'$. The Kronecker delta $\delta_{\mathbf{k}-\mathbf{p}-\mathbf{k}'}$ is one if $\mathbf{k} = \mathbf{p} + \mathbf{k}'$ and is zero otherwise. The square of a Kronecker delta has the same values, one or zero. We use it to eliminate the summation over **p**. We continue to use relativistic energies for the electron:

$$w = \frac{2\pi}{\hbar} N_e \left[\frac{2\pi e^2 \hbar}{\Omega m \sqrt{\omega}}\right]^2 \sum_{\mathbf{k}'\lambda'} \frac{(\hat{\xi}_f \cdot \hat{\xi}_i)^2}{\omega'}$$
$$\times \delta\left[mc^2 + \hbar c(k - k') - \sqrt{m^2c^4 + \hbar^2 c^2 (k^2 + k'^2 - 2\mathbf{k} \cdot \mathbf{k}')}\right] \tag{8.230}$$

where N_e is the number of electrons in the target box being scattered. Convert the summation over $\mathbf{k}'$ to the integral $d^3k' = k'^2 dk' d\Omega_{k'}$. The latter factor is the final-state solid angle. The integral over wave vector gives, with $z = k\lambda_C/2\pi$,

$$\int k' dk' \delta\left[mc^2 + \hbar c(k - k') - \sqrt{m^2c^4 + \hbar^2 c^2 (k^2 + k'^2 - 2\mathbf{k} \cdot \mathbf{k}')}\right]$$
$$= \frac{k'}{\hbar c}\left|1 + \frac{k' - k\cos\theta}{k - k' + 2\pi/\lambda_C}\right|^{-1} = \frac{k}{\hbar c}\frac{1 + z(1+z)(1-\cos\theta)}{[1 + z(1-\cos\theta)]^3} \tag{8.231}$$

Collecting all of the terms together gives the cross section for Compton scattering from a stationary electron:

$$\frac{d\sigma}{d\Omega} = r_0^2 \sum_{\lambda'} (\hat{\xi}_f \cdot \hat{\xi}_i)^2 \frac{1 + z(1+z)(1-\cos\theta)}{[1 + z(1-\cos\theta)]^3} \tag{8.232}$$

where $r_0 = e^2/(mc^2)$. The above formula is not correct. We used a hybrid theory with a nonrelativistic matrix element and a relativistic energy dispersion. In chapter 11 we revisit Compton scattering and also calculate the matrix elements relativistically. The final formula changes. For low-energy photons where $z = k\lambda_C/2\pi << 1$ the formula is

$$\frac{d\sigma}{d\Omega} \approx r_0^2 \sum_{\lambda'} (\hat{\xi}_f \cdot \hat{\xi}_i)^2 \tag{8.233}$$

which is very simple.

References

1. A. Aspect, P. Grangier and G. Roger, *Phys. Rev. Lett.* **47**, 460 (1981)
2. H.B.G. Casimir, *Proc. K. Ned. Akad. Wet.* **51**, 793 (1948)
3. A. Einstein, B. Podolsky, and N. Rosen, *Phys. Rev.* **47**, 777 (1935)
4. G. Greenstein and A.G. Zajonc, *The Quantum Challenge* (Jones & Bartlett, Boston, 2006)
5. J. J. Hopfield, *Phys. Rev.* **112**, 1555–1567 (1958)
6. U. Mohideen and A. Roy, *Phys. Rev. Lett.* **81**, 4549 (1998)

7. L.I. Schiff, *Quantum Mechanics*, 3rd ed. (McGraw-Hill, New York 1968)
8. W. Tittel, J. Brendel, H. Zbinden, N. Gisin, *Phys. Rev. Lett.* **81**, 3563 (1998)
9. G. Weihs, et al. *Phys. Rev. Lett.* **81**, 5039 (1998)

Homework

1. Evaluate the commutator of different times of the vector potential with itself. Derive the dependence on $\vec{R}=\mathbf{r}-\mathbf{r}'$, $s=t-t'$ by doing the integrals over wave vector:

$$D_{\mu\nu}(\vec{R}, s) = \langle [A_\mu(\mathbf{r}, t), A_\nu(\mathbf{r}', t')] \rangle \tag{8.234}$$

The left and right brackets are used to evaluate the commutators $\langle [a_\lambda, a_\lambda^\dagger] \rangle = 1$. Note that

$$\sum_\lambda \xi_\mu(\boldsymbol{\kappa}\lambda)\xi_\nu(\boldsymbol{\kappa}\lambda) = \delta_{\mu\nu} - \frac{k_\mu k_\nu}{k^2} \tag{8.235}$$

2. Calculate the optical absorption $\alpha(\omega)$ in one dimension of an electron initially bound in the one bound state of a delta-function potential. Calculate the frequency dependence of the absorption by doing the integral over a wave vector.

3. Calculate the oscillator strength, analytically and numerically, using the velocity formula for the transitions between the ground and the first excited state of an electron in the following potentials:

   a. An electron bound to an alpha-particle
   b. Three-dimensional harmonic oscillator
   c. One-dimensional box of length L and infinite walls

4. Repeat problem 3 using the length formula.

5. Consider the optical transition between an $(n=1, \ell=0)$ state and an $(n=2, \ell=1)$ state of an atom, where the latter has a small spin–orbit splitting.

   a. What fraction of the oscillator strength is contained in transitions to the $p_{\frac{1}{2}}$ state, and what fraction to the $p_{\frac{3}{2}}$ state?
   b. Does this fraction depend on the direction of the polarization of light?

6. The optical transition in atomic sodium $3S \to 3P$ has an energy of $\hbar\omega = 2.1$ eV and an oscillator strength of $f = 0.5$. What is the lifetime (in seconds) of an electron in the $3P$ state for radiative decay?

7. Derive eqn. (8.145) for the lifetime from magnetic dipole interaction.

8. Use eqn. (8.145) to calculate the radiative lifetime of Yb^{3+} between the f-levels split by the spin–orbit interaction: $\zeta_f = 2928$ cm^{-1}. Calculate the spin–orbit splitting to get

$\hbar\omega$. The upper state of the single f-hole is $|\frac{5}{2}, m\rangle$ and the lower state is $|\frac{7}{2}, m\rangle$. Find the excited-state lifetime for all values of m.

9. An argon atom has an isotropic polarizability of $\alpha = 1.64\,\text{Å}^3$. A continuous laser of wavelength $\lambda = 10.6\ \mu m$ and power of 1.0 watt is focused on a chamber of volume $1.0\,\text{cm}^3$ filled with argon at $P = 1.0$ bar and $T = 300\ K$. How many photons per second are Rayleigh scattered 90° and measured by a detector of area $A = 1\,\text{cm}^2$ at a distance of $L = 1$ meter away from the scattering chamber? Assume each atom scatters independently of the others.

10. Write down a Golden Rule for two photon absorption. In this process, the electronic transition requires the absorption of two photons and absorption of a single photon is not permitted. Assume there is only one incident light beam, so both photons have the same frequency. Specify the matrix element.

11. If $n \equiv n_\lambda$ is the number of photons of the prior problem, the absorption of two photons can be written as

$$\dot{n} = -\gamma n(n-1) \tag{8.236}$$

where γ is the result of the prior problem.

a. What is the solution of this differential equation, assuming $n(t=0) = n_0$?

b. Since $n >> 1$, we can write this equation also as

$$\dot{n} = -\gamma n^2 \tag{8.237}$$

What is the solution to this equation? Compare to Beer's law.

12. An x-ray with $\lambda = 10.0$ Å Compton scatters from a free electron initially at rest. What is the change in wavelength $\Delta\lambda$ of the x-ray? Derive an analytic formula for $\Delta\lambda$ as a function of λ and scattering angle and evaluate it for $\theta = 90°$.

13. Use the theory of Compton scattering to estimate the numerical value of the mean free path of a low-energy photon going through a plasma with an electron density of $n = 10^{20}\text{cm}^{-3}$.

9 Many-Particle Systems

9.1 Introduction

Consider quantum mechanical systems with many identical particles. A typical system is the many electrons in an atom. Similar statistics are found for many nucleons in a nuclei. Later we shall consider systems of many bosons such as superfluid ^{4}He. The nonrelativistic Hamiltonian for electrons in an atom contains three types of terms—(1) kinetic energy, (2) potential energy with the nucleus, and (3) electron–electron Coulomb interactions:

$$H = \sum_i \left[\frac{p_i^2}{2m} - \frac{Ze^2}{r_i} \right] + \sum_{i>j} \frac{e^2}{r_{ij}} \tag{9.1}$$

We work in the Coulomb gauge and neglect the role of photons. Since the electrons and nucleus are close together, the long-range forces from photons are not important. It is assumed the nucleus is fixed at the origin. Also note that the Hamiltonian does not depend on spin, although spin plays a large role in the solution.

The above Hamiltonian *is not a sum of one-particle Hamiltonians.* Electron–electron interactions prevent it from being written as $H = \sum_i H_i$. It is impossible to solve this Hamiltonian exactly when there is more than one electron. Approximate solutions are possible only by assuming that one can write it this way:

$$H = \sum_i H_i \tag{9.2}$$

$$H_i = \frac{p_i^2}{2m} - \frac{Ze^2}{r_i} + V_{ee}(\mathbf{r}_i) \tag{9.3}$$

The last term is an effective potential that simulates the role of electron–electron interactions. One goal of this chapter is to explain the nature of this self-consistent potential.

If one can write the Hamiltonian as a summation of individual Hamiltonians, one for each electron, then we solve each one separately. Since the electrons are identical, the H_i are all the same:

$$H_i\phi_n(\mathbf{r}) = E_n\phi_n(\mathbf{r}) \tag{9.4}$$

We assume that we can solve this problem and determine all of the eigenfunctions ϕ_n and eigenvalues E_n. Modern computers and expert codes make such solutions very accurate.

9.2 Fermions and Bosons

A fermion, such as an electron, proton, or neutron, will have a position $\mathbf{r}_i$ and spin s_i. The spin is usually given a quantum number α for spin-up and β for spin-down: "up" and "down" being along any axis of quantization. A wave function for two identical particles could be written as $\psi(\mathbf{r}_1, s_1; \mathbf{r}_2, s_2)$. Let P_x be an operator that exchanges the position of the two particles. Then

$$P_x\psi(\mathbf{r}_1, s_1; \mathbf{r}_2, s_2) = \psi(\mathbf{r}_2, s_2; \mathbf{r}_1, s_1) \tag{9.5}$$

$$P_x^2\psi(\mathbf{r}_1, s_1; \mathbf{r}_2, s_2) = \psi(\mathbf{r}_1, s_1; \mathbf{r}_2, s_2) \tag{9.6}$$

Operating twice by P_x returns the wave function to its original configuration. The operator P_x^2 has an eigenvalue equal to one, so that P_x must have an eigenvalue that is either plus one or minus one.

1. The eigenvalue of plus one gives

   $$P_x\psi(\mathbf{r}_1, s_1; \mathbf{r}_2, s_2) = \psi(\mathbf{r}_2, s_2; \mathbf{r}_1, s_1) = \psi(\mathbf{r}_1, s_1; \mathbf{r}_2, s_2) \tag{9.7}$$

   Particles that obey these statistics are called *bosons*. Photons are bosons. All bosons have integer spin, such as zero, one, or two.
2. The eigenvalue of minus one gives

   $$P_x\psi(\mathbf{r}_1, s_1; \mathbf{r}_2, s_2) = \psi(\mathbf{r}_2, s_2; \mathbf{r}_1, s_1) = -\psi(\mathbf{r}_1, s_1; \mathbf{r}_2, s_2) \tag{9.8}$$

   Particles that obey these statistics are called fermions. Examples are electrons, protons, and neutrons. Fermions have half-integer spins. All fermion wave functions of many particles have the feature that they are antisymmetric. Exchanging the positions and spins of any pair of particles must change the sign of the wave function. This feature puts a severe constraint on the nature of the many-particle wave function.

First exclusion principle: If two fermions are at the same point in space with the same spin, then the wave function must vanish. That is, if $(\mathbf{r}_1 = \mathbf{r}_2,\ s_1 = s_2)$, the above expression shows that $\psi(\mathbf{r}_1, s_1; \mathbf{r}_1, s_1)$ equals its negative. The only number that equals its negative is zero. Two fermions cannot be at the same point in space while having the same spin configuration.

A way of constructing a fermion wave function with the correct symmetry properties was proposed by John Slater. The Slater determinant for N-identical fermions is written in terms of single-particle wave functions $\psi_m(\xi_j)$, where m is the eigenvalue and $\xi_i = (\mathbf{r}_i, s_i)$:

$$\Psi_{1,2,3,\ldots,N}(\xi_1,\xi_2,\ldots,\xi_N)=\frac{1}{\sqrt{N!}}\begin{vmatrix}\psi_1(\xi_1) & \psi_1(\xi_2) & \psi_1(\xi_3) & \cdots & \psi_1(\xi_N)\\ \psi_2(\xi_1) & \psi_2(\xi_2) & \psi_2(\xi_3) & \cdots & \psi_2(\xi_N)\\ \psi_3(\xi_1) & \psi_3(\xi_2) & \psi_3(\xi_3) & \cdots & \psi_3(\xi_N)\\ \vdots & \vdots & \vdots & \ddots & \vdots\\ \psi_N(\xi_1) & \psi_N(\xi_2) & \psi_N(\xi_3) & \cdots & \psi_N(\xi_N)\end{vmatrix} \tag{9.9}$$

Each row has the same eigenvalue, while each column has the same value of ξ_i. Exchanging the ξ_i labels on two particles exchanges them on two columns. But exchanging two columns of a determinant changes the sign, in agreement with fermion statistics.

Second exclusion principle. If any two electrons are in the same state, the wave function vanishes. That is, if any two electrons have the same quantum numbers (m), then two rows of the determinant are identical, in which case it vanishes. This is called the *Pauli exclusion principle.*

For $N \geq 2$ the determinant has $N!$ terms, of which half are plus and half are minus. A modified determinant can be made by changing the minus signs to plus, so that every term has a plus sign. This modified determinant has the feature that exchanging two particles leaves the value unchanged. It is a suitable wave function for identical bosons.

9.2.1 Two Identical Particles

The statistics are well illustrated by considering the wave function for two identical particles. The subscript F is for fermions and B is for bosons:

$$\psi_F(\mathbf{r}_1,s_1;\mathbf{r}_2,s_2)=\frac{1}{\sqrt{2}}[\psi_1(\xi_1)\psi_2(\xi_2)-\psi_1(\xi_2)\psi_2(\xi_1)] \tag{9.10}$$

$$\psi_B(\mathbf{r}_1,s_1;\mathbf{r}_2,s_2)=\frac{1}{\sqrt{2}}[\psi_1(\xi_1)\psi_2(\xi_2)+\psi_1(\xi_2)\psi_2(\xi_1)] \tag{9.11}$$

Fermions have an antisymmetric combination of the two states, while bosons have the symmetric combination. The boson function was derived from the fermion function by changing the sign on the second term.

The fermion wave function has two factors: an orbital term $\phi_\lambda(\mathbf{r})$ and a spin factor χ_i. The spin part is either α_i or β_i. Two spin-$\frac{1}{2}$ states can be combined using Clebsch-Gordon coefficients into three spin-1 states and spin-0 state:

$$\chi_1(m)=\begin{cases} m=1 & \alpha_1\beta_1\\ m=0 & (\alpha_1\beta_2+\beta_1\alpha_2)/\sqrt{2}\\ m=-1 & \beta_1\beta_2\end{cases} \tag{9.12}$$

$$\chi_0(0)=\frac{1}{\sqrt{2}}(\alpha_1\beta_2-\beta_1\alpha_2) \tag{9.13}$$

The spin state $\chi_1(m)$ is symmetric in the two spin labels—exchanging 1 and 2 does not change the answer. So this spin state must be multiplied by an antisymmetric orbital state:

$$\Psi_{\lambda,\lambda',S=1}(\mathbf{r}_1,\mathbf{r}_2)=\frac{1}{\sqrt{2}}[\phi_\lambda(\mathbf{r}_1)\phi_{\lambda'}(\mathbf{r}_2)-\phi_\lambda(\mathbf{r}_2)\phi_{\lambda'}(\mathbf{r}_1)]\chi_1(m) \tag{9.14}$$

This state of two spin-$\frac{1}{2}$ fermions has the correct antisymmetry. It changes sign if one exchanges the two labels of (1, 2). The other spin state χ_0 is antisymmetric, and changes sign when exchanging labels. So the orbital part must be symmetric:

$$\Psi_{\lambda,\lambda',S=0}(\mathbf{r}_1,\mathbf{r}_2)=\frac{1}{\sqrt{2}}[\phi_\lambda(\mathbf{r}_1)\phi_{\lambda'}(\mathbf{r}_2)+\phi_\lambda(\mathbf{r}_2)\phi_{\lambda'}(\mathbf{r}_1)]\chi_0(0) \tag{9.15}$$

A special case is when both Fermions are in the same orbital state:

$$\Psi_{\lambda,\lambda,S=0}(\mathbf{r}_1,\mathbf{r}_2)=\phi_\lambda(\mathbf{r}_1)\phi_\lambda(\mathbf{r}_2)\chi_0(0) \tag{9.16}$$

Note that if $\lambda'=\lambda$ then the state $\Psi_{\lambda\lambda,S=1}(\mathbf{r}_1,\mathbf{r}_2)=0$. If two spin-$\frac{1}{2}$ fermions are in the same orbital state, they must be in the $S=0$ spin state. This rule is required by the antisymmetric nature of the fermion wave function.

As an example, consider the two electrons in the helium atom. The ground-state configuration has both electrons in 1s orbital states, so the spin configuration is $S=0$. However, consider the excited state where one electron is in the 1s state and the other is in the 2s state. Now there are four possible states: the three states with $S=1$ in (9.14) and the state (9.15) with $S=0$. The state with $S=1$ is called the *triplet state* and that with $S=0$ the *singlet state*. The triplet state has a very long radiative lifetime, since the usual optical transition does not flip electron spin. This is obvious for $m=\pm 1$, where the two spins are both up or both down. It is also true for the $S=1$, $m=0$ configuration.

9.3 Exchange Energy

9.3.1 *Two-Electron Systems*

The energy will be calculated for a two-electron system. Let $\phi_1(\mathbf{r})$, $\phi_2(\mathbf{r})$ be two different orbital states with eigenvalues E_1 and E_2. They could be atomic orbitals or plane waves; we need not yet specify their nature. Use them to calculate the expectation value of the Hamiltonian:

$$E(S)=\langle\Psi_S^\dagger|H|\Psi_S\rangle \tag{9.17}$$

$$H=\frac{p_1^2+p_2^2}{2m}+U(\mathbf{r}_1)+U(\mathbf{r}_2)+\frac{e^2}{|\mathbf{r}_1-\mathbf{r}_2|} \tag{9.18}$$

$$\Psi_S=\frac{1}{\sqrt{2}}[\phi_1(\mathbf{r}_1)\phi_2(\mathbf{r}_2)\pm\phi_1(\mathbf{r}_2)\phi_2(\mathbf{r}_1)]\chi_S(m) \tag{9.19}$$

The Hamiltonian does not involve spin, so the spin state gives unity when taking the expectation value $\langle\chi_{S'}(m')|\chi_S(m)\rangle=\delta_{SS'}\delta_{mm'}$. However, the value of S plays a role in the answer. The factor of ± 1 in front of the second orbital term depends on S: $+1$ for $S=0$ and -1 for $S=1$.

Now evaluate the orbital terms. First consider the central potential terms $U(\mathbf{r}_i)$. The integral has the form

$$\langle U\rangle = \int \frac{d^3r_1 d^3r_2}{2}[U(\mathbf{r}_1)+U(\mathbf{r}_2)]$$
$$\times\{|\phi_1(\mathbf{r}_1)\phi_2(\mathbf{r}_2)|^2+|\phi_1(\mathbf{r}_2)\phi_2(\mathbf{r}_1)|^2 \pm 2\Re[\phi_1(\mathbf{r}_1)\phi_2(\mathbf{r}_2)\phi_1^*(\mathbf{r}_2)\phi_2^*(\mathbf{r}_1)]\} \tag{9.20}$$

We assume the orbitals are orthogonal: $\int \phi_i^*\phi_j = \delta_{ij}$. The first term can be written as

$$I_1 = \int \frac{d^3r_1 d^3r_2}{2}[U(\mathbf{r}_1)+U(\mathbf{r}_2)]|\phi_1(\mathbf{r}_1)\phi_2(\mathbf{r}_2)|^2 \tag{9.21}$$
$$= \frac{1}{2}\left\{\int d^3r_1\, U(\mathbf{r}_1)|\phi_1(\mathbf{r}_1)|^2 \int d^3r_2 |\phi_2(\mathbf{r}_2)|^2\right. \tag{9.22}$$
$$\left. + \int d^3r_1 |\phi_1(\mathbf{r}_1)|^2 \int d^3r_2\, U(\mathbf{r}_2)|\phi_2(\mathbf{r}_2)|^2\right\} \tag{9.23}$$
$$= \frac{1}{2}\left\{\int d^3r_1\, U(\mathbf{r}_1)|\phi_1(\mathbf{r}_1)|^2 + \int d^3r_2\, U(\mathbf{r}_2)|\phi_2(\mathbf{r}_2)|^2\right\} \tag{9.24}$$
$$= \frac{1}{2}[U_1+U_2],\ U_i = \int d^3r U|\phi_i|^2 \tag{9.25}$$

The second term in brackets in (9.20), with $|\phi_1(\mathbf{r}_2)\phi_2(\mathbf{r}_1)|^2$ gives exactly the same thing, since d^3r_i are dummy variables of integration. The last factor gives zero since it involves $\int d^3r_i\phi_1^*(\mathbf{r}_i)\phi_2(\mathbf{r}_i) = 0$. So the final answer is

$$\langle U\rangle = U_1 + U_2 \tag{9.26}$$

Next consider the kinetic energy term:

$$\langle K\rangle = \langle \Psi_S^\dagger [\frac{p_1^2}{2m}+\frac{p_2^2}{2m}]\Psi_s\rangle \tag{9.27}$$
$$= K_1 + K_2,\ K_i = \int d^3r\phi_i^* \frac{p^2}{2m}\phi_i \tag{9.28}$$

The total kinetic energy is also the summation of the kinetic energy of each orbital. The kinetic energy and potential energy from U have this feature. These terms operate on only one spatial variable at a time, so the cross-terms play no role.

A very different result is found for the expectation value of the electron–electron interaction:

$$\langle V\rangle = \langle \Psi_S^\dagger | \frac{e^2}{|\mathbf{r}_1-\mathbf{r}_2|}|\Psi_S\rangle \tag{9.29}$$

The first two terms in $|\Psi|^2$ give a result called the *Hartree energy*:

$$V_H = \frac{e^2}{2}\int \frac{d^3r_1 d^3r_2}{|\mathbf{r}_1-\mathbf{r}_2|}[|\phi_1(\mathbf{r}_1)\phi_2(\mathbf{r}_2)|^2+|\phi_1(\mathbf{r}_2)\phi_2(\mathbf{r}_1)|^2] \tag{9.30}$$
$$= e^2\int \frac{d^3r_1 d^3r_2}{|\mathbf{r}_1-\mathbf{r}_2|}|\phi_1(\mathbf{r}_1)\phi_2(\mathbf{r}_2)|^2 \tag{9.31}$$

The two terms in brackets in eqn. (9.30) make equal contributions. The second is the same as the first after exchanging dummy variables $\mathbf{r}_1$ and $\mathbf{r}_2$. The Hartree energy is the one you would write down from classical consideration. $e|\phi_i(\mathbf{r})|^2 = \rho_i(\mathbf{r})$ is the density of charge in that electron state. The Hartree term is merely the Coulomb interaction between two charge distributions. The expression for $|\Psi_S|^2$ has a cross term, whose sign depends on S. This term is called the *exchange term* or the *Fock term*:

$$V_X = e^2\Re\{\int \frac{d^3r_1 d^3r_2}{|\mathbf{r}_1-\mathbf{r}_2|}\phi_1^*(\mathbf{r}_1)\phi_2^*(\mathbf{r}_2)\phi_2(\mathbf{r}_1)\phi_1(\mathbf{r}_2)\} \tag{9.32}$$

The exchange term has no classical analog. It arises from the first exclusion principle. The wave function vanishes if two electrons in the same spin configuration try to be at the same point in space. The two electrons have to stay apart. This exclusion, due to particle statistics, lowers the repulsive Coulomb interaction between them. The motion of the two electrons is correlated.

9.3.2 *Parahelium and Orthohelium*

Consider the helium atom with two electrons. If both are in the ground state, the $1s$ orbital, the spin must be in the singlet ($S = 0$) configuration. Its wave function was constructed in an earlier chapter using variational methods. There is no exchange term in this case.

Next consider the excited state with the lowest energy: One electron is in a $1s$ orbital and the other one is in the $2s$ orbital. Use the results of the prior subsection to deduce the energy for these two configurations.

1. $S = 0$ is called *parahelium* and has the energy

$$E_{S=0} = K_1 + K_2 + U_1 + U_2 + V_H + V_X \tag{9.33}$$

2. $S = 1$ is called *orthohelium* and has the energy

$$E_{S=1} = K_1 + K_2 + U_1 + U_2 + V_H - V_X \tag{9.34}$$

The exchange energy V_X is usually positive. The result of this exercise is that $E_{S=1} < E_{S=0}$. Orthohelium has the lowest energy. This is the triplet state, with spins parallel. It cannot readily decay to the ground state by emitting a photon, since that step does not permit flipping a spin. A spin flip is required to get from the triplet to the singlet state: the ground state is a spin singlet (χ_0), the excited state is a spin triplet (χ_1), and $\langle\chi_1|\chi_0\rangle = 0$. As a result, orthohelium is a metastable state and can last for a very long time. The most likely method of decay to the ground state is when two orthohelium atoms collide, mutually flip each others spin, followed by radiative decay.

9.3.3 *Hund's Rules*

The above results can be extended to systems of many electrons. In general, if the cost in energy is small, the electrons prefer to have their spins aligned to reduce the repulsive Coulomb repulsion between them. This important idea explains the order in which electrons fill up atomic shells.

Consider an s shell. If there is one electron in that shell, it can have either spin up or down. When there are two electrons in the shell, they must have the opposite spins to create the spin singlet. In s-shells the electrons cannot have parallel spins.

In p-shells, the angular momentum of $\ell = 1$ allows three orbital states with $m = 1, 0, -1$. When there is only one electron, it can have either spin state, say spin-up. It goes into the state with spin-up, and $m = 1$. When there are two electrons in the shell, they prefer to have the same spin. Because of the exclusion principle, they cannot both be in the same state, so they must have different values of $m = (1, 0)$. Three electrons also prefer to align themselves with the same spin, but must be in the states with different values of $m = (1, 0, -1)$. When there are three electrons, the configuration is to have three spins aligned ($S = \frac{3}{2}$) and all m states occupied. This arrangement lowers the repulsive electron–electron interaction among the electrons. The angular momentum is $L = \sum_i m_i$, which is zero for three electrons. For four electrons, one has to have one spin-down. Table 9.1 shows the actual order in which the states are filled. The filling of the first row of the periodic table is also shown.

All orbital states with $\ell \geq 1$ fill up in this fashion. This ordering is according to *Hund's rules*. For electrons in one atomic shell, the order of filling is governed by the following rules.

1. The electrons arrange into the maximum allowed value of total spin $S = \sum_i s_i$.
2. The electrons arrange to have the maximum value of angular momentum $L = \sum_i m_i$ consistent with rule 1.
3. The total angular momentum J is found according to the following rules:
   - $J = |L - S|$ when the shell is less than half-filled.
   - $J = S$, $L = 0$ when the shell is exactly half-filled.
   - $J = L + S$ when the shell is more than half-filled.

Table 9.1 shows this sequence. The last column shows standard notation in atomic physics for this configuration:

$$^{(2S+1)}L_J$$

The first superscript is the spin degeneracy $(2S + 1)$. The last subscript is J. The capital letter denotes angular momentum L according to

L	Notation
0	S
1	P
2	D
3	F
4	G
5	H

The notation is alphabetical after F.

Table 9.1 Hund's Rules for p-Shells.

n	S	L	J	Notation	Atom
1	$\frac{1}{2}$	1	$\frac{1}{2}$	$^2P_{1/2}$	B
2	1	1	0	$^3P_0$	C
3	$\frac{3}{2}$	0	$\frac{3}{2}$	$^4S_{3/2}$	N
4	1	1	2	$^3P_2$	O
5	$\frac{1}{2}$	1	$\frac{3}{2}$	$^2P_{3/2}$	F
6	0	0	0	$^1S_0$	Ne

Note. The First column is number of electrons in the atomic shell. The last column shows the configuration of the atoms in the first row of the periodic table.

9.4 Many-Electron Systems

9.4.1 *Evaluating Determinants*

The ideas developed for two electrons are easily extended for a many-electron system. We must learn to evaluate the determinant when there are a large number of electrons.

The square of the wave function gives the probability density for finding particles at positions $\mathbf{r}_j$ in the system and is called the diagonal density matrix:

$$\rho_N(\mathbf{r}_1, \mathbf{r}_2, \ldots, \mathbf{r}_N) = |\Psi(\mathbf{r}_1, \mathbf{r}_2, \ldots, \mathbf{r}_N)|^2 \tag{9.35}$$

The subscript N indicates that it applies to N particles. ρ_N is normalized so that the integral over all coordinates gives unity:

$$1 = \int d^3r_1 \cdots d^3r_N \rho_N(\mathbf{r}_1, \mathbf{r}_2, \ldots, \mathbf{r}_N) \tag{9.36}$$

The one-particle density matrix is obtained from ρ_N by integrating over all but one coordinate:

$$\rho_N(\mathbf{r}_1) = \int d^3r_2 d^3r_3 \cdots d^3r_N \rho_N(\mathbf{r}_1, \mathbf{r}_2, \ldots, \mathbf{r}_N) \tag{9.37}$$

$$1 = \int d^3r_1 \rho_N(\mathbf{r}_1) \tag{9.38}$$

The notation is a bit confusing: $\rho_N(\mathbf{r}_1)$ is the probability of one particle being at $\mathbf{r}_1$ in the N-particle system. Since all points are equivalent, $\rho_N(\mathbf{r}_1)$ is really independent of position—neglecting edges—so $\rho_N = 1/\Omega$, where Ω is the volume of the system.

Another useful quantity is the two-particle density matrix $\rho_N(\mathbf{r}_1, \mathbf{r}_2)$, which is obtained from ρ_N by integrating over all but two coordinates:

$$\rho_N(\mathbf{r}_1, \mathbf{r}_2) = \int d^3r_3 d^3r_4 \cdots d^3r_N \rho_N(\mathbf{r}_1, \mathbf{r}_2, \ldots, \mathbf{r}_N) \tag{9.39}$$

$$\frac{1}{\Omega} = \int d^3r_2 \rho_N(\mathbf{r}_1, \mathbf{r}_2) \tag{9.40}$$

Now $\rho_N(\mathbf{r}_1, \mathbf{r}_2)$ is the probability that any particle is at $\mathbf{r}_2$ if there is one at $\mathbf{r}_1$. In a homogeneous system it must depend only on the difference $\mathbf{r}_1 - \mathbf{r}_2$. It is proportional to a quantity $g(\mathbf{r})$ called the *pair distribution function*:

$$\rho_N(\mathbf{r}_1, \mathbf{r}_2) = \frac{1}{\Omega^2} g(\mathbf{r}_1 - \mathbf{r}_2) \tag{9.41}$$

Crystalline solid atoms are arranged in a regular order, which is indicated by structure in $g(\mathbf{r})$. Since the atoms are regularly spaced, $g(\mathbf{r})$ has large values where the atoms are located and is zero otherwise. Thermal vibrations smear out the answer. The two distribution functions $\rho_N(\mathbf{r}_1)$ and $\rho_N(\mathbf{r}_1, \mathbf{r}_2)$ are needed to evaluate the particle energy.

For a system of N fermions, one uses the Slater determinant to find the distribution functions. The number of electron pairs is $N_p = N(N-1)/2$. There is a theorem that the two-particle distribution function is exactly given by the summation of the square of the two-particle eigenfunction for all different pairs:

$$\rho_N(\mathbf{r}_1, \mathbf{r}_2) = \frac{1}{2N_p} \sum_{i>j} \begin{vmatrix} \psi_i(\xi_1) & \psi_i(\xi_2) \\ \psi_j(\xi_1) & \psi_j(\xi_2) \end{vmatrix}^2 \tag{9.42}$$

The notation $i > j$ means to count each pair (i, j) only once. If we expand the determinant and square it, we find

$$\rho_N(\mathbf{r}_1, \mathbf{r}_2) = \frac{1}{2N_p} \sum_{i>j} [|\psi_i(\xi_1)\psi_j(\xi_2)|^2 + |\psi_i(\xi_2)\psi_j(\xi_1)|^2$$
$$- 2\psi_i(\xi_1)\psi_j(\xi_2)\psi_i^*(\xi_2)\psi_j^*(\xi_1)] \tag{9.43}$$

Integrate this over all $d\xi_2$ and find

$$\rho_N(\mathbf{r}_1) = \frac{1}{2N_p} \sum_{i>j} [|\psi_i(\xi_1)|^2 + |\psi_j(\xi_1)|^2] \tag{9.44}$$

$$= \frac{1}{N} \sum_i |\psi_i(\xi_1)|^2 \tag{9.45}$$

These results for $\rho_N(\mathbf{r}_1)$ and $\rho_N(\mathbf{r}_1, \mathbf{r}_2)$ are used below to evaluate a number of expressions.

These results are proved in the following manner. Expand the original Slater determinant $\rho_N(\mathbf{r}_1, \mathbf{r}_2, \ldots, \mathbf{r}_N)$ in minors about the column that contains the variable ξ_N:

$$\Psi = \frac{1}{\sqrt{N!}} \sum_m \psi_m(\xi_N) M_{mN} \tag{9.46}$$

$$|\Psi|^2 = \frac{1}{N!} \sum_{mm'} \psi_{m'}^*(\xi_N)\psi_m(\xi_N) M_{m'N}^* M_{mN} \tag{9.47}$$

$$\int d^3r_N |\Psi|^2 = \frac{1}{N!} \sum_m |M_{mN}|^2 \tag{9.48}$$

Since the functions are orthogonal, the integral is zero for all of the cross terms and the result is a summation over the square of all of the minors of this column. Exactly the same result is obtained if we had chosen another variable d^3r_j to integrate. Since $j \neq 1, 2$, there are $(N-2)$ terms that give the same result.

The next step is to evaluate the square of the minors by performing another integration d^3r_n. In the same way, it has $(N-3)$ equivalent terms and reduces the dimension of the minor by one. Continue to do all of the integrals that define $\rho_N(\xi_1, \xi_2)$, and finally obtain eqn. (9.42).

9.4.2 Ground-State Energy

We wish to evaluate the energy of a Hamiltonian:

$$E = \langle \Psi^\dagger | H | \Psi \rangle \tag{9.49}$$

$$H = \sum_{i=1}^{N} \left[\frac{p_i^2}{2m} + U(\mathbf{r}_i) \right] + \sum_{i>j} \frac{e^2}{|\mathbf{r}_i - \mathbf{r}_j|} \tag{9.50}$$

The wave function $\Psi(\mathbf{r}_1, \mathbf{r}_2, \ldots, \mathbf{r}_N)$ has the form of a Slater determinant. Assume that the single-particle orbitals are orthogonal. They have a spin and orbital part. The orbital part is denoted as $\chi_j(m)$ for a single electron, where $\chi_j(\frac{1}{2}) = \alpha_j$, $\chi_j(-\frac{1}{2}) = \beta_j$. The subscript j denotes which electron, while m is the eigenvalaue for spin-up or spin-down:

$$\psi_m(\xi_j) = \phi_m(\mathbf{r}_j)\chi_j(m) \tag{9.51}$$

$$\int d^3r \phi_m(\mathbf{r})^* \phi_{m'}(\mathbf{r}) = \delta_{mm'} \tag{9.52}$$

$$\langle \chi_j(m) | \chi_{j'}(m') \rangle = \delta_{mm'} \delta_{jj'} \tag{9.53}$$

Using orthogonal eigenstates makes the evaluation of the determinant easy, as discussed above.

As a first step consider the evaluation of the potential energy term $\langle U \rangle$:

$$\langle U \rangle = \sum_{i=1}^{N} \int d^3r_1 d^3r_2 \cdots d^3r_N U(\mathbf{r}_i) |\Psi|^2 \tag{9.54}$$

Each term in the summation over i gives the same answer. That can be shown by just relabeling all of the variables of integration. So evaluate one term and multiply the result by the number of electrons N. This is the first step in canceling the $1/N!$ prefactor that results when we evaluate $|\Psi|^2$:

$$\langle U \rangle = N \int d^3r_1 U(\mathbf{r}_1) \rho_N(\mathbf{r}_1) \tag{9.55}$$

$$\rho_N(\mathbf{r}_1) = \int d^3r_2 \cdots d^3r_N |\Psi|^2 \tag{9.56}$$

$$= \frac{1}{N} \sum_{m=1}^{N} \rho_m(\mathbf{r}), \quad \rho_m(\mathbf{r}) = |\phi_m(\mathbf{r})|^2 \tag{9.57}$$

$$\langle U\rangle = \sum_{m=1}^{N} U_m, \quad U_m = \int d^3 r U(r)\rho_m(\mathbf{r}) \tag{9.58}$$

The quantity $N\rho_N(\mathbf{r})$ is the density of electrons in the system. It is the summation over all of the occupied states m of the density $|\phi_m(\mathbf{r})|^2$ provided by that orbital. The potential energy of the system of N electron is merely the summation of the potential energy of each occupied state. A similar result is found for the kinetic energy:

$$\langle K\rangle = \sum_{m=1}^{N} K_m, \; K_m = \int d^3 r \phi_m^*(\mathbf{r}) \frac{p^2}{2m} \phi_m(\mathbf{r}) \tag{9.59}$$

Notice that the spin plays no role in the evaluation of the kinetic and potential energy. For a single ψ_m one gets $\langle \chi_j(m)|\chi_j(m)\rangle = 1$ regardless of the values of m or j.

The last term to evaluate is the electron–electron interaction between all pairs of electrons. The summation over electron pairs is $(i > j)$. There are $N_p = N(N-1)/2$ pairs for N electrons. The integrals for each pair give the same results, since the identical particles are interchangable. So evaluate one pair and multiply by N_p:

$$\langle V\rangle = N_p \int d^3 r_1 d^3 r_2 \frac{e^2}{|\mathbf{r}_1 - \mathbf{r}_2|} \rho_N(\mathbf{r}_1, \mathbf{r}_2) \tag{9.60}$$

$$\rho_N(\mathbf{r}_1, \mathbf{r}_2) = \int d^3 r_3 d^3 r_4 \cdots d^3 r_N |\Psi|^2 \tag{9.61}$$

The two-particle distribution function $\rho_N(\mathbf{r}_1, \mathbf{r}_2)$ was evaluated above. The result is given by the summation over all pairs of two-particle distribution functions:

$$\langle V\rangle = \sum_{m>m'} V_{mm'} \tag{9.62}$$

$$V_{mm'} = \frac{e^2}{2} \int \frac{d^3 r_1 d^3 r_2}{|\mathbf{r}_1 - \mathbf{r}_2|} |\psi_m(\xi_1)\psi_{m'}(\xi_2) - \psi_m(\xi_2)\psi_{m'}(\xi_1)|^2$$

$V_{mm'}$ is the Coulomb energy between the two electron states m and m':

- If the spin states $\langle \chi_1(m)|\chi_2(m')\rangle = 0$, then

$$\begin{aligned} V_{mm'} &= \frac{e^2}{2} \int \frac{d^3 r_1 d^3 r_2}{|\mathbf{r}_1 - \mathbf{r}_2|} [|\phi_m(\mathbf{r}_1)\phi_{m'}(\mathbf{r}_2)|^2 + |\phi_m(\mathbf{r}_2)\phi_{m'}(\mathbf{r}_1)|^2] \\ &= e^2 \int \frac{d^3 r_1 d^3 r_2}{|\mathbf{r}_1 - \mathbf{r}_2|} |\phi_m(\mathbf{r}_1)\phi_{m'}(\mathbf{r}_2)|^2 = V_{H,mm'} \end{aligned} \tag{9.63}$$

- If the spin states $\langle \chi_1(m)|\chi_2(m')\rangle = 1$, then

$$\begin{aligned} V_{mm'} &= \frac{e^2}{2} \int \frac{d^3 r_1 d^3 r_2}{|\mathbf{r}_1 - \mathbf{r}_2|} |\phi_m(\mathbf{r}_1)\phi_{m'}(\mathbf{r}_2) - \phi_m(\mathbf{r}_2)\phi_{m'}(\mathbf{r}_1)|^2 \\ &= V_{H,\, mm'} - V_{X,\, mm'} \end{aligned} \tag{9.64}$$

$$V_{X,\, mm'} = e^2 \int \frac{d^3 r_1 d^3 r_2}{|\mathbf{r}_1 - \mathbf{r}_2|} \Re\{\phi_m^*(\mathbf{r}_1)\phi_{m'}^*(\mathbf{r}_2)\phi_m(\mathbf{r}_2)\phi_{m'}(\mathbf{r}_1)\} \tag{9.65}$$

The Hartree energy is $V_{H,mm'}$ and the exchange energy is $V_{X,\ mm'}$. The latter exists only between states with parallel spin. The summation over $(m,\ m')$ is over all pairs of occupied electron states.

The total expectation value of the Hamiltonian for this wave function is

$$\langle H\rangle = \sum_m (K_m + U_m) + \sum_{m>m'} (V_{H,mm'} - V_{X,mm'}\langle\chi(m)|\chi(m')\rangle)$$

Keep in mind that $\langle\chi(m)|\chi(m')\rangle = 0$ if the spins are antiparallel. This formula applies to any system with any form of potential energy $U(\mathbf{r})$.

The Slater determinant is never an exact eigenfunction of the Hamiltonian. One may select different orbital states $\phi_m(\mathbf{r})$ to use in setting up the determinant. In solving for electron states in molecules, chemists take a variety of different sets. The final wave function Ψ is a linear combination of such Slater determinants, each one with a different set of orbitals:

$$\Psi = \sum_{j=1}^{M} \frac{C_j}{\sqrt{N!}}|\cdots| \tag{9.66}$$

The coefficients C_j of these determinants are variational parameters that are varied to obtain the minimum energy. Such a procedure is called *configurational interaction*.

9.4.3 *Hartree-Fock Equations*

The beginning of this chapter discussed the need to find a simple Hamiltonian that described the behavior of a single electron. The correct Hamiltonian is for all of the electrons, but is cumbersome to solve. One possible approximate Hamiltonian for a single electron is called the Hartree-Fock approximation.

The previous subsection described the energy derived from a single Slater determinant. Think of this equation as a variational procedure. We could vary each orbital to find the minimum energy of the entire system. The way to do this is called a *functional derivative*. We vary each orbital $\delta\phi_m$:

$$0 = 2\int d^3r\delta\phi_m\left\{\frac{p^2}{2m}\phi_m + U(\mathbf{r})\phi_m \right. \tag{9.67}$$

$$\left. + e^2\int\frac{d^3r'}{|\mathbf{r}-\mathbf{r}'|}[\rho(\mathbf{r}')\phi_m(\mathbf{r}) - \sum_n \phi_n(\mathbf{r})\phi_n(\mathbf{r}')\phi_m(\mathbf{r}')\langle\chi(m)|\chi(m')\rangle]\right\}$$

$$\rho(\mathbf{r}) = \sum_n |\phi_n|^2 \tag{9.68}$$

The last two terms are from Hartree and exchange. The exchange term is nonzero only for states with parallel spin. For systems of two electrons we can form the singlet and triplet combinations. For more than two electrons, we cannot form singlet or triplet combinations from all pairs. So for more than two electrons, we abandon the singlet–triplet representation and just consider whether the orbital has a spin-up (α) or spin-down (β). Electrons of like spin have an exchange term; those with opposite spin do not.

The variational procedure has a constraint that the number of particles is fixed:

$$\int d^3r\delta\phi_m\phi_m = 0 \tag{9.69}$$

A variation with constraint has a Lagrange multiplier (λ), which is the eigenvalue. The above equations give

$$\lambda\phi_m = \frac{p^2}{2m}\phi_m + U(\mathbf{r})\phi_m + e^2\int\frac{d^3r'}{|\mathbf{r}-\mathbf{r}'|}[\rho(\mathbf{r}')\phi_m(\mathbf{r}) \\ -\sum_n \phi_n(\mathbf{r})\phi_n(\mathbf{r}')\phi_m(\mathbf{r}')\langle\chi(m)|\chi(n)\rangle] \tag{9.70}$$

This set of equations is called the *Hartree-Fock equations.*

The Hartree and exchange terms are of opposite sign and somewhat cancel. We have written both terms such that their sum over states n includes the term m. The term with $n = m$ cancels from the two terms: if the states are identical the spins are parallel. This convention has the advantage that the Hartree potential is the same for each electron. Another way to write these two terms is that the term $n = m$ is omitted from both terms. An obvious disadvantage of the H-F equations is that they are not effective Hamiltonians for a single electrons. The exchange term can be written as

$$-\sum_n V_{X,\,mn}(\mathbf{r})\phi_n(\mathbf{r})\langle\chi(m)|\chi(n)\rangle \tag{9.71}$$

$$V_{X,mn}(\mathbf{r}) = e^2\int\frac{d^3r'}{|\mathbf{r}-\mathbf{r}'|}\phi_n(\mathbf{r}')^*\phi_m(\mathbf{r}') \tag{9.72}$$

$$H_{H-F}\phi_m(\mathbf{r}) = \left[\frac{p^2}{2m} + U(\mathbf{r}) + V_H(\mathbf{r})\right]\phi_m - \sum_n V_{X,mn}(\mathbf{r})\phi_n\langle\chi(n)|\chi(m)\rangle$$

There is a summation n over all of the other occupied orbital states with the same spin. In an atom with a small number of different orbitals this summation includes only a few terms. For large molecules it becomes prohibitive. In every case, one has to numerically find a self-consistent solution. The effective potential acting on each electron includes Hartree and exchange. They, in turn, depend on the orbitals of the other occupied electrons states.

As an example, solve for the effective Hamiltonian for an electron in the $1s$ state of atomic helium. There are two electrons with opposite spins, so there is no exchange. There is a Hartree term of each electron having a Coulomb interaction on the other:

$$\varepsilon_{1s}\phi_{1s}(r) = \left[-\frac{\hbar^2\nabla^2}{2m} - \frac{2e^2}{r} + e^2\int\frac{d^3r'}{|\mathbf{r}-\mathbf{r}'|}|\phi_{1s}(\mathbf{r}')|^2\right]\phi_{1s}(r)$$

The equation is nonlinear and is usually solved by numerical iteration. The variational solution is easier.

The next case is to consider an excited state of the helium atom, with one electron in the ($1s$) state and the other in the ($2s$) state. There are three relevant potentials:

$$V_{1s}(r) = e^2\int\frac{d^3r'}{|\mathbf{r}-\mathbf{r}'|}\phi_{1s}(r')^2 \tag{9.73}$$

$$V_{2s}(r) = e^2 \int \frac{d^3r'}{|\mathbf{r}-\mathbf{r}'|} \phi_{2s}(r')^2 \tag{9.74}$$

$$V_X(r) = e^2 \int \frac{d^3r'}{|\mathbf{r}-\mathbf{r}'|} \phi_{2s}(r')\phi_{1s}(r') \tag{9.75}$$

There are two possible spin arrangements.

- Orthohelium has the spin triplet so the exchange term is negative:

$$\varepsilon_{1s}\phi_{1s}(r) = \left[-\frac{\hbar^2\nabla^2}{2m} - \frac{2e^2}{r} + V_{2s}(r)\right]\phi_{1s}(r) - V_X(r)\phi_{2s}(r)$$

$$\varepsilon_{2s}\phi_{2s}(r) = \left[-\frac{\hbar^2\nabla^2}{2m} - \frac{2e^2}{r} + V_{1s}(r)\right]\phi_{2s}(r) - V_X(r)\phi_{1s}(r)$$

- Parahelium has the spin singlet so the exchange term is positive:

$$\varepsilon_{1s}\phi_{1s}(r) = \left[-\frac{\hbar^2\nabla^2}{2m} - \frac{2e^2}{r} + V_{2s}(r)\right]\phi_{1s}(r) + V_X(r)\phi_{2s}(r)$$

$$\varepsilon_{2s}\phi_{2s}(r) = \left[-\frac{\hbar^2\nabla^2}{2m} - \frac{2e^2}{r} + V_{1s}(r)\right]\phi_{2s}(r) + V_X(r)\phi_{1s}(r)$$

The two orbitals ϕ_{1s}, ϕ_{2s} have different potential energies and are different orbitals for orthohelium and parahelium.

The main problem with H-F is that it is not an exact solution, even if one solves the numerical equations precisely. There are other energy terms, called correlation, that must also be included.

9.4.4 *Free Electrons*

Sommerfeld invented a simple model of the conduction electrons in a metal. He said they were "free": there was no potential energy. The actual model of a metal such as sodium or aluminum is that the valence electrons do indeed become free and wander around like electrons in zero potential. The free electrons model assumes that for N electrons, $N/2$ have spin-up and $N/2$ have spin-down. If the particles are in a cubic box of side L, the wave vectors are quantized according to $(k_{x,\ell} = 2\ell\pi/L, k_{y,m} = 2m\pi/L,\ k_{z,n} = 2n\pi/L)$. Each $\mathbf{k} = (k_x,\ k_y,\ k_z)$-state can be occupied by only two electrons, one with spin-up and one with spin-down. The states with lowest energy tend to fill up first. At zero temperature one can define a Fermi wave vector k_F such that states with $|\mathbf{k}| < k_F$ are occupied and states with $|\mathbf{k}| > k_F$ are unoccupied with electrons. The dominent energy term is the kinetic energy:

$$E(\mathbf{k}) = \frac{\hbar^2}{2m}(k_{x,\ell}^2 + k_{y,m}^2 + k_{z,n}^2) \tag{9.76}$$

The occupied states are a sphere of radius k_F in wave vector space. We use this model to calculate some of the energy terms, assuming that the wave function is a Slater determinant. The individual orbitals are $\phi_m(\mathbf{r}) = \exp(i\mathbf{k}\cdot\mathbf{r})/\sqrt{\Omega}$.

1. The number of electrons is found by summing over all of the **k** occupied states:

$$N_e = 2\sum_{\ell,m,n}\Theta(k_F - |\mathbf{k}|) = 2\Omega\int\frac{d^3k}{(2\pi)^3}\Theta(k_F - |\mathbf{k}|) \tag{9.77}$$

$$= \frac{2(4\pi)\Omega}{(2\pi)^3}\int_0^{k_F} k^2 dk = \Omega\frac{k_F^3}{3\pi^2} \tag{9.78}$$

$$n_e = \frac{N_e}{\Omega} = \frac{k_F^3}{3\pi^2} \tag{9.79}$$

The factor of two in front is from spin degeneracy. The electron density is n_e.

2. The average kinetic energy is found by

$$\langle K\rangle = 2\sum_{\mathbf{k}} K(\mathbf{k}),\; K(\mathbf{k}) = -\frac{\hbar^2}{2m\Omega}\int d^3r e^{-i\mathbf{k}\cdot\mathbf{r}}\nabla^2 e^{i\mathbf{k}\cdot\mathbf{r}} = \frac{\hbar^2k^2}{2m}$$

$$\langle K\rangle = 2\Omega\int\frac{d^3k}{(2\pi)^3}\Theta(k_F - |\mathbf{k}|)\frac{\hbar^2k^2}{2m} = \frac{\hbar^2\Omega}{2m\pi^2}\int_0^{k_F} k^4 dk \tag{9.80}$$

$$= \frac{3\hbar^2k_F^2}{10m}\Omega\frac{k_F^3}{3\pi^2} = N_e\frac{3\hbar^2k_F^2}{10m} \tag{9.81}$$

The average kinetic energy of an electron is

$$\frac{\langle K\rangle}{N_e} = \frac{3\hbar^2k_F^2}{10m} = \frac{3}{5}E_F,\; E_F = \frac{\hbar^2k_F^2}{2m} \tag{9.82}$$

E_F is called the *Fermi energy*. It is the energy of the electron with the maximum kinetic energy at zero temperature, where the zero of energy is measured from the bottom of the band ($\mathbf{k} = 0$).

3. The potential and Hartree energies are best treated together. The potential energy from $U(\mathbf{r})$ is from the interaction of the electrons with the positive ions. The Hartree energy V_H is from the electrons interacting with each other. The positive charge interacts with itself. These terms are given in terms of the electron charge density ρ_e and the ion charge density ρ_i as

$$\langle U + V_H\rangle = \frac{e^2}{2}\int\frac{d^3r_1 d^3r_2}{|\mathbf{r}_1 - \mathbf{r}_2|}[\rho_e(\mathbf{r}_1) - \rho_i(\mathbf{r}_1)][\rho_e(\mathbf{r}_2) - \rho_i(\mathbf{r}_2)]$$

The system must be neutral, on the average. The average electron charge must be equal to the average ion charge. This term is usually nonzero and rather large.

4. The most interesting term is from exchange. The evaluation of this term begins with $\rho_N(\mathbf{r}_1, \mathbf{r}_2)$ in eqn. (9.43). The summation over states includes wave vector states ($\mathbf{k} \geq \mathbf{k}'$) and spin states. In a paramagnetic system, the spin configurations for two electrons are (i) both up, (ii) both down, or (iii) one up and one down. Only the first two have an exchange energy. Since plane waves have $|\phi(\mathbf{k},\mathbf{r})|^2 = 1/\,\Omega$, the two-particle distribution for plane waves is

$$\rho_N(\mathbf{r}_1, \mathbf{r}_2) = \frac{2}{N_p\Omega^2}\sum_{\mathbf{k}>\mathbf{k}'}[2 - e^{i(\mathbf{k}-\mathbf{k}')\cdot(\mathbf{r}_1-\mathbf{r}_2)}] \tag{9.83}$$

The exchange term is from the last expression. A factor of two is used to change the double summation to all of ($\mathbf{k}$, $\mathbf{k}'$), which counts each pair twice:

$$\langle EE\rangle = \sum_{i>j} \int d^3r_1 \cdots d^3r_N \frac{e^2}{|\mathbf{r}_i - \mathbf{r}_j|} |\Psi|^2 \tag{9.84}$$

$$= N_p \int d^3r_1 d^3r_2 \frac{e^2}{|\mathbf{r}_1 - \mathbf{r}_2|} \rho_N(\mathbf{r}_1, \mathbf{r}_2) \tag{9.85}$$

Using the above expression for $\rho_N(\mathbf{r}_1, \mathbf{r}_2)$, the exchange energy is

$$E_X = -\frac{e^2}{\Omega^2} \sum_{\mathbf{k}_1, \mathbf{k}_2 < k_F} \int \frac{d^3r_1 d^3r_2}{|\mathbf{r}_1 - \mathbf{r}_2|} \exp\left[-i(\mathbf{k}_1 - \mathbf{k}_2) \cdot (\mathbf{r}_1 - \mathbf{r}_2)\right]$$

Evaluate the integral in center-of-mass coordinates: $\mathbf{r} = \mathbf{r}_1 - \mathbf{r}_2$, $\mathbf{R} = (\mathbf{r}_1 + \mathbf{r}_2)/2$. Then one can do $\int d^3R = \Omega$, which cancels one factor of volume. The remaining integral is just the Fourier transform of the Coulomb interaction:

$$E_X = -\frac{e^2}{\Omega} \sum_{\mathbf{k}_1 < k_F, \mathbf{k}_2 < k_F} \frac{4\pi}{|\mathbf{k}_1 - \mathbf{k}_2|^2} \tag{9.86}$$

The integral can be evaluated. It is useful to divide by the number of electrons N_e to have the exchange energy per electron. Also change variables to $\nu = \cos\theta$, where θ is the angle between the two wave vectors:

$$\frac{E_X}{N_e} = -\frac{(4\pi)^3 e^2}{n_e (2\pi)^6} \int_0^{k_F} k_1^2 dk_1 \int_0^{k_F} k_2^2 dk_2 \int_{-1}^{1} \frac{d\nu}{2} \frac{1}{k_1^2 + k_2^2 - 2k_1 k_2 \nu}$$

$$= -\frac{3e^2}{2\pi k_F^3} \int_0^{k_F} k_1 dk_1 \int_0^{k_F} k_2 dk_2 \ln\left|\frac{k_1 + k_2}{k_1 - k_2}\right| \tag{9.87}$$

$$= -\frac{3e^2 k_F}{4\pi} \tag{9.88}$$

The unit of $e^2 k_F$ is energy. If L is length, then e^2/L is energy. The only important length scale in the electron gas is k_F^{-1}. This energy term is also large.

The exchange energy makes a significant contribution to the binding energy of a metal. Since it is negative, it lowers the ground-state energy of the metallic state. The exchange energy results from the first exclusion principle that no two electrons with the same spin can be at the same point in space. This correlated motion reduces the repulsive electron–electron energy.

9.4.5 Pair Distribution Function

The pair distribution function $g(\mathbf{r}_1, \mathbf{r}_2)$ is defined as the probability that a particle is at $\mathbf{r}_2$ if there is also one at $\mathbf{r}_1$. It is normalized to go to one at large separation. In a fluid it can only depend on the separation of the two particles, $g(\mathbf{r}_1 - \mathbf{r}_2)$. It is found from the two-particle distribution, which was derived in eqn. (9.83) for the paramagnetic electron gas:

$$\rho_N(\mathbf{r}_1, \mathbf{r}_2) = \frac{1}{\Omega^2} g(\mathbf{r}_1 - \mathbf{r}_2) \tag{9.89}$$

$$g(r) = 1 - 2\Lambda(r)^2 \tag{9.90}$$

$$\Lambda(r) = \frac{1}{n_0}\int \frac{d^3k}{(2\pi)^3} e^{i\mathbf{k}\cdot\mathbf{r}}\Theta(k_F - k) \tag{9.91}$$

$$= \frac{3}{2k_F r} j_1(k_F r) = \frac{1}{2}[j_0(k_F r) + j_2(k_F r)] \tag{9.92}$$

Note $\Lambda(0) = \frac{1}{2}, g(0) = \frac{1}{2}$, and $\Lambda(\infty) = 0$, $g(\infty) = 1$. Half of the spins are up and half are down. If an electron is at one point, electrons of the same spin cannot be at that point, but electrons of opposite spin can be there. So $g(0) = \frac{1}{2}$. In a ferromagnetic system where all spins point in the same direction, $g(0) = 0$.

9.4.6 Correlation Energy

For the free electron gas, the following energy terms have been derived or discussed: kinetic energy, potential energy, and exchange energy. There are more energy terms. One way to derive them is by perturbation theory. The evaluation of $\langle \Psi_g | H | \Psi_g \rangle$ is equivalent to first-order perturbation theory. Here Ψ_g is the Slater determinant of the ground state. The next term is from second-order perturbation theory:

$$\Delta E = -\sum_j \frac{|\langle \Psi_j | H | \Psi_g \rangle|^2}{E_j - E_g} \tag{9.93}$$

The Slater determinants for excited states Ψ_j have one or more orbitals with $|\mathbf{k}| > k_F$. The above energy terms make a small but measurable contribution to the ground-state energy. Wigner coined the phrase *correlation energy* to describe the energy terms that arise in perturbation theory beyond Hartree-Fock. The exchange terms lower the total energy of the system by keeping apart electrons of parallel spin. The correlation energy applies to all electrons, but the largest effect is between electrons of opposite spin. Correlated motion, where particles of opposite spin avoid each other, also lowers the total energy of the system.

9.4.7 Thomas-Fermi Theory

Thomas-Fermi theory is called *Fermi-Thomas theory* about half of the time. As with most great ideas, it is built on a simple concept. Assume that the potential energy $V(r)$ varies slowly in space. The precise definition of "slowly" is deferred to later, where we learn that Thomas-Fermi theory works even for atoms. We introduce the chemical potential μ, which in equilibirium must be a constant in all space. Let the Fermi energy be $E_F = \hbar^2 k_F^2/2m$. If there is no potential then $\mu = E_F$. When there is a potential, and since the chemical potential is a constant, then

$$E_F + V(\mathbf{r}) = \mu \tag{9.94}$$

The above equation makes sense only if the kinetic energy also varies in space. The Fermi wave vector k_F must be a function of position:

$$\mu = \frac{\hbar^2 k_F(\mathbf{r})^2}{2m} + V(\mathbf{r}) \tag{9.95}$$

$$n_e(\mathbf{r}) = \frac{k_F(\mathbf{r})^3}{3\pi^2} = \left(\frac{2m}{\hbar^2}\right)^{3/2} \frac{[\mu - V(\mathbf{r})]^{3/2}}{3\pi^2} \tag{9.96}$$

Clearly the Thomas-Fermi theory is semiclassical. We define a local Fermi wave vector $k_F(\mathbf{r})$. Each point in space is assumed to be a local electron gas, with an electron density given by the above formula.

Introduce Poisson's equation for the potential from a charge distribution $\rho(\mathbf{r})$:

$$\nabla^2 V(\mathbf{r}) = 4\pi e^2 \rho(\mathbf{r}) \tag{9.97}$$

In the electron gas, the potential will be from the electrons as well as from the ion charge. We use the Thomas-Fermi expression for the electron density:

$$\nabla^2 V(\mathbf{r}) = 4\pi e^2 [\rho_i(\mathbf{r}) - n_e(\mathbf{r})] \tag{9.98}$$

$$\nabla^2 U(\mathbf{r}) = 4\pi e^2 \rho_i(\mathbf{r}) \tag{9.99}$$

The potential function $U(\mathbf{r})$ gives the interaction between the electrons and ions in a metal. Its source is the ion charge distribution $\rho_i(\mathbf{r})$, and $U(\mathbf{r})$ is derived from the second equation above. The potential function $V(\mathbf{r})$ is different from $U(\mathbf{r})$. We call $V(\mathbf{r})$ the *screened potential* or the *self-consistent potential*. As a result of the potential $U(\mathbf{r})$, the electrons all alter their motions. Since $U(\mathbf{r})$ is an attractive potential to electrons, they spend a little more time during their motion through the metal in the regions where $U(\mathbf{r})$ is large. When these motions are averaged, the electron density $n_e(\mathbf{r})$ is found to be higher in these regions. The potential $V(\mathbf{r})$ is the final potential, including the attractive part from $U(\mathbf{r})$ and the repulsive part from the additional electron density in these regions. The final point, which is important, is that the electrons feel the net potential $V(\mathbf{r})$ in their motion through the metal. It is found self-consistently. The algorithm for finding $V(\mathbf{r})$ is the following: *Starting from $U(\mathbf{r})$, $V(\mathbf{r})$ is the potential that acts on electrons to increase their density where U is large, such that the final potential is also V.*

Equation (9.98) is the Thomas-Fermi equation in three dimensions. In other dimensions $n(n = 1, 2)$, the equation must be written as

$$V(\mathbf{r}) = -e^2 \int \frac{d^n r'}{|\mathbf{r} - \mathbf{r}'|} [\rho_i(\mathbf{r}') - n_e(\mathbf{r}')] \tag{9.100}$$

In three dimensions, the operation by ∇^2 produces eqn. (9.98), but that is not true in other dimensions.

One way to solve the T-F equation is to assume that $V/\mu << 1$ and to expand the electron density in this ratio:

$$n_e(\mathbf{r}) = \frac{1}{3\pi^2} \left(\frac{2m\mu}{\hbar^2}\right)^{3/2} \left[1 - \frac{V(\mathbf{r})}{\mu}\right]^{3/2} \tag{9.101}$$

$$\approx n_0 \left[1 - \frac{3V}{2\mu} + \cdots\right] \tag{9.102}$$

$$n_0 = \frac{1}{3\pi^2}\left(\frac{2m\mu}{\hbar^2}\right)^{3/2} = \frac{k_F^3}{3\pi^2} \tag{9.103}$$

where n_0 is the equilibrium density. In this approximation, Poisson's equation is

$$[\nabla^2 - q_{T-F}^2]V(\mathbf{r}) = 4\pi e^2[\rho_i(\mathbf{r}) - n_0] \tag{9.104}$$

$$q_{T-F}^2 = \frac{6\pi e^2 n_0}{\mu} \tag{9.105}$$

The constant $q_{T\text{-}F}$ is the Thomas-Fermi screening length. For example, if the charge distribution $\rho_i = n_0 + Z\delta^3(\mathbf{r})$ is for a uniform background plus one point charge, the solution is

$$V(\mathbf{r}) = -\frac{Ze^2}{r}e^{-q_{T-F}r} \tag{9.106}$$

The potential has a Yukawa form. The bare potential $U = -Ze^2/r$ has been screened by the factor of $\exp(-q_{T\text{-}F}r)$. The above solution is approximate and is based on an expansion in V/μ.

There are many cases where one can solve the T-F equations without making such approximations. One case is for the electrons in a rare gas atom such as neon or argon. They are neutral, all electrons are in closed shells, and $[V(r), n(r)]$ are both spherically symmetric. Then one only has to solve the radial equation. Assuming the nucleus is a point charge,

$$\frac{1}{r}\frac{d^2}{dr^2}[rV(r)] = 4\pi e^2 Z\frac{\delta(r)}{r^2} - \frac{4\pi e^2}{3\pi^2}[\mu - V(r)]^{3/2}\left(\frac{2m}{\hbar^2}\right)^{3/2} \tag{9.107}$$

As r becomes large, the potential $V(r)$ must go to zero. That means the left-hand side of the above equation must vanish, and so must the right-hand side. However, if $V \to 0$, the right-hand side will vanish only if $\mu = 0$. Therefore, set the chemical potential to zero. Since $V < 0$, the argument of the fractional exponent is positive. Also, omit the term with $\delta(r)$ and replace it with the boundary condition that $V(r) \to -Ze^2/r$ as $r \to 0$.

Solve the equation in dimensionless units with $\rho = r/a_0$ (a_0 is Bohr radius) and

$$Q(\rho) = -\frac{r}{Ze^2}V(r) \tag{9.108}$$

$$\frac{d^2}{d\rho^2}Q(\rho) = \frac{2^{7/2}Z^{1/2}}{3\pi}\frac{Q(\rho)^{3/2}}{\sqrt{\rho}} \tag{9.109}$$

The initial condition is $Q(0) = 1$. The above equation can be solved using numerical methods. One obtains $V(r)$ and the effective electron density:

$$n(r) = \frac{1}{3\pi^2 a_0^3}\left(-\frac{V(r)}{E_{Ry}}\right)^{3/2} = \frac{(2Z)^{3/2}}{3\pi^2 a_0^3}\left(\frac{Q}{\rho}\right)^{3/2} \tag{9.110}$$

The equation can be put in universal dimensionless form by defining a new coordinate x as

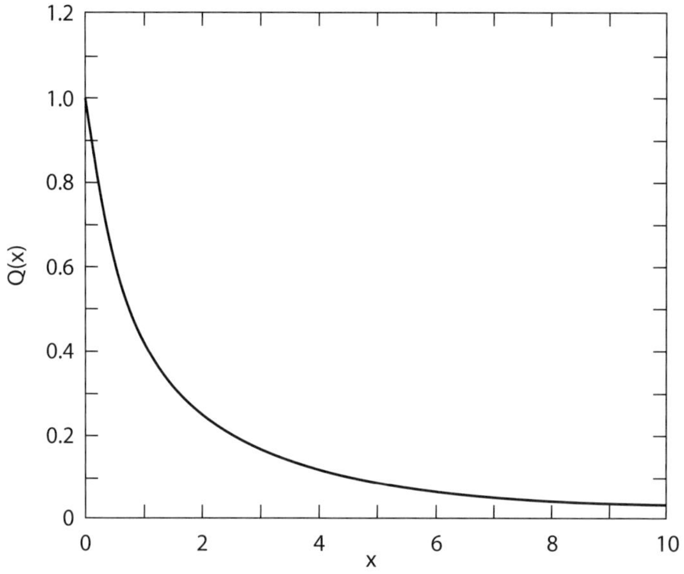

Figure 9.1. Thomas-Fermi function $Q(x)$ for a neutral atom.

$$x = \frac{\rho Z^{1/3}}{b}, b = \left(\frac{3\pi}{2^{7/2}}\right)^{2/3} \approx 0.885 \tag{9.111}$$

$$\sqrt{x}\frac{d^2 Q(x)}{dx^2} = Q(x)^{3/2} \tag{9.112}$$

There is a single equation that applies to all neutral atoms. Figure 9.1 shows a numerical solution of this equation, starting from $Q(0) = 1$. The curve falls to zero in a smooth way. The result is an electron density with the right general shape when graphed versus r. The shell structure of actual atoms is not present, since the Thomas-Fermi model gives only a single maximum.

9.4.8 Density Functional Theory

The goal of a many-electron system is to find an approximate Hamiltonian for a single electron. The Hamiltonian should have the form

$$H = \frac{p^2}{2m} + V(\mathbf{r}) \tag{9.113}$$

The potential energy $V(\mathbf{r})$ must include the potential $U(\mathbf{r})$ from the positive changes, as well as another potential that simulates the role of electron–electron interactions. John Slater made an important suggestion. His idea was to write for the exchange term

$$H = \frac{p^2}{2m} + U(\mathbf{r}) + V_X(\mathbf{r}) \tag{9.114}$$

$$V_X(\mathbf{r}) = -\alpha\frac{e^2 k_F(\mathbf{r})}{\pi} \tag{9.115}$$

$$k_F(\mathbf{r}) = [3\pi^2 n(\mathbf{r})]^{1/3} \tag{9.116}$$

$$n(\mathbf{r}) = \sum_m |\phi_m(\mathbf{r})|^2 \tag{9.117}$$

The exchange term is written in the form found for the electron gas $-e^2k_F/\pi$, where the Fermi wave vector k_F is, as in Thomas-Fermi, treated as a function of position. Instead of using the Thomas-Fermi definition of k_F, Slater used the electron gas definition $k_F(\mathbf{r})^3 = 3\pi^2 n(\mathbf{r})$. The electron density $n(\mathbf{r})$ is found by summing over all occupied orbital states. Slater's idea combined a bit of the electron gas with Thomas-Fermi. It is a self-consistent procedure, in that one has to know the density $n(\mathbf{r})$ to find $V_X(\mathbf{r})$, and one has to know V_X to find the orbitals that contribute to the density. Self-consistency is achieved by iterating the numerical equations.

Slater was unsure what coefficient should multiply e^2k_F/π. He put a dimensionless constant α in front and varied it to get good results. This technique was called $X\alpha$, where X stands for "exchange" and α is his constant.

$X\alpha$ was the first set of equations that could include exchange for a system with a large number of electrons. Using a local potential based on $k_F(\mathbf{r})$ is obviously an approximation. It has the desirable feature that exchange is large where there are many other electrons (large n) and small where the density of electrons is low. The $X\alpha$ method gave fairly good results for many atoms and molelcules. Yet the method was always suspect since it was obviously ad hoc. Also, the adjustable constant α destroyed any pretense of rigor. Nevertheless, $X\alpha$ was an important historical precedent to the present technology, which is called *density functional theory* (DFT).

Walter Kohn was the intellectual force behind DFT. He and Pierre Hohenberg proved a theorem that forms the basis of the entire technology: *The ground-state energy is a functional of only the density* $E_G(n)$. That is interesting, since it does not depend on any of the individual wave functions or their phases, but only on the entire density.

The second major idea is that the correct potential can be found by a functional derivative of this ground-state energy with respect to the electron density:

$$V(\mathbf{r}) = \frac{\delta E_G}{\delta n(\mathbf{r})} \tag{9.118}$$

All applications use the *local density approximation.* As with Slater, the dependence on density is assumed to be a local function of density such as $k_F[n(\mathbf{r})]$. Like Slater, they use the electron gas formulas to find exchange and correlation energies. For example, consider the exchange energy for the electron gas derived earlier in eqn. (9.88). Since $N_e = \int d^3r n(\mathbf{r})$ this expression can be rewritten as

$$E_X = -\int d^3 r n(\mathbf{r}) \frac{3e^2 k_F[n(\mathbf{r})]}{4\pi} \tag{9.119}$$

Since $k_F \sim n^{1/3}$ the right-hand side is $n^{4/3}$. Taking a functional derivative

$$\frac{\delta E_X}{\delta n} = V_X(n) = -\frac{e^2 k_F(n)}{\pi} \tag{9.120}$$

The right-hand side has the factor of $(\frac{3}{4})$ canceled by the exponent $(\frac{4}{3})$ from the derivative. The result is precisely Slater's potential, with $\alpha = 1$. DFT uses this formula for exchange. There is also another term from the correlation energy derived from electron–electron interactions:

$$H = \frac{p^2}{2m} + U(\mathbf{r}) + V_X[n(\mathbf{r})] + V_C[n(\mathbf{r})] \tag{9.121}$$

The result is an equation that looks like a Hamiltonian for a single electron and can be used to calculate self-consistently the density of electrons in a many-electron system. Density functional theory is used widely by physicists and chemists to calculate the properties of atoms, molecules, solids, liquids, and interfaces.

9.5 Second Quantization

Second quantization is needed to account for the fact that elementary particles occur in integer numbers. A system can have N electrons, where N is a nonnegative integer. If we have the eigenvalue equation $H\phi_n = \hbar\omega_n\phi_n$ the wave function is exactly given by

$$\psi(\mathbf{r}, t) = \sum_n a_n \phi_n(\mathbf{r}) e^{-i\omega_n t} \tag{9.122}$$

Normalize the wave function by setting

$$\int d^3r \psi^\dagger(\mathbf{r}, t)\psi(\mathbf{r}, t) = \sum_n |a_n|^2 = 1 \tag{9.123}$$

The values of $|a_n|^2$ are equal or less than one.

Writing the wave function this way does not require an integer number of particles. We do that by requiring that a_n is actually an operator! The expectation value is $\langle a_n^\dagger a_n \rangle = N_n$, where N_n is the average number of particles in state n. The above expression is now

$$\int d^3r \psi^\dagger(\mathbf{r}, t)\psi(\mathbf{r}, t) = \sum_n a_n^\dagger a_n \tag{9.124}$$

$$\left\langle \int d^3r \psi^\dagger(\mathbf{r}, t)\psi(\mathbf{r}, t) \right\rangle = \sum_n \langle a_n^\dagger a_n \rangle = \sum_n N_n = 1 \tag{9.125}$$

The average occupation number is one. However, the instantaneous value of $a_n^\dagger a_n$ is zero or one.

We use a Lagrangian-Hamiltonian approach to derive the properties of the operators a_n. The above discussion is general and applies to both fermion and boson particles.

9.5.1 *Bosons*

First consider boson particles. The method is often applied to atoms such as ^{4}He, which contain even numbers of fermions, so that it has boson-like properties. If the boson is in a potential $U(\mathbf{r})$, the one-particle Schrödinger equation is

$$i\hbar\dot{\psi}(\mathbf{r}) = H\psi(\mathbf{r}) = \left[-\frac{\hbar^2\nabla^2}{2m} + U(\mathbf{r})\right]\psi(\mathbf{r}) \tag{9.126}$$

This equation may be derived from the Lagrangian density:

$$L = i\hbar\psi^\dagger\dot{\psi} - \frac{\hbar^2}{2m}\nabla\psi^\dagger\cdot\nabla\psi - U(\mathbf{r},t)\psi^\dagger\psi \tag{9.127}$$

The wave function is complex, with real and imaginary parts. These two parts can be treated as independent variables. An alternate procedure is to treat $\psi(\mathbf{r})$ and $\psi^\dagger(\mathbf{r})$ as independent variables. Then the functional derivatives give

$$\frac{\partial L}{\partial\psi} = -U\psi^\dagger \tag{9.128}$$

$$\frac{\partial L}{\partial\psi_x} = -\frac{\hbar^2}{2m}\psi_x^\dagger \tag{9.129}$$

$$\frac{\partial L}{\partial\dot{\psi}} = i\hbar\psi^\dagger \tag{9.130}$$

Put these relations into Lagrange's equation:

$$0 = \frac{\partial L}{\partial\psi} - \sum_\mu \frac{\partial}{\partial x_\mu}\left(\frac{\partial L}{\partial(\partial\psi/\partial x_\mu)}\right) - \frac{\partial}{\partial t}\frac{\partial L}{\partial\dot{\psi}} \tag{9.131}$$

$$0 = -U\psi^\dagger + \frac{\hbar^2}{2m}\nabla^2\psi^\dagger - i\hbar\frac{\partial}{\partial t}\psi^\dagger \tag{9.132}$$

The latter is the Hermitian conjugate of Schrödinger's equation. In the Lagrangian formulation, the momentum that is conjugate to the variable ψ is

$$\pi = \frac{\partial L}{\partial\dot{\psi}} = i\hbar\psi^\dagger(\mathbf{r}) \tag{9.133}$$

The Hamiltonian density is given by

$$\mathcal{H} = \pi\dot{\psi} - L = \frac{\hbar^2}{2m}\nabla\psi^\dagger\cdot\nabla\psi + U\psi^\dagger\psi \tag{9.134}$$

Integrate over all volumes to obtain the Hamiltonian:

$$H = \int d^3r\mathcal{H} = \int d^3r\psi^\dagger\left(-\frac{\hbar^2}{2m}\nabla^2 + U\right)\psi \tag{9.135}$$

where the kinetic energy term was integrated by parts. Since π and ψ are conjugate variables, they obey commutation relations of the form

$$[\psi(\mathbf{r},t), \pi(\mathbf{r}',t)] = i\hbar\delta(\mathbf{r}-\mathbf{r}') \tag{9.136}$$

or, using (9.133),

$$[\psi(\mathbf{r},t), \psi^\dagger(\mathbf{r}',t)] = \delta(\mathbf{r}-\mathbf{r}') \tag{9.137}$$

This commutation relation is the fundamental basis of second quantization. Although it has been made plausible by the derivation from a Lagrangian, it is a basic premise.

These commutation relations may be satisfied by introducing creation and destruction operators for bosons. Let H have eigenstates and eigenvalues of the form

$$H\phi_\lambda = \varepsilon_\lambda \phi_\lambda \tag{9.138}$$

$$H = -\frac{\hbar^2}{2m}\nabla^2 + U(\mathbf{r}) \tag{9.139}$$

The wave function $\psi(\mathbf{r})$ and its conjugate $\psi^\dagger(\mathbf{r})$ are expanded in terms of this basis set:

$$\psi(\mathbf{r}) = \sum_\lambda a_\lambda(t)\phi_\lambda(\mathbf{r}) \tag{9.140}$$

$$\psi^\dagger(\mathbf{r}) = \sum_\lambda a_\lambda^\dagger(t)\phi_\lambda^*(\mathbf{r}) \tag{9.141}$$

The commutator (9.137) is satisfied if a and $a^\dagger$ operators have their own commutation relations:

$$\begin{aligned}
&[a_\lambda(t), a_{\lambda'}^\dagger(t)] = \delta_{\lambda\lambda'} \\
&[a_\lambda(t), a_{\lambda'}(t)] = 0 \\
&[a_\lambda^\dagger(t), a_{\lambda'}^\dagger(t)] = 0
\end{aligned} \tag{9.142}$$

The commutation relations for the field variables are

$$[\psi(\mathbf{r}, t), \psi^\dagger(\mathbf{r}', t)] = \sum_{\lambda\lambda'} [a_\lambda(t), a_{\lambda'}^\dagger(t)]\phi_\lambda(\mathbf{r})\phi_{\lambda'}^*(\mathbf{r}') \tag{9.143}$$

$$= \sum_\lambda \phi_\lambda(\mathbf{r})\phi_\lambda^*(\mathbf{r}') = \delta(\mathbf{r}-\mathbf{r}') \tag{9.144}$$

An operator of interest is the density operator:

$$\rho(\mathbf{r}) = \psi^\dagger(\mathbf{r})\psi(\mathbf{r}) = \sum_{\lambda\lambda'} a_\lambda^\dagger a_{\lambda'}\phi_\lambda^*(\mathbf{r})\phi_{\lambda'}(\mathbf{r}) \tag{9.145}$$

The integral of $\rho(\mathbf{r})$ is the number operator:

$$N = \int d^3r\rho(\mathbf{r}) = \sum_\lambda a_\lambda^\dagger a_\lambda = \sum_\lambda n_\lambda \tag{9.146}$$

Its thermal average is obtained by taking the thermal average of n_λ:

$$\langle N\rangle = \sum_\lambda \langle n_\lambda\rangle = \sum_\lambda N_\lambda \tag{9.147}$$

Any systems with these commutation relations behave as harmonic oscillators for each state λ. The eigenstates for each value of λ have a discrete set of occupation numbers $n_\lambda = 0, 1, 2, 3, \ldots$. All bosons have harmonic oscillator eigenstates. For phonons, the number n_λ is interpreted as the number of phonons in state λ. For particles, the interpretation is the same. The number n_λ tells how many particles in the system are in the same state λ. However, for particles, unlike phonons, the total number of particles is conserved. The many-particle wave function has the form

$$\Pi_\lambda \frac{(a_\lambda^\dagger)^{n_\lambda}}{\sqrt{n_\lambda!}}|0\rangle \tag{9.148}$$

In thermal equilibrium, the average number of particles in a state λ is given by the usual Boson occupation factor:

$$\langle n_\lambda\rangle = \frac{1}{e^{\beta(\varepsilon_\lambda-\mu)}-1} \equiv N_\lambda = n_B(\varepsilon_\lambda-\mu) \tag{9.149}$$

There is a chemical potential μ, which can vary with temperature and concentration. This equation serves as a definition of the chemical potential and determines its variations with temperature and particle number N.

9.5.2 *Fermions*

A similar method is used for second quantization for fermions. They are usually electrons, although the technique is also applied to holes, positrons, or ^{3}He atoms. Fermions have the property that any state may contain only zero or one particle, which is the exclusion principle. Jordan and Wigner (1928) discovered that Fermi statistics require that operators anticommute, which is represented by curly brackets:

$$\psi(\mathbf{r})\psi^\dagger(\mathbf{r}') + \psi^\dagger(\mathbf{r}')\psi(\mathbf{r}) \equiv \{\psi(\mathbf{r}), \psi^\dagger(\mathbf{r}')\} = \delta(\mathbf{r}-\mathbf{r}') \tag{9.150}$$

$$\{\psi(\mathbf{r}), \psi(\mathbf{r}')\} = 0 \tag{9.151}$$

$$\{\psi^\dagger(\mathbf{r}), \psi^\dagger(\mathbf{r}')\} = 0 \tag{9.152}$$

Expand these wave functions in a basis set $\phi_\lambda(\mathbf{r})$:

$$\psi(\mathbf{r}) = \sum_\lambda C_\lambda \phi_\lambda(\mathbf{r}) \tag{9.153}$$

$$\psi^\dagger(\mathbf{r}) = \sum_\lambda C_\lambda^\dagger \phi_\lambda^*(\mathbf{r}) \tag{9.154}$$

The coefficients $C_\lambda^\dagger$ and C_λ become creation and destruction operators, which obey anticommutation relations:

$$\{C_\lambda, C_{\lambda'}^\dagger\} = \delta_{\lambda\lambda'} \tag{9.155}$$

$$\{C_\lambda, C_{\lambda'}\} = 0 \tag{9.156}$$

$$\{C_\lambda^\dagger, C_{\lambda'}^\dagger\} = 0 \tag{9.157}$$

An example is $\{C_\lambda, C_\lambda\} = 2C_\lambda C_\lambda = 0$. The operator $C_\lambda C_\lambda$ acting on anything gives zero, since C_λ is a destruction operator, which destroys a particle from a state λ. A state may contain only zero or one particle, called $|0\rangle_\lambda$ and $|1\rangle_\lambda$, respectively:

$$\begin{aligned} &C_\lambda|1\rangle_\lambda = |0\rangle_\lambda, \quad C_\lambda^\dagger|1\rangle_\lambda = 0, \\ &C_\lambda|0\rangle_\lambda = 0, \quad C_\lambda^\dagger|0\rangle_\lambda = |1\rangle_\lambda \end{aligned} \tag{9.158}$$

So $C_\lambda C_\lambda$ acting on either $|1\rangle_\lambda$ or $|0\rangle_\lambda$ gives zero. Similarly, the combination $C_\lambda^\dagger C_\lambda^\dagger = 0$. It is zero because two particles cannot be created in the same state.

The thermodynamic average of the Fermion occupation number is

$$\langle C_\lambda^\dagger C_\lambda \rangle = n_\lambda = \frac{1}{e^{\beta(\varepsilon_\lambda - \mu)} + 1} = n_F(\varepsilon_\lambda) \tag{9.159}$$

There is always a chemical potential for fermions.

9.6 Bose-Einstein Condensation

Earth-bound nature has identified several superfluids. Among those are superconductors, the low temperature phase of liquid ^{4}He, and Bose-Einstein condensation of atoms in optical traps. Superfluids are macroscopic quantum states. The phase of all particles in the superfulid are tied together into one state that extends throughout the fluid. The phase coherence of the superfluid is similar to that found for the photons in a laser. Superfluids are unique environments to study macroscopic quantum phenomena.

The helium atom is found in two isotopes, called ^{4}He and ^{3}He. Both are found in the liquid state at very low temperatures. ^{3}He is a fermion, since it has an odd number of fermions: two electrons, two protons, and one neutron. ^{4}He is regarded as a boson, although it is actually a collection of an even number of fermions: two electrons, two protons, and two neutrons. ^{4}He undergoes a phase transition at $T_\lambda = 2.17$ K, which is believed to be Bose-Einstein condensation.

9.6.1 *Noninteracting Particles*

The easiest theory is for noninteracting particles. We ignore potential energy, and consider only the kinetic energy of the particles.

The traditional view of Bose-Einstein condensation is that there is a macroscopic occupation of the $\mathbf{k} = 0$ state. The number density n of spinless bosons in a state $\mathbf{k}$ is ($\beta = 1/k_B T$):

$$n_B(\mathbf{k}) = \frac{1}{e^{\beta(\varepsilon(k) - \mu)} - 1} \tag{9.160}$$

$$n = \int \frac{d^3k}{(2\pi)^3} n_B(\mathbf{k}) \tag{9.161}$$

$$\varepsilon(k) = \frac{\hbar^2 k^2}{2m} \tag{9.162}$$

where m is the mass of the helium atom. Boson occupation numbers have a chemical potential μ when the particle cannot be destroyed. Photon and phonon occupation numbers lack a chemical potential since one can easily create and destroy them.

At high temperatures, the chemical potential is less than zero. The zero of energy is the minimum value of the excitation energy $\varepsilon(k) = \hbar^2 k^2 / 2m$ at $\mathbf{k} = 0$. The long wavelength excitations of any fluid are sound waves. The density of sound excitations is $\{\exp[\beta\hbar\omega(k)]-1\}^{-1}$. However, in evaluating the number operator n, we want the number of quasiparticles of wave vector $\mathbf{k}$, which is not the same as the density of sound excitations.

Change integration variables to $x=\varepsilon(k)/(k_B T)$, set $y=\beta\mu$, and find

$$n=\frac{1}{4\pi^2}\left(\frac{2mk_BT}{\hbar^2}\right)^{3/2}F(\beta\mu) \tag{9.163}$$

$$F(y)=\int_0^\infty \frac{\sqrt{x}dx}{e^{x-y}-1} \tag{9.164}$$

The left-hand side (n) of (9.163) is a fixed number. As the temperature T becomes lower, the integral on the right becomes smaller due to the factor of $T^{3/2}$. The term $F(y)$ compensates by having y become smaller in magnitude (less negative). Eventually, $y=\beta\mu=0$. This condition defines the Bose-Einstein temperature:

$$k_BT_\lambda=\frac{\hbar^2}{2m}\left(\frac{4\pi^2 n}{F(0)}\right)^{2/3} \tag{9.165}$$

For $T<T_\lambda$, $\mu(T)=0$ and one has to replace eqn. (9.161) by

$$n=n_0+\int\frac{d^3k}{(2\pi)^3}\frac{1}{e^{\beta\varepsilon(k)}-1} \tag{9.166}$$

The quantity $n_0=f_0 n$ is the density of particles in the $\mathbf{k}=0$ state, which is usually represented by the fraction f_0 of all of the helium atoms. The definition of n_0 is

$$n_0=n-\frac{1}{4\pi^2}\left(\frac{2mk_BT}{\hbar^2}\right)^{3/2}F(0) \tag{9.167}$$

$$=n\left[1-\left(\frac{T}{T_\lambda}\right)^{3/2}\right] \tag{9.168}$$

$$f_0(T)=1-\left(\frac{T}{T_\lambda}\right)^{3/2} \tag{9.169}$$

The previous definition of T_λ has been used to write the condensate fraction as a $\frac{3}{2}$ power of the reduced temperature. For the noninteracting boson gas, all atoms are in the condensate at zero temperature [$f_0(0)=1$]. The condensate is the macroscopic quantum state. At $\mathbf{k}=0$ the particle wavelength extends throughout the fluid. In the next section we find a very different behavior for the actual ^{4}He superfluid.

9.6.2 *Off-Diagonal Long-Range Order*

The above derivation of BEC (Bose-Einstein condensation) treats the particles as noninteracting. That model is a poor approximation to ^{4}He. In the actual liquid, each atom has 6–8 near neighbors within its range of potential energy. Also, each atom has a large zero-point kinetic energy due to rapid motion within its small space in liquid.

Theories based on a plane-wave basis set give a poor description of the actual liquid. Much better results were obtained with wave functions based upon correlated basis functions.

Correlated basis functions (CBF) are also called Jastrow functions. The system has $N \sim 10^{23}$ identical bosons, which are spinless. Their ground state Ψ_0 is described by a many-particle wave function:

$$\Psi_0(\mathbf{r}_1, \mathbf{r}_2, \ldots, \mathbf{r}_N) = L_N \exp\left[-\sum_{i>j} u(\mathbf{r}_i - \mathbf{r}_j)\right] \tag{9.170}$$

where L_N is a normalization constant. The exponential has a summation over all pairwise interactions. The function $u(\mathbf{r})$ is designed to keep atoms from getting too close to each other, and has the generic form of $u(r) = (a/r)^n$. The parameters (a, n) are found by a variational solution to the ground-state energy, using large computer programs. This wave function is found to give a good description of particle energy and correlations in liquid helium.

Given this form of the wave function, how do we define the condensate, or zero-momentum state? What is the definition of "zero-momentum state" in a basis set that is not plane waves? What is meant by superfluidity and BEC in a system that is strongly interacting and highly correlated? The method of doing this was introduced by Penrose [3]. Yang [4] suggested the name of *off-diagonal long-range order* (ODLRO) for the type of ordering introduced by Penrose. The actual method of calculating this wave function, or at least some of its properties, is given in Mahan [2]. Here we summarize the main ideas.

Some insight into ODLRO is obtained by considering the techniques used for the weakly interacting systems. There a one-particle state function is defined in the plane-wave representation as

$$\Phi(\mathbf{r}) = \frac{1}{\sqrt{\Omega}} \sum_{\mathbf{k}} e^{i\mathbf{k}\cdot\mathbf{r}} C_{\mathbf{k}} \tag{9.171}$$

$$C_{\mathbf{k}} = \frac{1}{\sqrt{\Omega}} \int d^3 r e^{-i\mathbf{k}\cdot\mathbf{r}} \Phi(\mathbf{r}) \tag{9.172}$$

The number of particles in state $\mathbf{k}$ is

$$n(\mathbf{k}) = \langle C_{\mathbf{k}}^{\dagger} C_{\mathbf{k}} \rangle = \frac{1}{\Omega} \int d^3 r d^3 r' e^{-i\mathbf{k}\cdot(\mathbf{r}-\mathbf{r}')} \langle \Phi^{\dagger}(\mathbf{r}) \Phi(\mathbf{r}') \rangle \tag{9.173}$$

In a fluid, $\langle \Phi^{\dagger}(\mathbf{r}) \Phi(\mathbf{r}') \rangle$ must only be a function of $\mathbf{r} - \mathbf{r}'$. Change the integration to center-of-mass variables, $\vec{\tau} = \mathbf{r} - \mathbf{r}'$, $\vec{R} = (\mathbf{r} + \mathbf{r}')/2$. The integral $\int d^3 R = \Omega$. The integral $\int d^3 \tau$ gives the occupation number:

$$\langle \Phi^{\dagger}(r) \Phi(\mathbf{r}') \rangle = n R(\mathbf{r} - \mathbf{r}') = n R(\tau) \tag{9.174}$$

$$n(\mathbf{k}) = n \int d^3 \tau e^{-i\mathbf{k}\cdot\vec{\tau}} R(\tau) \equiv n \bar{R}(\mathbf{k}) \tag{9.175}$$

The dimensionless quantity $n(\mathbf{k})$ is the Fourier transform of $\langle \Phi^{\dagger}(\mathbf{r}) \Phi(\mathbf{r}') \rangle$ The quantity $n = N/\Omega$ is the density, with dimensional units of particles per volume. If $\mathbf{k}_0$ is the wave vector of the condensate (usually $\mathbf{k}_0 = 0$), then we need to have $n(\mathbf{k}_0) = N f_0$, where N is the total number of atoms. Then $\bar{R}(\mathbf{k}_0) = \Omega f_0$, where Ω is the volume. The only way that the above integral can be proportional to Ω is if

$$\lim_{\mathbf{r}\to\infty} R(\mathbf{r}) = e^{i\mathbf{k}_0\cdot\mathbf{r}} f_0 \tag{9.176}$$

$$n(\mathbf{k}) = N f_0 \delta_{\mathbf{k}=\mathbf{k}_0} + n \int d^3 r e^{-i\mathbf{k}\cdot\mathbf{r}} [R(\mathbf{r}) - e^{i\mathbf{k}_0\cdot\mathbf{r}} f_0] \tag{9.177}$$

The second term has an integral that is not proportional to Ω.

ODLRO is the feature that $R(\infty) \neq 0$. There is correlation in the wave functions of particles at large distances. This correlation is directly related to the condensate fraction f_0. In ^{4}He, it is found that $f_0(T=0) \sim 0.1$. The helium atoms spend only 10% of their time in the condensate.

The Penrose function $R(r)$ is related to the ground-state wave function Ψ_0. The square of the wave function gives the probability density for finding particles at positions $\mathbf{r}_j$ in the system and is called the diagonal density matrix:

$$\rho_N(\mathbf{r}_1, \mathbf{r}_2, \ldots, \mathbf{r}_N) = |\Psi_0(\mathbf{r}_1, \mathbf{r}_2, \ldots, \mathbf{r}_N)|^2 \tag{9.178}$$

The subscript N indicates that it applies to N particles. ρ_N is normalized so that the integral over all coordinates gives unity:

$$1 = \int d^3 r_1 \cdots d^3 r_N \rho_N(\mathbf{r}_1, \mathbf{r}_2, \ldots, \mathbf{r}_N) \tag{9.179}$$

The concept of ODLRO is contained in the function $R(\mathbf{r}_1, \mathbf{r}_1')$ obtained from $|\Psi_0|^2$ when all but one set of coordinates are equal and averaged over

$$R(\mathbf{r}_1 - \mathbf{r}_1') = \Omega \int d^3 r_2 d^3 r_3 \cdots d^3 r_N \Psi_0^*(\mathbf{r}_1, \mathbf{r}_2, \ldots, \mathbf{r}_N) \Psi_0(\mathbf{r}_1', \mathbf{r}_2, \mathbf{r}_3, \ldots, \mathbf{r}_N)$$

$$R(0) = 1 \tag{9.180}$$

The quantity $f_0 = R(\infty)$ is the fraction of time a particle spends in the condensate. Alternately, it is also the fraction of particles in this state at any one time.

References

1. P. Jordan and E. Wigner, *Z. Phys.* **47**, 631 (1951)
2. G.D. Mahan, *Many-Particle Physics*, 3rd ed. (Plenum-Kluwer, New York, 2000)
3. O. Penrose, *Philos. Mag.* **42**, 1373 (1951)
4. C.N. Yang, *Rev. Mod. Phys.*, **34**, 694 (1962)

Homework

1. Use Hund's rules to make a table of how *d*-shells ($\ell = 2$) fill up with electrons. Compare your results to the 3d series of atoms: Sc, Ti, ..., Zn.

2. Consider a system of three electrons: e.g., atomic lithium. Two have spin-up and one has spin down. The two spin-up electrons are in orbital states labeled 1 and 2, while the spin-down electron is in orbital state 3.

a. Write down the Slater determinant Ψ for this three-electron state.
b. Form the density matrix $\rho_{123}(\mathbf{r}_1, \mathbf{r}_2, \mathbf{r}_3) = |\Psi|^2$ and integrate it over d^3r_3.
c. Integrate the result of (b) over d^3r_2. What is the resulting function of $\mathbf{r}_1$?

3. Write out the Hartree-Fock equations for the electrons in the ground state of atomic Li: $(1s)^2(2s)$.

4. Write out the Hartree-Fock equations for the electrons in the ground state of atomic Be: $(1s)^2(2s)^2$.

5. Assume a Hamiltonian of the form below for a system of N bosons of spin-0:

$$H = \sum_i \left[\frac{p_i^2}{2m} + U(\mathbf{r}_i)\right] + \sum_{i>j} V(\mathbf{r}_i - \mathbf{r}_j) \tag{9.181}$$

a. Derive an expression for the ground-state energy assuming all of the orbital states are different for the bosons.
b. Derive an expression for the ground-state energy assuming all of the orbital states are identical—all of the bosons are in $\phi_1(\mathbf{r})$.

6. Three negative pions (spin-0) are bound to a lithium ($Z = 3$) nuclei. Use variational methods to estimate the ground-state binding energy.

7. Consider a system of N free electrons in three dimensions that interact with each other only through the delta-function potentials defined below. Define the interaction energy E_I as the sum of Hartree and exchange. Assume a paramagnetic state with equal number of spin-up and spin-down electrons. Solve for the two interactions below:

a. $V(\mathbf{r}_1 - \mathbf{r}_2) = I\delta^3(\mathbf{r}_1 - \mathbf{r}_2)$
b. $V(\mathbf{r}_1 - \mathbf{r}_2) = I(\vec{s}_1 \cdot \vec{s}_2)\delta^3(\mathbf{r}_1 - \mathbf{r}_2)$

8. Consider a ferromagnetic electron gas that has N free electrons in three dimensions with all spins pointing in the same direction, say "up."

a. Calculate the average kinetic energy per electron.
b. Calculate $\rho(\mathbf{r}_1, \mathbf{r}_2) = g(\mathbf{r}_1 - \mathbf{r}_2)/(\Omega^2)$ by summing over all pairs of electrons states.

9. Consider the two-particle wave function, where the two particles are in a d-state of an atom that has five orbital choices. How many states are available when:

a. The spin-0 particles are distinguishable?
b. The particles are spin-0 bosons?
c. The particles are electrons?

10. Consider the two-dimensional paramagnetic electron gas.

a. Derive the relationship between n (number of electrons per area) and k_F.
b. What is the average kinetic energy of each electron in the Fermi sea of the paramagnetic state (equal number of up and down spins).

11. In two dimensions, what is the Thomas-Fermi relation between $n(r)$ and $V(r)$?

12. A possible Thomas-Fermi equation is

$$\frac{d^2V(x)}{dx^2} = 4\pi n_0 e^2 \left\{ 1 + b\delta(x) - \left[1 - \frac{V(x)}{\mu} \right]^{3/2} \right\} \tag{9.182}$$

This equation applies in three dimensions if the term $\delta(x)$ is a sheet of charge. Solve this equation assuming $V/\mu << 1$.

13. Derive the formula for the Bose-Einstein transition temperature T_λ of a gas of free bosons ($\varepsilon(k) = \hbar^2k^2/2m$) in two dimensions. All integrals can be done exactly.

14. Consider a gas of boson atoms that has only a repulsive interaction $V(r_{ij})$ between atoms:

$$V(r_{ij}) = Ae^{-r_{ij}/b}, A >> \frac{\hbar^2k^2}{2m} \tag{9.183}$$

When using a correlated basis function for the ground-state wave function of the many atom gas, what is the best choice for the pair function $u(r_{ij})$?

15. Consider a one-dimensional system of two hard-sphere bosons confined to the interval $0 < x_i < 1$. By "hard sphere" is meant the wave function $\psi(x_1, x_2) = 0$ if $x_1 = x_2$. Also, the wave function must vanish if either particle gets to either end.

a. Show that the following is a suitable wave function

$$f(x_1, x_2) = \sin\left(\pi\frac{x_1}{x_2}\right)\sin\left(\pi\frac{1-x_2}{1-x_1}\right) \tag{9.184}$$

$$\psi(x_1, x_2) = \begin{cases} Af(x_1, x_2) & x_1 < x_2 \\ Af(x_2, x_1) & x_1 > x_2 \end{cases} \tag{9.185}$$

where A is a normalization constant.

b. Write down the expression for $\rho(x_1)$ obtained by integrating over dx_2. Specify all limits of integration.
c. Write down a similar wave function $\psi(x_1, x_2, x_3)$ for three hard-sphere bosons.

16. Write down the expression for the Penrose function $R(x_1 - x_1')$ for the above wave function with two hard bosons. Specify all of the limits of integration.

17. Calculate the Penrose function $R(\mathbf{r}_1 - \mathbf{r}_1')$ for a free *electron* gas at zero temperature. It will be a function of $x \equiv k_F|\mathbf{r}_1 - \mathbf{r}_1'|$. How does the answer differ for a paramagnetic gas compared to a ferromagnetic gas?

18. Since $S(q = 0) = 0$ there is an identity

$$1 = n_0 \int d^3r[1 - g(r)] \tag{9.186}$$

Someone proposes the following form for $g(r)$ based on the hard-core concept:

$$g(r) = \begin{cases} 0 & r < a \\ 2 & a < r < 2a \\ 1 & r > 2a \end{cases} \tag{9.187}$$

Does this form satisfy the above identity?

19. Assume a pair distribution function of the form

$$g(R) = 1 - e^{-\alpha R} \tag{9.188}$$

Adjust the value of α by using the sum rule of the problem 18. Then calculate $S(q)$.

10 Scattering Theory

The previous chapters have discussed scattering theory for electrons and photons. Usually the matrix elements were calculated in the first Born approximation, which is the Fourier transform of the potential energy. This chapter discusses scattering theory with more rigor. The exact matrix element will be obtained for a variety of systems.

The interaction of photons with charged particles is described well by perturbation theory. Perturbation theory for photons has a smallness parameter that is the fine-structure constant $\alpha_f \approx 1/137$. Each order of perturbation theory is smaller by this parameter. So the calculations in chapter 8 for photons are fine. In the present chapter the emphasis is scattering by particles such as electrons, neutrons, and protons.

10.1 Elastic Scattering

Elementary particles can be treated as waves or as particles. They are particles whose amplitudes obey a wave equation. It is useful to think about the scattering in terms of waves. Assume there is a target at the origin ($\mathbf{r} = 0$), which has a potential energy $V(\mathbf{r})$ with the particle being scattered. Represent the incoming particle as a wave with the initial wave function of

$$\phi_i(\mathbf{r}) = e^{i\mathbf{k}\cdot\mathbf{r}} \tag{10.1}$$

We are not including the factors for normalization. When this wave gets to the target, some of the wave amplitude is scattered in the form of a spherical outgoing wave. See figure 7.8b for a diagram.

Construct a set of equations that gives the same picture. The Hamiltonian is

$$H\psi = \left[-\frac{\hbar^2\nabla^2}{2m} + V(\mathbf{r})\right]\psi = \frac{\hbar^2 k^2}{2m}\psi \tag{10.2}$$

Rearrange this equation into the form

$$(\nabla^2 + k^2)\psi(\mathbf{r}) = \frac{2m}{\hbar^2} V(\mathbf{r})\psi(\mathbf{r}) \tag{10.3}$$

Introduce the classical *Green's function* $G(k; \mathbf{r}, \mathbf{r}')$ that obeys the equation

$$(\nabla^2 + k^2)G(k; \mathbf{r}, \mathbf{r}') = \frac{2m}{\hbar^2}\delta^3(\mathbf{r}-\mathbf{r}') \tag{10.4}$$

On the right is a three-dimensional delta-function. The Green's function provides a solution to (10.3):

$$\psi(\mathbf{r}) = e^{i\mathbf{k}\cdot\mathbf{r}} + \int d^3r' G(k; \mathbf{r}, \mathbf{r}')V(\mathbf{r}')\psi(\mathbf{r}') \tag{10.5}$$

$$(\nabla^2 + k^2)\psi(\mathbf{r}) = (\nabla^2 + k^2)e^{i\mathbf{k}\cdot\mathbf{r}} + \frac{2m}{\hbar^2}\int d^3r'\delta^3(\mathbf{r}-\mathbf{r}')V(\mathbf{r}')\psi(\mathbf{r}')$$

$$= \frac{2m}{\hbar^2} V(\mathbf{r})\psi(\mathbf{r}) \tag{10.6}$$

The first term on the right vanishes since $\nabla^2 e^{i\mathbf{k}\cdot\mathbf{r}} = -k^2 e^{i\mathbf{k}\cdot\mathbf{r}}$.

Equation (10.5) is an exact solution to the Hamiltonian equation (10.3). It is a self-consistent equation, in that the solution $\psi(\mathbf{r})$ depends on itself under the integral. The next step is to solve for the Green's function $G(k; \mathbf{r}, \mathbf{r}')$.

The right-hand-side of eqn. (10.4) is a function of only $\mathbf{R} = \mathbf{r} - \mathbf{r}'$, and so is the Green's function: $G(k; \mathbf{R})$. Take a Fourier transform of (10.4):

$$\tilde{G}(k; \mathbf{q}) = \int d^3 R e^{-i\mathbf{q}\cdot\mathbf{R}} G(k; \mathbf{R}) \tag{10.7}$$

$$\tilde{G}(k; \mathbf{q}) = -\frac{2m}{\hbar^2}\frac{1}{q^2-k^2} \tag{10.8}$$

$$G(k; \mathbf{R}) = \int \frac{d^3q}{(2\pi)^3} e^{i\mathbf{q}\cdot\mathbf{R}}\tilde{G}(k, \mathbf{q}) = -\frac{2m}{\hbar^2}\int \frac{d^3q}{(2\pi)^3} e^{i\mathbf{q}\cdot\mathbf{R}}\frac{1}{q^2-k^2}$$

The second equation is the Fourier transform of eqn. (10.4). The third equation is the definition of the inverse transform. Evaluate the integral and the calculation is over. First evaluate the angular integrals. Choose the direction of the vector $\mathbf{R}$ as the $\hat{z}$-direction, and then $\mathbf{q}\cdot\mathbf{R} = qR\nu$, $\nu = \cos(\theta)$. The angular integrals give

$$\int d\phi \int d\theta \sin(\theta) e^{iqR\cos\theta} = 2\pi \int_{-1}^{1} d\nu e^{iqR\nu} = \frac{2\pi}{iqR}(e^{iqR} - e^{-iqR}) \tag{10.9}$$

The integral over dq goes $0 < q < \infty$. In the term $\exp(-iqR)$ change $q' = -q$ and this integral goes $-\infty < q' < 0$. There is a sign change, so one gets

$$G(k; \mathbf{R}) = \frac{-m}{2\pi^2 iR\hbar^2}\int_{-\infty}^{\infty} qdq \frac{e^{iqR}}{q^2-k^2} \tag{10.10}$$

The integral over dq goes along the real axis. At the points where $q = \pm k$, the integrand diverges. The result for the Green's function depends on what is done at these points.

- If the integrand is treated as a principal part, the integral gives a standing wave:

$$G(k;\mathbf{R}) = -\frac{m}{2\pi R\hbar^2}\cos(kR) \tag{10.11}$$

- For the scattering problem, the Green's function should be an outgoing wave, since it describes a wave leaving the target. The outgoing wave is achieved by replacing the factor of $q^2 - k^2 \to q^2 - (k+i\eta)^2$. Here η is positive and infinitesimal. The poles of the integrand are now at $q = \pm(k+i\eta)$. The one at positive k is in the upper half-plane (UHP), and the one at $-k$ is in the lower half-plane (LHP). Treat q as a complex variable and close the contour in the UHP with a semicircle of infinite radius. According to Cauchy's theorem, the integral equals $2\pi i$ times the residue of the pole at $q = k + i\eta$:

$$G(k,\mathbf{R}) = -\frac{m}{2\pi\hbar^2}\frac{e^{ikR}}{R} \tag{10.12}$$

  This Green's function is used in scattering theory.

- Another possible choice is to use incoming waves. That is achieved by replacing $(k+i\eta)^2$ by $(k-i\eta)^2$. Now the pole in the UHP is at $q = -k + i\eta$ and the Green's function is

$$G(k,\mathbf{R}) = -\frac{m}{2\pi\hbar^2}\frac{e^{-ikR}}{R} \tag{10.13}$$

All three Green's functions are correct for some problem. For the scattering problem with its outgoing wave the correct choice is (10.12).

Let $\mathbf{k}_i$ be the initial wave vector, and equation (10.5) is now

$$\psi(\mathbf{k}_i,\mathbf{r}) = e^{i\mathbf{k}_i\cdot\mathbf{r}} - \frac{m}{2\pi\hbar^2}\int\frac{d^3r'}{|\mathbf{r}-\mathbf{r}'|}e^{ik|\mathbf{r}-\mathbf{r}'|}V(\mathbf{r}')\psi(\mathbf{k}_i,\mathbf{r}') \tag{10.14}$$

The potential energy $V(\mathbf{r})$ is assumed to be of short range—it is not a Coulomb potential. The integral over d^3r' extends only over the range of $V(\mathbf{r}')$, which is $O(\text{Å})$. If measurements of the scattering particle are being made a distance $r \sim$ meter from the target, then $r >> r'$. Expand the Green's function in this limit:

$$\lim_{r>>r'}|\mathbf{r}-\mathbf{r}'| = \lim_{r>>r'}\sqrt{r^2+(r')^2-2\mathbf{r}\cdot\mathbf{r}'} = r - \hat{r}\cdot\mathbf{r}' + O(r'^2/r)$$

$$\frac{\exp[ik|\mathbf{r}-\mathbf{r}'|]}{|\mathbf{r}-\mathbf{r}'|} = \frac{\exp[ikr - i\mathbf{k}_f\cdot\mathbf{r}']}{r}[1+O(r'/r)]$$

$$\psi(\mathbf{r}) = e^{i\mathbf{k}_i\cdot\mathbf{r}} + f(\mathbf{k}_f,\mathbf{k}_i)\frac{e^{ikr}}{r} \tag{10.15}$$

$$f(\mathbf{k}_f,\mathbf{k}_i) = -\frac{m}{2\pi\hbar^2}\int d^3r' e^{-i\mathbf{k}_f\cdot\mathbf{r}'}V(\mathbf{r}')\psi(\mathbf{k}_i,\mathbf{r}')$$

$$\mathbf{k}_f \equiv k\hat{r},\quad v_f = \frac{\hbar k_f}{m} \tag{10.16}$$

The wave vector $\mathbf{k}_f$ for the final state is the magnitude of k taken in the direction of $\hat{r}$. The quantity $f(\mathbf{k}_f, \mathbf{k}_i)$ is the scattering amplitude. It has the dimensions of length.

The flux of scattered particles is found from the current operator:

$$\vec{j}(\mathbf{r}) = \frac{\hbar}{2mi}[\psi^*\vec{\nabla}\psi - \psi\vec{\nabla}\psi^*] \tag{10.17}$$

$$\vec{\nabla}\left[\frac{e^{ikr}}{r}\right]=\hat{r}\left(ik-\frac{1}{r}\right)\frac{e^{ikr}}{r} \tag{10.18}$$

$$\vec{j}(\mathbf{r})=\frac{\vec{v}_f}{r^2}|f(\mathbf{k}_f,\mathbf{k}_i)|^2[1+O(1/r)] \tag{10.19}$$

At the point $\mathbf{r}$, where the measurement is done, the flux through the area $dA=r^2d\Omega$ is $r^2jd\Omega$. The scattered flux is divided by the incident flux. The incident flux comes from the term $e^{i\mathbf{k}i\cdot\mathbf{r}}$ and is v_i. The ratio of these two fluxes gives the scattering cross section:

$$\frac{d\sigma}{d\Omega}=\frac{v_f}{v_i}|f(\mathbf{k}_f,\mathbf{k}_i)|^2=\frac{v_f}{v_i}\left(\frac{m}{2\pi\hbar^2}\right)^2|T(\mathbf{k}_f,\mathbf{k}_i)|^2 \tag{10.20}$$

$$T(\mathbf{k}_f,\mathbf{k}_i)=\int d^3r' e^{-i\mathbf{k}_f\cdot\mathbf{r}'}V(\mathbf{r}')\psi(\mathbf{k}_i,\mathbf{r}') \tag{10.21}$$

$$f(\mathbf{k}_f,\mathbf{k}_i)=\frac{m}{2\pi\hbar^2}T(\mathbf{k}_f,\mathbf{k}_i) \tag{10.22}$$

For elastic scattering, $v_i=v_f$ and the initial prefactor is omitted.

The last two lines introduce the *T-matrix* for scattering. The *T*-matrix is the exact matrix element. Our derivation contained no approximation. The only assumption we made was that the measurement apparatus was far from the center of scattering. The *T*-matrix is a matrix element of the potential $V(\mathbf{r})$. One wave function is the exact one $\psi(\mathbf{k}_i, \mathbf{r})$, while the other is a plane wave $\exp(-i\mathbf{k}_f\cdot\mathbf{r})$. In an earlier chapter we derived the Born approximation to the *T*-matrix, where $\psi(\mathbf{k}_i, \mathbf{r})$ was approximated as a plane wave $\exp(i\mathbf{k}_i\cdot\mathbf{r})$ and the *T*-matrix is approximated as the Fourier transform of the potential $T(\mathbf{k}_f, \mathbf{k}_i)\rightarrow \tilde{V}(\mathbf{k}_i-\mathbf{k}_f)$. A guess at the correct answer might have been

$$\int d^3r\psi^*(\mathbf{k}_f,\mathbf{r})V(\mathbf{r})\psi(\mathbf{k}_i,\mathbf{r}) \tag{10.23}$$

in which an exact eigenstate is on both sides of the potential. This guess is reasonable but incorrect. The correct result (10.21) has one side with an exact eigenfunction and the other side with a plane wave.

10.1.1 *Partial Wave Analysis*

In this section the scattering potential $V(\mathbf{r})$ is assumed to be spherically symmetric: it depends only on the magnitude of vector $\mathbf{r}$, and is written $V(r)$. A plane wave can be expanded as

$$e^{i\mathbf{k}\cdot\mathbf{r}}=\sum_{\ell=0}^{\infty}(2\ell+1)i^\ell j_\ell(kr)P_\ell(\cos\theta) \tag{10.24}$$

where $j_\ell(z)$ is a spherical Bessel function, $P_\ell(\cos\theta)$ is a Legendre polynomial, and $\cos(\theta)=\hat{k}\cdot\hat{r}$. In a similar way, assume that the exact eigenfunction can be expanded as

$$\psi(\mathbf{k},\mathbf{r})=\sum_{\ell=0}^{\infty}(2\ell+1)i^\ell R_\ell(kr)P_\ell(\cos\theta) \tag{10.25}$$

This form is valid for spherically symmetric potentials. If the potential $V(r)$ is short-ranged, we can express the radial function in terms of phase shifts when $kr >> 1$:

$$\lim_{kr\to\infty} j_\ell(kr) = \frac{\sin(kr - \pi\ell/2)}{kr} \tag{10.26}$$

$$\lim_{kr\to\infty} R_\ell(kr) = \frac{\sin[kr + \delta_\ell(k) - \pi\ell/2]}{kr} \tag{10.27}$$

Examples of phase shifts were given in previous chapters. The form for $R_\ell(kr)$ needs to be changed to be appropriate for an outgoing wave. For example, the Bessel function is

$$\lim_{kr>>1} j_\ell(kr) = \frac{1}{2ikr}\left(e^{i(kr-\pi\ell/2)} - e^{-i(kr-\pi\ell/2)}\right) \tag{10.28}$$

The first term on the right has the exponent of an outgoing wave. The second exponent is an incoming wave. In a scattering experiment, the incoming wave has not yet experienced the potential and is not phase shifted. The outgoing wave is phase-shifted twice: once on the incoming path and once on the outgoing path. Write the actual wave function, at large distances, as

$$\lim_{kr>>1} R_\ell(kr) = \frac{1}{2ikr}(e^{2i\delta_\ell}e^{i(kr-\pi\ell/2)} - e^{-i(kr-\pi\ell/2)}) \tag{10.29}$$

$$= \frac{e^{i\delta_\ell}}{kr}\sin[kr + \delta_\ell(k) - \frac{\pi\ell}{2}] \tag{10.30}$$

Outgoing wave boundary conditions have added a phase factor to the radial function. Equation (10.29) can be rewritten exactly as

$$\lim_{kr>>1} R_\ell(kr) = \frac{\sin(kr - \pi\ell/2)}{kr} + \frac{(e^{2i\delta_\ell} - 1)}{2ikr}e^{i(kr-\pi\ell/2)} \tag{10.31}$$

Put this form in eqn. (10.25). The first term sums to the incoming wave, and the second term is the scattered wave:

$$\psi(\mathbf{k}, \mathbf{r}) = e^{i\mathbf{k}\cdot\mathbf{r}} + f\frac{e^{ikr}}{r} \tag{10.32}$$

$$f(k, \theta) = \frac{1}{2ik}\sum_\ell (2\ell+1)(e^{2i\delta_\ell(k)} - 1)P_\ell(\cos\theta) \tag{10.33}$$

$$= \frac{1}{k}\sum_\ell (2\ell+1)e^{i\delta_\ell(k)}\sin[\delta_\ell(k)]P_\ell(\cos\theta) \tag{10.34}$$

$$T(k, \theta) = \frac{2\pi\hbar^2}{m}f(k, \theta) = \frac{\pi\hbar^2}{ikm}\sum_\ell (2\ell+1)(e^{2i\delta_\ell(k)} - 1)P_\ell(\cos\theta)$$

where now $\cos(\theta) = \hat{k}_f \cdot \hat{k}_i$ since $\hat{r} = \hat{k}_f$. The last formula is exact for elastic scattering. The scattering amplitude f is expressed as a function of k and $\cos\theta = \hat{k}_i \cdot \hat{k}_f$. The phase shifts play a major role in the scattering amplitude.

The differential cross section for elastic scattering is

$$\frac{d\sigma}{d\Omega} = |f|^2 = \frac{1}{4k^2}\left|\sum_\ell (2\ell+1)(e^{2i\delta_\ell(k)} - 1)P_\ell(\cos\theta)\right|^2 \tag{10.35}$$

The total cross section is obtained by integrating over solid angle $d\Omega = d\phi\ \sin\theta d\theta$. Since the Legendre polynomials are orthogonal, this integral forces $\ell = \ell'$ in the double summation:

$$\sigma(k) = \frac{2\pi}{4k^2}\sum_{\ell\ell'}(2\ell+1)(2\ell'+1)(e^{2i\delta}-1)(e^{-2i\delta'}-1)I_{\ell\ell'} \tag{10.36}$$

$$I_{\ell\ell'} = \int_0^\pi d\theta \sin\theta P_\ell(\cos\theta)P_{\ell'}(\cos\theta) = \delta_{\ell\ell'}\frac{2}{2\ell+1} \tag{10.37}$$

$$\sigma(k) = \frac{4\pi}{k^2}\sum_{\ell=0}^{\infty}(2\ell+1)\sin^2[\delta_\ell(k)] \tag{10.38}$$

The final formula gives the total cross section as a function of the scattering phase shifts. This formula is exact for elastic scattering from potentials that are not Coulombic.

We can immediately derive the *optical theorem*. The imaginary part of the scattering amplitude is

$$\Im\{f(k,\theta)\} = \frac{1}{k}\sum_\ell(2\ell+1)\sin^2[\delta_\ell(k)]P_\ell(\cos\theta) \tag{10.39}$$

At $\theta = 0$ then $P_\ell(\cos 0) = P_\ell(1) = 1$ and the above summation becomes

$$\sigma(k) = \frac{4\pi}{k}\Im\{f(k,0)\} \tag{10.40}$$

which is the optical theorem. The imaginary part of the forward scattering amplitude is proportional to the total cross section.

Another interesting result is obtained by equating two expressions for the scattering amplitude f. First start with

$$f(\mathbf{k}_f,\mathbf{k}_i) = \frac{m}{2\pi\hbar^2}\int d^3r' e^{-i\mathbf{k}_f\cdot\mathbf{r}'}V(\mathbf{r}')\psi(\mathbf{k}_i,\mathbf{r}') \tag{10.41}$$

Expand two factors of the integrand in Legendre functions and do the integral over angle

$$e^{-i\mathbf{k}_f\cdot\mathbf{r}'} = \sum_{\ell'=0}^{\infty}(2\ell'+1)(-i)^{\ell'}j_{\ell'}(kr')P_{\ell'}(\hat{k}_f\cdot\hat{r}') \tag{10.42}$$

$$\psi(\mathbf{k}_i,\mathbf{r}') = \sum_{\ell=0}^{\infty}(2\ell+1)i^\ell R_\ell(kr')P_\ell(\hat{k}_i\cdot\hat{r}') \tag{10.43}$$

$$\frac{4\pi}{2\ell+1}\delta_{\ell\ell'}P_\ell(\hat{k}_i\cdot\hat{k}_f) = \int_0^{2\pi}d\phi\int_0^\pi d\theta'\sin(\theta')P_{\ell'}(\hat{k}_f\cdot\hat{r}')P_\ell(\hat{k}_i\cdot\hat{r}') \tag{10.44}$$

$$f(\mathbf{k}_f,\mathbf{k}_i) = -\frac{4\pi m}{2\pi\hbar^2}\sum_\ell(2\ell+1)P_\ell(\hat{k}_i\cdot\hat{k}_f)\int_0^\infty r'^2dr'j_\ell(kr')V(r')R_\ell(kr')$$

Equation (10.44) is related to the addition theorem for Legendre functions. Compare this last formula with eqn. (10.33) and deduce that

$$e^{i\delta_\ell}\sin[\delta_\ell(k)] = -\frac{2mk}{\hbar^2}\int_0^\infty r'^2dr'j_\ell(kr')V(r')R_\ell(kr') \tag{10.45}$$

This expression is exact for elastic scattering from potentials of short range. The integral on the right appears complicated, but it is given by the simple expression on the left of the equal sign. The phase factor $\exp[i\delta_\ell]$ is in $R_\ell(kr')$.

The above formula can be used to derive the behavior of the phase shift at small values of wave vector. As $k \to 0$, then $j_\ell(kr) \to (kr)^\ell$. Since the potential has a short range, only small values of r are needed. One finds as $k \to 0$ that

$$\sin(\delta_\ell) \to -k^{\ell+1} a_\ell \tag{10.46}$$

$$a_\ell = \frac{2^\ell \ell!}{(2\ell+1)!} \frac{2m}{\hbar^2} R_\ell(0) \int_0^\infty (r')^{\ell+2} dr' V(r') \tag{10.47}$$

The constant a_ℓ is called *the scattering length*, and is an important experimental quantity. For *s*-wave scattering ($\ell = 0$), one finds for the cross section at low energy:

$$\lim_{k \to 0} \frac{d\sigma}{d\Omega} = a_0^2, \ \lim_{k \to 0} \sigma = 4\pi a_0^2 \tag{10.48}$$

The differential cross section is a constant area given by a_0^2.

The other important information is the value of the phase shift at $k = 0$. *Levinson's theorem* states that

$$\delta_\ell(0) = N_\ell \pi \tag{10.49}$$

where N_ℓ is the number of bound states in the potential with angular momentum ℓ. This number is finite for a potential of short range.

The definition of the phase shift usually has the form $\tan(\delta_\ell) = f(k)$. If $f(k = 0) = 0$, the tangent of the phase shift is zero. Since $\tan(N\pi) = 0$, how do we know the value of N? The unique answer is obtained by graphing the phase shift as a function of k. We demand that this graph has two characteristics:

- The phase shift vanishes at very large values of k.
- The phase shift is a continuous function of k.

These two conditions guarantee that one knows the unique value of the phase shift at $k = 0$.

10.1.2 *Scattering in Two Dimensions*

The same approach to elastic scattering can be applied to two dimensions. The particles are confined to a plane, and polar coordinates are used. The wave function has an incident term plus a term scattering outward from a potential, as in eqn. (10.5):

$$\psi(\mathbf{k}, \mathbf{r}) = e^{i\mathbf{k}\cdot\mathbf{r}} + \int d^2r' G(k; \mathbf{r} - \mathbf{r}') V(\mathbf{r}') \psi(\mathbf{k}, \mathbf{r}') \tag{10.50}$$

$$[\nabla^2 + k^2] G(k, R) = \frac{2m}{\hbar^2} \delta^2(\mathbf{R}) \tag{10.51}$$

The important step is to find the Green's function in two dimensions. For three dimensions, the Green's function was found using a Fourier transform method. Here we use another method to provide variety.

At $R \neq 0$ the Green's function obeys the wave equation for a free particle. The solutions are Bessel functions $J_n(kR)$, $Y_n(kR)$. Both are needed since the solution is away from the origin. The combination of these two functions that forms an outgoing wave is called a Hankel function:

$$H_n^{(1)}(z) = J_n(z) + iY_n(z) \tag{10.52}$$

$$H_n^{(2)}(z) = J_n(z) - iY_n(z) \tag{10.53}$$

$H_n^{(1)}(z)$ is an outgoing wave, while $H_n^{(2)}(z)$ is an incoming wave. We need the circularly symmetric solution $n=0$, so just guess the solution is

$$G(k, R) = CH_0^{(1)}(kR) \tag{10.54}$$

The only remaining step is to find the constant C. It is determined by the delta-function at the origin. First integrate eqn. (10.51) over $d\theta$. This step gives 2π on the left and eliminates the angular term in the delta-function on the right:

$$2\pi\left[\frac{1}{R}\frac{d}{dR}R\frac{d}{dR} + k^2\right]CH_0^{(1)}(kR) = \frac{2m}{\hbar^2}\frac{\delta(R)}{R} \tag{10.55}$$

Multiply the entire equation by $R/2\pi$ and then integrate each term by

$$\lim_{\varepsilon\to 0}\int_0^{\varepsilon} dR \tag{10.56}$$

On the right, the integral gives unity, independent of ε. On the left we get

$$\lim_{\varepsilon\to 0}\varepsilon\frac{d}{d\varepsilon}CH_0^{(1)}(k\varepsilon) = \frac{m}{\pi\hbar^2} \tag{10.57}$$

The Bessel function $J_0(z) = 1 + O(z^2)$ and the operation by

$$\lim_{\varepsilon\to 0}\varepsilon\frac{d}{d\varepsilon}J_0(k\varepsilon) = 0 \tag{10.58}$$

The Neumann function at small argument goes as $Y_0(z) = (2/\pi)\ln(z/2)$ and its derivative is

$$\lim_{\varepsilon\to 0}\varepsilon\frac{d}{d\varepsilon}Y_0(k\varepsilon) = \frac{2}{\pi} \tag{10.59}$$

$$iC\frac{2}{\pi} = \frac{m}{\pi\hbar^2} \tag{10.60}$$

$$C = -i\frac{m}{2\hbar^2} \tag{10.61}$$

$$G(k, R) = -i\frac{m}{2\hbar^2}H_0^{(1)}(kR) \tag{10.62}$$

The scattering part of the wave is found from the asymptotic form of the Hankel function:

$$\lim_{z>>1} H_0^{(1)}(z) = \sqrt{\frac{2}{i\pi z}} e^{iz} \tag{10.63}$$

$$\psi(\mathbf{k},\mathbf{r}) = e^{i\mathbf{k}\cdot\mathbf{r}} + f(\mathbf{k}_f,\mathbf{k}_i)\frac{e^{ikr}}{\sqrt{r}} \tag{10.64}$$

$$f = -\frac{m}{\hbar^2}\sqrt{\frac{i}{2\pi k}}\int d^2r' e^{-i\mathbf{k}_f\cdot\mathbf{r}'} V(r')\psi(\mathbf{k}_i,\mathbf{r}') \tag{10.65}$$

$$\frac{d\sigma}{d\theta} = |f(\mathbf{k}_f,\mathbf{k}_i)|^2 \tag{10.66}$$

The scattering amplitude f has the units of (length)$^{1/2}$, and the cross section in two dimensions has the units of length.

Another expression for the scattering amplitude is found by a partial wave analysis:

$$e^{i\mathbf{k}\cdot\mathbf{r}} = \sum_n i^n e^{in\theta} J_n(kr) \tag{10.67}$$

$$\psi(\mathbf{k}_i,\mathbf{r}) = \sum_n i^n e^{in\theta} R_n(kr) \tag{10.68}$$

At large values of (kr) the asymptotic forms are

$$\lim_{z>>1} J_n(z) = \sqrt{\frac{2}{\pi z}}\cos\left(z - \frac{n\pi}{2} - \frac{\pi}{4}\right) \tag{10.69}$$

$$\lim_{z>>1} R_n(z) = \sqrt{\frac{2}{\pi z}} e^{i\delta_n}\cos\left[z + \delta_n(k) - \frac{n\pi}{2} - \frac{\pi}{4}\right] \tag{10.70}$$

$$\lim_{z>>1} R_n(z) = \sqrt{\frac{2}{\pi z}}\left[\cos\left(z - \frac{n\pi}{2} - \frac{\pi}{4}\right) + \frac{(e^{2i\delta_n}-1)}{2} e^{i(z-n\pi/2-\pi/4)}\right]$$

Put the latter result in eqn. (10.68) and get back eqn. (10.64) with a new expression for the scattering amplitude:

$$f(\mathbf{k}_f,\mathbf{k}_i) = \frac{1}{\sqrt{2\pi i k}}\sum_n e^{in\theta}(e^{2i\delta_n}-1) \tag{10.71}$$

$$\sigma_T(k) = \frac{4}{k}\sum_{n=-\infty}^{\infty}\sin^2[\delta_n(k)] \tag{10.72}$$

The final formula is the total cross section for elastic scattering in two dimensions.

10.1.3 *Hard-Sphere Scattering*

Consider the scattering of two classical hard spheres, say, two bowling balls or two cue balls in pool. Both have the same diameter d and radius $a = d/2$. The scattering is elastic and there are no bound states. All that is needed for the scattering theory are the phase shifts. They can be found by the solution of the two-sphere Hamiltonian in relative coordinates:

$$H = -\frac{\hbar^2\nabla^2}{2\mu} + V(r),\ V(r) = \infty \text{ if } r < d \tag{10.73}$$

where the reduced mass is $\mu = m/2$. The potential is zero if $r > d$ and is infinite if $r < d$. This is equivalent to setting the radial function equal to zero at $r = d$: $R_\ell(kd) = 0$. Since the potential vanishes for $r > d$, the solution in this region are plane waves. Since solutions are only needed away from the origin, the most general solution has both Bessel functions, $j_\ell(kr)$ and $y_\ell(kr)$:

$$R_\ell(kr) = \cos(\delta_\ell)e^{i\delta_\ell}\{j_\ell(kr) - \tan[\delta_\ell(k)]y_\ell(kr)\} \tag{10.74}$$

$$\tan[\delta_\ell(k)] = \frac{j_\ell(kd)}{y_\ell(kd)} \tag{10.75}$$

The second line is the definition of the phase shift. This choice does fulfill the boundary condition that $R_\ell(kd) = 0$. Also, using the asymptotic forms for the two spherical Bessel functions gives the right behavior for the radial function:

$$\lim_{z>>1} j_\ell(z) = \frac{\sin(z - \pi\ell/2)}{z}\left[1 + O\left(\frac{1}{z}\right)\right] \tag{10.76}$$

$$\lim_{z>>1} y_\ell(z) = -\frac{\cos(z - \pi\ell/2)}{z}\left[1 + O\left(\frac{1}{z}\right)\right] \tag{10.77}$$

$$\lim_{z>>1} R_\ell(z) = \frac{e^{i\delta_\ell}}{z}\left[\cos(\delta_\ell)\sin\left(z - \frac{\pi\ell}{2}\right) + \sin(\delta_\ell)\cos\left(\frac{z - \pi\ell}{2}\right)\right]\left[1 + O\left(\frac{1}{z}\right)\right]$$

$$= \frac{e^{i\delta_\ell}}{z}\sin\left(z + \delta_\ell - \frac{\pi\ell}{2}\right)\left[1 + O\left(\frac{1}{z}\right)\right] \tag{10.78}$$

The last equation is the correct asymptotic formula for the radial function, in terms of the phase shift.

The total cross section for the scattering of two hard spheres is given by

$$\sigma = \frac{4\pi}{k^2}\sum_{\ell=0}(2\ell + 1)\sin^2[\delta_\ell(k)] \tag{10.79}$$

$$\sin^2[\delta_\ell(k)] = \frac{\tan^2[\delta_\ell(k)]}{1 + \tan^2[\delta_\ell(k)]} = \frac{j_\ell(kd)^2}{j_\ell(kd)^2 + y_\ell(kd)^2} \tag{10.80}$$

$$\sigma = \frac{4\pi}{k^2}\sum_{\ell=0}(2\ell + 1)\frac{j_\ell(kd)^2}{j_\ell(kd)^2 + y_\ell(kd)^2} \tag{10.81}$$

The last equation is the exact result for the total cross section for the elastic scattering of two hard spheres. There is no overt Planck's constant (h) in the expression, but the wave vector $k = \sqrt{2mE}/\hbar$ of the particle is a quantum concept.

The *s*-wave phase shift has a simple form $\delta_0(k) = -kd$, which is derived from eqn. (10.75), which gives that $\tan(\delta_0) = -\tan(kd)$. An alternative derivation is to write out eqn. (10.74) and find

$$R_0(kr) = \frac{e^{i\delta_0}}{kr}\sin(kr + \delta_0) \tag{10.82}$$

$$= \frac{e^{i\delta_0}}{kr}\sin[k(r - d)] \tag{10.83}$$

Writing the argument of the sine function as $\sin[k(r - d)]$ guarantees that it vanishes at $r = d$. This choice determines that $\delta_0 = -kd$.

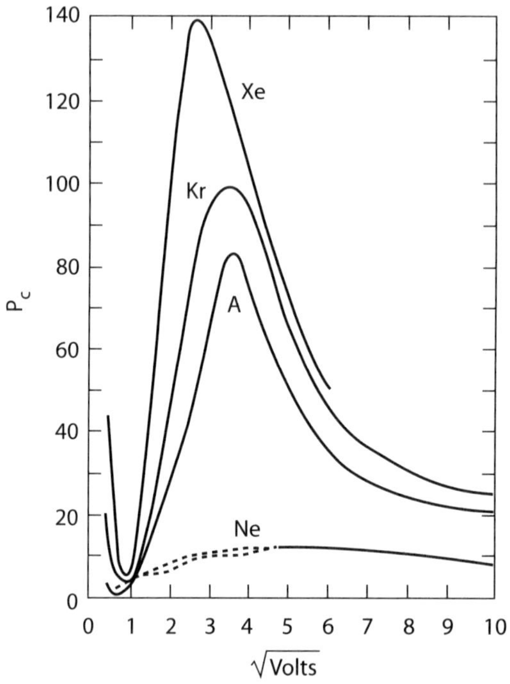

FIGURE 10.1. Experimental probability of an electron scattering from the atoms Ne, Ar, Kr, Xe. From Brode, *Rev. Mod. Phys.* 5, 257(1933).

An interesting result is the limit of long wavelength ($k \rightarrow 0$). Since there are no bound states, the phase shifts vanish according to $\delta\ell \rightarrow k^{\ell+1}a_\ell$ as shown above. Therefore, the limit of $\delta_\ell/k = k^\ell = 0$ for all values of ℓ except for s-waves. In the limit that $k \rightarrow 0$ the exact cross section becomes

$$\lim_{k=0} \sigma = 4\pi d^2 \tag{10.84}$$

$$\lim_{k=0} \frac{d\sigma}{d\Omega} = d^2 \tag{10.85}$$

Since only s-waves contribute to the scattering, the angular dependence is isotropic. Note that it is four times the classical area of πd^2.

10.1.4 *Ramsauer-Townsend Effect*

Electron scattering from atomic helium or neon has a cross section $\sigma(k)$ that is rather featureless. However, electron scattering from the other rare gas atoms—argon, krypton, and xenon—show an interesting phenomenon. Figure 10.1 shows a pronounced minimum in the cross section $\sigma(k)$ for electron kinetic energies around 0.7–1.0 eV. This minimum is called the *Ramsauer-Townsend effect*. The minimum in the cross section occurs because the phase shifts have the feature that $\delta_0 = \pi$, $\sin(\delta_0) = 0$, while the other phase shifts $[\delta_\ell(k), \ell \geq 1]$ are small. It is a quantum effect and is caused by the wave nature of electrons.

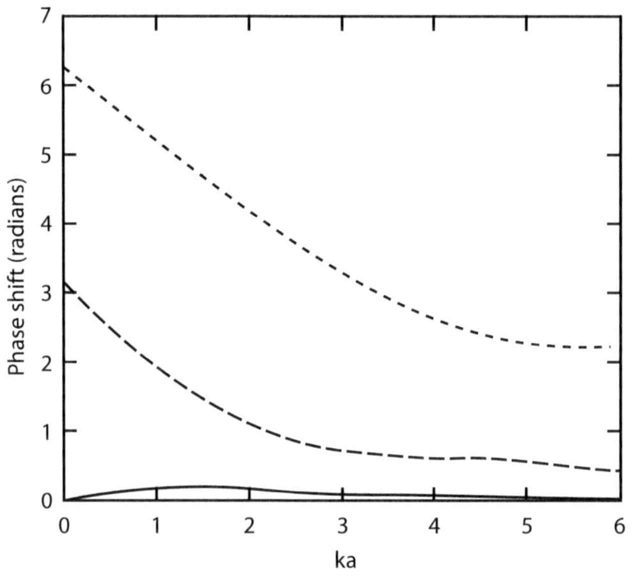

FIGURE 10.2. s-wave phase shift $\delta_0(k)$ as a function of ka for the spherical square well for three values of k_0a: (1) $\pi/4$, (2) $3\pi/4$, (3) $7\pi/4$. Case 1 has no bound states, case 2 has one bound state, and case 3 has two bound states. The Ramsauer-Townsend effect occurs for case 3, when the phase shift passes through π

The s-wave phase shifts at small wave vector behave as $\delta_0(k) = N_0\pi - ka$. Here a is the scattering length. If $N_0 = 1$, then $\sin(\delta_0) = \sin(ka) \approx ka$ and the cross section at small k is $4\pi a^2$. Similarly, if $N_0 = 2$, the scattering cross section is $4\pi a^2$ at small wave vectors. However, if $N_0 = 2$, the phase shift has a value of $\delta_0(0) = 2\pi$ at the origin. It declines in value with increasing wave vector and becomes small at large values of wave vector. As shown in figure 10.2, if it goes between 2π and 0, then it passes through π at some nonzero value of wave vector, here at $ka \approx 3$. If it passes through π then $\sin(\pi) = 0$ and the scattering by s-waves vanishes at this value of wave vector.

The Ramsauer-Townsend effect is due to bound states in the potential $V(r)$ between an electron and these atoms. Bound states are required to have $N_0 \neq 0$. The potential between an electron and helium or neon is largely repulsive, with no bound states. However, an attractive potential occurs between an electron and the other rare gases, and bound states exist. The negative ions Ar^-, Xe^- do exist. The reason for the bound state is due to the polarizability of the atoms. At large distance r, the electron of charge e creates an electric field $E = e\vec{r}/r^3$ at the atom. This electric field induces a dipole moment $p = \alpha E$ on the atom, which acts back on the electron. The long-range potential between the electron and the atom is

$$\lim_{r>>a_0} V(r) = -\frac{e^2\alpha}{2r^4} \tag{10.86}$$

The polarizability α increases with atomic number, so the above interaction is larger for Ar, Kr, Xe. As the scattering electron nears the electron clouds in the atom, exchange and correlation alter the above formula.

A simple potential that demonstrates the above phenomena is the spherical square well, where $V(r) = -V_0$ for $r < a$ and is zero elsewhere. The radial function for s-waves is

$$\chi_0(r) = \begin{cases} A\sin(pr) & r < a \\ \sin(kr + \delta_0) & r > a \end{cases} \tag{10.87}$$

$$p^2 = k^2 + k_0^2,\ k_0^2 = \frac{2mV_0}{\hbar^2} \tag{10.88}$$

Matching the two solutions at $r = a$ gives the equation for the phase shift:

$$\tan(ka + \delta_0) = \frac{k}{p}\tan(pa) \tag{10.89}$$

$$\delta_0(k) = -ka + \arctan\left(\frac{k}{p}\tan(pa)\right) \tag{10.90}$$

This formula is plotted in figure 10.2 with the values $k_0 a = (1)\ \pi/4$, (2) $3\pi/4$, (3) $7\pi/4$. Case 1 has no bound states, case 2 has one bound state, and case 3 has two bound states. The Ramsauer-Townsend effect occurs for case 3, when the phase shift passes through π. Another interesting feature of eqn. (10.90) is that the differential cross section at small wave vector is

$$\frac{d\sigma}{d\Omega} = a^2\left(1 - \frac{\tan(k_0 a)}{k_0 a}\right)^2 \tag{10.91}$$

The expression in parentheses can be either greater or smaller than one, depending on the value of $\tan(k_0 a)$.

10.1.5 *Born Approximation*

For any potential function $V(r)$, modern computers can be used to calculate the phase shifts $\delta_\ell(k)$ as a function of energy $\varepsilon(k)$. There is very little need for approximate methods. Before the age of computers, approximate methods were very useful for calculating the phase shifts. One popular method is called the *Born approximation*. The exact expression for the eigenfunction can be evaluated by iteration:

$$\psi(\mathbf{k},\mathbf{r}) = e^{i\mathbf{k}\cdot\mathbf{r}} + \int d^3r' G(k;\mathbf{r},\mathbf{r}')V(\mathbf{r}')\psi(\mathbf{k},\mathbf{r}') \tag{10.92}$$

$$= e^{i\mathbf{k}\cdot\mathbf{r}} + \int d^3r' G(k;\mathbf{r},\mathbf{r}')V(\mathbf{r}')e^{i\mathbf{k}\cdot\mathbf{r}'} \tag{10.93}$$

$$+ \int d^3r' G(k;\mathbf{r},\mathbf{r}')V(\mathbf{r}') \int d^3r' G(k;\mathbf{r}',\mathbf{r}')V(\mathbf{r}')e^{i\mathbf{k}\cdot\mathbf{r}'} + \cdots$$

Similarly, the integral for the T-matrix is

$$T(\mathbf{k}_f,\mathbf{k}_i) = \int d^3r' e^{-i\mathbf{k}_f\cdot\mathbf{r}'}V(\mathbf{r}')\psi(\mathbf{k}_i,\mathbf{r}') \tag{10.94}$$

$$= \int d^3r' e^{-i\mathbf{k}_f\cdot\mathbf{r}'}V(\mathbf{r}')\left[e^{i\mathbf{k}_i\cdot\mathbf{r}'} + \int d^3r' G(k;\mathbf{r}',\mathbf{r}')V(\mathbf{r}')e^{i\mathbf{k}_i\cdot\mathbf{r}'} + \cdots\right]$$

The terms in the series have names.

1. Keeping only the first term is called the *first Born approximation*:

$$T^{(1)}(\mathbf{k}_f, \mathbf{k}_i) = \int d^3r' e^{-i\mathbf{k}_f \cdot \mathbf{r}'} V(\mathbf{r}') e^{i\mathbf{k}_i \cdot \mathbf{r}'} = \bar{V}(\mathbf{k}_f - \mathbf{k}_i) \tag{10.95}$$

   The matrix element for scattering in the first Born approximation is merely the Fourier transform of the potential energy function. This approximation is best at high energy, where the kinetic energy of the incoming particle is much higher than any binding energy of the potential.

2. The *second Born approximation* includes the second term in the series:

$$T^{(2)}(\mathbf{k}_f, \mathbf{k}_i) = \int d^3r' e^{-i\mathbf{k}_f \cdot \mathbf{r}'} V(\mathbf{r}') \int d^3r'' G(k; \mathbf{r}', \mathbf{r}'') V(\mathbf{r}'') e^{i\mathbf{k}_i \cdot \mathbf{r}''}$$

$$T \approx T^{(1)} + T^{(2)} \tag{10.96}$$

$$\frac{d\sigma}{d\Omega} = \left(\frac{m}{2\pi\hbar^2}\right)^2 |T|^2 \tag{10.97}$$

   This approximation is more accurate than keeping only the first term, but the integrals can often be difficult to evaluate. The integrals are often easier if one Fourier transforms everything and starts from

$$T^{(2)}(\mathbf{k}_f, \mathbf{k}_i) = -\frac{2m}{\hbar^2} \int \frac{d^3q}{(2\pi)^3} \frac{\bar{V}(\mathbf{k}_f - \mathbf{q}) \bar{V}(\mathbf{q} - \mathbf{k}_i)}{k^2 - q^2} \tag{10.98}$$

10.2 Scattering of Identical Particles

The previous treatment of scattering considered that a particle scattered from a potential $V(\mathbf{r})$. In reality, the scattering target is a collection of other particles such as atoms or nuclei. This section provides two examples of a particle scattering from an identical particle. In one case the target particle is free, and in the other case it is bound.

10.2.1 *Two Free Particles*

A simple example is the scattering of a particle from another particle of the same kind. Assume both particles are free. It is logical to work the problem in the center of mass, so that one particle initially enters the scattering region with a wave function of $e^{i\mathbf{k}\cdot\mathbf{r}}$, while the other particle enters with a wave function of $e^{-i\mathbf{k}\cdot\mathbf{r}}$. Let the two particles be at positions $\mathbf{r}_1$, $\mathbf{r}_2$ and the relative separation is $\mathbf{R} = \mathbf{r}_1 - \mathbf{r}_2$.

If the two particles are identical spinless bosons, the symmetric wave function for the two particle system is

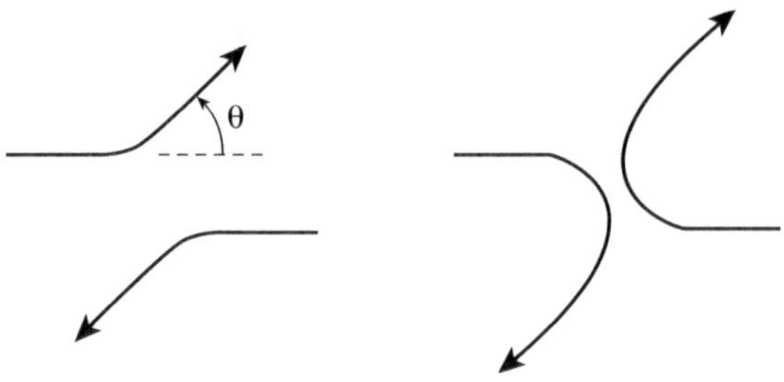

FIGURE 10.3. Direct (*left*) and exchange scattering (*right*) of two identical particles.

$$\Psi_i(\mathbf{r}_1,\mathbf{r}_2)=\frac{1}{\sqrt{2}}[\psi(\mathbf{k},\mathbf{r}_1)\psi(-\mathbf{k},\mathbf{r}_2)+\psi(-\mathbf{k},\mathbf{r}_1)\psi(\mathbf{k},\mathbf{r}_2)] \tag{10.99}$$

$$\psi(\mathbf{k},\mathbf{r})=e^{i\mathbf{k}\cdot\mathbf{r}} \tag{10.100}$$

$$\Psi_i(\mathbf{r}_1,\mathbf{r}_2)=\frac{1}{\sqrt{2}}\left[e^{i\mathbf{k}\cdot\mathbf{R}}+e^{-i\mathbf{k}\cdot\mathbf{R}}\right] \tag{10.101}$$

Write down a similar wave function for the final state (using $\mathbf{k}_f$) and then calculate the scattering matrix element in the first Born approximation:

$$\begin{aligned}M(\mathbf{k}_f,\mathbf{k}_i)&=\frac{1}{2}\int d^3R[e^{-i\mathbf{k}_f\cdot\mathbf{R}}+e^{i\mathbf{k}_f\cdot\mathbf{R}}]V(R)[e^{i\mathbf{k}_i\cdot\mathbf{R}}+e^{-i\mathbf{k}_i\cdot\mathbf{R}}]\\&=\tilde{V}(\mathbf{k}_i-\mathbf{k}_f)+\tilde{V}(\mathbf{k}_i+\mathbf{k}_f)\end{aligned} \tag{10.102}$$

There are two scattering terms. One has a final state with wave vector $\mathbf{k}_f$, while the other has a final state with the wave vector $-\mathbf{k}_f$. These two cases are shown in figure 10.3. In the first figure the two particles each scatter by an angle θ. In the second event they both scatter by the angle of $\pi-\theta$. Since the particles are identical, these two scatter processes cannot be distinguished. The matrix elements are added when evaluating the Golden Rule:

$$w=\frac{2\pi}{\hbar}\sum_{\mathbf{k}_f}|\tilde{V}(\mathbf{k}_i-\mathbf{k}_f)+\tilde{V}(\mathbf{k}_i+\mathbf{k}_f)|^2\delta[\varepsilon(\mathbf{k}_i)-\varepsilon(\mathbf{k}_f)] \tag{10.103}$$

The above discussion is in the Born approximation. The extension to an exact theory is simple. The incident and scattered wave functions have the form

$$\Psi(\mathbf{k},\mathbf{R})=\left[e^{i\mathbf{k}_i\cdot\mathbf{R}}+e^{-i\mathbf{k}_i\cdot\mathbf{R}}\right]+\frac{e^{ikR}}{R}[f(k,\theta)+f(k,\pi-\theta)] \tag{10.104}$$

$$\frac{d\sigma}{d\Omega}=|f(k,\theta)+f(k,\pi-\theta)|^2 \tag{10.105}$$

The scattering amplitude $f(k,\theta)$ is the one when the two particles are distinguishable.

It is also interesting to express the result in terms of phase shifts. The angular dependence is contained in the Legendre polynomials, which have the property that $\cos(\pi-\theta)=-\cos(\theta)$, $P_\ell(\pi-\theta)=(-1)^\ell P_\ell(\theta)$. When ℓ is an even integer, the terms $P_\ell(\pi-\theta)$ and $P_\ell(\theta)$ just add and double the answer. When ℓ is an odd integer, the two terms have opposite sign and cancel. The net scattering amplitude for bosons is

$$f(k,\theta)+f(k,\pi-\theta)=\frac{2}{2ik}\sum_{\ell\text{even}}(2\ell+1)\left(e^{2i\delta_\ell(k)}-1\right)P_\ell(\theta)$$

The summation over angular momentum contains only even integers. At small wave vector, if $\delta_0 \approx ka$, $\sin(\delta_0) \approx ka$, the differential cross section is

$$\lim_{k\to 0}\frac{d\sigma}{d\Omega}=4a^2 \tag{10.106}$$

It is increased by a factor of four over the result obtained when the two particles were not identical bosons.

A similar result is obtained when the two particles are identical fermions. If the two spin-$\frac{1}{2}$ fermions are in a spin singlet, the orbital terms must be symmetric and the same result is found as given for bosons. If the two fermions are in a spin triplet state, the orbital part must be antisymmetric. In this case the result is

$$\frac{d\sigma}{d\Omega}=|f(k,\theta)-f(k,\pi-\theta)|^2 \tag{10.107}$$

$$f(k,\theta)-f(k,\pi-\theta)=\frac{2}{2ik}\sum_{\ell\text{odd}}(2\ell+1)(e^{2i\delta_\ell(k)}-1)P_\ell(\theta)$$

Now the summation over angular momentum contains only odd integers. At small wave vectors the differential cross section actually goes to zero since it does not have an *s*-wave contribution.

10.2.2 *Electron Scattering from Hydrogen*

Another example of exchange during electron scattering is to consider the scattering of an electron from a hydrogen atom. Instead of viewing the scattering center as a simple potential, now view it as a proton plus an electron. The electron on the atom and the one scattering form a two-electron system. The scattering cross section depends on whether the spins are in a singlet or in a triplet state. The proton is a different particle, and can be treated as a simple potential acting on the electron.

Divide the Hamiltonian of the system as

$$H=H_{01}+H_{02}+V \tag{10.108}$$

$$H_{01}=\frac{p_1^2}{2m} \tag{10.109}$$

$$H_{02}=\frac{p_2^2}{2m}-\frac{e^2}{r_2} \tag{10.110}$$

$$V(\mathbf{r}_1,\mathbf{r}_2)=-\frac{e^2}{r_1}+\frac{e^2}{|\mathbf{r}_1-\mathbf{r}_2|} \tag{10.111}$$

The eigenfunction of $H_{01}+H_{02}$ is given by either of the two forms:

$$\Psi_0(\mathbf{k};\mathbf{r}_1,\mathbf{r}_2)=\frac{1}{\sqrt{2}}[e^{i\mathbf{k}\cdot\mathbf{r}_1}\phi_{1s}(r_2)+e^{i\mathbf{k}\cdot\mathbf{r}_2}\phi_{1s}(r_1)]\chi_0 \tag{10.112}$$

$$\Psi_{1m}(\mathbf{k};\mathbf{r}_1,\mathbf{r}_2)=\frac{1}{\sqrt{2}}[e^{i\mathbf{k}\cdot\mathbf{r}_1}\phi_{1s}(r_2)-e^{i\mathbf{k}\cdot\mathbf{r}_2}\phi_{1s}(r_1)]\chi_1(m) \tag{10.113}$$

The wave function has the correct antisymmetry. The hydrogen wave function ϕ_{1s} occurs with both arguments of r_1 and r_2. We must revise our thinking on the form of the scattering potential V. Instead of the above form for V, write it as $H-E$, where the energy is $E=\hbar^2k^2/2m-E_{\mathrm{Ry}}$. Then the effective potential is

$$(H-E)\Psi_S=\frac{1}{\sqrt{2}}[V(\mathbf{r}_1,\mathbf{r}_2)e^{i\mathbf{k}\cdot\mathbf{r}_1}\phi_{1s}(r_2)\pm V(\mathbf{r}_2,\mathbf{r}_1)e^{i\mathbf{k}\cdot\mathbf{r}_2}\phi_{1s}(r_1)]\chi_S$$

Assume the scattering is elastic and the electron goes from $\mathbf{k}_i$ to $\mathbf{k}_f$, where $k=|\mathbf{k}_i|=|\mathbf{k}_f|$. Assume the spin does not flip during scattering; the scattering leaves the spin states unchanged. In the Born approximation, the matrix element for scattering is

$$M_S=\langle\Psi_S|H-E|\Psi_S\rangle \tag{10.114}$$

$$=\frac{1}{2}\int d^3r_1d^3r_2[e^{-i\mathbf{k}_f\cdot\mathbf{r}_1}\phi_{1s}(r_2)\pm e^{-i\mathbf{k}_f\cdot\mathbf{r}_2}\phi_{1s}(r_1)]$$
$$\times[V(\mathbf{r}_1,\mathbf{r}_2)e^{i\mathbf{k}_i\cdot\mathbf{r}_1}\phi_{1s}(r_2)\pm V(\mathbf{r}_2,\mathbf{r}_1)e^{i\mathbf{k}_i\cdot\mathbf{r}_2}\phi_{1s}(r_1)] \tag{10.115}$$

There are four terms in the above integral and they are grouped into two pairs. Each pair is alike after interchanging dummy variables of integration $\mathbf{r}_1$, $\mathbf{r}_2$. One finally has

$$M_S=M_H\pm M_X \tag{10.116}$$

$$M_H=\int d^3r_1d^3r_2V(\mathbf{r}_1,\mathbf{r}_2)e^{-i(\mathbf{k}_f-\mathbf{k}_i)\cdot\mathbf{r}_1}\phi_{1s}(r_2)^2 \tag{10.117}$$

$$=\frac{4\pi e^2}{|\mathbf{k}_f-\mathbf{k}_i|^2}[1-F_{1s}(\mathbf{k}_f-\mathbf{k}_i)] \tag{10.118}$$

$$F_n(\mathbf{q})=\int d^3re^{i\mathbf{q}\cdot\mathbf{r}}\phi_n^2(r) \tag{10.119}$$

$$M_X=e^2\int d^3r_1d^3r_2\phi_{1s}(r_1)\phi_{1s}(r_2)e^{-i\mathbf{k}_f\cdot\mathbf{r}_1+i\mathbf{k}_i\cdot\mathbf{r}_2}\left[\frac{1}{r_2}-\frac{1}{|\mathbf{r}_1-\mathbf{r}_2|}\right]$$

The integral for the exchange term M_X cannot be done in a simple manner. There are too many vectors in the integrand, which makes the angular integrals complicated. The Hartree part of the matrix element M_H is the same term derived in a prior chapter. It contains the Fourier transform of the Coulomb interaction and the Fourier transform of the $1s$ charge distribution. The exchange term is due to the fact that events are possible where the two electrons change roles during the scattering process. The incoming electron gets bound and the intially bound electron gets unbound. The two electrons switch places in the bound state. Of course, that can happen, since the particles are identical, so a matrix element is needed, which is M_X.

This calculation is a simple example of electron scattering from atoms. If the atom is further up the periodic table and contains many bound electrons, the exchange processes become more complicated. The exchange terms become smaller as the wave vector k increases. Experiments are usually done with electrons of large initial energy, which reduces the size of the exchange terms. The Hartree term depends only on $\mathbf{q}=\mathbf{k}_f-k_i$, $q^2=2k^2(1-\cos\theta)$, which can be small at small angles θ even when k^2 is large. However, the exchange term depends on the magnitude of k and becomes small at large energy at all angles. High-energy electron scattering can reveal the charge distribution of electrons in atoms by making exchange effects negligible.

10.3 T-Matrices

The *T-matrix* was defined earlier as the exact scattering matrix:

$$T(\mathbf{k}_f, \mathbf{k}_i) = \int d^3 r e^{-i\mathbf{k}_f \cdot \mathbf{r}} V(\mathbf{r}) \psi(\mathbf{k}_i, \mathbf{r}) \tag{10.120}$$

Here the definition of the T-matrix is expanded to make it an operator. The above integral is really the expectation value of the operator. Define an operator $T(\mathbf{r})$ such that

$$T(\mathbf{k}_f, \mathbf{k}_i) = \int d^3 r e^{-i\mathbf{k}_f \cdot \mathbf{r}} T(\mathbf{r}) e^{i\mathbf{k}_i \cdot \mathbf{r}} \tag{10.121}$$

The goal is to derive $T(\mathbf{r})$.

Start with the outgoing wave definition of the wave function:

$$\psi(\mathbf{k}_i, \mathbf{r}) = \phi(\mathbf{k}_i, \mathbf{r}) + \int d^3 r' G_0(k, \mathbf{r}, \mathbf{r}') V(\mathbf{r}') \psi(\mathbf{k}_i, \mathbf{r}') \tag{10.122}$$

$$G_0(\mathbf{k}, \mathbf{r}, \mathbf{r}') = \sum_{\mathbf{k}'} \frac{\phi(\mathbf{k}', \mathbf{r}) \phi^*(\mathbf{k}', \mathbf{r}')}{\varepsilon(k) - \varepsilon(k') + i\eta} \tag{10.123}$$

$$\phi(\mathbf{k}, \mathbf{r}) = \frac{e^{i\mathbf{k} \cdot \mathbf{r}}}{\sqrt{\Omega}} \tag{10.124}$$

The symbol G_0 denotes a Green's function.

The first formal manipulation is to replace, in the denominator, $\varepsilon(k')$ by the Hamiltonian H_0:

$$H_0 \phi(\mathbf{k}', \mathbf{r}) = \varepsilon(k') \phi(\mathbf{k}', \mathbf{r}) \tag{10.125}$$

$$G_0(\mathbf{k}, \mathbf{r}, \mathbf{r}') = \frac{1}{\varepsilon(k) - H_0 + i\eta} \sum_{\mathbf{k}'} \phi(\mathbf{k}', \mathbf{r}) \phi^*(\mathbf{k}', \mathbf{r}') \tag{10.126}$$

$$= \frac{1}{\varepsilon(k) - H_0 + i\eta} \delta^3(\mathbf{r} - \mathbf{r}') \tag{10.127}$$

$$\psi(\mathbf{k}_i, \mathbf{r}) = \phi(\mathbf{k}_i, \mathbf{r}) + \frac{1}{\varepsilon(k) - H_0 + i\eta} V(\mathbf{r}) \psi(\mathbf{k}_i, \mathbf{r}) \tag{10.128}$$

The last equation is the basis for the T-matrix expansion. By comparing the two equations (10.120) and (10.121), an identity is

$$V(\mathbf{r}) \psi(\mathbf{k}, \mathbf{r}) = T(\mathbf{r}) \phi(\mathbf{k}, \mathbf{r}) \tag{10.129}$$

Multiply eqn. (10.128) by $V(\mathbf{r})$ from the left, and use the above identity. We then derive

$$T(\mathbf{r}) = V(\mathbf{r}) + V(\mathbf{r}) \frac{1}{\varepsilon - H_0 + i\eta} T(\mathbf{r}) \tag{10.130}$$

This expression is called the *T-matrix equation*. The *T*-matrix is a function of the energy ε and the position **r**.

The *T*-matrix equation can be iterated to get

$$T = V + VG_0V + VG_0VG_0V + VG_0VG_0VG_0V + \cdots \tag{10.131}$$

$$G_0 = \frac{1}{\varepsilon - H_0 + i\eta} \tag{10.132}$$

The *T*-matrix is the summation of the infinite series that comes from perturbation theory. Two other exact expressions for the *T*-matrix are

$$T(\mathbf{r}) = V(\mathbf{r}) + T(\mathbf{r})\frac{1}{\varepsilon - H_0 + i\eta}V(\mathbf{r}) \tag{10.133}$$

$$T = V + V\frac{1}{\varepsilon - H + i\eta}V, \quad H = H_0 + V \tag{10.134}$$

The first can be proved using the series equation (10.131). There are many ways to prove the second expression. One way is write the denominator as

$$\begin{aligned} G &= \frac{1}{\varepsilon - H + i\eta} = \frac{1}{[\varepsilon - H_0 + i\eta](1 - G_0V)} = \frac{G_0}{1 - G_0V} \\ &= G_0 + G_0VG_0 + G_0VG_0VG_0 + \cdots \end{aligned} \tag{10.135}$$

This series generates the same solution found in (10.131).

One way to evaluate eqn. (10.134) is to use $H\psi(\mathbf{k}, \mathbf{r}) = \varepsilon(\mathbf{k})\,\psi(\mathbf{k}, \mathbf{r})$:

$$G\delta^3(\mathbf{r} - \mathbf{r}') = G\sum_{\mathbf{k}'}\psi(\mathbf{k}', \mathbf{r})\psi^*(\mathbf{k}', \mathbf{r}') = \sum_{\mathbf{k}'}\frac{\psi(\mathbf{k}', \mathbf{r})\psi^*(\mathbf{k}', \mathbf{r}')}{\varepsilon - \varepsilon(\mathbf{k}') + i\eta}$$

Taking the matrix element of eqn. (10.134) gives

$$\begin{aligned} T(\mathbf{k}_f, \mathbf{k}_i) &= \int d^3r e^{-i\mathbf{k}_f\cdot\mathbf{r}}V(\mathbf{r})\left[1 + \int d^3r'\sum_{\mathbf{k}'}\frac{\psi(\mathbf{k}', \mathbf{r})\psi^*(\mathbf{k}', \mathbf{r}')}{\varepsilon - \varepsilon(\mathbf{k}') + i\eta}V(\mathbf{r}')\right]e^{i\mathbf{k}_i\cdot\mathbf{r}'} \\ &= \tilde{V}(\mathbf{k}_f - \mathbf{k}_i) + \frac{1}{\Omega}\sum_{\mathbf{k}'}\frac{T(\mathbf{k}_f, \mathbf{k}')T(\mathbf{k}', \mathbf{k}_i)}{\varepsilon - \varepsilon(\mathbf{k}') + i\eta} \end{aligned} \tag{10.136}$$

Set $\mathbf{k}_f = \mathbf{k}_i = \mathbf{k}$ and take the imaginary part of this expression. The Fourier transform of the potential $\tilde{V}(0)$ is real and drops out:

$$-2\Im\{T(\mathbf{k}, \mathbf{k})\} = 2\pi\int\frac{d^3k'}{(2\pi)^3}|T(\mathbf{k}, \mathbf{k}')|^2\delta[\varepsilon(\mathbf{k}) - \varepsilon(\mathbf{k}')] \tag{10.137}$$

This identity is another version of the optical theorem. In deriving this expression we have used the identity

$$\begin{aligned} &T(\mathbf{k}_i, \mathbf{k}_f) = T(\mathbf{k}_f, \mathbf{k}_i)^\dagger \\ &T(\mathbf{k}_f, \mathbf{k}_i) = \int d^3r\phi^*(\mathbf{k}_f, \mathbf{r})V(\mathbf{r})\psi(\mathbf{k}_i, \mathbf{r}) = \int d^3r\tilde{\psi}^*(\mathbf{k}_f, \mathbf{r})V(\mathbf{r})\phi(\mathbf{k}_i, \mathbf{r}) \end{aligned} \tag{10.138}$$

This latter identity can be proven by writing $V\psi(\mathbf{k}, \mathbf{r}) = T\phi(\mathbf{k}, \mathbf{r})$ and taking the Hermitian conjugate of this expression:

$$\phi^*(\mathbf{k}_f,\mathbf{r})V\psi(\mathbf{k}_i,\mathbf{r}) = \phi^*(\mathbf{k}_f,\mathbf{r})T\phi(\mathbf{k}_i,\mathbf{r}) = [T^\dagger\phi(\mathbf{k}_f,\mathbf{r})]^\dagger\phi(\mathbf{k}_i,\mathbf{r})$$
$$= [V\tilde{\psi}(\mathbf{k}_f,\mathbf{r})]^\dagger\phi(\mathbf{k}_i,\mathbf{r}) = \tilde{\psi}(\mathbf{k}_f,\mathbf{r})^* V\phi(\mathbf{k}_i,\mathbf{r})$$

The function $\tilde{\psi}(\mathbf{k}_f, \mathbf{r})^*$ is the complex conjugate of $\tilde{\psi}$. The tilde has a special meaning. We defined $T\phi = V\psi$ as an outgoing wave for ψ. However, the conjugate$T^\dagger$ has all of the factors of $i\eta$ in the denominators of the Green's functions changed to $-i\eta$. Earlier we stated that a Green's function with an imaginary part of $-i\eta$ was an incoming wave. The wave ψ is an outgoing wave, but the wave $\tilde{\psi}$ is an incoming wave. The incoming wave is the time reverse of the outgoing wave.

The conjugate T-matrix obeys the equation

$$T = V + VG_0T = V + TG_0V \tag{10.139}$$

$$T^\dagger = V + T^\dagger G_0^\dagger V = V + VG_0^\dagger T^\dagger \tag{10.140}$$

The second equation is the Hermitian conjugate of the first. The Green's function $G_0^\dagger$ generates incoming waves. The incoming wave function is defined as $T^\dagger\phi = V\tilde{\psi}$, so that

$$T^\dagger\phi = V\phi + VG_0^\dagger T^\dagger\phi \tag{10.141}$$

$$V\tilde{\psi} = V\phi + VG_0^\dagger V\tilde{\psi} \tag{10.142}$$

Take out a factor of V from the left and get

$$\tilde{\psi} = \phi + G_0^\dagger V\tilde{\psi} \tag{10.143}$$

$$\tilde{\psi}^* = \phi^* + \tilde{\psi}^* VG_0 \tag{10.144}$$

The second equation is the Hermitian conjugate of the previous one. These results are useful below.

The results are illustrated by a partial wave expansion. The Green's function for an incoming wave is

$$G_0^\dagger = -\frac{m}{2\pi\hbar^2|\mathbf{r}-\mathbf{r}'|}e^{-ik|\mathbf{r}-\mathbf{r}'|} \to -\frac{m}{2\pi\hbar^2 r}e^{-i(kr-\mathbf{k}_f\cdot\mathbf{r}')} \tag{10.145}$$

The conjugate wave function is

$$\tilde{\psi} = e^{i\mathbf{k}_i\cdot\mathbf{r}} + \tilde{f}\frac{e^{-ikr}}{r} \tag{10.146}$$

$$\tilde{f}(\mathbf{k}_f,\mathbf{k}_i) = -\frac{m}{2\pi\hbar^2}\int d^3r' e^{i\mathbf{k}_f\cdot\mathbf{r}'}V(\mathbf{r}')\tilde{\psi}(\mathbf{k}_i,\mathbf{r}') \tag{10.147}$$

If we do the usual partial wave analysis, we find

$$\tilde{\psi}(\mathbf{k}_i,\mathbf{r}) = \sum_\ell (2\ell+1)i^\ell \tilde{R}_\ell(kr)P_\ell(\hat{k}_i\cdot\hat{r}) \tag{10.148}$$

$$\tilde{R}_\ell(kr) = R_\ell^*(kr) \to e^{-i\delta_\ell}\sin(kr+\delta_\ell-\pi\ell/2)/kr \tag{10.149}$$

$$\tilde{f}(k,\theta) = \frac{1}{2ik}\sum_\ell (2\ell+1)\left(e^{-2i\delta_\ell}-1\right)P_\ell(\cos\theta) \tag{10.150}$$

The conjugate eigenfunction is used in the next section on distorted wave scattering.

10.4 Distorted Wave Scattering

Distorted wave scattering is denoted as DWS. Start with a Hamiltonian that contains two terms in the potential

$$H = \frac{p^2}{2m} + V_1(\mathbf{r}) + V_2(\mathbf{r}) = H_1 + V_2(\mathbf{r}) \tag{10.151}$$

$$H_1 = \frac{p^2}{2m} + V_1(\mathbf{r}) \tag{10.152}$$

The potential $V_1(\mathbf{r})$ is too large to treat by perturbation theory, and is included in H_1. The smaller potential V_2 is treated in the DWS. The DWS is based on the eigenstates of H_1, which are not pure plane waves. The DWS is a theory of scattering where the initial eigenstates are not plane waves, but are eigenstates of a particle in the potential V_1.

For example, in nuclear physics, the scattering of a proton by a nucleus has two potentials: V_1 is the strong nucleon–nucleon potential that the proton has inside of the nucleus. In this case, V_2 is the "weak" Coulomb potential between the proton and nucleus when the proton is outside of the nucleus. Other examples are given below.

There are three possible Hamiltonians, each with a different eigenfunction:

$$\frac{p^2}{2m}\phi(\mathbf{k},\mathbf{r}) = \varepsilon(\mathbf{k})\phi(\mathbf{k},\mathbf{r}), \quad \phi(\mathbf{k},\mathbf{r}) = \frac{e^{i\mathbf{k}\cdot\mathbf{r}}}{\sqrt{\Omega}} \tag{10.153}$$

$$\left[\frac{p^2}{2m} + V_1\right]\psi^{(1)}(\mathbf{k},\mathbf{r}) = \varepsilon(\mathbf{k})\psi^{(1)} \tag{10.154}$$

$$\left[\frac{p^2}{2m} + V_1 + V_2\right]\psi(\mathbf{k},\mathbf{r}) = \varepsilon(\mathbf{k})\psi \tag{10.155}$$

All of the states have the same energy $\varepsilon(\mathbf{k}) = \hbar^2k^2/2m$ in the continuum. There might be bound states of H_1 or H, but they do not participate in scattering. We also can write some T-matrix equations:

$$\psi = \phi + G_0(V_1 + V_2)\psi \tag{10.156}$$

$$\psi^{(1)} = \phi + G_0 V_1 \psi^{(1)} \tag{10.157}$$

$$\tilde{\psi}^{(1)} = \phi + G_0^{\dagger} V_1 \tilde{\psi}^{(1)} \tag{10.158}$$

The last equation is the incoming wave discussed in the previous section.

These various definitions are used to prove the following theorem.

THEOREM: *The exact T-matrix can be written as*

$$T(\mathbf{k}_f,\mathbf{k}_i) = \int d^3r\phi^*(\mathbf{k}_f,\mathbf{r})[V_1 + V_2]\psi(\mathbf{k}_i,\mathbf{r}) = T^{(1)}(\mathbf{k}_f,\mathbf{k}_i) + \delta T(\mathbf{k}_f,\mathbf{k}_i)$$

$$T^{(1)}(\mathbf{k}_f,\mathbf{k}_i) = \int d^3r\phi^*(\mathbf{k}_f,\mathbf{r})V_1\psi^{(1)}(\mathbf{k}_i,\mathbf{r}) \tag{10.159}$$

$$\delta T(\mathbf{k}_f,\mathbf{k}_i) = \int d^3r\tilde{\psi}^{(1)*}(\mathbf{k}_f,\mathbf{r})V_2\psi(\mathbf{k}_i,\mathbf{r}) \tag{10.160}$$

The first term $T^{(1)}(\mathbf{k}_f, \mathbf{k}_i)$ is the exact T-matrix for scattering from the potential V_1 alone, if V_2 were absent. The second term is the contribution of V_2. It is the matrix element of V_2 with the incoming part of $\psi^{(1)}$ and the exact wave function ψ.

Proof: Take the Hermitian conjugate of eqn. (10.158) and then rearrange it:

$$\phi^*(\mathbf{k}_f) = \tilde{\psi}^{(1)*}[1 - V_1 G_0] \tag{10.161}$$

$$T(\mathbf{k}_f, \mathbf{k}_i) = \int d^3r \tilde{\psi}^{(1)*}[1 - V_1 G_0][V_1 + V_2]\psi(\mathbf{k}_i, \mathbf{r}) \tag{10.162}$$

The last factor can be simplified using eqn. (10.156):

$$G_0(V_1 + V_2)\psi = \psi - \phi \tag{10.163}$$

$$T(\mathbf{k}_f, \mathbf{k}_i) = \int d^3r \tilde{\psi}^{(1)*}\{[V_1 + V_2]\psi - V_1(\psi - \phi)\} \tag{10.164}$$

$$= \int d^3r \tilde{\psi}^{(1)*}\{V_1\phi + V_2\psi\} \tag{10.165}$$

The last line proves the theorem.

So far no approximation is made. The *distorted wave born approximation* (DWBA) comes from approximating δT with a Born approximation:

$$\delta T(\mathbf{k}_f, \mathbf{k}_i) \approx \int d^3r \tilde{\psi}^{(1)*}(\mathbf{k}_f, \mathbf{r}) V_2 \psi^{(1)}(\mathbf{k}_i, \mathbf{r}) \tag{10.166}$$

For the incoming wave, the exact function ψ is replaced by $\psi^{(1)}$. The basic idea is quite simple. One first solves the eigenstates of the Hamiltonian H_1. This solution produces its eigenstates $\psi^{(1)}$ and also the exact T-matrix $T^{(1)}$. Then one does the Born approximation to evaluate the contribution of V_2, using the eigenfunctions $\psi^{(1)}$.

The first example of the DWBA is a potential step and delta-function in one dimension:

$$V_1(x) = -V_0\Theta(x) \tag{10.167}$$

$$V_2(x) = \lambda\delta(x - a), \quad a > 0 \tag{10.168}$$

The potential is shown in figure 10.4a.

First consider the solution to $H_1 = p^2/2m + V_1(x)$. A wave enters from the left with kinetic energy $E = \hbar^2k^2/2m$. On the right of the origin, its energy is $E = \hbar^2p^2/2m - V_0$, $p^2 = k^2 + 2mV_0/\hbar^2$. The wave function is

$$\psi^{(1)}(x) = \begin{cases} e^{ikx} + Re^{-ikx} & x < 0 \\ Se^{ipx} & x > 0 \end{cases} \tag{10.169}$$

Matching the wave function at $x = 0$ gives

$$R = -\frac{p - k}{p + k}, \quad S = \frac{2k}{k + p} \tag{10.170}$$

In one dimension the Green's function is

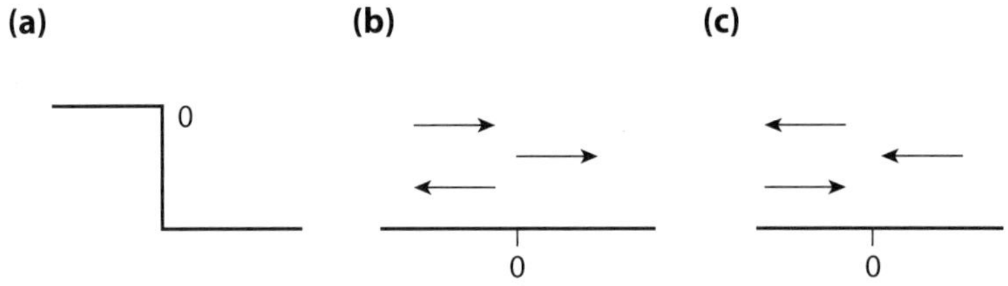

FIGURE 10.4. (a) Potential V(x). (b) Incoming, reflected, and transmitted waves. (c) Time-reversed state.

$$G(x,x') = -\frac{2m}{\hbar^2}\int \frac{dq}{2\pi}\frac{e^{iq(x-x')}}{q^2-k^2-i\delta} = -\frac{im}{\hbar^2 k}e^{ik|x-x'|} \tag{10.171}$$

$$\psi^{(1)}(k_i,x) = e^{ikx} + \int dx' G(x,x')V_1(x')\psi^{(1)}(k_i,x') = e^{ikx} + Re^{-ikx} \text{for } x<0$$

$$R = -\frac{im}{\hbar^2 k}T^{(1)}, \quad T^{(1)} = i\frac{\hbar^2 k}{m}R = -i\frac{\hbar^2 k}{m}\frac{p-k}{p+k} \tag{10.172}$$

$$T^{(1)} = \int_{-\infty}^{\infty} dx' e^{ikx'}V_1(x')\psi^{(1)}(x') = -V_0 S\int_0^{\infty} dx e^{ix(k+p)} \tag{10.173}$$

$$= -i\frac{V_0 2k}{(p+k)^2} \tag{10.174}$$

The two expressions for $T^{(1)}$ are identical when using the identity $V_0 = \hbar^2(p^2-k^2)/2m$.

In the DWBA, the outgoing wave function $\psi^{(1)}$ is given in (10.169). The incoming wave is simply the complex conjugate of the outgoing wave, which makes it the time-reversed state:

$$\tilde{\psi}^{(1)}(x) = \begin{cases} e^{-ikx} + R^* e^{ikx} & x<0 \\ S^* e^{-ipx} & x>0 \end{cases} \tag{10.175}$$

These two are shown in figure 10.4. The additional T-matrix from V_2 in the DWBA is

$$\delta T = \int_{-\infty}^{\infty} dx \tilde{\psi}^{(1)*}(x)V_2(x)\psi^{(1)} = \lambda S^2 e^{2ipa} \tag{10.176}$$

It is easy to derive the exact phase shift for back scattering and to show it gives this result in the limit that λ is small.

The second example of the DWBA is in three dimensions. For spherically symmetric potentials, one can expand the various eigenfunctions in angular momentum:

$$e^{i\mathbf{k}\cdot\mathbf{r}} = \sum_\ell (2\ell+1)i^\ell j_\ell(kr)P_\ell(\cos\theta) \tag{10.177}$$

$$\psi^{(1)}(\mathbf{k},\mathbf{r}) = \sum_\ell (2\ell+1)i^\ell R_\ell^{(1)}(kr)P_\ell(\cos\theta) \tag{10.178}$$

$$\tilde{\psi}^{(1)}(\mathbf{k},\mathbf{r}) = \sum_\ell (2\ell+1)i^\ell \tilde{R}_\ell^{(1)}(kr)P_\ell(\cos\theta) \tag{10.179}$$

$$\psi(\mathbf{k},\mathbf{r}) = \sum_\ell (2\ell+1)i^\ell R_\ell(kr)P_\ell(\cos\theta) \tag{10.180}$$

Keep in mind that R has a built in factor of $\exp[i\delta_\ell(k)]$, while $\tilde{R}$ has the complex conjugate of this factor. Also note that $\tilde{\psi}^{(1)*} \neq \psi^{(1)}$. In the DWBA the two terms in the T-matrix are $[\cos\theta = \hat{k}_f \cdot \hat{k}_i]$:

$$T^{(1)}(\mathbf{k}_f, \mathbf{k}_i) = \frac{\pi\hbar^2}{ikm} \sum_\ell (2\ell+1)\left(e^{2i\delta_\ell^{(1)}(k)} - 1\right) P_\ell(\cos\theta) \tag{10.181}$$

$$\delta T(\mathbf{k}_f, \mathbf{k}_i) = 4\pi \sum_\ell (2\ell+1) P_\ell(\cos\theta) \int_0^\infty r^2 dr \tilde{R}_\ell^{(1)2}(kr) V_2(r)$$

The exact T-matrix is

$$T(\mathbf{k}_f, \mathbf{k}_i) = \frac{\pi\hbar^2}{ikm} \sum_\ell (2\ell+1)\left(e^{2i\delta_\ell(k)} - 1\right) P_\ell(\cos\theta) \tag{10.182}$$

In the DWS the phase shift is divided into two terms: $\delta_\ell(k) = \delta_\ell^{(1)}(k) + \Delta\delta_\ell(k)$, where $\delta_\ell^{(1)}(k)$ is the contribution from V_1, and $\Delta\delta_{\ell(k)}$ is the contribution from V_2. Expanding the above result gives

$$\left(e^{2i\delta_\ell(k)} - 1\right) = \left(e^{2i[\delta_\ell^{(1)}(k) + \Delta\delta_\ell(k)]} - 1\right) \tag{10.183}$$

$$\approx \left(e^{2i\delta_\ell^{(1)}(k)} - 1\right) + 2ie^{2i\delta_\ell^{(1)}(k)} \Delta\delta_\ell(k) \tag{10.184}$$

The first term on the right gives $T^{(1)}$. The second term on the right gives δT in the DWBA. Note that the second term is always multiplied by the factor $\exp[2i\delta_\ell^{(1)}(k)]$ that comes from $\tilde{R}^2$ in the integrand.

10.5 Scattering from Many Particles

So far the discussion has considered a single-particle scattering from a single target. Here we consider the situation where a single particle scatters from a collection of identical targets. The collection of targets might be the atoms in a solid or liquid.

For inelastic scattering, the target absorbs an arbitrary amount of phase, which is transferred to the scattered particle. Then the target acts as an incoherent source: each target particle is emitting scattered particles with a random phase. The scattering probability is proportional to the number of targets.

The more interesting case is when the scattering is elastic. The coherence of the incident particle is transferred to coherently scattered particles. One can get a variety of effects.

10.5.1 *Bragg Scattering*

Consider the elastic scattering theory of the prior section. The N target particles are located at the points $\mathbf{R}_j$. Assume the target is a solid, so the atoms are in a regular arrangement. A crystal has reciprocal lattice vectors ($\vec{G}_\alpha$) that are chosen such that for every lattice point $\mathbf{R}_j$:

$$e^{i\mathbf{G}_\alpha \cdot \mathbf{R}_j} = 1 \tag{10.185}$$

There are many $\mathbf{G}_\alpha$ that satisfy this condition. A fundamental theory of Fourier analysis is that a periodic potential can be expanded in a Fourier series that contains only the wave vectors $\mathbf{G}_\alpha$:

$$\mathcal{V}(\mathbf{r}) = \sum_j V(\mathbf{r}-\mathbf{R}_j) = \sum_\alpha e^{i\mathbf{G}_\alpha \cdot \mathbf{r}} \tilde{V}_\alpha \tag{10.186}$$

$$\tilde{V}_\alpha = \int d^3 r e^{i\vec{G}_\alpha \cdot r} V(\mathbf{r}) \tag{10.187}$$

Here $V(\mathbf{r}-\mathbf{R}_j)$ is the potential of one atom centered at point $\mathbf{R}_j$, and the crystal target is a regular array of such targets.

The outgoing wave function is

$$\begin{aligned}\psi(\mathbf{k}_i, \mathbf{r}) &= e^{i\mathbf{k}_i \cdot \mathbf{r}} + \sum_{j=1}^{N} \int d^3 r' G_k(\mathbf{r}-\mathbf{r}') V(\mathbf{r}-\mathbf{R}_j) \psi(\mathbf{k}_i, \mathbf{r}') \\ &= e^{i\mathbf{k}_i \cdot \mathbf{r}} + \sum_\alpha \tilde{V}_\alpha \int d^3 r' G_k(\mathbf{r}-\mathbf{r}') e^{i\mathbf{G}_\alpha \cdot \mathbf{r}'} \psi(\mathbf{k}_i, \mathbf{r}')\end{aligned}$$

The main result is that for elastic scattering the wave vector of the scattered particle must change by G_α:

$$\mathbf{k}_f = \mathbf{k}_i + \mathbf{G}_\alpha \tag{10.188}$$

The elastic scattering is called *Bragg scattering*. The lattice wave vectors $\mathbf{G}_\alpha$ are fairly large ($G \sim 2\pi/a$) where a is the separation between atoms. The wave vector $\mathbf{k}_i$ must be larger in value than the smallest G_α. The scattered particles are either photons ("x-ray scattering"), electrons, or neutrons. The elastically scattered particles go in particular directions, so they appear as "spots" on the measurement apparatus. The intensity of each spot provides information on the value of $|\tilde{V}_\alpha|^2$. A knowledge of where the spots are located, which gives the values of G_α, and the intensity of each spot permits the atomic arrangments in the solids to be determined.

X-rays scatter from electrons, so the x-ray pattern gives the density of electrons. Neutrons scatter from the nucleus, and their scattering gives the arrangement of nuclei. Usually the two methods give the same structure. In molecules containing hydrogen, the location of the protons can be found only by neutron scattering since there are not enough electrons attached to the proton to measure using x-rays.

10.5.2 *Scattering by Fluids*

Fluids denote liquids or gases. The targets are atoms in an arrangment that has short-range order but not long-range order. First consider the case that the scattering is weak.

The incident particle (photon, electron, or neutron) scatters only once. Use the first Born approximation and write

$$\psi(\mathbf{k}_i,\mathbf{r}) = e^{i\mathbf{k}_i\cdot\mathbf{r}} + \sum_{j=1}^{N}\int d^3r' G_k(\mathbf{r}-\mathbf{r}'-\mathbf{R}_j)V(\mathbf{r}')\phi(\mathbf{k}_i,\mathbf{r}'+\mathbf{R}_j)$$

Expanding the Green's function for outgoing waves gives

$$\psi(\mathbf{k}_i,\mathbf{r}) = e^{i\mathbf{k}_i\cdot\mathbf{r}} + \frac{m}{2\pi\hbar^2}\tilde{V}(\mathbf{k}_f-\mathbf{k}_i)\frac{e^{ikr}}{r}\sum_{j=1}^{N}e^{i\mathbf{R}_j\cdot(\mathbf{k}_i-\mathbf{k}_f)} \tag{10.189}$$

The last factor is contains the phases caused by the targets in different positions. All of the other factors are identical to the scattering by a single target. So we can write

$$\left(\frac{d\sigma}{d\Omega}\right)_N = \left(\frac{d\sigma}{d\Omega}\right)_1 \Big|\sum_j e^{i\mathbf{R}_j\cdot(\mathbf{k}_f-\mathbf{k}_i)}\Big|^2 \tag{10.190}$$

$$\left(\frac{d\sigma}{d\Omega}\right)_1 = \left(\frac{m}{2\pi\hbar^2}\right)^2 \tilde{V}(\mathbf{k}_f-\mathbf{k}_f)^2 \tag{10.191}$$

where $(d\sigma/d\Omega)_N$ is the scattering by N-particles. The first term on the right is the scattering by a single particle. Define the *structure factor* $S(q)$ as

$$\Big|\sum_j e^{i\mathbf{R}_j\cdot\mathbf{q}}\Big|^2 = N[S(q)+N\delta_q] = \sum_j\Big[1+\sum_{m\neq j}e^{i\mathbf{q}\cdot(\mathbf{R}_m-\mathbf{R}_j)}\Big]$$

If $\mathbf{q}=0$, then the summation over N gives N and the square gives N^2, so that $S(q=0)\ =0$. In a liquid of identical atoms, each term in the summation over j will be the same, since each atom sees the same average environment. We ignore edge effects. There are N terms all alike. So

$$S(q)+N\delta_{\mathbf{q}=0} = 1+\sum_{m\neq 0}e^{i\mathbf{q}\cdot\mathbf{R}_m} = 1+n_0\int d^3R\, g(R)e^{i\mathbf{q}\cdot\mathbf{R}}$$

The summation over $\mathbf{R}_m$ assumes a particle is at the origin, and excludes that particle. The summation is changed to a continuous integral. If there is a particle at the origin, then the probability of having one in volume d^3R is $n_0 g(R)d^3R$, where $g(R)$ is the pair distribution function and n_0 is the density.

Rearrange the above terms. Move the "1" to the left of the equal sign, and the $\delta_{\mathbf{q}=0}$ to the right, while changing it to

$$N\delta_{\mathbf{q}=0} = n_0\int d^3R\, e^{i\mathbf{q}\cdot\mathbf{R}} \tag{10.192}$$

$$S(q)-1 = n_0\int d^3R[g(R)-1]e^{i\mathbf{q}\cdot\mathbf{R}} \tag{10.193}$$

$$g(R)-1 = \frac{1}{n_0}\int\frac{d^3q}{(2\pi)^3}[S(q)-1]e^{-i\mathbf{q}\cdot\mathbf{R}} \tag{10.194}$$

The above equation is the important relationship between the pair distribution function and the structure factor. Both $S(q)$ and $g(R)$ are defined so that they vanish at small

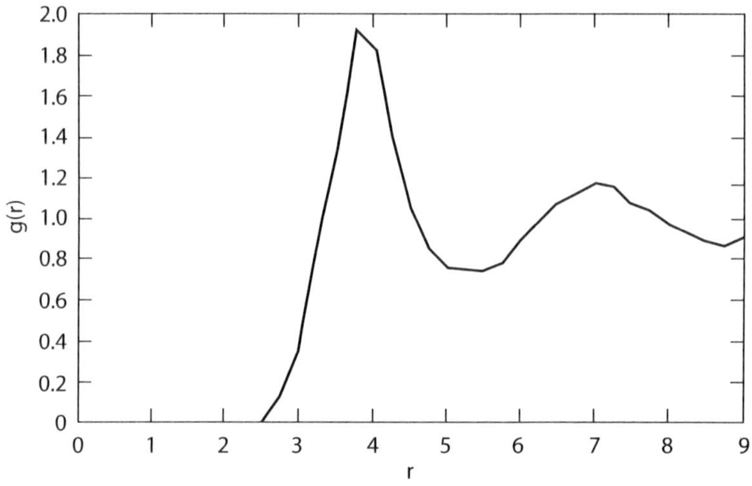

FIGURE 10.5. $g(r)$ in liquid sodium, where r is in units of Å. Data from Tasasov and Warren, *J. Chem. Phys.* 4, 236 (1936).

arguments and go to unity at large values of their arguments. Therefore, $[g(R)-1]$ and $[S(q)-1]$ are both functions that are nonzero only for small values of R or q, respectively. Their mutual Fourier transforms exist. Since $S(q=0)=0$ there is an identity

$$1 = n_0 \int d^3R[1-g(R)] \tag{10.195}$$

that is useful for checking proposed forms for the pair distribution function.

Experiments using x-ray and neutron scattering on liquids measure the differential cross sections at small values of $q=|\mathbf{k}_f-\mathbf{k}_i|$. This gives a direct measurement of $S(q)$. The experimental values of $S(q)-1$ are Fourier transformed to give $g(R)-1$. This procedure gives the experimental value for the pair distribution function. This process works only if the Born approximation is valid, that is, if there is only a single scattering in the fluid. Results for liquid metallic sodium are shown in figure 10.5, where $g(r)$ is the Na–Na correlation function. $g(r)=0$ for $r<2$ Å since the Na^+ ions do not interpenetrate. The peak at 4 Å is the first shell of nearest neighbors.

10.5.3 *Strong Scattering*

Another situation is when the scattering is quite strong. One example is the scattering of visible light by liquids such as water. The photons scatter from the individual water molecules. The *Huygens principle* is that the scattered waves add coherently into a new wave front that is going in the forward direction. The scattering slows down the photons and is represented by a refractive index. Thus, it appears as if light goes through water without scattering. In fact, it scatters strongly, but the scattered waves add coherently and continue along in the same direction as the incident light beam. A detector sitting at 90° to the incident beams sees no signal.

A very different case is when light is shown on a glass of milk. The light scatters strongly from the fat molecules, and the scattered waves do not add coherently. Instead, the light random walks through the milk. Its properties obey the diffusion equation. An incident light beam scatters many times and exits in all different directions. Light in water obeys the wave equation, but light in milk obeys the diffusion equation. The difference seems to be the ratio of $k\ell$, where ℓ is the scattering mean free path. If $k\ell > 1$ the light random walks, while if $k\ell < 1$ it obeys Huygens' principle. For more detail see Sheng's book [3].

10.6 Wave Packets

Most scattering experiments are done with pulsed beams. Particles are generated or accelerated in groups. Then it is useful to think of the scattering experiment in terms of wave packets. The packet is a local region in space that contains one or more particles. The packet approaches the target, gets scattered by it, and some part of the packet continues unscattered, while another part becomes an outgoing wave. The packet picture is a good visualization of this process. Chapter 2 discussed wave packets in one dimension. For scattering theory, we need to extend the treatment to three-dimensional packets.

10.6.1 *Three Dimensions*

A realistic description of a scattering experiment requires a three-dimensional wave packet. If the particle is going in the z-direction, it will still have a nonzero width in the transverse direction. It could even miss the target if poorly aimed!

$$\Xi(\mathbf{k}_0, \mathbf{r}, t) = A \int d^3k C(\mathbf{k}-\mathbf{k}_0) \exp[i(\mathbf{k}\cdot\mathbf{r}-t\omega(\mathbf{k}))] \tag{10.196}$$

where again $\omega(\mathbf{k}) = \hbar k^2/2m$ for nonrelativistic particles. If the width function is another Gaussian, then the three-dimensional packet is the product of three one-dimnsional ones, one in each direction:

$$C(\mathbf{k}-\mathbf{k}_0) = \exp\left[-\frac{\Delta^2}{2}(\mathbf{k}-\mathbf{k}_0)^2\right] \tag{10.197}$$

$$\Xi(\mathbf{k}_0, \mathbf{r}, t) = \Psi(k_{0x}, x, t)\Psi(k_{0y}, y, t)\Psi(k_{0z}, z, t) \tag{10.198}$$

If the particle is going in the $\hat{z}$-direction, then $k_{0z} = k_0$, $k_{0x} = 0 = k_{0y}$. In this case the three-dimensional wave packet is

$$\Xi(\mathbf{k}_0, \mathbf{r}, t) = G_3(\mathbf{r}-v_0 t) \exp[i(k_0 z-\omega_0 t)] \tag{10.199}$$

$$G_3(\mathbf{r}-v_0 t) = G(x)G(y)G(z-v_0 t) \tag{10.200}$$

where $G(x)$ are the one-dimensional packets.

The theory is not limited to Gaussian packets. Another way to express the answer is to change variables in the integral in eqn. (10.196) to $\mathbf{q} = \mathbf{k}-\mathbf{k}_0$ and find

$$\Xi(\mathbf{k}_0, \mathbf{r}, t) = Ae^{i(\mathbf{k}_0\cdot\mathbf{r}-\omega_0 t)} \int d^3 q C(\mathbf{q}) e^{i\mathbf{q}\cdot(\mathbf{r}-\vec{v}_0 t)} e^{it\hbar q^2/2m} \tag{10.201}$$

We can often neglect the last phase factor of $\exp[it\hbar q^2/2m]$. Since $q \sim O(\Delta)$, this neglect is similar to omitting the spreading term tu for the packets in one dimension. If this term can be neglected, then the above expression becomes

$$\Xi(\mathbf{k}_0, \mathbf{r}, t) = e^{i(\mathbf{k}_0\cdot\mathbf{r}-\omega_0 t)} \tilde{C}(\mathbf{r}-\vec{v}_0 t) \tag{10.202}$$

$$\tilde{C}(\mathbf{r}) = A \int d^3 q C(q) e^{i\mathbf{q}\cdot\mathbf{r}} \tag{10.203}$$

where $\tilde{C}$ is the Fourier transform of $C(\mathbf{q})$. This way of writing the answer applies to all packet functions, including Gaussian and Lorentzian.

10.6.2 Scattering by Wave Packets

The plane wave in the argument of eqn. (10.196) is replaced by the wave function for scattering:

$$\Xi(\mathbf{k}_0, \mathbf{r}, t) = A \int d^3 k C(\mathbf{k}-\mathbf{k}_0) \psi(\mathbf{k}, \mathbf{r}) \exp[-it\omega(\mathbf{k})] \tag{10.204}$$

$$\psi(\mathbf{k}_i, \mathbf{r}) = e^{i\mathbf{k}_i\cdot\mathbf{r}} + \int d^3 r' G(k, \mathbf{r}-\mathbf{r}') V(\mathbf{r}') \psi(\mathbf{k}, \mathbf{r}') \tag{10.205}$$

The first term in $\psi(\mathbf{k}_i, \mathbf{r})$ gives the incoming wave packet:

$$\Xi(\mathbf{k}_0, \mathbf{r}, t) = \Xi_i(\mathbf{k}_0, \mathbf{r}, t) + \Xi_S(\mathbf{k}_0, \mathbf{r}, t) \tag{10.206}$$

$$\Xi_i(\mathbf{k}_0, \mathbf{r}, t) = e^{i(\mathbf{k}_0\cdot\mathbf{r}-\omega_0 t)} \tilde{C}(\mathbf{r}-\vec{v}_0 t) \tag{10.207}$$

Evaluate the scattering packet in the Born approximation:

$$\Xi_S(\mathbf{k}_0, \mathbf{r}, t) = -\frac{mA}{2\pi\hbar^2 r} \int d^3 k C(\mathbf{k}-\mathbf{k}_0) e^{ikr - i\omega_k t} \tag{10.208}$$
$$\times \int d^3 r' V(r') e^{i\mathbf{r}'\cdot(\mathbf{k}-k\hat{r})}$$

Make the variable change $\mathbf{q} = \mathbf{k} - \mathbf{k}_0$, and neglect terms of $O(qr')$ in the integral. The distance r' is of the order of an atom, while $1/q$ is the size of the packet, which is usually millimeters or more. Also neglect the spreading terms. Then the integrals give

$$\Xi_S(\mathbf{k}_0, \mathbf{r}, t) = -\frac{m}{2\pi\hbar^2 r} e^{ik_0 r - i\omega_0 t} \tilde{C}[\hat{k}_0(r-v_0 t)] \tilde{V}(\mathbf{k}_0 - \mathbf{k}_{f0}) \tag{10.209}$$

$$\mathbf{k}_{f0} = k_0 \hat{r} \tag{10.210}$$

The cross section is

$$\frac{d\sigma}{d\Omega} = \left[\frac{d\sigma}{d\Omega}\right]_0 |\tilde{C}[\hat{k}_0(r-v_0 t)]|^2 \tag{10.211}$$

$$\left[\frac{d\sigma}{d\Omega}\right]_0 = \left(\frac{m}{2\pi\hbar^2}\right)^2 |\tilde{V}(\mathbf{k}_0 - \mathbf{k}_{f0})|^2 \tag{10.212}$$

The last line is the usual definition of cross section in the Born approximation. The prior line gives the expression in terms of wave packets. The factor of $|\tilde{C}[\hat{k}_0(r - v_0 t)]|^2$ ensures that a signal is measured only when the packet is going by the counter, which is a distance of r from the target.

References

1. R.B. Brode, *Reviews of Modern Physics*, **5**, 257 (1933): "The Quantitative Study of the Collision of Electrons with Atoms"
2. N.S. Gingrich, *Reviews of Modern Physics* **15**, 90 (1943): "The Diffraction of X-Rays by Liquid Elements"
3. P. Sheng, *Introduction to Wave Scattering, Localization, and Mesoscopic Phenomena* (Academic Press, New York, 1995)

Homework

1. In one dimension, consider the scattering of a particle from the potential $V(x) = -\lambda\delta(x)$.

a. For a wave coming from the left,

$$\psi = \begin{cases} e^{ikx} + Re^{-ikx} & x<0 \\ Te^{ikx} & x>0 \end{cases} \tag{10.213}$$

Find R and T by matching wave functions at the origin.

b. Derive the same result from the scattering equation:

$$\psi(x) = e^{ikx} + \int dx' G(k, x-x') V(x') \psi(x') \tag{10.214}$$

Use the form of the Green's function in one dimension.

2. For the spherical square-well potential $V(r) = -V_0$ for $r<a$, verify the expression below by doing the integral for $\ell = 0$:

$$e^{i\delta_\ell} \sin(\delta_\ell) = -\frac{2m}{\hbar^2}\int_0^\infty r^2 dr j_\ell(kr) V(r) R_\ell(kr) \tag{10.215}$$

3. For bound states of negative energy, The Green's function obeys the equation

$$[\nabla^2 - \alpha^2] G(\alpha, \mathbf{r}-\mathbf{r}') = \frac{2m}{\hbar^2}\delta^3(\mathbf{r}-\mathbf{r}') \tag{10.216}$$

$$\alpha^2 = \frac{2m}{\hbar^2}E_B > 0 \tag{10.217}$$

a. Solve this equation and find $G(\alpha, R)$, $\mathbf{R} = \mathbf{r} - \mathbf{r}'$.
b. Show that bound states obey the integral equation

$$\phi_\alpha(\mathbf{r}) = \int d^3r' G(\alpha, \mathbf{r}-\mathbf{r}') V(\mathbf{r}') \phi_\alpha(\mathbf{r}') \tag{10.218}$$

c. Explicitly verify this equation by doing the integrals on the right for the hydrogen 1*s*-state.

4. For the scattering of two hard spheres, we found the *s*-wave phase shift is $\delta_0(k) = -kd$. Find the similar expression for $\delta_1(k)$.

5. Make a graph of the θ-dependence of the scattering cross section $(d\sigma/d\Omega)$ for the mutual scattering of two hard spheres of diameter d. Assume $kd = 1.0$, where k is the relative wave vector. For this value, only *s*- and *p*-wave phase shifts have nonnegligible value.

6. An electron scatters from a spherical square well of depth V_0 and radius $d = 5.0$ Å.

a. What is the value of V_0 that just binds an electron with $\ell = 0$?
b. Increase that value of V_0 by 10%. What is the differential cross section as $k \rightarrow 0$?

7. From the small binding energy of the deuteron (n + p), the nuclear potential between a proton and a neutron is approximated as a spherical square well of depth $V_0 = 36$ MeV and radius $a = 2.0 \times 10^{-15}$ meters. Estimate the numerical value of the total cross section for low-energy $(E \rightarrow 0)$ scattering of a proton with a neutron.

8. Derive the formula in two dimensions for the total cross section of the elastic scattering of two "hard circles" (e.g., two checkers).

9. Calculate the differential cross section in the first Born approximation for elastic scattering of a proton by a nuclei of charge Z, where the potential is from the Coulomb interaction. Show that this gives the Rutherford formula.

10. Derive the differential cross section for the inelastic scattering of an electron by a hydrogen atom in the first Born approximation. The initial state has an electron in $\mathbf{k}_i$ and H in 1*s*, and the final state has an electron in $\mathbf{k}_f$ and H in 2*s*. Include exchange, and discuss the matrix elements.

11. Solve the *s*-wave scattering in the DWBA using the two potentials:

$$V_1(r) = \begin{cases} \infty & 0 < r < a \\ 0 & a < r \end{cases} \tag{10.219}$$

$$V_2(r) = \begin{cases} -V_0 & a < r < b \\ 0 & b < r \end{cases} \tag{10.220}$$

The potential V_1 is a hard sphere that forces the wave function to vanish at $r = a$.

12. Consider a particle scattering from a strong square-well potential and a weak delta-function potential:

$$V_1(r) = \begin{cases} -V_0 & r < a \\ 0 & a < r \end{cases} \tag{10.221}$$

$$V_2(r) = \lambda\delta^3(r) \tag{10.222}$$

Solve the *s*-wave scattering amplitude in the distorted wave Born approximation.

11 Relativistic Quantum Mechanics

The previous chapters have been concerned with nonrelativistic quantum mechanics. The basic equation is Schrödinger's equation. It does not apply when the particles have kinetic energies of the same size as their rest energy mc^2. Then a full relativistic equation is required. In this chapter we discuss two different relativistic equations: Klein-Gordon and Dirac. The K-G (Klein-Gordon) equation applies to spinless particles such as mesons. The Dirac equation applies to spin-1/2 fermions, such as electrons, nucleons, or neutrinos.

11.1 Four-Vectors

In special relativity, the energy of a free particle is given by

$$E(p) = \sqrt{m^2c^4 + c^2p^2} \tag{11.1}$$

where m is the mass, p is the momentum, and c is the speed of light in vacuum. An important feature of relativity is that time plays a role on the same footing as space. Instead of using three vectors such as $\mathbf{r} = (x, y, z)$, the theory requires the use of a *four-vector* $(\mathbf{r}, ct) = (x, y, z, ct)$. This space–time four-vector is denoted as x_μ, $(\mu = 1, 2, 3, 4)$, where $x_1 = x$, $x_2 = y$, $x_3 = z$, $x_4 = ct$. The invariant quantity in relativity is $r^2 - (ct)^2$. When we multiply together two four-vectors, we must change the sign of the last term. This is done by introducing the matrix

$$g_{\mu\nu} = \begin{pmatrix} 1 & 0 & 0 & 0 \\ 0 & 1 & 0 & 0 \\ 0 & 0 & 1 & 0 \\ 0 & 0 & 0 & -1 \end{pmatrix} \tag{11.2}$$

$$x \cdot x = \sum_{\mu\nu} x_\mu g_{\mu\nu} x_\nu = r^2 - (ct)^2 \tag{11.3}$$

The tensor $g_{\mu\nu}$ is diagonal, with $+1$ in the first three diagonal elements and -1 in the fourth. Different authors use different notation. Some put the time component first, $x_\mu = (ct, x, y, z)$, and then the first element of $g_{\mu\nu}$ is -1.

There are other important four-vectors. In listing them, the subscript μ or ν always denotes four components. The subscript i denotes the three spatial components, e.g., $r_i = (x, y, z)$.

1. The momentum is $p_\mu = (p_x, p_y, p_z, E/c)$, where E is the energy.
2. The vector potential is $A_\mu = (A_x, A_y, A_z, \phi)$, where ϕ is the scalar potential.
3. The current is $j_\mu = (j_x, j_y, j_z, c\rho)$, where ρ is the charge density.
4. The derivative is

$$\frac{\partial}{\partial x_\mu} = \left(\frac{\partial}{\partial x}, \frac{\partial}{\partial y}, \frac{\partial}{\partial z}, -\frac{\partial}{c\partial t} \right) \tag{11.4}$$

Note the sign change in the last term. The sign change comes because we have

$$p_\mu = \frac{\hbar}{i} \frac{\partial}{\partial x_\mu}, \quad E = i\hbar \frac{\partial}{\partial t} \tag{11.5}$$

where i is in the denominator in **p** and is in the numerator in E. This change in the location of i is the change in sign.

The product of two four-vectors is usually an important equation. For example, the product of the momentum with itself is an invariant:

$$p \cdot p = \sum_{\mu\nu} p_\mu g_{\mu\nu} p_\nu = p^2 - \frac{E^2}{c^2} = -(mc)^2 \tag{11.6}$$

The latter identity follows from eqn. (11.1). Following are other examples of four-vector products.

1. The product of the derivative and the current gives the equation of continuity:

$$\frac{\partial}{\partial x} \cdot j = \sum_{\mu\nu} \frac{\partial}{\partial x_\mu} g_{\mu\nu} j_\nu = \vec{\nabla} \cdot \vec{j} + \frac{\partial \rho}{\partial t} = 0 \tag{11.7}$$

This equation is valid in the relativistic limit.

2. The product of the derivative and the vector potential gives the Lorentz gauge condition in electromagnetism:

$$\frac{\partial}{\partial x} \cdot A = \sum_{\mu\nu} \frac{\partial}{\partial x_\mu} g_{\mu\nu} A_\nu = \vec{\nabla} \cdot \vec{A} + \frac{\partial \phi}{c\partial t} = 0 \tag{11.8}$$

3. The product of the derivative with itself gives the wave equation:

$$\frac{\partial}{\partial x} \cdot \frac{\partial}{\partial x} = \sum_{\mu\nu} \frac{\partial}{\partial x_\mu} g_{\mu\nu} \frac{\partial}{\partial x_\nu} = \nabla^2 - \frac{1}{c^2} \frac{\partial^2}{\partial t^2} \tag{11.9}$$

Any product of four-vectors is invariant under a Lorentz transformation. Most of the basic equations can be written as products of four-vectors. This idea provides the guidance needed to write down a relativistically invariant Hamiltonian for various particles.

11.2 Klein-Gordon Equation

11.2.1 *Derivation*

The derivation of the K-G equation proceeds in several steps.

1. The first is to start with (11.6):

$$\left[p^2 + (mc)^2\right]\phi(x) = \frac{E^2}{c^2}\phi(x) \tag{11.10}$$

2. The second step is to insert potentials. The potentials are generated by either a vector or scalar potential. In nonrelativistic physics, the vector potential was inserted using

$$\mathbf{p} \to \mathbf{p} - \frac{e}{c}\mathbf{A} \tag{11.11}$$

We follow the same procedure, but it must be done with four-vectors. For the three vector components we still use eqn. (11.11). But the same substitution must be done in the fourth component:

$$\frac{E}{c} \to \frac{E - e\phi}{c} \tag{11.12}$$

$$\left[c^2(\mathbf{p} - \frac{e}{c}\mathbf{A})^2 + (mc^2)^2\right]\phi(x) = (E - V)^2\phi(x) \tag{11.13}$$

where $V(\mathbf{r}) = e\phi(\mathbf{r})$ is the usual scalar potential.

3. The last step is to replace the four momenta with their derivatives in eqn. (11.5):

$$\left[c^2(\frac{\hbar}{i}\vec{\nabla} - \frac{e}{c}\mathbf{A})^2 + (mc^2)^2\right]\phi(x) = \left(i\hbar\frac{\partial}{\partial t} - V\right)^2\phi(x) \tag{11.14}$$

Equation (11.14) is the Klein-Gordon equation.

11.2.2 *Free Particle*

The easiest solution is for a free particle, where all vector and scalar potentials are zero:

$$[-c^2\hbar^2\nabla^2 + (mc^2)^2]\psi(\mathbf{r}, t) = -\hbar^2\frac{\partial^2}{\partial t^2}\psi(\mathbf{r}, t) \tag{11.15}$$

The obvious solution is

$$\psi(\mathbf{r}, t) = \frac{1}{\sqrt{\Omega}}\exp[i\mathbf{k}\cdot\mathbf{r} - iEt/\hbar] \tag{11.16}$$

$$E(k)^2 = (\hbar ck)^2 + (mc^2)^2 \tag{11.17}$$

The plane wave gives the correct relativistic dispersion for a free particle.

The K-G equation applies to spinless particles such as pions. Below we solve this equation with a potential in order to find standard solutions such as the eigenfunctions in a Coulomb potential. Before proceeding, it is necessary to comment on a peculiar feature of the K-G equation.

11.2.3 Currents and Densities

The solutions to the K-G equation obey the equation of continuity. This section derives the expressions for the current operator and the charge density. The charge density has a peculiar form in the K-G equation.

The derivation is more efficient using four-vector notation. Write the K-G equation as

$$(cp-eA)^2\psi=-(mc^2)^2\psi \tag{11.18}$$

The product of two four-vectors has the definition of (11.13). Multiply this equation from the left by ψ^*:

$$\psi^*(cp-eA)^2\psi=-(mc^2)^2|\psi|^2 \tag{11.19}$$

Then take the complex conjugate (not the Hermitian conjugate) of this expression and subtract the two equations. Terms without derivatives, such as the right-hand side of the above equation, cancel:

$$\psi(cp^*-eA)^2\psi^*=-(mc^2)^2|\psi|^2 \tag{11.20}$$

$$0=c^2[\psi^*p\cdot p\psi-\psi p\cdot p\psi^*]-2ec[\psi^*p\psi+\psi p\psi^*]\cdot A$$

Since p_μ in eqn. (11.5) contains i, then $p^*=-p$, which explains the sign change in the last term. Also note that the Lorentz gauge condition is $p\cdot A=0$ so the order of these two quantities does not matter. The above expression can also be written as

$$0=p\cdot\left[\psi^*p\psi-\psi p\psi^*-\frac{2e}{c}A|\psi|^2\right] \tag{11.21}$$

Since the equation of continuity is $p\cdot j=0$, we identify the quantity in brackets as the definition of current, which means we must multiply by $e/2m$:

$$j=\frac{e}{2m}\left[\psi^*p\psi-\psi p\psi^*-\frac{2e}{c}A|\psi|^2\right] \tag{11.22}$$

Since $j_\mu=(\vec{j},c\rho)$, we obtain an explicit definition of both the current and charge density:

$$\vec{j}=\frac{e\hbar}{2mi}(\psi^*\vec{\nabla}\psi-\psi\vec{\nabla}\psi^*)-\frac{e^2}{mc}\mathbf{A}|\psi|^2 \tag{11.23}$$

$$\rho=\frac{ie\hbar}{2mc^2}\left(\psi^*\frac{\partial}{\partial t}\psi-\psi\frac{\partial}{\partial t}\psi^*\right)-\frac{e^2}{mc^2}\phi|\psi|^2 \tag{11.24}$$

The current has exactly the same form as found in nonrelativistic physics. The expression for the charge density is not the same and is odd indeed. We might expect that $\rho\sim e|\psi|^2$, but that is not correct for the K-G equation. The correct expression is (11.24).

A simple example is given by the plane wave, using the wave function in (11.16). Since $\mathbf{A}=0,\ V=0$, we find

$$\rho = \frac{eE(p)}{mc^2\Omega} \tag{11.25}$$

At least this answer is a positive number times e. When the four-vector $A \neq 0$ sometimes ρ is a negative number times e. It is not possible to think of ρ as a particle density multiplied by a charge. Instead, ρ is the charge density, which has no simple relation to the particle density.

11.2.4 Step Potential

A simple example of the charge density is provided by a particle that scatters from a step potential. In one dimension, set $\vec{A}=0,\ e\phi = V(x) = V_0\Theta(x)$. A particle approaches this step potential from the left with energy $E > V_0 + mc^2$. The eigenfunction in the two regions is

1. $x<0$ has

$$k^2 = \frac{E^2-(mc^2)^2}{(\hbar c)^2} > 0 \tag{11.26}$$

$$\psi(x) = Ie^{ikx} + Re^{-ikx} \tag{11.27}$$

2. $x>0$ has

$$p^2 = \frac{(E-V_0)^2-(mc^2)^2}{(\hbar c)^2} > 0 \tag{11.28}$$

$$\psi(x) = Te^{ipx} \tag{11.29}$$

Since the K-G equation has two x-derivatives, one matches the eigenfunction and its first derivative at $x=0$:

$$I+R=T,\ ik(I-R)=ipT \tag{11.30}$$

$$R=I\frac{k-p}{k+p},\quad T=I\frac{2k}{k+p} \tag{11.31}$$

This matching is identical to the nonrelativistic case. The interesting result occurs when calculating the charge density:

$$\rho(x) = \frac{e[E-V(x)]}{mc^2}|\psi|^2 \tag{11.32}$$

The charge density is not continuous at $x=0$.

The K-G equation has a formula for the charge density that makes predictions that are very strange. They are not necessarily wrong, but are not intuitive.

11.2.5 Nonrelativistic Limit

It is always useful to derive the nonrelativistic limit of a relativisitic equation, to ensure that the proper limit is obtained. For the K-G equation, let the wave function $\psi(\mathbf{r},t)$ be given by

$$\psi(\mathbf{r},\ t) = \Psi(\mathbf{r},\ t)e^{-imc^2t/\hbar} \tag{11.33}$$

Take various time derivatives:

$$\frac{\partial\psi}{\partial t} = e^{-imc^2t/\hbar}\left[\frac{\partial\Psi}{\partial t} - i\frac{mc^2}{\hbar}\Psi\right] \tag{11.34}$$

$$\frac{\partial^2\psi}{\partial t^2} = e^{-imc^2t/\hbar}\left[\frac{\partial^2\Psi}{\partial t^2} - i\frac{2mc^2}{\hbar}\frac{\partial\Psi}{\partial t} - \frac{m^2c^4}{\hbar^2}\Psi\right] \tag{11.35}$$

These derivatives are inserted into the K-G equation (11.14):

$$0 = \left[-\hbar^2\nabla^2 - \frac{2e\hbar}{ic}\mathbf{A}\cdot\vec{\nabla} + \frac{e^2}{c^2}A^2 + \frac{\hbar^2}{c^2}\frac{\partial^2}{\partial t^2} - 2im\left(\hbar\frac{\partial}{\partial t} + iV\right) - \frac{V^2}{c^2} + 2i\frac{V\hbar}{c^2}\frac{\partial}{\partial t}\right]\Psi \tag{11.36}$$

Divide the entire equation by $2m$, and rearrange terms:

$$i\hbar\frac{\partial\Psi}{\partial t} = \left(-\frac{\hbar^2\nabla^2}{2m} - \frac{e\hbar}{imc}\mathbf{A}\cdot\vec{\nabla} + \frac{e^2}{2mc^2}A^2 + V\right)\Psi - \frac{1}{mc^2}\left(V^2 - \hbar V\frac{\partial}{\partial t} - \hbar^2\frac{\partial^2}{\partial t^2}\right)\Psi \tag{11.37}$$

On the right, the second line contains the terms that are divided by mc^2. All of these terms are neglected in the nonrelativistic limit. The first line on the right has the familiar terms

$$H = \frac{1}{2m}(\mathbf{p} - \frac{e}{c}\mathbf{A})^2 + V \tag{11.38}$$

The transformation (11.33) does produce the standard nonrelativistic limit for a spinless particle.

How does the charge density behave in the nonrelativistic limit? Inserting the time derivatives into (11.24) gives

$$\rho = \frac{ie\hbar}{2mc^2}\left[\Psi^*\left(\frac{\partial\Psi}{\partial t} - i\frac{mc^2}{\hbar}\Psi\right) - \Psi\left(\frac{\partial\Psi^*}{\partial t} + i\frac{mc^2}{\hbar}\Psi^*\right)\right] - \frac{eV}{mc^2}|\Psi|^2 = e|\Psi|^2\left[1 + O\left(\frac{E' - V}{mc^2}\right)\right] \tag{11.39}$$

where $E' = E - mc^2$ is the nonrelativistic energy. The first term gives the correct nonrelativistic limit. The current operator is unchanged in the nonrelativistic limit since it has no time derivatives.

11.2.6 π-*Mesonic Atoms*

If negative pions or muons are injected into matter, they are attracted to the positive nucleus by Coulomb's law. They form hydrogenic bound states that can be observed in x-ray spectra. They are much more tightly bound than electrons due to the much larger mass of the pion. The pions are inside all of the electron orbits, so they are unscreened. Of course, the pions live only a short time, but the hydrogenic bound states are observed before the pion decay process. The K-G equation is solved for the Coulomb potential to give a relativistic theory for the pion bound states.

The K-G equation (11.14) is solved with the following features:

1. The vector potential $\mathbf{A} = 0$
2. The scalar potential $V = -Ze^2/r$
3. The bound states are stationary, so set $\psi(\mathbf{r},t) = \psi(\mathbf{r})\exp(-itE/\hbar)$ where $E < mc^2$ for bound states.

   These features yield the equation

$$0 = \left[c^2\hbar^2\nabla^2 + \left(E + \frac{Ze^2}{r}\right)^2 - (mc^2)^2 \right]\psi(\mathbf{r}) \tag{11.40}$$

This equation is solved in spherical coordinates. Assume the solution is a product of a radial function and a spherical harmonic:

$$\psi(\mathbf{r}) = R_{n\ell}(r)Y_\ell^m(\theta, \phi) \tag{11.41}$$

$$0 = \left[\hbar^2c^2\left(\frac{1}{r}\frac{\partial^2}{\partial r^2}r - \frac{\ell(\ell+1)}{r^2}\right) + E^2 - (mc^2)^2 + \frac{2Ze^2E}{r} + \frac{Z^2e^4}{r^2}\right]R_{n\ell}(r)$$

Next, define $\chi(r) = rR_{n\ell}$ and divide the above equation by $(\hbar c)^2$, yielding

$$0 = \left[\frac{\partial^2}{\partial r^2} - \frac{\ell(\ell+1) - Z^2\alpha^2}{r^2} + \frac{2Z\alpha E}{\hbar cr} + \frac{E^2 - m^2c^4}{\hbar^2c^2}\right]\chi(r)$$

where α is the fine structure constant. Note that the factor of $(Z\alpha)$ occurs in several terms. These terms represent the relativistic corrections.

There are four terms in the differential equation. The first is the second derivative. There is a constant term, one divided by r, and one divided by r^2. The same four types of terms are found when solving for the hydrogen atom in chapter 5. The coefficients in the numerator are different than in the nonrelativistic case. We use the same method of solving this equation. Use the constant term to define a unit of length r_0, and then define a dimensionless distance $\rho = r/r_0$:

$$\frac{1}{r_0^2} = 4\frac{m^2c^4 - E^2}{\hbar^2c^2} \tag{11.42}$$

$$0 = \left[\frac{d^2}{d\rho^2} + \frac{\lambda}{\rho} - \frac{\ell(\ell+1) - Z^2\alpha^2}{\rho^2} - \frac{1}{4}\right]\chi(\rho) \tag{11.43}$$

$$\lambda = \frac{2Z\alpha Er_0}{\hbar c} = \frac{Z\alpha E}{\sqrt{(mc^2)^2 - E^2}} \tag{11.44}$$

where the last equation for λ uses the definition of r_0. Equation (11.43) is solved by making the usual ansatz:

$$\chi(\rho) = \rho^{s+1} W(\rho) e^{-\rho/2} \tag{11.45}$$

$$s(s+1) = \ell(\ell+1) - (Z\alpha)^2 \tag{11.46}$$

$$s = -\frac{1}{2} + \sqrt{(\ell+1/2)^2 - (Z\alpha)^2} \tag{11.47}$$

The ansatz is chosen to give the correct behavior in the limits of small and of large ρ. Putting (11.45) into (11.43) gives a differential equation for $W(\rho)$:

$$0 = \left[\rho \frac{d^2 W}{d\rho^2} + (2s+2-\rho)\frac{dW}{d\rho} - (s+1-\lambda) W\right] \tag{11.48}$$

This equation has the form for the confluent hypergeometric function $F(a, b, \rho)$, where $a = s+1-\lambda$, $b = 2(s+1)$. The eigenvalue equation for bound states is found by setting a equal to a negative integer n_r:

$$\chi(\rho) = \rho^{s+1} e^{-\rho/2} F(s+1-\lambda,\ 2s+2,\ \rho) \tag{11.49}$$

$$-n_r = s+1-\lambda,\ n_r = 0,\ 1,\ 2,\ \ldots \tag{11.50}$$

$$\lambda = s+1+n_r \tag{11.51}$$

Use (11.44) for λ, and solve this equation for E:

$$E_{\ell,n_r} = \frac{mc^2}{\sqrt{1+(Z\alpha/\lambda)^2}} \tag{11.52}$$

$$\lambda = n_r + 1/2 + \sqrt{(\ell+1/2)^2 - (Z\alpha)^2} \tag{11.53}$$

Equation (11.52) is the final solution for the eigenvalue. It depends on the quantity λ, which depends on the two quantum numbers: the angular momentum ℓ and the radial quantum number n_r. Note for bound states that $0 < E < mc^2$.

The following are some sample eigenvalues.

1. The lowest eigenvalue ("ground state") is when $\ell = 0$, $n_r = 0$. Then

$$\lambda = \frac{1}{2} + \sqrt{\frac{1}{4} - (Z\alpha)^2} \tag{11.54}$$

Note that the argument of the square root is negative when $Z\alpha > \frac{1}{2}$. That is experimentally possible, since $\alpha \approx \frac{1}{137}$, while Z can be any integer up to 92, or higher. This unphysical result is caused by assuming the nucleus is a point charge. When $Z\alpha$ is large, the orbits becomes as small as the nucleus, and the finite size of the nucleus has to be taken into account. That removes the unphysical result of imaginary values of λ. It also makes a significant change in the eigenvalue.

For the ground state,

$$s = \sqrt{\frac{1}{4} - (Z\alpha)^2} - \frac{1}{2} < 0 \tag{11.55}$$

That s is negative does not cause problems, since $s+1 > 0$.

2. The next set of energy levels correspond to the $n = 2$ state of the hydrogen atom. Here the two levels do not have the same energy, unlike the nonrelativistic case:

   - The s-state has $\ell = 0$, $n_r = 1$ and

$$\lambda = \frac{3}{2} + \sqrt{\frac{1}{4} - (Z\alpha)^2} \tag{11.56}$$

   - The p-state has $\ell = 1, n_r = 0$

$$\lambda = \frac{1}{2} + \sqrt{\frac{9}{4} - (Z\alpha)^2} \tag{11.57}$$

3. The next set of three levels corresponds to the $n = 3$ state of hydrogen. Here they all have different energies and correspond to the case $(\ell = 0, n_r = 2)$, $(\ell = 1, n_r = 1)$, $(\ell = 2, n_r = 0)$.

It is useful to take the nonrelativistic limit of the eigenvalue in (11.52). Assume that $Z\alpha <,1$ and expand the numerator:

$$E = mc^2 \left[1 - \frac{1}{2} \left(\frac{Z\alpha}{\lambda} \right)^2 + \frac{3}{8} \left(\frac{Z\alpha}{\lambda} \right)^4 + \cdots \right] \tag{11.58}$$

The first term is the rest energy mc^2. In the next term, note that $mc^2\alpha^2 = 2E_{\rm Ry}$, where $E_{\rm Ry} = me^2/2\hbar^2$ is the Rydberg energy for the pion mass. Next expand λ in powers of $(Z\alpha)^2$. Remember that the principle quantum number is $n = \ell + 1 + n_r$:

$$E = mc^2 - \frac{Z^2 E_{\rm Ry}}{\lambda^2} \left[1 - \frac{3}{4} \frac{Z^2\alpha^2}{\lambda^2} + \cdots \right] \tag{11.59}$$

$$\lambda = n_r + \ell + 1 - \frac{Z^2\alpha^2}{2\ell + 1} + O(Z^4\alpha^4) \tag{11.60}$$

$$\approx n - \frac{Z^2\alpha^2}{2\ell + 1} \tag{11.61}$$

$$\frac{1}{\lambda} \approx \frac{1}{n} + \frac{Z^2\alpha^2}{n^2(2\ell + 1)} + O(Z^4\alpha^4) \tag{11.62}$$

$$E = mc^2 - \frac{Z^2 E_{\rm Ry}}{n^2} - \frac{Z^4 E_{\rm Ry}\alpha^2}{n^3} \left(\frac{2}{2\ell + 1} - \frac{3}{4n} \right) + O(Z^6\alpha^6) \tag{11.63}$$

The third term on the right is the first relativistic correction. If relativistic corrections are important, it is better to use the full result (11.52).

Table 11.1 shows some experimental values of the 1s to 2p transition in a π-mesonic atom, compared with the predictions of the K-G equation. The predictions get worse as the value of $Z\alpha$ increases, due to the finite size of the nucleus.

11.3 Dirac Equation

The Dirac equation is a relativistic Hamiltonian that describes the properties of particles with spin$-\frac{1}{2}$. Here we write down this equation and try to make it plausible. One

Table 11.1 $1s \rightarrow 2p$ **energies for pions bound to nuclei, compared to the predictions of the Klein-Gordon equation**

Atom	Z	Exp. (keV)	K-G (keV)
Li	3	23.8	24.5
Be	4	42.1	43.9
B	5	65.2	68.7
C	6	92.6	99.3

Source. From M.B. Stearns, *Prog. Nuclear Phys.* **6**, 108–137 (1957).

cannot really derive a fundamental equation of physics. One writes it down as a good guess and examines its properties. If it agrees with experiment, it was a good guess indeed.

11.3.1 Derivation

Dirac was motivated to derive a relativistic Hamiltonian equation with a single time derivative. The double time derivative of the Klein-Gordon equation leads to the strange formula for the charge density. The charge density will have a simple form if the Hamiltonian has a single time derivative:

$$i\hbar \frac{\partial}{\partial t}\psi(\mathbf{r}, t) = H\psi(\mathbf{r}, t) \tag{11.64}$$

The discussion of four-vectors at the beginning of this chapter emphasized that equations which are relativistic invariant are composed of four-vectors. The left-hand side of (11.64) is equivalent to the energy E, which is part of the four-vector of momentum. The single power of E requires a single power of the momentum $\mathbf{p}$. Furthermore, the Hamiltonian must contain the rest mass energy mc^2. So make a guess that the Hamiltonian has to have the form

$$H = c\vec{\alpha} \cdot \mathbf{p} + \beta mc^2 \tag{11.65}$$

Here $\vec{\alpha} = (\alpha_x, \alpha_y, \alpha_z)$ is a new use of the symbol α, and β has none of its prior meanings. What must be the form of $\vec{\alpha}, \beta$ if eqns. (11.64) and (11.65) are a relativistically invariant Hamiltonian?

1. $\vec{\alpha}$, β must be independent of $(\mathbf{r}, t, \mathbf{p}, E)$.
2. Since H has to be Hermitian, then so must $\vec{\alpha}, \beta$.
3. All particles obey the Klein-Gordon equation. The K-G equation is derived from (11.64) by taking

$$\left(i\hbar\frac{\partial}{\partial t}\right)^2\psi(\mathbf{r},t)=H^2\psi(\mathbf{r},t)=(c\vec{\alpha}\cdot\mathbf{p}+\beta mc^2)^2\psi$$
$$=[c^2p^2+(mc^2)^2]\psi \tag{11.66}$$

If we compare the last two equations, the Dirac equation gives the K-G equation provided

$$\alpha_i^2=1 \tag{11.67}$$

$$\beta^2=1 \tag{11.68}$$

$$\alpha_i\alpha_j+\alpha_j\alpha_i=2\delta_{ij} \tag{11.69}$$

$$\alpha_i\beta+\beta\alpha_i=0 \tag{11.70}$$

where (i, j) are the three space components.

Since scalars always commute, the last two equations require that α_i, β are not scalars. They must be matrices. The Dirac equation (11.64) is a matrix equation and the eigenfunction $\psi(\mathbf{r}, t)$ is actually a vector (sometimes called a *spinor*). The spinor denotes the spin states of the fermion.

As an aside, physics students are familiar with differential equations written in matrix form, although this feature is often disguised. For example, one of the equations of Maxwell is written in conventional and then matrix form:

$$\vec{\nabla}\times\vec{H}=\frac{1}{c}\frac{\partial}{\partial t}\vec{E} \tag{11.71}$$

$$\begin{bmatrix} 0 & -\partial/\partial z & \partial/\partial y \\ \partial/\partial z & 0 & -\partial/\partial x \\ -\partial/\partial y & \partial/\partial x & 0 \end{bmatrix}\begin{bmatrix} H_x \\ H_y \\ H_z \end{bmatrix}=\frac{\partial}{\partial(ct)}\begin{bmatrix} E_x \\ E_y \\ E_z \end{bmatrix} \tag{11.72}$$

The Dirac equation is often written in a form that disguises its matrix nature.

The Dirac matrices are actually 4×4. It is shown below that this is the lowest dimension of a matrix that satisfies the above requirements.

Two dimensions. In two dimensions the four Hermitian matrices that span the space are the three Pauli matrices and the identity matrix:

$$\sigma_x=\begin{pmatrix} 0 & 1 \\ 1 & 0 \end{pmatrix},\quad \sigma_y=\begin{pmatrix} 0 & -i \\ i & 0 \end{pmatrix} \tag{11.73}$$

$$\sigma_z=\begin{pmatrix} 1 & 0 \\ 0 & -1 \end{pmatrix},\quad I=\begin{pmatrix} 1 & 0 \\ 0 & 1 \end{pmatrix} \tag{11.74}$$

The three Pauli matrices do anticommute:

$$\{\sigma_i,\sigma_j\}=\sigma_i\sigma_j+\sigma_j\sigma_i=2\delta_{ij} \tag{11.75}$$

and they could serve as the three components of the vector matrices $\vec{\alpha}$. However, the Dirac equation requires a fourth matrix (β) that anticommutes with these three. The identity matrix commutes with everything and anticommutes with nothing. Two-dimensional space does not have four Hermitian matrices that anticommute.

Three dimensions. There is a simple proof that the matrices have to have a dimension that is an even integer. Odd dimensions, such as 3, 5, 7, are not possible. The proof starts out with any of the anticommutation relations, such as

$$\alpha_x \beta = -\beta \alpha_x = (-I)\beta \alpha_x \tag{11.76}$$

On the right we have put in the minus sign as the matrix $(-I)$, which has all elements along the diagonal equal to -1 and all other elements zero. For dimension three it is

$$-I = \begin{pmatrix} -1 & 0 & 0 \\ 0 & -1 & 0 \\ 0 & 0 & -1 \end{pmatrix} \tag{11.77}$$

Then we use a theorem that

$$\det[AB] = \det[A]\det[B] \tag{11.78}$$

$$\det[\alpha_x \beta] = \det[\alpha_x]\det[\beta] = \det[-I]\det[\beta]\det[\alpha_x] \tag{11.79}$$

Since determinants are c-numbers, they can be canceled from the two sides of the equation. The determinant of the matrix $(-I)$ is $(-1)^N$, where N is the dimension:

$$1 = (-1)^N \tag{11.80}$$

The last equation requires that N be an even integer, such as 2, 4, etc. We have shown that the dimension of the matrix N cannot be 1, 2, 3, 5. The choice of $N = 4$ works.

There are many possible representations for the four matrices $(\alpha_x, \alpha_y, \alpha_z, \beta)$. Several possible choices are given as homework assignments. One very popular choice is

$$\alpha_k = \begin{pmatrix} 0 & \sigma_k \\ \sigma_k & 0 \end{pmatrix}, \quad \beta = \begin{pmatrix} I & 0 \\ 0 & -I \end{pmatrix} \tag{11.81}$$

Here we have introduced an important shorthand for the Dirac matrices. The three Pauli matrices σ_k are of dimension 2×2, while α_k is 4×4. For example,

$$\alpha_x = \begin{pmatrix} 0 & 0 & 0 & 1 \\ 0 & 0 & 1 & 0 \\ 0 & 1 & 0 & 0 \\ 1 & 0 & 0 & 0 \end{pmatrix} \equiv \begin{pmatrix} 0 & \sigma_x \\ \sigma_x & 0 \end{pmatrix} \tag{11.82}$$

The shorthand notation in (11.81) is an easier way of writing the 4×4 matrix. In (11.81) the matrix I is the 2×2 identity matrix.

In summary, the Dirac equation is a matrix equation:

$$i\hbar \frac{\partial}{\partial t}\psi_\nu = c \sum_{k=x,y} \sum_{\mu=1}^{4} \alpha_{\nu\mu}^{(k)} p_k \psi_\mu + mc^2 \sum_{\mu=1}^{4} \beta_{\nu\mu}\psi_\mu \tag{11.83}$$

where $\nu = 1, 2, 3, 4$. Note that we have written $\alpha_{\nu\mu}^{(k)}$ for the 4×4 matrix of vector component k.

11.3.2 *Current and Charge Density*

A primary motivation for the Dirac equation is to have a simple expression for the charge density. It is derived here and indeed the result is simple and logical.

Multiply eqn. (11.83) by ψ_ν^* and sum over all four values of ν. Also use the gradient formula for the three-momentum:

$$i\hbar\sum_\nu \psi_\nu^* \frac{\partial}{\partial t}\psi_\nu = c\sum_{k=x,y,z}\sum_{\mu,\nu=1}^{4}\frac{\hbar}{i}\psi_\nu^*\alpha_{\nu\mu}^{(k)}\frac{\partial}{\partial x_k}\psi_\mu + mc^2\sum_{\mu,\nu=1}^{4}\psi_\nu^*\beta_{\nu\mu}\psi_\mu$$

Next take the complex conjugate of this expression (not the Hermitian conjugate) . Then subtract these two equations:

$$i\hbar\sum_\nu \frac{\partial}{\partial t}(\psi_\nu^*\psi_\nu) = c\sum_{k=x,y,z}\sum_{\mu,\nu=1}^{4}\frac{\hbar}{i}\left(\psi_\nu^*\alpha_{\nu\mu}^{(k)}\frac{\partial}{\partial x_k}\psi_\mu + \psi_\nu\alpha_{\nu\mu}^{(k)*}\frac{\partial}{\partial x_k}\psi_\mu^*\right)$$
$$+ mc^2\sum_{\mu,\nu=1}^{4}(\psi_\nu^*\beta_{\nu\mu}\psi_\mu - \psi_\nu\beta_{\nu\mu}^*\psi_\mu^*) \tag{11.84}$$

Recall that the Dirac matrices are Hermitian, which is

$$\beta_{\nu\mu}^* = \beta_{\mu\nu} \tag{11.85}$$

$$\alpha_{\nu\mu}^{(k)*} = \alpha_{\mu\nu}^{(k)} \tag{11.86}$$

Then the last term in eqn. (11.84) cancels to zero, and the other term on the right has a simple form:

$$i\hbar\sum_\nu \frac{\partial}{\partial t}(\psi_\nu^*\psi_\nu) = c\sum_{k=x,y,z}\sum_{\mu,\nu=1}^{4}\frac{\hbar}{i}\frac{\partial}{\partial x_k}(\psi_\nu^*\alpha_{\nu\mu}^{(k)}\psi_\mu) \tag{11.87}$$

If we cancel out the common factor of $i\hbar$ from both terms, what we have left is the equation of continuity:

$$\frac{\partial}{\partial t}\rho = -\sum_{k=x}\frac{\partial}{\partial x_k}j_k \tag{11.88}$$

$$\rho(\mathbf{r}, t) = e\sum_\nu(\psi_\nu^*\psi_\nu) \tag{11.89}$$

$$j_k(\mathbf{r}, t) = ec\sum_{\mu,\nu=1}^{4}(\psi_\nu^*\alpha_{\nu\mu}^{(k)}\psi_\mu) \tag{11.90}$$

The particle density has the desired form as the square of the absolute magnitude of the spinor wave function. The current has the interesting form as the matrix element of the $\alpha^{(k)}$ matrix. Unlike the nonrelativistic case, it does not contain any derivatives.

11.3.3 *Gamma Matrices*

Equation (11.83) is a popular way to write the Dirac equation. We shall use it many times. However, it is useful to have a formula that explicitly employs a four-vector formalism. The four-vector method uses matrices represented by the symbol γ_μ:

$$\left[\sum_{\mu,\nu} \gamma_\mu g_{\mu\nu} p_\nu + mc\right]\psi = 0 \tag{11.91}$$

This way of writing the equation is a bit confusing. It is still a 4×4 matrix equation, but all matrix indices have been suppressed. For example, the term mc should be multiplied by the 4×4 identity matrix. The wave function ψ is still a spinor. The γ-matrices should have three indices: two representing the matrix components and a third representing the four-vector components. Here we just use the latter. This way of writing the equation is conventional. An even more shorthand method is

$$[\gamma \cdot p + mc]\psi = 0 \tag{11.92}$$

Equation (11.91) is derived from eqn. (11.83) by multiplying the latter equation from the left by the matrix β. The first three gamma-matrices are defined by $\gamma_i = \beta\alpha^{(i)}$:

$$i\hbar\beta\frac{\partial}{\partial t}\psi = [c\vec{\gamma}\cdot\vec{p} + Imc^2]\psi \tag{11.93}$$

where I is the 4×4 identity matrix. Rearrange terms in the above equation to get

$$0 = \left[c\vec{\gamma}\cdot\vec{p} - i\hbar\beta\frac{\partial}{\partial t} + Imc^2\right]\psi \tag{11.94}$$

The four-vector $\gamma \cdot p$ is

$$\gamma \cdot p = \vec{\gamma}\cdot\vec{p} - \gamma_4 p_4 \tag{11.95}$$

For the last term on the right to agree with (11.94),

$$p_4 = -\frac{\hbar}{ic}\frac{\partial}{\partial t} \tag{11.96}$$

$$\gamma_4 = \beta,\ \gamma_\mu = (\vec{\gamma}, \beta) \tag{11.97}$$

This choice of gamma-matrices is conventional.

We examine some properties of gamma-matrices. Use $\alpha_k \equiv \alpha^{(k)}$ to denote the three alpha-matrices.

1. Take the Hermitian conjugate: $\gamma_k^\dagger = (\beta\alpha_k)^\dagger = \alpha_k\beta = -\beta\alpha_k = -\gamma_k$. Since β and α_k are both Hermitian and anticommute, we find that γ_k is skew-Hermitian: it changes sign under a Hermitian conjugate.
2. The fourth component is Hermitian: $\gamma_4^\dagger = \beta^\dagger = \beta = \gamma_4$.
3. The anticommutator of any of the three-components is

$$\{\gamma_i, \gamma_j\} = (\beta\alpha_i)(\beta\alpha_j) + (\beta\alpha_j)(\beta\alpha_i) \tag{11.98}$$

$$= -\beta^2\{\alpha_i, \alpha_j\} = -2\delta_{ij} \tag{11.99}$$

   The sign change comes from commuting β to the left.
4. $\{\gamma_4, \gamma_i\} = \beta(\beta\,\alpha_i) + (\beta\alpha_i)\beta = \beta^2\alpha_i(1-1) = 0$.
5. $\gamma_4^2 = \beta^2 = I$.

All of these results are summarized by the single equation

$$\{\gamma_\mu, \gamma_\nu\} = -2g_{\mu\nu} \tag{11.100}$$

where $g_{\mu\nu}$ is the tensor introduced in the section on four-vectors.

An advantage of writing the Dirac equation in four-vector notation is that we can easily introduce external fields as the four-vector potential:

$$0 = \left[\gamma \cdot \left(p - \frac{e}{c}A\right) + mc\right]\psi \tag{11.101}$$

Equation (11.101) can be used to find how potentials enter the Dirac Hamiltonian. By reversing the steps to derive the above equation, one finds

$$\left\{c\vec{\alpha} \cdot (\vec{p} - \frac{e}{c}\vec{A}) + \beta mc^2 - I[E - V(\mathbf{r})]\right\}\psi = 0 \tag{11.102}$$

This latter equation is the basis for many calculations, such as the relativistic hydrogen atom.

The interaction between fermions and photons is given by the usual interaction:

$$H_{\text{int}} = -\frac{1}{c}\int d^3r j(r) \cdot A(r) \tag{11.103}$$

$$j \cdot A = c[e(\psi^* \vec{\alpha}\psi) \cdot \vec{A} - \psi^* V(r)\psi] \tag{11.104}$$

The interaction is converted to γ-notation by defining

$$\bar{\psi} = \beta\psi,\ \bar{\psi}^\dagger = \psi^\dagger\beta^\dagger = \psi^\dagger\beta \tag{11.105}$$

Since $\beta^2 = I$, this factor is inserted in the middle of the matrix element, which gives

$$j \cdot A = c[e(\psi^* \beta^2 \vec{\alpha}\psi) \cdot \vec{A} - \psi^* \beta^2 V(r)\psi] \tag{11.106}$$

$$j \cdot A = c[e(\bar{\psi}^* \vec{\gamma}\psi) \cdot \vec{A} - \bar{\psi}^* \gamma_4 V(r)\psi] \tag{11.107}$$

$$H_{\text{int}} = e\int d^3r(\bar{\psi}^\dagger \gamma_\mu \psi) g_{\mu\nu} A_\nu(r) \tag{11.108}$$

Writing the interaction this way, in terms of γ-matrices, is quite conventional.

11.3.4 *Free-Particle Solutions*

We consider the solution for a free particle that has no external potential (A = 0). Generally, we write the solutions as

$$\psi_\nu(\mathbf{r}, t) = \Psi_\nu \exp[i(\mathbf{k} \cdot \mathbf{r} - tE(k)/\hbar)] \tag{11.109}$$

The space and time dependence has the usual form for a free particle. What is the form of the spinor function Ψ_ν?

We use the form of the Dirac equation in (11.81). In keeping with representing 4×4 states as two 2×2, we define the spinor as

$$\Psi_\nu = \begin{bmatrix} \Psi_1 \\ \Psi_2 \\ \Psi_3 \\ \Psi_4 \end{bmatrix} = \begin{bmatrix} \hat{\phi} \\ \hat{\chi} \end{bmatrix} \tag{11.110}$$

$$\hat{\phi} = \begin{bmatrix} \Psi_1 \\ \Psi_2 \end{bmatrix}, \ \hat{\chi} = \begin{bmatrix} \Psi_3 \\ \Psi_4 \end{bmatrix} \tag{11.111}$$

In terms of this notation, the stationary Dirac equation is

$$H\Psi_\mu = \begin{bmatrix} mc^2 & c\hbar\vec{\sigma}\cdot\vec{k} \\ c\hbar\vec{\sigma}\cdot\vec{k} & -mc^2 \end{bmatrix} \begin{bmatrix} \hat{\phi} \\ \hat{\chi} \end{bmatrix} = E \begin{bmatrix} \hat{\phi} \\ \hat{\chi} \end{bmatrix} \tag{11.112}$$

It separates into two equations, which are each 2×2 matrix equations:

$$0 = (mc^2 - E(k))\hat{\phi} + c\hbar\vec{\sigma}\cdot\vec{k}\hat{\chi} \tag{11.113}$$

$$0 = c\hbar\vec{\sigma}\cdot\vec{k}\hat{\phi} - (mc^2 + E(k))\hat{\chi} \tag{11.114}$$

$$\vec{\sigma}\cdot\vec{k} = \begin{pmatrix} k_z & k_x - ik_y \\ k_x + ik_y & -k_z \end{pmatrix} \tag{11.115}$$

The eigenvalue is found by setting to zero the determinant of the above equations:

$$0 = E(k)^2 - (mc^2)^2 - (\hbar c)^2(\vec{\sigma}\cdot\vec{k})^2 \tag{11.116}$$

$$(\vec{\sigma}\cdot\vec{k})^2 = k_x^2\sigma_x^2 + k_y^2\sigma_y^2 + k_z^2\sigma_z^2 + k_xk_y\{\sigma_x, \sigma_y\} \tag{11.117}$$

$$+ k_xk_z\{\sigma_x, \sigma_z\} + k_yk_z\{\sigma_y, \sigma_z\}$$

$$= k^2 \tag{11.118}$$

Since $\{\sigma_i, \sigma_j\} = 2\delta_{ij}$, only the diagonal terms are nonzero. The eigenvalue is

$$E(k)^2 = (mc^2)^2 + (\hbar ck)^2 \tag{11.119}$$

which is the correct result for a free particle.

Equation (11.114) is solved to give

$$\hat{\chi} = \frac{\hbar c\vec{\sigma}\cdot\vec{k}}{E(k) + mc^2}\hat{\phi} \tag{11.120}$$

Since $E > mc^2$, the denominator in the above expression is always larger than the numerator:

- $\hat{\phi}$ is called the *large component*
- $\hat{\chi}$ is called the *small component*

Write the spinor function as

$$\Psi_\nu = N_k \begin{pmatrix} \hat{\phi} \\ \frac{\hbar c \vec{\sigma}\cdot\vec{k}}{E(k)+mc^2}\hat{\phi} \end{pmatrix} \tag{11.121}$$

$$\hat{\phi} = \begin{pmatrix} \phi_1 \\ \phi_2 \end{pmatrix} \tag{11.122}$$

$$1 = \hat{\phi}^\dagger\hat{\phi} = \phi_1^2 + \phi_2^2 \tag{11.123}$$

The normalization constant N_k is found from

$$1 = \Psi^\dagger\Psi = N_k^2\left[\hat{\phi}^\dagger\hat{\phi} + \frac{(\hbar c)^2}{(E(k)+mc^2)^2}\hat{\phi}^\dagger(\mathbf{k}\cdot\vec{\sigma})^2\hat{\phi}\right] \tag{11.124}$$

$$= N_k^2\left[1 + \frac{(\hbar ck)^2}{(E(k)+mc^2)^2}\right] = N_k^2\frac{2E(k)}{E(k)+mc^2} \tag{11.125}$$

$$N_k = \sqrt{\frac{E(k)+mc^2}{2E(k)}} \tag{11.126}$$

These results are collected into the final expression for the eigenfunction of a free particle:

$$\psi_\nu = N_k \begin{pmatrix} \hat{\phi} \\ \frac{\hbar c \vec{\sigma}\cdot\vec{k}}{E(k)+mc^2}\hat{\phi} \end{pmatrix} \exp[i(\mathbf{k}\cdot\mathbf{r} - E(k)t/\hbar)] \tag{11.127}$$

There are no factors of volume, since we are using delta-function normalization. Using this eigenfunction, the charge density is

$$\rho = e|\psi|^2 = e\Psi^\dagger\Psi = e \tag{11.128}$$

The result is a constant, since a particle in a plane wave is equally likely to be anywhere. The current operator is more interesting:

$$j_\ell = ec(\psi^\dagger\alpha^{(\ell)}\psi) = ec(\Psi^\dagger\alpha^{(\ell)}\Psi) \tag{11.129}$$

$$= ecN_k^2[\hat{\phi}^\dagger, \hat{\chi}^\dagger]\begin{pmatrix} 0 & \sigma_\ell \\ \sigma_\ell & 0 \end{pmatrix}\begin{pmatrix} \hat{\phi} \\ \hat{\chi} \end{pmatrix} \tag{11.130}$$

$$j_\ell = ecN_k^2[\hat{\phi}^\dagger\sigma_\ell\hat{\chi} + \hat{\chi}^\dagger\sigma_\ell\hat{\phi}] \tag{11.131}$$

The second term in brackets is the Hermitian conjugate of the first. The first one is evaluated using eqn. (11.120) for the small component:

$$\hat{\phi}^\dagger\sigma_\ell\hat{\chi} = \frac{\hbar c}{E(k)+mc^2}(\hat{\phi}^\dagger\sigma_\ell\vec{\sigma}\cdot\vec{k}\hat{\phi}) \tag{11.132}$$

To evaluate the last expression, use the identity that, if $\vec{A} = (A_x, A_y, A_z)$ and $\vec{B} = (B_x, B_y, B_z)$ are ordinary vectors, then

$$(\vec{A}\cdot\vec{\sigma})(\vec{B}\cdot\vec{\sigma}) = \vec{A}\cdot\vec{B}I + i\vec{\sigma}\cdot(\vec{A}\times\vec{B}) \tag{11.133}$$

where I is the 2×2 identity matrix. The last term can be arranged into

$$(\vec{A}\cdot\vec{\sigma})(\vec{B}\cdot\vec{\sigma}) = \vec{A}\cdot\vec{B}I + i\vec{A}\cdot(\vec{B}\times\vec{\sigma}) \tag{11.134}$$

$$= \vec{A}\cdot[\vec{B}I + i\vec{B}\times\vec{\sigma}] \tag{11.135}$$

Since $\vec{A}$ is an arbitrary vector, eliminate it and find

$$\vec{\sigma}(\vec{B}\cdot\vec{\sigma}) = \vec{B}I + i\vec{B}\times\vec{\sigma} \tag{11.136}$$

Set $\vec{B} = \vec{k}$ and we have the expression in eqn. (11.132) :

$$\hat{\phi}^\dagger \sigma_\ell \hat{\chi} = \frac{\hbar c}{E(k) + mc^2}\{k_\ell|\hat{\phi}|^2 + i[\hat{\phi}^\dagger(\vec{k}\times\vec{\sigma})_\ell\hat{\phi}]\} \tag{11.137}$$

Equation (11.131) has the above term plus its Hermitian conjugate. When we take the Hermitian conjugate in the above equation, the last term changes sign. This term cancels. The term N_k^2 cancels much of the prefactor:

$$j_\ell = \frac{ec^2\hbar k_\ell}{E(k)} = \frac{e}{\hbar}\frac{dE(k)}{dk_\ell} = ev_\ell(k) \tag{11.138}$$

The current is given by the charge times the relativistic velocity.

11.3.5 Spin-Projection Operators

The four spinor components give the orientation of the electron spin. The two large components in $\hat{\phi}$ give the spin of the Dirac particle. The other two spinor components relate to the spin of the negative energy states. Here we first discuss the spin states of the large component $\hat{\phi}$. In nonrelativistic physics, we usually say they are spin-up (α) or spin-down (β). For a free particle, down or up relate to the direction of motion. The natural way to define the direction of spin is along the direction of motion.

Define a 4×4 vector matrix:

$$\Sigma_k = \begin{pmatrix} \sigma_k & 0 \\ 0 & \sigma_k \end{pmatrix} \tag{11.139}$$

$$\vec{B}\cdot\vec{\Sigma} = \begin{pmatrix} \vec{B}\cdot\vec{\sigma} & 0 \\ 0 & \vec{B}\cdot\vec{\sigma} \end{pmatrix} \tag{11.140}$$

THEOREM: $\vec{p}\cdot\vec{\Sigma}$ *commutes with the Dirac Hamiltonian.*
Proof: The Dirac Hamiltonian has two terms

$$H = c\vec{\alpha}\cdot\vec{p} + \beta mc^2 \tag{11.141}$$

We must show that our matrix $\vec{p}\cdot\vec{\Sigma}$ commutes with both terms. Use the representation in eqn. (11.81):

$$\alpha_k = \begin{pmatrix} 0 & \sigma_k \\ \sigma_k & 0 \end{pmatrix}, \quad \beta = \begin{pmatrix} I & 0 \\ 0 & -I \end{pmatrix} \tag{11.142}$$

The matrix $(\vec{p}\cdot\vec{\Sigma})$ commutes with β since they are both diagonal matrices:

$$\begin{pmatrix} \sigma_k & 0 \\ 0 & \sigma_k \end{pmatrix}\begin{pmatrix} I & 0 \\ 0 & -I \end{pmatrix}=\begin{pmatrix} \sigma_k & 0 \\ 0 & -\sigma_k \end{pmatrix} \tag{11.143}$$

$$=\begin{pmatrix} I & 0 \\ 0 & -I \end{pmatrix}\begin{pmatrix} \sigma_k & 0 \\ 0 & \sigma_k \end{pmatrix} \tag{11.144}$$

The other term in the Dirac Hamiltonian has the commutator

$$[\vec{p}\cdot\vec{\Sigma}, \vec{p}\cdot\vec{\alpha}] \tag{11.145}$$

$$=\begin{pmatrix} \vec{p}\cdot\vec{\sigma} & 0 \\ 0 & \vec{p}\cdot\vec{\sigma} \end{pmatrix}\begin{pmatrix} 0 & \vec{p}\cdot\vec{\sigma} \\ \vec{p}\cdot\vec{\sigma} & 0 \end{pmatrix}-\begin{pmatrix} 0 & \vec{p}\cdot\vec{\sigma} \\ \vec{p}\cdot\vec{\sigma} & 0 \end{pmatrix}\begin{pmatrix} \vec{p}\cdot\vec{\sigma} & 0 \\ 0 & \vec{p}\cdot\vec{\sigma} \end{pmatrix}$$

$$=\begin{pmatrix} 0 & (\vec{p}\cdot\vec{\sigma})^2(1-1) \\ (\vec{p}\cdot\vec{\sigma})^2(1-1) & 0 \end{pmatrix}=0 \tag{11.146}$$

which proves the theorem.

Define the *spin projection operator* as

$$S=\frac{1}{2}\hat{p}\cdot\vec{\Sigma} \tag{11.147}$$

The symbol S stands for a 4×4 matrix. Note that we use the unit vector $\hat{p}$ so that S is dimensionless. The matrix S commutes with the Dirac Hamiltonian.

If two operators commute, they can have the same set of eigenfunctions, albeit with different eigenvalues for each operator. An eigenstate of the Dirac Hamiltonian can also be an eigenstate of the spin-projection operator:

$$H\psi(\mathbf{k}, \mathbf{r}) = E(k)\psi(k, \mathbf{r}) \tag{11.148}$$

$$S\psi(\mathbf{k}, \mathbf{r}) = \mu\psi(\mathbf{k}, \mathbf{r}) \tag{11.149}$$

The eigenvalue $\mu=\pm\frac{1}{2}$: $+\frac{1}{2}$ denotes a spin direction along $\hat{p}$, while $\mu=-\frac{1}{2}$ denotes a spin along $-\hat{p}$.

A simple example is provided by a free particle going in the $\hat{z}$-direction: $\vec{k}=k\hat{z}$. The eigenfunction is

$$\psi(\mathbf{k}, \mathbf{r}) = N_k\begin{bmatrix} \hat{\phi} \\ \lambda_k\sigma_z\hat{\phi} \end{bmatrix}\exp\left[i(kz-\omega_k t)\right] \tag{11.150}$$

$$\lambda_k=\frac{\hbar ck}{E(k)+mc^2} \tag{11.151}$$

The spin-projection operator is

$$S=\frac{\Sigma_z}{2}=\frac{1}{2}\begin{bmatrix} \sigma_z & 0 \\ 0 & \sigma_z \end{bmatrix} \tag{11.152}$$

For $\mu=\frac{1}{2}$, set

$$\hat{\phi} = \begin{pmatrix} 1 \\ 0 \end{pmatrix} \tag{11.153}$$

$$\Psi = N_k \begin{bmatrix} 1 \\ 0 \\ \lambda_k \\ 0 \end{bmatrix} \tag{11.154}$$

$$S\Psi = \frac{1}{2}\Psi \tag{11.155}$$

For $\mu = -\frac{1}{2}$, set

$$\hat{\phi} = \begin{pmatrix} 0 \\ 1 \end{pmatrix} \tag{11.156}$$

$$\Psi = N_k \begin{bmatrix} 0 \\ 1 \\ 0 \\ -\lambda_k \end{bmatrix} \tag{11.157}$$

$$S\Psi = -\frac{1}{2}\Psi \tag{11.158}$$

These results can be generalized to an arbitrary direction of $\vec{p}$. First find the form of the 2-spinor $\hat{\phi}$. When $\mu = \frac{1}{2}$ it obeys the matrix equation:

$$\vec{p} \cdot \vec{\sigma}\hat{\phi} = p\hat{\phi} \tag{11.159}$$

$$\begin{bmatrix} p_z - p & p_x - ip_y \\ p_x + ip_y & -(p_z + p) \end{bmatrix} \begin{bmatrix} \phi_1 \\ \phi_2 \end{bmatrix} = 0 \tag{11.160}$$

The spinor that obeys this matrix equation is

$$\begin{bmatrix} \phi_1 \\ \phi_2 \end{bmatrix} = \frac{1}{\sqrt{2p(p+p_z)}} \begin{bmatrix} p + p_z \\ p_x + ip_y \end{bmatrix} \tag{11.161}$$

which can be verified by direct substitution. A similar matrix can be found for $\mu = -\frac{1}{2}$ that obeys the equation

$$\vec{p} \cdot \vec{\sigma}\hat{\phi} = -p\hat{\phi} \tag{11.162}$$

Solving this equation is a homework assignment.

11.3.6 *Scattering of Dirac Particles*

Spin plays a role in the scattering of fermions that obey the Dirac Hamiltonian. Here we consider a simple case of a fermion undergoing elastic scattering from a fixed potential $V(\mathbf{r})$. We solve for the differential cross section in the Born approximation. The same calculation was done in chapter 9 for nonrelativistic particles. Here we evaluate the role of (1) relativity, and (2) spins.

A potential $V(\mathbf{r})$ is located at the origin. An incoming plane wave has momentum $\mathbf{p}_i$ and energy $E(p_i) = E_i$. The wave encounters the potential, and part of the wave scatters outward with final momentum $\mathbf{p}_f$, $E_f = E(p_f)$. The transition rate in the Born approximation is

$$w = \frac{2\pi}{\hbar^4}\int \frac{d^3p_f}{(2\pi)^3}|M|^2\delta[E(p_i)-E(p_f)] \tag{11.163}$$

$$E(p) = c\sqrt{(mc)^2 + p^2} \tag{11.164}$$

The integral over final momentum is evaluated in spherical coordinates $d^3p_f = E_f \sqrt{E_f^2 - m^2c^4}\, dE_f\, d\Omega_f/c^3$, where $E_f = E(p_f)$, $E_i = E(p_i)$. The integral over energy

$$\int dE_f\, F(E_f)\delta(E_i - E_f) = F(E_i) \equiv F(E) \tag{11.165}$$

$$\frac{d\sigma}{d\Omega} = \frac{1}{v_f}\frac{dw}{d\Omega} = \left(\frac{E}{2\pi c^2\hbar^2}\right)^2 |M|^2 \tag{11.166}$$

The prefactor reduces to the nonrelativistic limit when $E \to mc^2$.

The scattering center is from the scalar potential, and the vector potential $\mathbf{A} = 0$. Let Ψ_i be the four-spinor of the initial state, and Ψ_f be that of the final state. In the Born approximation, the matrix element is

$$M_{fi} = \langle\Psi_f|\Psi_i\rangle \int d^3r V(\mathbf{r}) \exp[-i\mathbf{r}\cdot(\mathbf{p}_f - \mathbf{p}_i)/\hbar] \tag{11.167}$$

$$= \langle\Psi_f|\Psi_i\rangle \tilde{V}(\mathbf{k}_f - \mathbf{k}_i) \tag{11.168}$$

$$\frac{d\sigma}{d\Omega} = \left(\frac{E}{2\pi c^2\hbar^2}\right)^2 \tilde{V}(\mathbf{k}_f - \mathbf{k}_i)^2 |\langle\Psi_f|\Psi_i\rangle|^2 \tag{11.169}$$

where $\mathbf{p}_{i,f} = \hbar\mathbf{k}_{i,f}$. The matrix element contains the Fourier transform of the scattering potential. This factor is familiar from nonrelativistic scattering in the Born approximation. The above formula, without the spinor factor, is also valid for relativistic particles that obey the K-G equation. The additional factor is due to the matrix element of the initial and final spinor function.

Assume that the initial state is in the $\hat{z}$-direction ($\mathbf{k}_i = k\hat{z}$) with spin projection $\mu = \frac{1}{2}$ so that

$$\hat{\phi} = \begin{pmatrix} 1 \\ 0 \end{pmatrix} \tag{11.170}$$

$$\Psi_i = N_k \begin{bmatrix} 1 \\ 0 \\ \lambda_k \\ 0 \end{bmatrix} \tag{11.171}$$

There are two possible spinors in the final state:

1. For $\mu_f = \frac{1}{2}$, the large component is (11.170) and the small component contains the factor

$$\mathbf{k} \cdot \vec{\sigma}\hat{\phi} = \begin{pmatrix} k_z \\ k_x + ik_y \end{pmatrix} = k \begin{pmatrix} \cos(\theta) \\ \sin(\theta)e^{i\phi} \end{pmatrix} \tag{11.172}$$

$$\Psi_{f1} = N_k \begin{bmatrix} 1 \\ 0 \\ \lambda_k \cos(\theta) \\ \lambda_k \sin(\theta)e^{i\phi} \end{bmatrix}$$

$$\langle \Psi_{f1} | \Psi_i \rangle = N_k^2[1 + \lambda_k^2 \cos(\theta)] = 1 - \Lambda(k) \sin^2(\theta/2)$$

$$\Lambda(k) = 2\lambda_k^2 N_k^2 = \frac{\hbar^2 c^2 k^2}{E(k)[E(k) + mc^2]} \tag{11.174}$$

where (θ, ϕ) are the scattering angles in spherical coordinates. Since the scattering is assumed to be elastic, the wave vector k is the same in the initial and final state.

2. For $\mu_f = -\frac{1}{2}$,

$$\hat{\phi} = \begin{pmatrix} 0 \\ 1 \end{pmatrix} \tag{11.175}$$

$$\Psi_{f2} = N_k \begin{bmatrix} 0 \\ 1 \\ \lambda_k \sin(\theta)e^{-i\phi} \\ -\lambda_k \cos(\theta) \end{bmatrix} \tag{11.176}$$

$$\langle \Psi_{f2} | \Psi_i \rangle = N_k^2 \lambda_k^2 \sin(\theta)e^{-i\phi} = \frac{1}{2}\Lambda(k) \sin(\theta)e^{-i\phi} \tag{11.177}$$

If the spin is measured in the final state, then the values of $|\langle \Psi_{fj} | \Psi_i \rangle|^2$ give the cross section for the two spin arrangements. A more common experiment is not to measure the final spin, but to have a detector that measures all particles regardless of spin. Then the differential cross section has the factor of

$$\sum_{j=1}^{2} |\langle \Psi_{fj} | \Psi_i \rangle|^2 = [1 - \Lambda \sin^2(\theta/2)]^2 + \Lambda^2 \sin^2(\theta/2) \cos^2(\theta/2)$$

$$= 1 - \sin^2(\theta/2)[2\Lambda - \Lambda^2] \tag{11.178}$$

$$= 1 - \left(\frac{v_k}{c}\right)^2 \sin^2(\theta/2), \quad v_k = \frac{\hbar c^2 k}{E(k)} \tag{11.179}$$

The spin components make this factor. At relativistic velocities $v_k \sim c$ it makes an important change in the angular cross section. Back scattering $(\theta \sim \pi)$ is much less probable.

11.4 Antiparticles and Negative Energy States

The eigenvalue for the free-particle Dirac Hamiltonian was found to be

$$E^2 = E(k)^2 = (mc^2)^2 + (\hbar ck)^2 \tag{11.180}$$

$$E = \pm E(k) = \pm\sqrt{(mc^2)^2 + (\hbar ck)^2} \tag{11.181}$$

So far, all of the solutions to the Dirac Hamiltonian have assumed that the energy is $E = +E(k)$. These are called the positive energy states. They are thought to describe fermions such as electrons or neutrons. The negative energy states have $E = -E(k) = -\sqrt{(mc^2)^2 + (\hbar ck)^2}$. They are related to eigenfunctions for the antiparticles.

Go back to eqns. (11.113) and (11.114) and change the sign of the energy:

$$0 = (mc^2 + E(k))\hat{\phi} + c\hbar\vec{\sigma}\cdot\vec{k}\hat{\chi} \tag{11.182}$$

$$0 = c\hbar\vec{\sigma}\cdot\vec{k}\hat{\phi} - (mc^2 - E(k))\hat{\chi} \tag{11.183}$$

From eqn. (11.182) we deduce

$$\hat{\phi} = -\frac{c\hbar\vec{\sigma}\cdot\vec{k}}{mc^2 + E(k)}\hat{\chi} \tag{11.184}$$

Since $E(k) \sim mc^2$, it is inconvenient to use (11.183), which would produce an energy denominator of $(E(k) - mc^2)$, which could be small. Using (11.182) ensures that the energy denominator $(E(k) + mc^2) > 2mc^2$ is a large energy.

For negative energy states, $\hat{\chi}$ is the "large component" while $\hat{\phi}$ is the "small component." The eigenfunction for a negative energy plane wave is

$$\psi = N_k \begin{bmatrix} -\frac{c\hbar\vec{\sigma}\cdot\vec{k}}{mc^2 + E(k)}\hat{\chi} \\ \hat{\chi} \end{bmatrix} \exp\left[i\left(\mathbf{k}\cdot\mathbf{r} + \frac{E(k)t}{\hbar}\right)\right] \tag{11.185}$$

The lower two components of the 4-spinor are the largest.

The negative energy states are related to the states of the antiparticle. To be specific, we discuss the Dirac equation for electrons, which have a charge $q_e < 0$. Its antiparticle state is the positron, which has a charge $q_p > 0$. Dirac derived his equation before the discovery of the positron or other antiparticles. Before then, no one knew why the spinor had four components, since the electron spin has only two configurations. Only after the discovery of the positron, which had been predicted all along by Dirac's Hamiltonian, was his theory finally understood and accepted.

The electron is an elementary particle with negative charge and positive mass. The positron is an elementary particle with positive charge and the same positive mass. These two particles are antiparticles, since they can mutually annihilate and conserve energy and momentum by emitting two photons (γ-rays).

The Dirac Hamiltonian for electrons and positrons is $e \equiv q_e = -q_p$, $e < 0$:

$$\left\{\gamma\cdot\left(p-\frac{e}{c}A\right)+mc\right\}\psi^{(e)}=0 \tag{11.186}$$

$$\left\{\gamma\cdot\left(p+\frac{e}{c}A\right)+mc\right\}\psi^{(p)}=0 \tag{11.187}$$

What is the relationship between these two equations? Equation (11.187) is related to the negative energy states of eqn. (11.186). We describe a sequence of operations that converts (11.186) into (11.187). These steps show that the above two equations are not independent, but are the same equation in two different forms.

Take the complex conjugate of eqn. (11.186). The complex conjugate gives $p^* = -p$ for all four components. Also, $A^* = A$:

$$\left\{-\gamma^*\cdot\left(p+\frac{e}{c}A\right)+mc\right\}\psi^{(e)*}=0 \tag{11.188}$$

$$\left\{-\vec{\gamma}^*\cdot\left(\vec{p}+\frac{e}{c}\vec{A}\right)+\gamma^*{}_4\left(p_4+\frac{e\phi}{c}\right)+mcI\right\}\psi^{(e)*}=0 \tag{11.189}$$

In the second equation we have written the four-vectors in terms of the three-vector and fourth component. Examine the behavior of γ^*:

1. $\gamma_1^*=\gamma_1$ since

$$\gamma_1=\beta\alpha^{(x)}=\begin{pmatrix} I & 0\\ 0 & -I\end{pmatrix}\begin{pmatrix} 0 & \sigma_x\\ \sigma_x & 0\end{pmatrix} \tag{11.190}$$

$$=\begin{pmatrix} 0 & \sigma_x\\ -\sigma_x & 0\end{pmatrix}=\gamma_1^* \tag{11.191}$$

2. $\gamma_2^*=-\gamma_2$ since

$$\gamma_2=\beta\alpha^{(y)}=\begin{pmatrix} I & 0\\ 0 & -I\end{pmatrix}\begin{pmatrix} 0 & \sigma_y\\ \sigma_y & 0\end{pmatrix} \tag{11.192}$$

$$=\begin{pmatrix} 0 & \sigma_y\\ -\sigma_y & 0\end{pmatrix}=-\gamma_2^* \tag{11.193}$$

3. $\gamma_3{}^*=\gamma_3$ since

$$\gamma_3=\beta\alpha^{(z)}=\begin{pmatrix} I & 0\\ 0 & -I\end{pmatrix}\begin{pmatrix} 0 & \sigma_z\\ \sigma_z & 0\end{pmatrix} \tag{11.194}$$

$$=\begin{pmatrix} 0 & \sigma_z\\ -\sigma_z & 0\end{pmatrix}=\gamma_3^* \tag{11.195}$$

4. $\gamma_4=\beta=\gamma_4^*$

Three of the γ-matrices are unchanged by the complex conjugation, while γ_2 changes sign. The difference is that, of the three Pauli matrices, σ_x and σ_z are real, while σ_y is imaginary.

THEOREM: $\gamma_2\gamma_\mu^*\gamma_2=\gamma_\mu$

Proof: The proof uses two properties of γ functions: (i) $\gamma_2^2=-I$ and (ii) $\gamma_\nu\gamma_2=-\gamma_2\gamma_\nu$ if $\nu\neq 2$.

- If $\mu \neq 2$, then $\gamma_\mu^* = \gamma_\mu$, and

$$\gamma_2 \gamma_\mu^* \gamma_2 = \gamma_2 \gamma_\mu \gamma_2 = -\gamma_2 \gamma_2 \gamma_\mu = \gamma_\mu \tag{11.196}$$

- If $\nu = 2$, then $\gamma_2^* = -\gamma_2$, and

$$\gamma_2 \gamma_2^* \gamma_2 = -\gamma_2 \gamma_2 \gamma_2 = \gamma_2 \tag{11.197}$$

This theorem tells us how to handle the feature that only the γ_2-matrix changes sign under the operation of complex conjugation:

1. In eqn. (11.189), insert $\gamma_2^2 = -I$ between the right bracket "}" and the eigenfunction $\psi^{(e)*}$.
2. Multiply the entire equation from the left side by the matrix γ_2.
3. In the γ_2^2 in step 1, one factor of γ_2 multiplies the eigenfunction, while the other is combined with the γ_2 from step 2 to change the sign so that $\gamma_2 \gamma_\mu^* \gamma_2 = \gamma_\mu$. This step makes the term $mcI\gamma_2^2 = -mcI$.

These steps give the new equation from (11.189):

$$\left\{ -\vec{\gamma} \cdot \left(\vec{p} + \frac{e}{c}\vec{A} \right) + \gamma_4 \left(p_4 + \frac{e\phi}{c} \right) - mcI \right\} \gamma_2 \psi^{(e)*} = 0 \tag{11.198}$$

Change the sign of the entire equation. Also recall that $e < 0$ so $-e = |e|$ and we find

$$\left\{ \vec{\gamma} \cdot \left(\vec{p} - \frac{|e|}{c}\vec{A} \right) - \gamma_4 \left(p_4 - \frac{|e|\phi}{c} \right) + mcI \right\} \gamma_2 \psi^{(e)*} = 0 \tag{11.199}$$

This equation is just (11.187). The relationship between the electron and positron eigenfunctions is

$$\psi^{(p)} = -i\gamma_2 \psi^{(e)*} \tag{11.200}$$

The phase factor of $-i$ is added for later convenience: just multiply eqn. (11.199) by $-i$. The convenient feature is that $-i\gamma_2$ is a real matrix:

$$-i\gamma_2 = -i \begin{bmatrix} 0 & \sigma_y \\ -\sigma_y & 0 \end{bmatrix} = \begin{bmatrix} 0 & 0 & 0 & -1 \\ 0 & 0 & 1 & 0 \\ 0 & 1 & 0 & 0 \\ -1 & 0 & 0 & 0 \end{bmatrix} \tag{11.201}$$

The matrix $-i\gamma_2$, when operating on $\psi^{(e)*}$, inverts the order of the spinor components. The end components change sign:

$$-i\gamma_2 \psi^{(e)*} = \begin{bmatrix} 0 & 0 & 0 & -1 \\ 0 & 0 & 1 & 0 \\ 0 & 1 & 0 & 0 \\ -1 & 0 & 0 & 0 \end{bmatrix} \begin{bmatrix} \psi_1^* \\ \psi_2^* \\ \psi_3^* \\ \psi_4^* \end{bmatrix} = \begin{bmatrix} -\psi_4^* \\ \psi_3^* \\ \psi_2^* \\ -\psi_1^* \end{bmatrix} \tag{11.202}$$

The lower two components of the electron four-spinor are just the large components of the positron.

As an example, consider the negative energy state of the electron in (11.185):

$$\psi^{(e)} = N_k \begin{bmatrix} -\frac{c\hbar\vec{\sigma}\cdot\vec{k}}{mc^2 + E(k)}\hat{\chi} \\ \hat{\chi} \end{bmatrix} \exp\left[i\left(\mathbf{k}\cdot\mathbf{r} + \frac{E(k)t}{\hbar}\right)\right] \tag{11.203}$$

Set $\vec{k} = k\hat{z}$, $S = 1/2$, so the large spinor is $\hat{\chi}^{\dagger} = (1, 0)$, and the eigenfunction is

$$\psi^{(e)} = N_k \begin{bmatrix} -\lambda_k \\ 0 \\ 1 \\ 0 \end{bmatrix} \exp\left[i\left(kz + \frac{E(k)t}{\hbar}\right)\right] \tag{11.204}$$

$$\psi^{(e)*} = N_k \begin{bmatrix} -\lambda_k \\ 0 \\ 1 \\ 0 \end{bmatrix} \exp\left[-i\left(kz + \frac{E(k)t}{\hbar}\right)\right] \tag{11.205}$$

Multipying by $-i\gamma_2$ gives the positron state:

$$\psi^{(p)} = N_k \begin{bmatrix} 0 \\ 1 \\ 0 \\ \lambda_k \end{bmatrix} \exp\left[i\left(-k)z - i\frac{E(k)t}{\hbar}\right)\right] \tag{11.206}$$

Compare the positron eigenfunction with the earlier ones for the electron. The eigenfunction describes a free particle of

1. Positive energy E_k
2. Wave vector $(-k)$
3. Spin with $\mu = -\frac{1}{2}$

An electron state of negative energy, wave vector **k**, and spin projection $\mu = +\frac{1}{2}$, is a positron state of positive energy, wave vector $-$**k**, and spin projection $\mu = -\frac{1}{2}$.

11.5 Spin Averages

In the measurement of the cross section, often the particle spins are not measured. The detector measures spins with both values of $\mu_f = \pm\frac{1}{2}$. Similarly, often the initial values of the spin polarization are not known, and one also has to average over both initial spin eigenvalues $\mu_i = \pm\frac{1}{2}$. In the calculation of a particle scattering from a potential, we write these two averages as

$$\Xi = \frac{1}{2} \sum_{\mu_i = \pm 1/2} \sum_{\mu_f = \pm 1/2} |\langle \Psi^*(\varepsilon_f, \mu_f)\Psi(\varepsilon_i, \mu_i)\rangle|^2 \tag{11.207}$$

$$= \frac{1}{2}\mathrm{Tr}\{\Xi^{(i)}\Xi^{(f)}\} \tag{11.208}$$

$$\Xi^{(i)}_{\alpha\beta} = \sum_{\mu_i} \Psi_\alpha(\varepsilon_i, \mu_i)\Psi_\beta{}^*(\varepsilon_i, \mu_i) \tag{11.209}$$

The matrices $\Xi^{(i,f)}$ can be found easily. The spinor part of the eigenfunction $\Psi_\alpha(\pm\,\varepsilon,\,\mu)$ forms a complete set of states over the four-dimensional space. The set is complete, and

$$\sum_{\mu_i}\left[\Psi_\alpha(\varepsilon_p,\,\mu_i)\Psi_\beta{}^*(\varepsilon_p,\,\mu_i)+\Psi_\alpha(-\varepsilon_p,\,\mu_i)\Psi_\beta{}^*(-\varepsilon_p,\,\mu_i)\right]=\delta_{\alpha\beta} \tag{11.210}$$

Operate on the above equation by the projection operator:

$$\mathcal{P}=\frac{H+\varepsilon_p I}{2\varepsilon_p} \tag{11.211}$$

where H is the Hamiltonian matrix. Since $H\psi\,(\varepsilon,\,\mu)\;=\;\varepsilon\psi(\varepsilon,\mu)$ and $H\psi(-\varepsilon,\,\mu)=-\varepsilon\psi(-\varepsilon,\,\mu)$, the projection operator selects only the states of positive energy:

$$\mathcal{P}\delta_{\alpha\beta}=\mathcal{P}\sum_{\mu_i}\left[\Psi_\alpha(\varepsilon_p,\,\mu_i)\Psi_\beta{}^*(\varepsilon_p,\,\mu_i)+\Psi_\alpha(-\varepsilon_p,\,\mu_i)\Psi_\beta{}^*(-\varepsilon_p,\,\mu_i)\right]$$

$$\frac{1}{2\varepsilon_p}\left[H+\varepsilon_p I\right]_{\alpha\beta}=\sum_{\mu_i}\Psi_\alpha(\varepsilon_p,\,\mu_i)\Psi_\beta{}^*(\varepsilon_p,\,\mu_i)=\frac{1}{2\varepsilon_p}[\hbar c\vec{\alpha}\cdot\mathbf{p}+\beta mc^2+\varepsilon_p I]$$

$$\Xi^{(i)}_{\alpha\beta}=\frac{1}{2\varepsilon_p}[\hbar c\vec{\alpha}\cdot\mathbf{p}+\beta mc^2+\varepsilon_p I]_{\alpha\beta} \tag{11.212}$$

Returning to the original problem, eqn. (11.208), we have

$$\Xi=\frac{1}{8\varepsilon_k\varepsilon_p}\mathrm{Tr}\{[\hbar c\vec{\alpha}\cdot\mathbf{k}+\beta mc^2+\varepsilon_k I][\hbar c\vec{\alpha}\cdot\mathbf{p}+\beta mc^2+\varepsilon_p I]\} \tag{11.213}$$

where $\mathbf{p}$ and $\mathbf{k}$ are the initial and final scattering wave vector. Since $\alpha_i^2=I,\ \beta^2=I$, their trace equals four. All other combinations have a zero trace. The above expression reduces to

$$\Xi=\frac{1}{2\varepsilon_k\varepsilon_p}[\hbar^2c^2\mathbf{k}\cdot\mathbf{p}+m^2c^4+\varepsilon_k\varepsilon_p] \tag{11.214}$$

This formula gives the same result as evaluating the eigenfunctions directly. For example, in the case of elastic scattering from a potential, then $\varepsilon_k\;=\;\varepsilon_p$, and $\mathbf{k}\cdot\mathbf{p}\;=\;k^2\,\cos(\theta)$. Writing $\cos(\theta)\;=\;1-2\,\sin^2(\theta/2)$ gives

$$\Xi=\frac{1}{2\varepsilon^2}\left\{\varepsilon^2+m^2c^4+\hbar^2c^2k^2\left[1-2\sin^2\left(\frac{\theta}{2}\right)\right]\right\} \tag{11.215}$$

$$=1-\frac{\hbar^2c^2k^2}{\varepsilon^2}\sin^2\left(\frac{\theta}{2}\right)=1-\left(\frac{v_k}{c}\right)^2\sin^2\left(\frac{\theta}{2}\right) \tag{11.216}$$

which is the spinor contribution to the scattering cross section.

Eventually we are going to formulate scattering theory entirely using γ-matrices. Then it turns out that the summation we want is

$$\Xi^{(i)}_{\alpha\beta}=\sum_{\mu_i}\Psi_\alpha(\varepsilon_p,\,\mu_i)\bar{\Psi}_\beta(\varepsilon_p,\,\mu_i) \tag{11.217}$$

where $\bar{\Psi}\equiv\Psi^*\beta$. This means that

$$\Xi^{(i)}_{\alpha\delta} = \Xi^{(i)}_{\alpha\varepsilon}\beta_{\varepsilon\delta} = \frac{1}{2\varepsilon_p}[\hbar c\vec{\alpha}\cdot\mathbf{p} + \beta mc^2 + \varepsilon_p I]_{\alpha\varepsilon}\beta_{\varepsilon\delta} \tag{11.218}$$

$$= \frac{c}{2\varepsilon_p}[mc - \gamma\cdot p]_{\alpha\delta} \tag{11.219}$$

since $\vec{\alpha}\beta = -\beta\vec{\alpha} = -\vec{\gamma}$, etc. These spin-averaging methods are used in the next sections.

11.6 Nonrelativistic Limit

The Dirac Hamiltonian has a nonrelativistic limit for the case that the mass is nonzero. The rest energy $mc^2 \equiv \Omega$ becomes a large quantity and we expand the Hamiltonian in a series in powers of $O(1/\Omega)$. The relativistic eigenvalue E is set equal to $E = E' + mc^2$, where E' is the nonrelativisitic eigenvalaue.

11.6.1 *First Approximation*

The nonrelativistic limit of the Dirac equation has several levels of approximation, depending on how many terms in $(1/\Omega)$ are retained. The first case is to keep a minimal number of terms. Write eqn. (11.102), which includes the potentials, in the standard 2×2 notation:

$$0 = -(E' - V)\hat{\phi} + c\vec{\sigma}\cdot(\mathbf{p} - \frac{e}{c}\vec{A})\hat{\chi} \tag{11.220}$$

$$0 = c\vec{\sigma}\cdot(\mathbf{p} - \frac{e}{c}\vec{A})\hat{\phi} - (E' + 2mc^2 - V)\hat{\chi} \tag{11.221}$$

Solve the second equation for the small component and insert the results into the first equation. Neglect V and E' compared to mc^2:

$$\hat{\chi} = \frac{c\vec{\sigma}\cdot(\mathbf{p} - \frac{e}{c}\vec{A})}{2mc^2 + E' - V}\hat{\phi} \approx \frac{c\vec{\sigma}\cdot(\mathbf{p} - \frac{e}{c}\vec{A})}{2mc^2}\hat{\phi} \tag{11.222}$$

$$0 = \left[-E' + V + \frac{c^2}{2mc^2}\vec{\sigma}\cdot\left(\mathbf{p} - \frac{e}{c}\vec{A}\right)\vec{\sigma}\cdot\left(\mathbf{p} - \frac{e}{c}\vec{A}\right)\right]\hat{\phi} \tag{11.223}$$

The last expression is evaluated using

$$(\vec{\sigma}\cdot\vec{B})(\vec{\sigma}\cdot\vec{C}) = \vec{B}\cdot\vec{C} + i\vec{\sigma}\cdot(\vec{B}\times\vec{C}) \tag{11.224}$$

where $\vec{B} = \vec{C} = \mathbf{p} - e\mathbf{A}/c$. Since $\mathbf{p}\times\mathbf{p} = 0$, $\mathbf{A}\times\mathbf{A} = 0$, the last term has the only nonzero cross product:

$$\vec{\sigma}\cdot\left(\mathbf{p} - \frac{e}{c}\vec{A}\right)\vec{\sigma}\cdot\left(\mathbf{p} - \frac{e}{c}\vec{A}\right) = \left(\mathbf{p} - \frac{e}{c}\mathbf{A}\right)^2 - i\frac{e}{c}\vec{\sigma}\cdot(\mathbf{p}\times\mathbf{A} + \mathbf{A}\times\mathbf{p})$$

$$\mathbf{p}\times\mathbf{A} + \mathbf{A}\times\mathbf{p} = \frac{\hbar}{i}\vec{\nabla}\times\mathbf{A} = -i\hbar\vec{B} \tag{11.225}$$

The effective Hamiltonian is

$$H'\hat{\phi} = E'\hat{\phi} \tag{11.226}$$

$$H' = \frac{1}{2m}(\mathbf{p} - \frac{e}{c}\mathbf{A})^2 + V(\mathbf{r}) - \mu_0\vec{\sigma}\cdot\vec{B},\ \mu_0 = \frac{e\hbar}{2mc} \tag{11.227}$$

These manipulations derive Schrödinger's equation with the usual form for the vector and scalar potentials. The last term is the Pauli interaction between the particle spin and the external magnetic field $\vec{B}$. It has a single power of c in the denominator, and should be viewed as a relativistic correction term of $O(1/\sqrt{\Omega})$. It is the only correction term of this order, and further terms are of $O(1/\Omega)$. Deriving them takes far more effort.

11.6.2 Second Approximation

Some effort is saved in deriving the next terms in $O(1/mc^2)$ by neglecting the vector potential $\vec{A} = 0$. The scalar potential $V(\mathbf{r}) = e\phi(\mathbf{r})$ is retained and produces some important relativistic corrections. Rewrite eqns. (11.220) and (11.221) without the vector potential:

$$0 = -(E' - V)\hat{\phi} + c\vec{\sigma}\cdot\mathbf{p}\hat{\chi} \tag{11.228}$$

$$0 = c\vec{\sigma}\cdot\mathbf{p}\hat{\phi} - (E' + 2mc^2 - V)\hat{\chi} \tag{11.229}$$

Solve for the small component more accurately than before:

$$\hat{\chi} = \frac{1}{2mc[1 + (E' - V)/2mc^2]}\vec{\sigma}\cdot\mathbf{p}\hat{\phi} \tag{11.230}$$

$$\approx \frac{1}{2mc}\left[1 - \frac{(E' - V)}{2mc^2}\right]\vec{\sigma}\cdot\mathbf{p}\hat{\phi} \tag{11.231}$$

Put the last equation into (11.228):

$$0 = \left[-E' + V + \frac{1}{2m}\vec{\sigma}\cdot\mathbf{p}\left(1 - \frac{E' - V}{2mc^2}\right)\vec{\sigma}\cdot\mathbf{p}\right]\hat{\phi} \tag{11.232}$$

Recall that $(\vec{\sigma}\cdot\mathbf{p})^2 = p^2$. The other term takes more work:

$$\vec{\sigma}\cdot\mathbf{p}V(\mathbf{r})\vec{\sigma}\cdot\mathbf{p} = V(\vec{\sigma}\cdot\mathbf{p})^2 + \vec{\sigma}\cdot\mathbf{p},\ V]\vec{\sigma}\cdot\mathbf{p} \tag{11.233}$$

$$= Vp^2 + \frac{\hbar}{i}(\vec{\sigma}\cdot\vec{\nabla}V)\vec{\sigma}\cdot\mathbf{p} \tag{11.234}$$

$$= Vp^2 - i\hbar(\vec{\nabla}V)\cdot\mathbf{p} + \hbar\vec{\sigma}\cdot(\vec{\nabla}V\times\mathbf{p}) \tag{11.235}$$

The last term is the spin–orbit interaction. Put the above expression into eqn. (11.232) and find that

$$E'\left(1 + \frac{p^2}{4m^2c^2}\right)\hat{\phi} = \left[V\left(1 + \frac{p^2}{4m^2c^2}\right) + \frac{p^2}{2m} - i\frac{\hbar}{4m^2c^2}(\vec{\nabla}V)\cdot\mathbf{p} + \frac{\hbar}{4m^2c^2}\vec{\sigma}\cdot(\vec{\nabla}V\times\mathbf{p})\right]\hat{\phi} \tag{11.236}$$

Two more steps remain to find an equation of the form

$$E'\hat{u} = H_{\rm nr}\hat{u} \tag{11.237}$$

The first step is to normalize the eigenfunction. In nonrelativistic physics we choose $\hat{u}^\dagger\hat{u} = 1$, while in relativistic physics

$$|\Psi|^2 = 1 = |\hat{\phi}|^2 + |\hat{\chi}|^2 \tag{11.238}$$

Treating p^2 as a scalar, the above transformation gave, to lowest order,

$$\hat{\chi} \approx \frac{\vec{\sigma}\cdot\mathbf{p}}{2mc}\hat{\phi} \tag{11.239}$$

$$1 = \left[1 + \frac{p^2}{4m^2c^2}\right]|\hat{\phi}|^2 \tag{11.240}$$

So a properly normalized nonrelativistic spinor is

$$\hat{u} = \left[1 + \frac{p^2}{8m^2c^2}\right]\hat{\phi} \tag{11.241}$$

$$\hat{\phi} = \left[1 - \frac{p^2}{8m^2c^2}\right]\hat{u} \tag{11.242}$$

We only retain terms of $O(1/mc^2)$, which is why the above two relationships are both valid. Rewrite eqn. (11.236) using the eigenfunction $\hat{u}$ and find

$$E'\left(1 + \frac{p^2}{8m^2c^2}\right)\hat{u} = \left[V\left(1 + \frac{p^2}{8m^2c^2}\right) + \frac{p^2}{2m}\left(1 - \frac{p^2}{8m^2c^2}\right)\right.$$
$$\left. - i\frac{\hbar}{4m^2c^2}(\vec{\nabla}V)\cdot\mathbf{p} + \frac{\hbar}{4m^2c^2}\vec{\sigma}\cdot(\vec{\nabla}V\times\mathbf{p})\right]\hat{u} \tag{11.243}$$

Finally, multiply every term by $\left(1 - \frac{p^2}{8m^2c^2}\right)$, which eliminates the factor on the left of the equal sign. Now the equation has the form of eqn. (11.237) with a nonrelativistic Hamiltonian:

$$H_{\rm nr} = \tilde{V} + \frac{p^2}{2m}\left(1 - \frac{p^2}{4m^2c^2}\right) \tag{11.244}$$

$$-i\frac{\hbar}{4m^2c^2}(\vec{\nabla}V)\cdot\mathbf{p} + \frac{\hbar}{4m^2c^2}\vec{\sigma}\cdot(\vec{\nabla}V\times\mathbf{p}) \tag{11.245}$$

$$\tilde{V} = \left(1 - \frac{p^2}{8m^2c^2}\right)V\left(1 + \frac{p^2}{8m^2c^2}\right) \tag{11.246}$$

$$= V + \frac{\hbar^2}{8m^2c^2}\nabla^2 V + \frac{i\hbar}{4m^2c^2}\vec{\nabla}V\cdot\mathbf{p} \tag{11.247}$$

Note that the two terms in $\vec{\nabla}V\cdot\mathbf{p}$ cancel. The result to $O(1/mc^2)$ is

$$H_{\rm nr} = \frac{p^2}{2m} + V(\mathbf{r}) + \frac{1}{mc^2}\left[-\frac{p^4}{8m^2} + \frac{\hbar^2}{8m}\nabla^2 V + \frac{\hbar}{4m}\vec{\sigma}\cdot(\vec{\nabla}V\times\mathbf{p})\right]$$

The first two terms are the standard Schrödinger equation. The three corrections of $O(1/mc^2)$ have names.

1. The p^4 term is the correction to the kinetic energy, which is derived from the expansion

$$\sqrt{c^2p^2+(mc^2)^2}=mc^2\left[1+\frac{p^2}{m^2c^2}\right]^{1/2} \tag{11.248}$$

$$=mc^2\left[1+\frac{1}{2}\left(\frac{p^2}{m^2c^2}\right)-\frac{1}{8}\left(\frac{p^2}{m^2c^2}\right)^2+\cdots\right]$$

$$=mc^2+\frac{p^2}{2m}-\frac{p^4}{8m^3c^2}+\cdots \tag{11.249}$$

2. The term in $\nabla^2 V$ is called the *Darwin term*. If the potential is Coulombic, then

$$\nabla^2\frac{1}{r}=-4\pi\delta^3(\mathbf{r}) \tag{11.250}$$

A delta-function potential at the origin affects only *s*-states, since only *s*-states have $R(r=0)\neq 0$.

3. The last term is called the *spin–orbit interaction*. For atoms it is written in a slightly different form. If $V(\mathbf{r})$ is a central potential, then

$$\vec{\nabla}V=\frac{\mathbf{r}}{r}\frac{dV}{dr} \tag{11.251}$$

Since $\vec{\ell}=\vec{r}\times\vec{p}$ is angular momentum and the spin is $\vec{s}=\hbar\vec{\sigma}/2$, the spin–orbit term can be written as

$$V_{\text{so}}=\frac{\hbar}{4m^2c^2}\vec{\sigma}\cdot(\vec{\nabla}V\times\mathbf{p})=\frac{\zeta(r)}{\hbar^2}\vec{s}\cdot\vec{\ell} \tag{11.252}$$

$$\zeta(r)=\frac{\hbar^2}{2m^2c^2r}\frac{dV}{dr} \tag{11.253}$$

which is the way it is written for atoms.

11.6.3 *Relativistic Corrections for Hydrogenic States*

The three relativistic correction terms are evaluated for an electron in the Coulomb potential of the nucleus. We define the problem as

$$H=H_0+\frac{1}{mc^2}\mathcal{L} \tag{11.254}$$

$$H_0=-\left[\frac{\hbar^2}{2m}\nabla^2+\frac{Ze^2}{r}\right] \tag{11.255}$$

$$\mathcal{L}=-\frac{p^4}{8m^2}+\frac{\hbar^2}{8m}\nabla^2V+\frac{\hbar}{4m}\vec{\sigma}\cdot(\vec{\nabla}V\times\mathbf{p}) \tag{11.256}$$

The eigenvalues of H_0 are $E_n=-E_{\text{Ry}}Z^2/n^2$, where E_{Ry} is the Rydberg of energy. The eigenfunctions are given in chapter 5. For the lowest-order corrections, we use first-order perturbation theory:

$$\Delta E_{n\ell} = \frac{1}{mc^2}\langle n\ell|\mathcal{L}|n\ell\rangle \tag{11.257}$$

The three term in $\mathcal{L}$ give the following.

- The Darwin term gives $\nabla^2 V = 4\pi Z e^2 \delta(\vec{r})$ and the matrix element has nonzero values only for s-states:

$$\Delta E_{n\ell,1} = \frac{\hbar^2}{8m^2c^2}\langle n\ell|\nabla^2 V|n\ell\rangle \tag{11.258}$$

$$= \frac{Z\hbar^2 e^2}{8m^2c^2} R_{n0}(0)^2 = \frac{E_{\text{Ry}}^2}{2mc^2} Z a_0^3 R_{n0}(0)^2 \tag{11.259}$$

The result depends on the magnitude of the radial eigenfunction at the origin. Note that $R_{n0}(0)^2 = 4Z^3/(a_0^3 n^3)$, and

$$\frac{E_{\text{Ry}}}{mc^2} = \frac{e^4}{2\hbar^2 c^2} = \frac{1}{2}\alpha_f^2 \tag{11.260}$$

$$\Delta E_{n\ell,1} = E_{\text{Ry}}\frac{\alpha_f^2 Z^4}{n^3} \tag{11.261}$$

where the fine structure constant is $\alpha_f = e^2/\hbar c \approx \frac{1}{137}$.

- The kinetic energy correction is evaluated by defining $p^2 = 2m(E_n - V)$ and the energy contribution is

$$\Delta E_{n\ell,2} = -\frac{1}{2mc^2}\langle n\ell|\left(E_n + \frac{Ze^2}{r}\right)^2|n\ell\rangle \tag{11.262}$$

In evaluating this expression, we employ the earlier formulas for the expectation values of $\langle 1/r^n\rangle$ given in chapter 5:

$$\langle n\ell|\frac{1}{r}|n\ell\rangle = \frac{Z}{n^2 a_0} \tag{11.263}$$

$$\langle n\ell|\frac{1}{r^2}|n\ell\rangle = \frac{Z^2}{n^3 a_0^2(\ell+\frac{1}{2})} \tag{11.264}$$

$$\Delta E_{n\ell,2} = E_{\text{Ry}}\frac{Z^4\alpha_f^2}{n^3}\left[\frac{3}{4n} - \frac{2}{2\ell+1}\right] \tag{11.265}$$

- The spin–orbit interaction is

$$\vec{\nabla} V = \frac{Ze^2}{r^3}\vec{r},\ \vec{\nabla} V \times \vec{p} = \frac{Ze^2}{r^3}\vec{\ell} \tag{11.266}$$

$$V_{\text{so}} = \frac{Ze^2}{2m^2c^2}\left\langle n\ell|\frac{\vec{\ell}\cdot\vec{s}}{r^3}|n\ell\right\rangle \tag{11.267}$$

$$\left\langle n\ell|\frac{\vec{\ell}\cdot\vec{s}}{r^3}|n\ell\right\rangle = \frac{Z^3\hbar^2}{2a_0^3 n^3 \ell(\ell+\frac{1}{2})(\ell+1)}[j(j+1)-\ell(\ell+1)-s(s+1)]$$

$$\Delta E_{n\ell,3} = \ell E_{\text{so}}, \quad j = \ell + \frac{1}{2} \tag{11.268}$$

$$= -(\ell+1) E_{\text{so}}, \quad j = \ell - \frac{1}{2} \tag{11.269}$$

$$E_{\text{so}} = \frac{E_{\text{Ry}} Z^4 \alpha_f^2}{2n^3 \ell(\ell + \frac{1}{2})(\ell+1)} \tag{11.270}$$

Note that all three terms are of $O(E_{\text{Ry}} Z^4 \alpha_f^2/n^3)$ times various dimensionless factors. They are all of similar value. If we add them all together we get

$$\Delta E_{n\ell} = E_{\text{Ry}} \frac{Z^4 \alpha_f^2}{n^3} \left[\frac{3}{4n} - \frac{2}{2j+1} \right] \tag{11.271}$$

11.7 Relativistic Interactions

Many scattering experiments are done at accelerators, where the particles have a high energy. It is necessary to do many scattering calculations relativistically. This section also serves as an introduction to field theory.

The prior sections discussed scattering from a potential $V(\mathbf{r})$. Potentials are actually generated by other particles. A more fundamental view of scattering is that particles scatter from other particles. The particle could scatter from a collection of particles: examples are proton scattering from a nucleus and electron scattering from an atom.

To discuss particle–particle scattering, we must learn to calculate Green's functions. They are quantum mechanical solutions to particle motion in four-dimensional space–time. We need them for two kinds of particles: photons and Dirac fermions. Green's functions come in a variety of forms: time-ordered, retarded, advanced, etc. For scattering theory we need only the time-ordered Green's functions. The others are discussed by Mahan [1].

11.7.1 Photon Green's Function

The derivation of the Green's functions for photons starts with eqns. (8.19) and (8.20):

$$\nabla^2 \psi + \frac{1}{c} \frac{\partial \mathbf{A}}{\partial t} = -4\pi\rho \tag{11.272}$$

$$\nabla \times (\nabla \times \mathbf{A}) + \frac{1}{c^2} \frac{\partial^2}{\partial t^2} \mathbf{A} + \nabla \frac{\partial}{c \partial t} \psi = \frac{4\pi}{c} \mathbf{j} \tag{11.273}$$

where $\mathbf{A}$ is the vector potential, and ψ is the scalar potential. We adopt the Lorentz gauge,

$$\nabla \cdot \mathbf{A} + \frac{\partial}{c \partial t} \psi = 0 \tag{11.274}$$

which simplifies these equations. We also use the standard form of $\mathbf{\nabla}\times(\mathbf{\nabla}\times\mathbf{A}) = \mathbf{\nabla}(\mathbf{\nabla}\cdot\mathbf{A}) - \nabla^2\mathbf{A}$ to get

$$\left[\nabla^2 - \frac{1}{c^2}\frac{\partial^2}{\partial t^2}\right]\psi = -4\pi\rho \tag{11.275}$$

$$\left[\nabla^2 - \frac{1}{c^2}\frac{\partial^2}{\partial t^2}\right]A_i = -\frac{4\pi}{c}j_i \tag{11.276}$$

These two wave equations can be combined into a four-vector notation:

$$\left[\nabla^2 - \frac{1}{c^2}\frac{\partial^2}{\partial t^2}\right]A_\mu = -\frac{4\pi}{c}j_\mu \tag{11.277}$$

$$A_\mu = [A_i,\ \psi],\quad j_\mu = [j_i,\ c\rho] \tag{11.278}$$

Define a type of Green's function $D(x)$ according to

$$\left[\nabla^2 - \frac{1}{c^2}\frac{\partial^2}{\partial t^2}\right]D(x) = \delta^4(x) \tag{11.279}$$

which has a solution in terms of a four-dimensional Fourier transform:

$$D(x) = -\int \frac{d^4q}{(2\pi)^4}\frac{e^{-iq\cdot x}}{q\cdot q + i\eta} \tag{11.280}$$

where η is an infinitesimal positive constant to make the photons go the right direction in time. Recall that $q\cdot q = q^2 - \omega^2/c^2$, and $d^4q = d^3q d\omega/c$. A solution to eqn. (11.277) is

$$A_\mu(x) = -\frac{4\pi}{c}\int d^4y D(x-y)j_\mu(y) \tag{11.281}$$

The four-vector interaction is

$$-\frac{1}{c}\int d^3r j_\mu A_\mu = -\frac{1}{c}\int d^3r \mathbf{j}\cdot\mathbf{A} + \int d^3r\rho\psi \tag{11.282}$$

Integrating this expression over time defines the S-matrix for the scattering process:

$$S_{fi} = \frac{1}{\hbar c}\int d^4x j_\mu(x)A_\mu(x) \tag{11.283}$$

Using the equation for $A_\mu(x)$ allows us to rewrite the S-matrix as

$$S_{fi} = -\frac{4\pi}{\hbar c^2}\int d^4x \int d^4y j_\mu^{(a)}(x)D(x-y)j_\mu^{(b)}(y) \tag{11.284}$$

Two particles interact by exchanging a photon. The superscripts (a, b) denote the separate particles. This process is shown in figure 11.1. The two solid lines are the fermions and the dashed line is the photon. This is a virtual process: actual photons are neither emitted nor absorbed. One particle has a current $j_\mu^{(b)}(y)$, which emits the photon, while the other particle immediately absorbs it through the current $j_\mu^{(a)}(x)$. This interaction can also be written in Fourier space using eqn. (11.280):

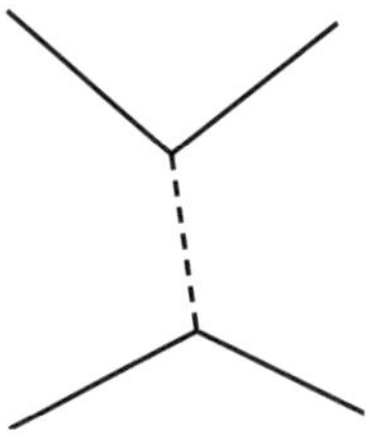

FIGURE 11.1 Two fermions scatter by exchanging a photon.

$$S_{fi} = \frac{4\pi}{\hbar c^2} \int \frac{d^4q}{(2\pi)^4} \frac{\bar{j}_\mu^{(a)}(q)\bar{j}_\mu^{(b)}(q)}{q \cdot q + i\eta} \tag{11.285}$$

$$\bar{j}_\mu^{(a)}(q) = \int d^4x e^{iq\cdot x} j_\mu^{(a)}(x) = ec \int d^4x e^{iq\cdot x} \left(\bar{\Psi}^{(a)}\gamma_\mu \Psi^{(a)}\right) \tag{11.286}$$

This expression will be the basis for the matrix element for scattering of two Dirac particles.

The photon Green's function is defined in terms of the vector potential as

$$\mathcal{D}_{\mu\nu}(x_1 - x_2) = -i\langle TA_\mu(x_1)A_\nu(x_2)\rangle \tag{11.287}$$

$$A_\mu(x) = c\sum_{\mathbf{k}\lambda} \sqrt{\frac{2\pi\hbar}{\omega(k)\Omega}} \xi_\mu(\mathbf{k}, \lambda)\left[a_{\mathbf{k}\lambda}e^{i\mathbf{k}\cdot x} + a^\dagger_{\mathbf{k}\lambda}e^{-i\mathbf{k}\cdot x}\right] \tag{11.288}$$

where Ω is the volume of the room. The symbol T is the time-ordering operator. The four-vector dot product is $k \cdot x = \mathbf{k}\cdot\mathbf{r} - \omega(k)t$. The sum over polarizations gives

$$\sum_\lambda \xi_\mu(\mathbf{k}, \lambda)\xi_\nu(\mathbf{k}, \lambda) = \delta_{\mu\nu} \tag{11.289}$$

In evaluating the Green's function at zero temperature we get expressions such as

$$\langle Ta_{\mathbf{k}\lambda}(t_1)a_{\mathbf{k}'\lambda'}(t_2)\rangle = 0 \tag{11.290}$$

$$\langle Ta_{\mathbf{k}\lambda}(t_1)a^\dagger_{\mathbf{k}'\lambda'}(t_2)\rangle = \delta_{\mathbf{k}\mathbf{k}'}\delta_{\lambda\lambda'}\Theta(t_1 - t_2)e^{i\omega(k)(t_1 - t_2)} \tag{11.291}$$

$$\langle Ta^\dagger_{\mathbf{k}\lambda}(t_1)a^\dagger_{\mathbf{k}'\lambda'}(t_2)\rangle = 0 \tag{11.292}$$

since $|\rangle$ denotes the vacuum state. Therefore, the photon Green's function is

$$\mathcal{D}_{\mu\nu}(x_1 - x_2) = -i\delta_{\mu\nu}c^2 \sum_{\mathbf{k}} \frac{2\pi\hbar}{\omega(k)\Omega} e^{i\mathbf{k}\cdot(\mathbf{r}_1 - \mathbf{r}_2) - i\omega(k)|t_1 - t_2|} \tag{11.293}$$

One can write the above expression as

$$\mathcal{D}_{\mu\nu}(x_1 - x_2) = 4\pi\hbar\delta_{\mu\nu} \int \frac{d^4q}{(2\pi)^4} \frac{e^{iq\cdot(x_1 - x_2)}}{q \cdot q + i\eta} \tag{11.294}$$

$$= 4\pi\hbar\delta_{\mu\nu} D(x_1 - x_2) \tag{11.295}$$

where $q_\mu = (\mathbf{k}, \omega/c)$. The photon Green's function $\mathcal{D}_{\mu\nu}(x)$ is proportional to the scalar Green's function $D(x)$ used to derive the Coulomb interaction between fermions. We can also write this interaction using the photon Green's function

$$S_{fi} = -\frac{1}{\hbar^2 c^2}\int d^4x \int d^4y j_\mu^{(a)}(x)\mathcal{D}_{\mu\nu}(x-y)j_\nu^{(b)}(y) \tag{11.296}$$

11.7.2 *Electron Green's Function*

The electron Green's function is defined as

$$G_{\alpha\beta}(x_1 - x_2) = -i\langle T\psi_\alpha(x_1)\bar{\psi}_\beta(x_2)\rangle \tag{11.297}$$

where $x_j = (\mathbf{r}_j, t_j)$ are four vectors. The symbols $\psi_\alpha(x)$ denote eigenfunctions of the free-particle Dirac equation, and $\bar{\psi} = \psi^*\beta$. The subscripts (α, β) are spinor components, so that $G_{\alpha\beta}$ is a tensor. The symbol T denotes time ordering—the function with the earliest time goes to the right. The symbols $\langle\cdots\rangle$ denote an average over the system. For scattering, we take this average at zero temperature.

First we give a short derivation. The functions $\psi_\alpha(x)$ are solutions to the Dirac equation, and so is the Green's function:

$$[\gamma\cdot p + mc]\psi = 0 \tag{11.298}$$

$$[\gamma\cdot p_1 + mc]_{\alpha\beta}G_{\beta\delta}(x_1 - x_2) = \delta_{\alpha\delta}\delta^4(x_1 - x_2) \tag{11.299}$$

Multiply both sides of this equation by $[\gamma\cdot p_1 - mc]_{\varepsilon\alpha}$ and get

$$[\gamma\cdot p_1 - mc]_{\varepsilon\alpha}[\gamma\cdot p_1 + mc]_{\alpha\beta}G_{\beta\delta}(x_1 - x_2) = [\gamma\cdot p_1 - mc]_{\varepsilon\delta}\delta^4(x_1 - x_2)$$

$$[(\gamma\cdot p_1)^2 - (mc)^2]_{\varepsilon\beta}G_{\beta\delta}(x_1 - x_2) = [\gamma\cdot p_1 - mc]_{\varepsilon\delta}\delta^4(x_1 - x_2)$$

$$-[(p_1)^2 + (mc)^2]_{\varepsilon\beta}G_{\beta\delta}(x_1 - x_2) = [\gamma\cdot p_1 - mc]_{\varepsilon\delta}\delta^4(x_1 - x_2)$$

Now take the four-dimensional Fourier transform of this equation and get

$$G_{\alpha\beta}(q) = -\frac{[\gamma\cdot q - mc]_{\alpha\beta}}{q\cdot q + m^2c^2} \tag{11.300}$$

This equation is the electron's Green's function for a free Dirac particle.

11.7.3 *Boson Exchange*

Equation (11.284) describes the interaction of two Dirac particles due to the exchange of a virtual photon. The nonrelativistic limit of this interaction is Coulomb's law $-e_1e_2/r_{12}$, where $e_{1,2} = \pm e$ are the charges of the particles. All interactions between particles are due to the exchange of a boson particle between them. The $1/r$ potential is due to the exchange of a massless boson such as the photon. Most other bosons have a mass. Yukawa first proposed that the nuclear force between nucleons was due to the exchange of a boson with a mass. He predicted the existence of the pion (π-meson) before its discovery in particle accelerators. For a boson with mass, eqn. (11.285) is modified to have the form

$$S_{fi} = \frac{4\pi}{\hbar c^2}\int \frac{d^4q}{(2\pi)^4}\frac{\bar{j}_\mu^{(a)}(q)\bar{j}_\mu^{(b)}(q)}{q\cdot q + m^2c^2} \tag{11.301}$$

where the factor of $(mc)^2$ has been added to the denominator. This form is similar to that of the Green's function for the electron. The coupling need not be only by γ-matrices. These matrices are 4×4. There are 16 Hermitian matrices needed to span this space, and the identity (I) and γ-matrices are only five. There are 11 other matrices, which can be found in books on field theory. If $\mathcal{M}, \mathcal{M}'$ are two such matrices, the most general interaction is

$$S_{fi} = \frac{4\pi}{\hbar c^2}\int \frac{d^4q}{(2\pi)^4}\frac{\mathcal{M}^{(a)}(q)\mathcal{M}^{(b)'}(q)}{q\cdot q + m^2c^2} \tag{11.302}$$

$$\mathcal{M}^{(j)}(q) = \int d^4x \langle \bar{\Psi}^{(j)}\mathcal{M}\Psi^{(j)}\rangle \tag{11.303}$$

For example, consider the β-decay of the neutron:

$$n \to p + e^- + \bar{\nu}_e \tag{11.304}$$

Angular correlation experiments have determined the matrix element is

$$\mathcal{M}(x)\mathcal{M}(y) = \langle \bar{\psi}^{(p)}(x)\gamma_\mu(1-\lambda\gamma_5)\Psi^{(n)}(x)\rangle\langle \bar{\psi}^{(e)}(y)\gamma_\mu(1-\gamma_5)\Psi^{(\nu)}(y)\rangle$$

$$\gamma_5 = -i\gamma_1\gamma_2\gamma_3\gamma_4 \tag{11.305}$$

and $\lambda = 1.25\pm 0.02$. Note that the nucleons (proton and neutron) are paired together, as are the leptons (electrons and neutrino). γ_5 is one of the other 11 matrices. The combination $\gamma_5\gamma_\mu$ is called *axial vector.*

In the nonrelativistic limit, a boson with mass gives a real-space potential function:

$$V(r) = 4\pi C\int \frac{d^3q}{(2\pi)^3}\frac{e^{i\mathbf{q}\cdot\mathbf{r}}}{q^2 + (mc/\hbar)^2} \tag{11.306}$$

$$= \frac{C}{r}e^{-r/b},\ b = \frac{\hbar}{mc} \tag{11.307}$$

Potential functions with this form are called *Yukawa potentials.* The potential function is dominated by the exponential. The range of the force b is given by the boson mass.

11.8 Scattering of Electron and Muon

The μ-meson (muon) is a Dirac particle with spin-$\frac{1}{2}$ and a heavy mass. The scattering of an electron and a muon is a simple problem. The scattering of two electrons is complicated by the exchange interaction. No such exchange exists between different particles.

The calculation of the cross section always proceeds in two steps. The first is kinematics, and the second is the matrix element.

The kinematics are done first. The basic variables are

- The electron of mass m has initial variables: $\mathbf{k}_i$, $\varepsilon_i = \sqrt{m^2c^4 + \hbar^2 c^2 k_i^2}$, μ_i
- The electron has final variables: $\mathbf{k}_f$, ε_f, μ_f
- The muon of mass M has initial variables: $\mathbf{p}_i = 0$, $E_i = Mc^2$, U_i
- The muon has final variables: $\mathbf{p}_f$, E_f, U_f

where (μ, U) are spin projection eigenvalues. The electron scatters from the muon, which is initially at rest. Momentum and energy conservation are

$$\mathbf{k}_i = \mathbf{k}_f + \mathbf{p}_f \tag{11.308}$$

$$\varepsilon_i + Mc^2 = \varepsilon_f + E_f \tag{11.309}$$

Suppose that the experiment measures the final trajectory of the electrons, so that the direction k_f is determined, as is the scattering angle $k_i \cdot k_f = \cos(\theta)$. Then setting $\mathbf{p}_f = \mathbf{k}_i - \mathbf{k}_f$ gives

$$\varepsilon_i + Mc^2 - \varepsilon_f = \sqrt{M^2c^4 + \hbar^2 c^2 (\mathbf{k}_i - \mathbf{k}_f)^2} \tag{11.310}$$

Squaring this equation and canceling like terms gives

$$\begin{aligned} Mc^2(\varepsilon_f - \varepsilon_i) - m^2c^4 + \varepsilon_i\varepsilon_f &= \hbar^2 c^2 k_i k_f \cos(\theta) \\ &= \cos(\theta)\sqrt{\varepsilon_i^2 - m^2c^4}\sqrt{\varepsilon_f^2 - m^2c^4} \end{aligned} \tag{11.311}$$

Square this equation and collect terms according to their power of ε_f:

$$0 = A\varepsilon_f^2 - 2B\varepsilon_f + C \tag{11.312}$$

$$A = [M^2 + m^2\cos^2(\theta)]c^4 + 2\varepsilon_i Mc^2 + \varepsilon_i^2 \sin^2(\theta) \tag{11.313}$$

$$B = \varepsilon_i c^4 (m^2 + M^2) + \varepsilon_i^2 Mc^2 + m^2 Mc^6 \tag{11.314}$$

$$C = c^4 \varepsilon_i^2 [M^2 + m^2 \cos^2(\theta)] + m^4 c^8 \sin^2(\theta) + 2m^2 Mc^6 \varepsilon_i \tag{11.315}$$

$$\varepsilon_f = \frac{1}{A}\left[B \pm \sqrt{B^2 - CA}\right] \tag{11.316}$$

This determines the final electron energy ε_f as a function of the incident energy ε_i and the scattering angle θ.

The scattering rate is given by the Golden Rule, where one sums over all final-state momentum variables:

$$w = \frac{(2\pi)^4}{\hbar}\int \frac{d^3k_f}{(2\pi)^3}\int \frac{d^3p_f}{(2\pi)^3} |\mathcal{M}|^2 \delta(\varepsilon_i + Mc^2 - \varepsilon_f - E_f)\delta^3(\mathbf{k}_i - \mathbf{k}_f - \mathbf{p}_f)$$

The formula is written this way to treat all final variables equally. We use the momentum delta-function to eliminate the integral d^3p_f:

$$1 = \int d^3p_f \delta^3(\mathbf{k}_i - \mathbf{k}_f - \mathbf{p}_f) \tag{11.317}$$

In spherical coordinates $d^3k_f = d\Omega_f k_f^2 dk_f$. The solid angle part $d\Omega_f$ is moved to the left of the equal sign to get the differential cross section for electron scattering:

$$\frac{d\sigma}{d\Omega_f} = \frac{w}{v_i d\Omega_f} \tag{11.318}$$

$$= \frac{1}{(2\pi)^2 \hbar v_i} \int k_f^2 dk_f |\mathcal{M}|^2 \delta(\varepsilon_i + Mc^2 - \varepsilon_f - E_f) \tag{11.319}$$

The integral is

$$\int k_f^2 dk_f \delta(\varepsilon_i + Mc^2 - \varepsilon_f - E_f) = \frac{k_f^2}{\hbar^2 c^2 |k_f/\varepsilon_f + (k_f - k_i\cos(\theta))/E_f|}$$
$$= \frac{k_f \varepsilon_f (Mc^2 + \varepsilon_i - \varepsilon_f)}{\hbar^2 c^2 |\varepsilon_i + Mc^2 - (\varepsilon_f k_i/k_f)\cos(\theta)|}$$

where energy conservation is used to eliminate E_f. The final formula for the differential cross section is

$$\frac{d\sigma}{d\Omega} = \frac{\varepsilon_i \varepsilon_f k_f}{(2\pi\hbar^2 c^2)^2 k_i} |\mathcal{M}|^2 \frac{\varepsilon_i + Mc^2 - \varepsilon_f}{|\varepsilon_i + Mc^2 - (\varepsilon_f k_i/k_f)\cos(\theta)|} \tag{11.320}$$

In the nonrelativistic limit where $E_f \approx Mc^2$ the last factor becomes unity.

The matrix element for this process is derived from the S-matrix:

$$\mathcal{M} = \frac{4\pi e^2}{q \cdot q} \langle \bar{\Psi}^{(e)*}(k_f)\gamma_\nu \Psi^{(e)}(k_i)\rangle \langle \bar{\Psi}^{(\mu)*}(p_f)\gamma_\nu \Psi^{(\mu)}(p_i)\rangle \tag{11.321}$$

where the brackets $\langle \cdots \rangle$ denote a spinor product. Here $q_\mu = [\mathbf{k}_f - \mathbf{k}_i, (\varepsilon_f - \varepsilon_i)/\hbar c]$ is the four-vector momentum transfer between the two currents. The first matrix element is from the electron current and the second is from the muon current. The eigenfunctions are also functions of the spin projection eigenvalues (μ_i, μ_f, U_i, U_f).

The next step in the derivation depends on the nature of the measuring apparatus. If the detector determines the spin of the scattered electron, then we are done. We calculate the above matrix elements for those values of μ, U. However, in the usual case, they are not determined. Then sum over all final spin operators and average over the initial ones. Thus, evaluate

$$\Xi = \frac{1}{4} \sum_{\mu_i \mu_f U_i U_f} |\langle \bar{\Psi}^{(e)}(k_f)\gamma_\nu \Psi^{(e)}(k_i)\rangle \langle \bar{\Psi}^{(\mu)}(p_f)\gamma_\nu \Psi^{(\mu)}(p_i)\rangle|^2 \tag{11.322}$$

$$= \sum_{\mu\nu} \Xi^{(e)}_{\mu\nu} \Xi^{(\mu)}_{\nu\mu} \tag{11.323}$$

$$\Xi^{(e)}_{\mu\nu} = \frac{1}{2} \sum_{\mu_i \mu_j} \langle \bar{\Psi}^{(e)}(k_f)\gamma_\nu \Psi^{(e)}(k_i)\rangle \langle \bar{\Psi}^{(e)}(k_i)\bar{\gamma}_\mu \Psi^{(e)}(k_f)\rangle \tag{11.324}$$

$$\bar{\gamma} = \beta\gamma^\dagger\beta = \beta[\beta\alpha_\mu]^\dagger \beta = \beta\alpha_\mu\beta^2 = \gamma_\mu \tag{11.325}$$

We sum the spins using the matrix method:

$$\sum_{\mu_i} \Psi^{(e)}(k_i)\bar{\Psi}^{(e)}(k_i) = \frac{c}{2\varepsilon_i}[mc - \gamma \cdot k_i] \tag{11.326}$$

$$\sum_{\mu_f} \Psi^{(e)}(k_f)\bar{\Psi}^{(e)}(k_f) = \frac{c}{2\varepsilon_f}[mc - \gamma \cdot k_f] \tag{11.327}$$

The term is now reduced to a trace:

$$\Xi^{(e)}_{\mu\nu} = \frac{c^2}{8\varepsilon_i\varepsilon_f}\mathrm{Tr}\{\gamma_\nu[mc - \gamma \cdot k_i]\gamma_\mu[mc - \gamma \cdot k_f]\} \tag{11.328}$$

The trace contains up to four γ-matrices. For two γ-matrices, start with the anticommutator $\{\gamma_\mu, \gamma_\nu\} = -2g_{\mu\nu}$. Recall from chapter 10 that the tensor $g_{\mu\nu}$ is diagonal. It is $+1$ for the vector components and -1 for the fourth component. Next prove

$$\mathrm{Tr}\{\gamma_\mu\gamma_\nu\} = -4g_{\mu\nu} \tag{11.329}$$

- If $\mu \neq \nu$, then $\gamma_\mu\gamma_\nu = -\gamma_\nu\gamma_\mu$, and

  $$\mathrm{Tr}\{\gamma_\mu\gamma_\nu\} = -\mathrm{Tr}\{\gamma_\nu\gamma_\mu\} = \mathrm{Tr}\{\gamma_\nu\gamma_\mu\} \tag{11.330}$$

  The first equality is because the operators anticommute. The second equality comes from the cyclic properties of a trace: the trace of $\mathrm{Tr}[\mathcal{ABC}] = \mathrm{Tr}[\mathcal{BCA}]$. Since the last two terms are equal except for sign, the trace is zero. This result is independent of representation.

- If $\mu = \nu$, then $\gamma_\mu^2 = -g_{\mu\nu}$. If $\mu = 4$ then $\gamma_\mu^2 = \beta^2 = I$ and its trace is four. If $\mu = i = (x, y, z)$, then $\gamma_i^2 = \beta\alpha_i\beta\alpha_i = -\beta^2\alpha_i^2 = -I$ and the trace is -4.

These results are summarized in eqn. (11.329). Note that one *does not* take the trace of $g_{\mu\nu}$. The trace of an odd number of γ-matrices vanishes:

$$\mathrm{Tr}\{\gamma_\mu\} = 0 \tag{11.331}$$

$$\mathrm{Tr}\{\gamma_\mu\gamma_\nu\gamma_\delta\} = 0 \tag{11.332}$$

The trace of four γ-matrices is obtained from all possible pairings:

$$\mathrm{Tr}\{\gamma_\mu\gamma_\nu\gamma_\delta\gamma_\epsilon\} = 4[g_{\mu\nu}g_{\delta\epsilon} + g_{\mu\epsilon}g_{\nu\delta} - g_{\mu\delta}g_{\nu\epsilon}] \tag{11.333}$$

Therefore, we get from eqn. (11.328)

$$\Xi^{(e)}_{\mu\nu} = \frac{c^2}{2\varepsilon_i\varepsilon_f}[-g_{\mu\nu}(m^2c^2 + \hbar^2 k_i \cdot k_f) + \hbar^2 g_{\mu\alpha}k_{i\alpha}g_{\nu\beta}k_{f\beta} + \hbar^2 g_{\mu\alpha}k_{f\alpha}g_{\nu\beta}k_{i\beta}]$$

$$\Xi^{(\mu)}_{\mu\nu} = \frac{c^2}{2E_iE_f}[-g_{\mu\nu}(M^2c^2 + \hbar^2 p_i \cdot p_f) + \hbar^2 g_{\mu\alpha}p_{i\alpha}g_{\nu\beta}p_{f\beta} + \hbar^2 g_{\mu\alpha}p_{f\alpha}g_{\nu\beta}p_{i\beta}]$$

We have written the identical expression for $\Xi^{(\mu)}$. Multiply these together, and then sum over (μ,ν). This gives,

$$\Xi = \frac{c^4}{4\varepsilon_i \varepsilon_f E_i E_f}[4(m^2c^2 + \hbar^2 k_i \cdot k_f)(M^2c^2 + \hbar^2 p_i \cdot p_f) + 2\hbar^4(k_i \cdot p_i)(k_f \cdot p_f).$$
$$2\hbar^4(k_i \cdot p_f)(k_f \cdot p_i) - 2\hbar^2 p_i \cdot p_f(m^2c^2 + \hbar^2 k_i \cdot k_f)$$
$$- 2\hbar^2 k_i \cdot k_f(M^2c^2 + \hbar^2 p_i \cdot p_f)] \tag{11.334}$$

After canceling of like terms,

$$\Xi = \frac{c^4}{2\varepsilon_i \varepsilon_f E_i E_f}[2m^2M^2c^4 + M^2c^2\hbar^2 k_i \cdot k_f + m^2c^2\hbar^2 p_i \cdot p_f$$
$$+ \hbar^4(k_i \cdot p_i)(k_f \cdot p_f) + \hbar^4(k_i \cdot p_f)(k_f \cdot p_i)] \tag{11.335}$$

$$k_i \cdot k_f = \mathbf{k}_i \cdot \mathbf{k}_f - \frac{\varepsilon_i \varepsilon_f}{\hbar^2 c^2}, \quad p_i \cdot p_f = -\frac{Mc^2 E_f}{\hbar^2 c^2} \tag{11.336}$$

$$k_i \cdot p_f = \mathbf{k}_i \cdot \mathbf{p}_f - \frac{\varepsilon_i E_f}{\hbar^2 c^2}, \quad p_i \cdot k_f = -\frac{Mc^2 \varepsilon_f}{\hbar^2 c^2} \tag{11.337}$$

The above result can be simplified further using $E_f = Mc^2 + \varepsilon_i - \varepsilon_f$.

References

1. G.D. Mahan, *Many-Particle Physics*, 3rd ed. (Plenum-Kluwer, New York, 2000)

Homework

1. Consider a π-mesonic atom for cadmium ($Z = 48$).

a. Find the value of the 1s to 2p transition energy nonrelativistically.
b. Find the value of the 1s to 2p transition energy relativistically.

2. Solve the K-G equation in one dimension for the lowest bound state of an attractive square well $V(x) = -V_0\Theta(a - |x|)$. Then derive an expression for the charge density $\rho(x)$ of this bound state and discuss whether it is continuous at the edge of the potential step.

3. Solve the Klein-Gordon equation in three dimensions for a spherical square-well potential of depth V_0 and radius b.

a. Find the eigenvalue equation for general angular momentum ℓ.
b. Give the explicit expression for the case $\ell = 0$ in terms of b, V_0, m_π.

What value of V_0 gives the first bound state (with zero binding energy)?

4. Solve for the eigenvalue and eigenvector of the K-G equation for a charged pion in a constant magnetic field in two dimensions: the field is in the z-direction and the pion is confined to the (x, y)-plane. Assume: $p_z = 0$, $V(\mathbf{r}) = 0$, $\mathbf{A} = B(0, x, 0)$.

5. Discuss the Zeeman effect for the $n = 2$ state of a π-mesonic atom.

6. Another way to take the nonrelativisitic limit of the energy for π-mesonic atoms is to write

$$E(k) = \sqrt{m^2c^4 + c^2p^2} = mc^2 + \frac{p^2}{2m} - \frac{p^4}{8m^3c^2} + \cdots \tag{11.338}$$

$$H_0 = mc^2 + \frac{p^2}{2m} - \frac{Ze^2}{r} \tag{11.339}$$

$$V = -\frac{p^4}{8m^3c^2} \tag{11.340}$$

Treat V as a perturbation on H_0 and evaluate its contribution in first-order of perturbation theory. You should get the same answer found by expanding the exact relativistic solution.

7. Start from the Klein-Gordon Hamiltonian with a zero vector potential but a nonzero potential energy:

$$[-\hbar^2c^2\nabla^2 + m^2c^4 - (E - V)^2]\phi(r) = 0 \tag{11.341}$$

Derive the WKBJ condition for bound states in three dimensions, which is the relativistic version of the Bohr-Sommerfeld law. Use it to find the relativistic eigenvalues of the hydrogen atom in three dimensions, and compare your result with the exact solution.

8. Examine the proposed representations of the Dirac matrices and determine whether they are correct representations:

a. $$\beta = \begin{pmatrix} 0 & I \\ -I & 0 \end{pmatrix}, \quad \alpha^{(k)} = \begin{pmatrix} \sigma_k & 0 \\ 0 & \sigma_k \end{pmatrix} \tag{11.342}$$

b. $$\beta = \begin{pmatrix} 0 & I \\ I & 0 \end{pmatrix}, \quad \alpha^{(k)} = \begin{pmatrix} \sigma_k & 0 \\ 0 & -\sigma_k \end{pmatrix} \tag{11.343}$$

c. $$\beta = \begin{pmatrix} I & 0 \\ 0 & -I \end{pmatrix}, \quad \alpha^{(k)} = \begin{pmatrix} 0 & i\sigma_k \\ -i\sigma_k & 0 \end{pmatrix} \tag{11.344}$$

9. Consider a relativistic electron in one dimension of energy E that is incident from the left on a step barrier at the origin of height $V_0 > E > mc^2$.

a. Calculate the intensity of the reflected and transmitted waves assuming the spin is initially in the z-direction and does not flip.

b. For $V_0 > E + mc^2$ show that the reflected current exceeds the incident current, due to a positron current in the barrier.

10. Solve the Dirac equation in one dimension for the lowest bound state of an attractive square well $V(z) = V_0\Theta(a - |z|)$. You will get a complicated eigenvalue equation, which you need not solve.

11. Find the 2-spinor that obeys eqn. (11.162).

12. Derive the equivalent of eqn. (11.172) for the case that the incident particle has $\mu = -\frac{1}{2}$.

13. Solve for the eigenvalue and eigenvector of the Dirac equation for an electron in a constant magnetic field in two dimensions. The field is in the z-direction and the pion is confined to the (x, y)-plane. Assume $p_z = 0$, $V(\mathbf{r}) = 0$, $\mathbf{A} = B(0,x,0)$.

14. Another Lorentz invariant way to introduce the electromagnetic fields into the Dirac equation is through the term

$$H_3 = -\frac{g}{2}\mu_0 \sum_{\mu\nu} F_{\mu\nu}\sigma_{\mu\nu} \tag{11.345}$$

$$F_{\mu\nu} = \frac{\partial A_\mu}{\partial x_\nu} - \frac{\partial A_\nu}{\partial x_\mu} \tag{11.346}$$

$$\sigma_{\mu\nu} = \frac{i}{2}[\gamma_\mu\gamma_\nu - \gamma_\nu\gamma_\mu] \tag{11.347}$$

where g is a dimensionless constant and μ_0 is the Bohr magneton. Take the nonrelativisitic limit of this interaction and show it contributes to the magnetic moment.

15. Consider two relativistic corrections to an electron in a three-dimensional harmonic oscillator: $V(r) = Kr^2/2$. Evaluate (a) the Darwin term, and (b) the spin–orbit interaction.

16. Another gamma-matrix is defined as

$$\gamma_5 = -i\gamma_1\gamma_2\gamma_3\gamma_4 \tag{11.348}$$

a. Show that γ_5 is real.
b. Use the Pauli form for the gamma-matrices and show that

$$\gamma_5 = \begin{pmatrix} 0 & I \\ I & 0 \end{pmatrix} \tag{11.349}$$

c. What is $\gamma_5^2 = ?$
d. What is $\{\gamma_5, \gamma_\mu\} = ?$

17. Suppose a Dirac particle was scattering from a fixed potential $V(r)$, and that the matrix element had the form

$$M_{fi} = \left(\bar{u}(\mathbf{p}_f, \varepsilon_f, \mu_f)\gamma_5 u(\mathbf{p}_i, \varepsilon_i, \mu_i)\right)\tilde{v}(\mathbf{p}_f - \mathbf{p}_i) \tag{11.350}$$

where $\tilde{v}(\mathbf{q})$ is the usual Fourier tranform of $V(r)$. Determine the differential cross section in the Born approximation by averaging over initial spins and summing over final spins. Note $\bar{u} = \beta u$.

Index